Bonfiglioli Riduttori S.p.A.

Bonfiglioli Riduttori S.p.A. (Eds.)

Gear Motor Handbook

With contributions by
D. W. Dudley, J. Sprengers,
D. Schröder, H. Yamashina

With 325 Figures

 Springer

Bonfiglioli Riduttori S.p.A. (Eds.)

Via Giovanni XXIII, 7/A
Lippo di Calderara di Reno,
Bologna / Italy

Darle W. Dudley

17777 Camino Murrillo,
San Diego, CA 92128, USA

Prof. Hon.-Eng. Jacques Sprengers

Rue E. Mahaim 98,
B-4100 Seraing, Belgium

Prof. Dierk Schröder

Technische Universität München,
Lehrstuhl für Elektrische Antriebstechnik,
Arcisstraße 21,
D-80333 München, FRG

Prof. Hajime Yamashina

Uji-Shi
Nanryo-Cho
5 Chome 7
Kyoto, Japan

ISBN 3-540-58988-0 Springer-Verlag Berlin Heidelberg NewYork

This work is subject to copyright. All rights are reserved, whether the whole or part of the material is concerned, specifically the rights of translation, reprinting, re-use of illustrations, recitation, broadcasting, reproduction on microfilms or in other ways, and storage in data banks. Duplication of this publication or parts thereof is only permitted under the provisions of the German Copyright Law of September 9, 1965, in its current version, and a copyright fee must always be paid.

Springer-Verlag Berlin Heidelberg 1995
Printed in Germany

The use of registered names, trademarks, etc. in this publication does not imply, even in the absence of a specific statement, that such names are exempt from the relevant protective laws and regulations and therefore free for general use.

Product liability: The publishers cannot guarantee the accuracy of any information about dosage and application contained in this book. In every individual case the user must check such information by consulting the relevant literature.

Typesetting: camera-ready by author
SPIN: 10495689   62/3020 - 5 4 3 2 1   - Printed on acid-free paper

Darle W. Dudley, Manager of Gear Technology
Development with Dudley Technical Group, Inc.;
Honorary Member of AGMA

Jacques Sprengers, President ISO/TC 60;
Chevalier de l'Ordre de Léopold, Commandeur
de l'Ordre de Léopold II, Insigne d'argent de
Lauréat du Travail

Prof. Dr.-Ing. Dr.-Ing. h.c. Dierk Schröder,
Chair for Power Engineering and Electric Drives,
Technical University of Munich

Prof. Hajime Yamashina, Chair for Production
Eng. at the Department of Precision Mechanics,
Kyoto University, and Affiliate Professor
at London Business School

Preface

In these years of constant growth and further development for our company, research and development has become more and more important, and has allowed us to be at the forefront in our business sector, where innovation is the obvious and decisive factor.

It has therefore been consistent with our everyday business philosophy to involve ourselves deeply in writing and printing this handbook, which is designed to recognize the capacity and hard work of all employees working successfully in the Bonfiglioli Group.

The book is intended to be a concrete contribution by Bonfiglioli Riduttori S.p.A. to the development and application of power transmissions.

The book is addressed to all who have technical dealings with power transmissions, from university students to engineers active in the workplace. For this reason we have invited the cooperation of four prestigious professionals - Darle W. Dudley, Jacques Sprengers, Dierk Schröder, and Hajime Yamashina - in the knowledge that only through the cooperation of the leading specialists in the field of power transmissions could we develop a truly useful and helpful handbook.

It has been hard work, but we are sure the reader's appreciation will amply reward our efforts.

Clementino Bonfiglioli
President

Table of Contents

Part I
Editor: Darle W. Dudley

1 Introduction 3

    1.1 History of the Art of Gear Making 3
    1.2 Development of Gear Technology 7
    1.3 Development of Gear Manufacturing Organizations 7
    1.4 Developments at Bonfiglioli Riduttori.................................. 8
    1.5 The content of This Technical Handbook............................ 11

Part II
Editor: Jacques Sprengers

1 The Dynamics of Solids 21

    1.1 Introduction 21
    1.2 Gravity Centre of Solids 21
    1.3 Movements of Solids 23
    1.4 Rectilinear Translating Movement 23
    1.5 Movement of Rotation 23
    1.6 Mechanical Energy, Work and Power 23
        1.6.1 Introduction... 23
        1.6.2 Potential Energy.. 23
        1.6.3 Kinetic Energy 24
        1.6.4 Work in the Uniform Rectilinear Movement 26
        1.6.5 Work in the Uniform Movement of Rotation 26
        1.6.6 Power in the Uniform Rectilinear Movement 27
        1.6.7 Power in the Movement of Rotation 27
    1.7 Inertia.. 28
        1.7.1 Introduction... 28
        1.7.2 Inertia of Solids During Translation....................... 28
        1.7.3 Balance of Rotors. Imbalances................................ 30
        1.7.4 Inertia During Accelerated Rotation. Moments of Inertia of Solids 31

1.8 Friction 34
    1.8.1 Fundamental Equation 34
    1.8.2 Angular Value of the Coefficient of Friction. 34
    1.8.3 Wedge Effect 35
    1.8.4 Friction of a Roller on its Axis 37
    1.8.5 Friction of a Belt around its Pulley 38
    1.8.6 Adhesion 40
    1.8.7 Friction on disks 41
    1.8.8 Values of some Coefficients of Friction 42
1.9 Rolling Resistance 43
1.10 Torque Necessary for a Roller of Translation 44
1.11 Cranking Torque 44
1.12 Kinematic Chain 45
    1.12.1 Definition. 45
    1.12.2 Torques 45
    1.12.3 Moments of Inertia 45
    1.12.4 Elastic Constants of Shafts 46
    1.12.5 Application of a Lifting Winch on Bridge Crane 47

2 Strength of Materials 48

2.1 Single Traction or Compression 48
2.2 Shearing 49
2.3 Simple Bending 50
    2.3.1 Plane Bending 50
    2.3.2 Left Bending 52
2.4 Torsion 53
2.5 Buckling 54
2.6 Composition of Simple Stresses 56

Part III
Editor: Jacques Sprengers

1 Stresses and Materials 61

1.1 Introduction 61
1.2 Stresses 61
1.3 Materials: Properties and Features 62
    1.3.1 Permissible Static Stress 62
    1.3.2 Fatigue. Endurance Limit 63
1.4 Materials 69

1.4.1 General Features	69
1.4.2 Ferrous Materials	69
1.4.3 Bronzes	73

2 Geometry and Gears — 74

2.1 General Features	74
2.2 Involute to a Circle	75
2.3 Elementary Geometric Wheel. Meshing Theory	77
2.4 Standard Basic Rack Tooth Profile	80
2.5 Geometrical Wheel - Basic Rack Profile	80
2.6 Tooth Thickness	82
2.7 Influence of the Rack Shift Factor. Undercut and Meshing Interference	84
2.8 Reference Centre Distance. Modified Centre Distance	86
2.9 Contact Ratio	88
2.10 Calculation of an Angle Starting from its Involute	91

3 Mechanical Gears — 92

3.1 General Spur Features	92
3.2 Cylindrical Wheels	92
3.2.1 Definition	92
3.2.2 Geometry	92
3.2.3 Standardized Basic Profile and Modules	93
3.2.4 Facewidth	94
3.2.5 Undercut	94
3.2.6 Tooth Thickness	94
3.2.7 Internal Toothing	95
3.3 Parallel Spur Gears	96
3.4 Helical Gears. Involute Helicoid	98
3.4.1 Helical Wheel	100
3.4.2 Helical Parallel Gear Pairs	105
3.4.3 Sliding	110
3.4.4 Forces	115
3.4.5 Gear Ratio - Speed Ratio	117
3.4.6 Normal Pressure	118
3.4.7 Rack Shift Factor	119
3.4.8 Gears that Reduce Speed/Gears that Increase Speed	120
3.4.9 Tooth Modifications	121
3.4.10 Mechanical Losses in Gears	123
3.5 Bevel Gear Pair	124
3.5.1 Definition	124
3.5.2 Equivalent Gears. Trestgold's Hypothesis	127
3.5.3 Constant Tooth Depth Bevel Gear	129

        3.5.4 Constant Tooth Depth Gear 130
        3.5.5 Forces on Bevel Gear Pairs 131
        3.5.6 Dimensions of the Blanks Forming the Gears 132
    3.6 Worms and Worm Wheels 135
        3.6.1 Definition 135
        3.6.2 Main Dimension 135
        3.6.3 Efficiency of Worms and Worm Wheels. Reversing 138
        3.6.4 Tooth Profile 139
        3.6.5 Forces 140
    3.7 Single Planetary Gear Train 142
        3.7.1 Description 142
        3.7.2 Kinematic Relation 143
        3.7.3 Forces 144
        3.7.4 Advantages Derived from the Use of Planetary Train Gears 145

4 Gear Measuring 146

    4.1 Measurement of the Tooth Thickness 146
        4.1.1 Base Tangent Measurement 146
        4.1.2 Normal Chordal Tooth Thickness Measurement 148
        4.1.3 Measuring by Balls and Pins 150
    4.2 Pitch Measurement 151
    4.3 Profile Measurements 154
    4.4 Measurement of the Helix Deviation 157
    4.5 Composite Tangential Deviation 158
    4.6 Composite Radial Deviation (Measurement on the two Flanks) 159
    4.7 Eccentricity 161
    4.8 Precision in Gear Blank 162
    4.9 Centre Distance Deviation and Axis Parallelism 163
    4.10 Information on Standards 163
    4.11 Backlash 164
    4.12 Contact Pattern Measurement 167

5 Gear Calculation 168

    5.1 Aspects of Teeth After Operation 168
    5.2 Parallel Gear and Bevel Gear Pair Load Capacity 169
        5.2.1 Prevention of Pitting (Contact Stress) 169
        5.2.2 Stress Applied to the Tooth Root 170
    5.3 Standard ISO 6336 (or DIN 3990) 172
        5.3.1 Introduction 172
        5.3.2 Influence Factors 173
        5.3.3 Contact Stress Calculation 180
        5.3.4 Calculation of the Tooth Bending Strength 185
        5.3.5 Allowable Stress Number 191

5.4 Standard ISO 10300 (Bevel Gear Pairs)		191
5.4.1 Introduction		191
5.4.2 Influence Factors		192
5.4.3 Contact Area		193
5.4.4 Contact Stress Calculation		195
5.4.5 Calculation of Tooth Bending Strength		198
5.4.6 Materials		199
5.5 Calculation with a Load Spectrum		200
5.5.1 Complex Calculation		200
5.5.2 Approximate Calculation		200
5.6 Worm and Worm Wheel Calculation		201
5.6.1 General Features		201
5.6.2 Load Capacity for Contact Stress		201
5.6.3 Wear Prevention		203
5.6.4 Tooth Strength		204
5.6.5 Standard Information		205
5.6.6 Reversing and Efficiency		205

6 Shafts 207

6.1 Definition and Function	207
6.2 Shafts and Forces in Gears	207
6.3 Study of the Theoretical Stresses	207
6.3.1 Bending	207
6.3.2 Torsion	212
6.3.3 Normal Stress	213
6.3.4 External Compression	213
6.4 Real Stresses (Stress Concentration Factor - Form Factor)	213
6.5 Static Allowable Stress	214
6.5.1 Static Breaking Stress due to Traction	214
6.5.2 Elastic Limit	214
6.5.3 Allowable Static Stress	215
6.6 Allowable Stress for an Unlimited Duration	215
6.6.1 Endurance (Endurance Factor Y_C)	215
6.6.2 Factor Modifying Endurance	216
6.6.3 Allowable Stress for Unlimited Life Duration	217
6.7 Allowable Stress	217
6.8 Stress Concentration Factor	218
6.8.1 Definition	218
6.8.2 Notch Sensibility Factor	219
6.8.3 Notch Gradient	219
6.8.4 Notch Factor	219
6.8.5 Values of the Allowable Stress Concentration Coefficient and Notch Gradient	220

6.8.6 Static Stress Concentration Factor	224
6.9 Form Factor	224
6.10 Real Stress	224
6.11 Safety Factors	225
6.11.1 Simple Stress	225
6.11.2 Stress Reduced to Reference Stress	225
6.11.3 Composition of Stresses Following the same Direction	226
6.11.4 Stresses Having any Direction	226
6.11.5 Stresses Generated by a Spectrum of Loads	228
6.12 Stress Concentration Importance	229
6.13 Shaft Designing	230
6.14 Distortion of the Shaft	231

7 Fits Between the Shaft and the Hub and Fits Between the Shaft and the Shaft 234

7.1 ISO System Concerning Tolerances	234
7.2 Measurement of Roughness	237
7.3 Fixing by means of Prismatic Key	238
7.4 Interference Mounting	241
7.5 Fretting Corrosion	243
7.6 Splines	245
7.7 Couplings - Generalities	247
7.8 Rigid Couplings	248
7.9 Elastic Couplings	250
7.10 Gear Toothing Couplings	251
7.11 Clutch Couplings	251
7.12 Couplings by Means of Shrink Disks	252

8 Bearings 254

8.1 Generalities	254
8.2 Classification of Bearings on the Basis of the Rolling Elements	254
8.3 Classification of Bearings on the Basis of their Positioning	255
8.4 Calculation of Bearings	255
8.5 Variable Loads	257
8.6 Static Capacity	257
8.7 Dimensions of Bearings	257
8.8 Radial Deep Groove Ball Bearings	258
8.9 Double Row Self-Aligning Ball Bearings	259
8.10 Cylindrical Roller Bearings	260
8.11 Double Row Self-Aligning Roller Bearings	261
8.12 Angular Contact Ball Bearings	262
8.13 Taper Roller Bearings	263

8.14 Selection of Bearings	265
8.15 Bearing Fixings	266
8.16 Dismounting of Bearings	269
8.17 Maximum Speed and Number of Cycles	270
8.18 Oil Seals	270

9 Lubricants and Lubrication ... 273

9.1 Function of Lubrication	273
9.2 Lubricants	273
9.3 Mineral Lubricants	273
9.4 Greases	274
9.5 Synthetic Lubricants	274
9.6 Viscosity	275
9.6.1 Definition	275
9.6.2 Variation of Viscosity by Temperature	276
9.6.3 Viscosity Variation due to Pressure	277
9.6.4 Measurement of Viscosity	277
9.6.5 Standardized Designation of Viscosity (ISO)	278
9.7 Other Properties of Lubricants	278
9.8 Causes of Deterioration	279
9.9 Selection of Lubricants	279
9.10 Working at Low Temperatures	280
9.11 Working at High Temperatures	280
9.12 Modalities of Lubrication of Gearing	281
9.13 Elasto-Hydro-Dynamic State (EHD)	281
9.14 Scoring	282

10 Housings ... 284

10.1 Definitions and Function	284
10.2 Materials	284
10.3 Primary Function: Support the Bearings	284
10.4 Calculation with Finite Elements	285
10.5 Second Function: Seal	288
10.6 Third Function: External Fixing	288
10.7 Accessories	289
10.8 Working and Precision of Housings	289
10.9 Mounting Screwed Elements (Bolts and Nuts)	290
10.10 Economic Conditions Influencing the Selection of Housings	294

11 Standard Gear Units ... 295

11.1 Definition and Fields of Application	295
11.2 Design Principle	296

XVI   Table of Contents

11.3 Types of Gearboxes 297
11.4 Selection of a Gear Unit 298
11.5 Service Factor 298
11.6 Equivalent Power 299
11.7 Example of Calculation 299
11.8 Factors that Influence the Operation Factor 300
11.9 Particular Precautions 300
11.10 Gear Motors 301

12 Vibrations and Noises 302

12.1 Natural Vibration 302
12.2 Forced Vibration 303
12.3 Vibration Modalities 303
12.4 Flexion Vibration 304
12.5 Torsion Vibration 305
12.6 Sinusoidal Vibration Function 306
12.7 Periodic Vibration Function (Fourier) 307
12.8 Modal Anlysis 307
12.9 Measurement of Vibrations 307
12.10 Causes of Vibrations in Gearboxes. Frequency of Meshing 308
12.11 Effects of Vibrations 309
12.12 Noise 310
12.13 Measurement of the Noise 310
12.14 Well-Considered Scales (dB_A) 311
12.15 Measurement of the Noise Produced by Gear Units 312

13 Thermal Power of the Gearing 314

13.1 Definition 314
13.2 Losses in Gearboxes 314
13.3 Losses During the Meshing 315
13.4 Losses due to Dipping 316
13.5 Losses in Bearings 317
13.6 Losses in Sealings 320
13.7 Total Losses 320
13.8 Dissipation of Heat 320
13.9 Efficiency of Gearboxes 322
13.10 Measurement of the Efficiency of Gearboxes 322
   13.10.1 Open Circuit Measurement 322
   13.10.2 Back to Back Measurement 323
13.11 Comparison Between the Different Types of Gear Units 325

14 The Manufacturing of Gearboxes ... 326

14.1 Introduction ... 326
14.2 The Raw Material ... 327
14.3 The Processing of Carters ... 328
14.4 The Processing of Shafts ... 330
14.5 The Working of Gear Blanks ... 331
14.6 Cutting Cylindrical Spur and Helical Gears ... 332
    14.6.1 Particulars ... 332
    14.6.2 Processing by Milling with Form Tools ... 332
    14.6.3 Processing by Generation. Principle ... 332
    14.6.4 Cutting by Rack-Tool ... 333
    14.6.5 Cutting by Pinion-Tool ... 335
    14.6.6 Processing by Lead-Miller (Hob) ... 336
    14.6.7 Processing by a Tool of Carbide or Fired Clay ... 337
14.7 The Processing of Bevel Gears ... 338
14.8 The Processing of Worm and Worm Wheels ... 338
14.9 The Finishing of Cylindrical Gears ... 340
    14.9.1 Grinding ... 340
    14.9.2 Shaving ... 342
    14.9.3 Honing ... 343
    14.9.4 Skiving ... 343
14.10 Finishing of Bevel Gears ... 343
14.11 Grinding of Worms ... 344
14.12 Assembly ... 344

15 Applications ... 345

15.1 The Application of Gears ... 345
15.2 Hoisting Instruments ... 346
    15.2.1 Bridge Cranes ... 346
    15.2.2 Cranes ... 347
15.3 Conveyors ... 348
    15.3.1 Belt Conveyors ... 348
    15.3.2 Chain Conveyors ... 349
    15.3.3 Roller Conveyors ... 350
    15.3.4 Screw Conveyors ... 351
15.4 Waste and Treatment of Used Water ... 351
    15.4.1 Waste Treatment ... 351
    15.4.2 The Treatment of Used Waters ... 352
15.5 Special Vehicles and Farm Machines ... 353
15.6 Equipments Used for Intermittent Operation ... 354
15.7 Precision Positioners ... 354

XVIII   Table of Contents

Appendix. Hydraulic Motors and Variators 357

    A.1 Hydraulic Motors 357
    A.2 Speed Variators 359
        A.2.1 Belt Variators 359
        A.2.2 Planetary Variators 360

The ISO Standards Relating to Gears (TC 60) 362

Bibliography 365

Part IV
Editor: Dierk Schröder

1   Electric Machines - Standards and Definitions 369

    1.1 Introduction and Standards 369
    1.2 Duty and Rating 371
    1.3 Machines with Different Duty Types 380
    1.4 Operating Conditions: Altitude, Ambient Temperatures and Coolant
        Temperatures 381
    1.5 Electrical Conditions 382
    1.6 System: Production Machine-Driving Machine 386
        1.6.1 Stationary Behaviour of the Production Machine 386
        1.6.2 Stationary Behaviour of the Driving Machine: $t_M = f(n,\varphi)$ 388
        1.6.3 Static Stability at the Operating Point 390
        1.6.4 Rating of the Electrical Drive System 392

2   Electric Machines 395

    2.1 Direct-Current Machine 395
    2.2 Rotating-Field Machines 397
        2.2.1 Fundamental Principle of the Rotating-Field Machines 400
        2.2.2 Synchronous Machines 403

3   Direct-Current Machine 406

    3.1 Signal-Flow Graph of Separately Excited DC-Machine Armature
        Circuit 406
        3.1.1 Normalization 409
        3.1.2 Field Circuit Excitation Circuit 413
    3.2 Transfer-Functions Transient Responses 414
        3.2.1 Variable Command Behaviour and Variable Command Transfer
            Function 414

  3.2.2 Load Behaviour and Disturbance Transfer-Function 418
  3.2.3 Influence of φ on *n* (Field Weakening) 418
 3.3 Open-Loop Speed Control 420
  3.3.1 Open-Loop Speed Control by Means of Armature Voltage 420
  3.3.2 Open-Loop Control Through Field Weakening 422
  3.3.3 Control Through Armature and Field-Voltage 423
  3.3.4 Control with a Series-Resistor on the Armature Circuit 423

4 Converters and DC-Machine Control 426

 4.1 DC-DC-Converter 426
  4.1.1 DC-DC-Converter Principle 426
  4.1.2 Control Strategies for DC-DC-Converters 429
  4.1.3 DC-DC-Converters for one or Multiple Quadrant Operation
    of DC-DC-Motors 434
 4.2 Line Commutated Converters 439
  4.2.1 Three Phase "Y" Half Wave Converter 440
  4.2.2 Single-Phase Converters 445
  4.2.3 Three-Phase Bridge-Converter 448
  4.2.4 Operating Limits Converter and Machine 450
  4.2.5 Torque Reversal of DC-Machines Supplied by Converters 454
 4.3 Control of DC-Drives (Current-Control, Speed Control) 458
  4.3.1 Control of the Armature Current 459
  4.3.2 Speed Control 463

5 Three-Phase AC-Machines Signal Flow Graphs 466

 5.1 Space-Vector Theory -Phasors 466
 5.2 General Rotating Field Machine 469
 5.3 Induction Machine in the Steady State Supplied from the Mains 477
  5.3.1 Mains and Converter Power Supply 477
 5.4 Single-Phase Equivalent Circuit in Steady-State Operation 483
  5.4.1 Equivalent Electrical Circuit for the IM 483
 5.5 Induction Machine Supplied by Inverter 485
  5.5.1 Control Strategies Based on Constant Stator Flux 486
  5.5.2 Control Strategies Based on Constant Rotor Flux 490

6 Synchronous Machine 495

 6.1 Synchronous Machine with Salient Pole Rotor 496
  6.6.1 Machine with Salient Pole Rotor with Impressed Stator
    Voltage-Signal Flow Graph 496
 6.2 Synchronous Machine with Non-Salient Pole Rotor 502
 6.3 Non-Salient Pole Rotor Synchronous Machine Control 504

7 Inverter Drives ... 509

7.1 Cycloconverter ... 509
7.2 Wound Rotor Induction Machine with Slip-Power Recovery (Subsynchronous Cascade; Scherbius Drive) ... 511
7.3 Double Fed Induction Machine ... 516
7.4 The Load-Commutated Inverter Drive ... 516
7.5 Forced Commutation Current Source Inverter with Thyristors ... 522
7.6 Voltage Source Inverter with Intermediate Direct Voltage (VSI) ... 527
    7.6.1 VSI with Variable DC-Voltage ... 528
    7.6.2 VSI with Constant Intermediate Direct Voltage ... 530

8 Basic Considerations for the Control of Rotating Field Machines ... 534

8.1 Decoupling ... 535
8.2 Field Orientation ... 540

9 Permanent Magnet Machines ... 543

9.1 Permanent Magnet Synchronous Machine ... 543
9.2 Brushless DC-Machine ... 547

Part V
Editor: Haijme Yamashina

1 Outline of Total Quality Management ... 551

1.1 Definition of Quality ... 551
1.2 Brief History of Total Quality Management in Japan ... 552
    1.2.1 Introduction of SQC (Statistical Quality Control) ... 552
    1.2.2 Necessity of Quality Assurance ... 554
    1.2.3 Toward Total Quality Management ... 555
1.3 Importance of Production Techniques to Build-in Quality in New Innovative Product Development ... 556
    1.3.1 Product and Process Technology ... 556
    1.3.2 A Closer Look at the Production Process ... 558
    1.3.3 Matching Manufacturing to Market Changes ... 560
    1.3.4 Costs of Diversification ... 560
    1.3.5 Mobile Competitive Edge ... 562

2 The QC Viewpoint ... 565

2.1 Customer Orientation ... 565
2.2 Customer's Quality Requirements ... 566
    2.2.1 Quality Requirements and Substitutional Characteristics ... 566
    2.2.2 How to Express Quality ... 566
2.3 Management through Quality to Satisfy Customers ... 569
    2.3.1 Problems in Conventional Management ... 569
    2.3.2 Management by Rotating the Plan-Do-Check-Act Cycle ... 569

3 Quality Assurance ... 575

3.1 Basic Quality Assurance Policy ... 575
3.2 Quality Assurance Organization ... 575
3.3 Quality Assurance Process ... 578
3.4 Quality Assurance System Diagram ... 579

4 Main Quality Control Activities ... 583

4.1 Quality Control at the Product Design Stage ... 583
    4.1.1 Quality Engineering ... 583
    4.1.2 FMEA and FTA ... 585
4.2 Quality Control at the Preparation Stage for Production ... 585
    4.2.1 Quality Engineering ... 585
    4.2.2 Process FMEA ... 586
    4.2.3 Process Capability Assessment ... 586
    4.2.4 Fool-Proof System ... 589
    4.2.5 Quality Control of Purchased Goods ... 590
4.3 Quality Control at the Daily Production Stage ... 590
    4.3.1 Statistical Process Control (SPC) by Using the Control Chart ... 590
    4.3.2 Every Aspect of Inspection ... 592
    4.3.3 QA Network ... 595
    4.3.4 Establishment of the Standard Operation ... 596
    4.3.5 Role of the Foreman ... 597
    4.3.6 Quality Maintenance ... 597
    4.3.7 QC Circle Activities ... 598
    4.3.8 Quality Control of Purchased Goods ... 598
4.4 Necessity of Education and Training and Training on the QC-Tools ... 600

Author- and Subject Index ... 603

General Notes and Symbols

Part II, III

Not all the symbols are listed below, only the values of the main symbols and signs used. Complex symbols will be easily reconstructed through the indications given. If some symbols are missing from the list, it is because they are used only in a specific paragraph within their definition.

For gears, symbols of the ISO 701 are followed. For calculation of gears, symbols of standard ISO 6336 or ISO 10300 are followed. For tolerances, the symbols are those of ISO 1328-1.

a	centre distance,	mm
b	tooth width,	mm
c	teeth elasticity,	N/mm.µm
d	diameter,	mm
f	deviations,	µm
f	frequency,	Hz
g	path of contact,	mm
h	teeth height,	mm
i	speed ratio,	–
j	backlash,	µm
k	shortening,	–
l	shaft length,	mm
m	module,	mm
n	speed of rotation (revolutions),	min^{-1}
p	pitch, pressure,	mm
r	radius,	mm
s	teeth thickness,	mm
t	time,	sec
u	gear ratio,	–
v	linear speed, reference speed,	m/s
x	rackshift factor,	–
z	teeth number,	–
E	coefficient of elasticity,	MPa
F	force, cumulated deviations,	µm
J	inertial moments,	kg.m^2
K	factor relative to the load,	–
M	moment of a force,	N.m
N	number of cycles,	–
P	power,	kW

XXIV General Notes and Symbols

Q	shear force,	N
R	cone generator,	mm
S	safety factor,	–
T	torque,	N.m
W	base tangent length,	mm
X	factor related to scuffing,	–
Y	factor related to tooth root benmding,	–
Z	factor relative to sewing,	–
IT	space of tolerance,	μm
es	upper deviation for a shaft,	μm
ei	lower deviation for a shaft,	μm
ES	upper deviation for a bore,	μm
EI	lower deviation for a bore,	μm
α	pressure angle, clearance angle,	° or –
β	helix angle,	° or –
γ	cone angle, helix slope,	° or –
δ	density,	kg/m³
ε	contact ratio	–
κ	ratio of the minimum stress to the maximum stress during a periodical stimulus,	
ν	coefficient of Poisson,	
σ	normal stress,	MPa
τ	tangential stress,	MPa
ω	pulsation	s⁻¹

Main indexes:

a	relative to the tooth tip, amplitude,
b	relative to the basic circle, bending,
d	relative to the diameter,
f	relative to the tooth root,
k	relative for b, t or n,
m	medium,
n	real, equivalent, normal,
p	relative to pitch,
t	transverse, of shearing,
v	virtual,
w	working,
x	axial,
y	any point,
D	relative to resistance,
E	relative to the limit of elasticity,

F	relative to the tooth root stress,
H	relative to the contact stress,
N	relative to the number of cycles,
P	permissible,
R	of breaking,
lim	limit,
max	aximum,
min	mnimum,
stat	static,
α	relative to the profile,
β	relative to the facewidth,
γ	total,
1	relative to the driving wheel, to the pinion, to the sungear,
2	relative to the driven wheel, to the wheel, to the planetgear,
3	relative to the ringwheel of a planetary train.

Index
* reduced to the module = 1
  reduced to the unitary width

PART IV

Capital letters: non-normalized variables and signals

Lower-case letters: normalized variables and originals

Index S: stator fixed coordinate system

Index K: optional coordinate system

Index L: rotor fixed coordinate system

a:	switching ratio
$\underline{a}$:	spatial displacement operator
B:	acceleration
$\bar{B}$:	complex spatial magnetic flux density phasor
C_E:	machine constant (electrical) of the DC-machine
C_M:	machine constant (mechanical) of the DC-machine
Cv:	thermal capacity
d:	damping factor second order system
d_x:	relative inductive voltage drop in converters
D_F:	freewheeling diode in DC-DC converters
D_x:	inductive voltage drop in converter
E_A:	induced electromotive force
F:	general: frequency, force
F_N:	mains frequency

$G(s)$:	general: transfer function
$G_r(s)$:	transfer function in the feedback channel
$G_V(S)$:	transfer function in the feed-forward channel
$G_R(S)$:	transfer function of the controller
i_{adm}:	current load allowed
I_A:	armature current of the DC-machine
I_B:	load current
I_E:	excitation current
I_K:	short-circuit current
I_{circle}:	circulating current in four quadrant converters
I_{max}:	maximum current
I_V:	user current, load current
I^*:	reference value of the current
L:	general: inductance
L_A:	armature inductance
L_K:	circulating current inductance
T_K:	pull out torque
M_M:	motor torque
M_{MI}:	air gap torque
M_{MR}:	motor friction torque
M_W:	load torque
M_{12}, M_{21}:	mutual inductances
N^*:	speed
N_{ON}:	reference speed (no load)
N:	speed reference
p:	pulse number
P:	general: power
P_v:	dissipation
R:	radius, resistance
R_A:	armature resistor in dc-machines
R_v:	additional series resistor
$sgn(...)$:	step function
s:	slip, operator Laplace domain
s_K:	break down slip
t_E:	switch on-time of switches
t_b:	operating time
t_p:	pause time
t_a:	starting time; switch off-time of switches
t_{Br}:	braking time
t_s:	cycle time, release time
T_A:	armature circuit, time constant
T_E:	excitation circuit, time constant
T_{equi}:	equivalent time constant
T_R:	regulator time constant
$T_{\Theta N}$:	normalized inertia time constant
$u_K\%$:	relative short-circuit voltage
$ü$:	commutation interval, transformer transmission ratio

U_A:	armature voltage of dc-machine
U_E:	excitation voltage
U_d:	DC-voltage of converters
U_{dia}:	DC- voltage controlled by α
U_Q:	source voltage
U:	line to line voltage
V:	velocity, losses
V_{FE}:	iron losses
V_w:	copper losses
V_R:	friction losses, proportional gain of controller
V_{str}:	static converter gain
X_d:	control error
Z_p:	number of pole pairs in electric machines
a:	firing angle in converters
a_{max}:	maximum control angle
β:	first derivate of torque at the operating point
β_K:	phase displacement of the system K with respect to the stator
β_L:	phase displacement of the rotor with respect to the stator
γ:	protection time angle
η_{el}:	electrical yield
η_{mech}:	mechanical yield
$\delta(t)$:	dirac function
$\sigma(t)$:	unit step function
σ:	Blondel coefficient
T:	normalized time
ωd:	resonant circuit frequency
Θ:	mass inertia moment
Θtot:	total mass inertia moment
θ:	general: polar wheel angle, temperature
ϕ:	angle of rotation
Ψ:	interlinked flux
Ω:	angular frequency
Ω_{on}:	reference angular frequency
Ω_m:	mechanical angular frequency
Ω_{el}:	electrical angular frequency
Ω_l:	angular stator frequency
Ω_2:	angular rotor frequency
Ω_k:	angular frequency of optional coordinate system K
Ω_{syn}:	synchronous angular frequency
Ω_o:	angular frequency at the operating point

Part I

Introduction

Darle W. Dudley
Professional Engineer, Honorary Member AGMA

1 Introduction

This handbook is designed to help customers and gear specialists have the appropriate technical information in regard to the gear products now available at Bonfiglioli Riduttori. Several kinds of gear units are in high volume production and Bonfiglioli serves a rapidly growing, world wide market. This technical handbook is intended to be broadly useful to those involved with geared machinery.

Those concerned with gearing need to have some background on the generalities of gear manufacture. This involves the history of technology and manufacture of gears. These items will be discussed briefly in this introduction before getting into specific comments on the Bonfiglioli organisation and the contents of this handbook.

1.1 History of the Art of Gear Making

Gears are about as old as any machinery of mankind. So far as we know, the only "machine" that is older is the potter's wheel.

At some time over 3000 years ago, primitive gears first meshed with each other and transmitted rotary motion. Ancient data from Greece shows that both wooden and metal gears were used. The later Roman Empire used wooden gears in grist mills and metal gears in many small devices.

An outline of the history of gears up to the 20th century is given below.

1. The early Chinese used gears and developed a very ingenious South Pointing Chariot.
2. The earliest gears were wooden. The teeth were really engaging pins.

3. The early Greeks made some use of metal gears with wedge-shaped teeth. They also used wooden gears.
4. The Romans made considerable use of gears in their mills.
5. In the Middle Ages water power mills used wooden gears. Stone gears were used at one location in Sweden.
6. Metal clock gearing became important (started in the 13th century).
7. Dutch windmills all used wooden gears to pump the water out of the Netherlands (18th century).
8. Wooden gears were used extensively in mills all over the world in the 18th century (grist mills, textile mills, steel mills).
9. The wide scale use of electric motors and steam engines created a considerable need for good metal gears. This happened in the latter part of the 19th century (railroads, steamboats, factories, power stations, etc).

Figure 1.1 shows a schematic sketch of a water powered Roman grist mill. This kind of mill was in limited use at the time of Julius Caesar. At the end of the Roman Empire over 300 years later, there had been a considerable increase in the use of gearing for mills and for a variety of other devices.

Mechanical clock gearing started in Europe in about 1285. It was invented by monks and first used to strike hours (timing of church activity like early morning prayers).

The first mechanical clocks had an escapement device and one set of toothed gearing. A suspended set of "verge" blades on a vertical shaft developed a back and forth rotary motion of a few degrees. This primitive clock developed over the next 400 years to have three sets of gears and a hand and dial so that hours and minutes could be read. The clock gears were usually made of metal and had wedge shaped teeth.

Astronomers and physicists needed a more accurate clock. In 1656 a Dutch scientist, Christiaan Huygens, invented the pendulum clock. The swinging pendulum, as we all know, is suspended from a horizontal shaft. Pendulum clocks were soon developed to have high accuracy. Very large ones were made for public buildings. As many as ten or more sets of gears were used. Some of these gears were as large as a meter in diameter. The technology and manufacture of gearing took a big step forward with the advent of the pendulum clocks. Figure 1.2 shows a schematic of the early pendulum clock.

The water-powered mill was still the work horse of industry in the 18th century and early 19th century. Besides grist mills, other mills were powered by water and sawed wood, machined metal or wood, made from iron and steel rods, pumped water, etc. In a few places, like the Netherlands, wind power (instead of water power) drove a mill with wooden gearing. Figure 1.3 shows a schematic sketch of a typical grist mill in operation in the United States in the latter part of the 18th century.

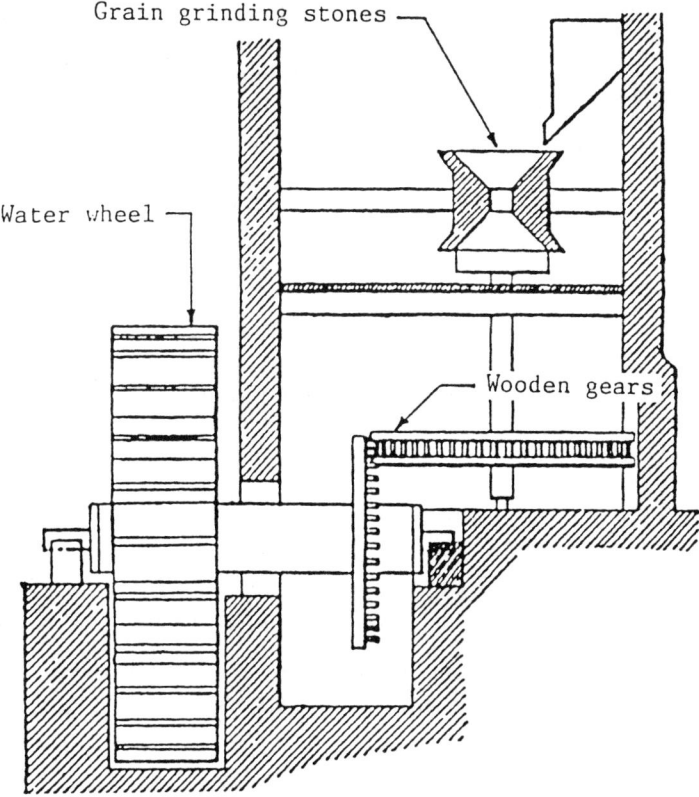

Fig. 1.1. Schematic of an early Roman grist mill. In the first century B.C. these mills were in limited use. Most of the flour used was ground by human or animal power.

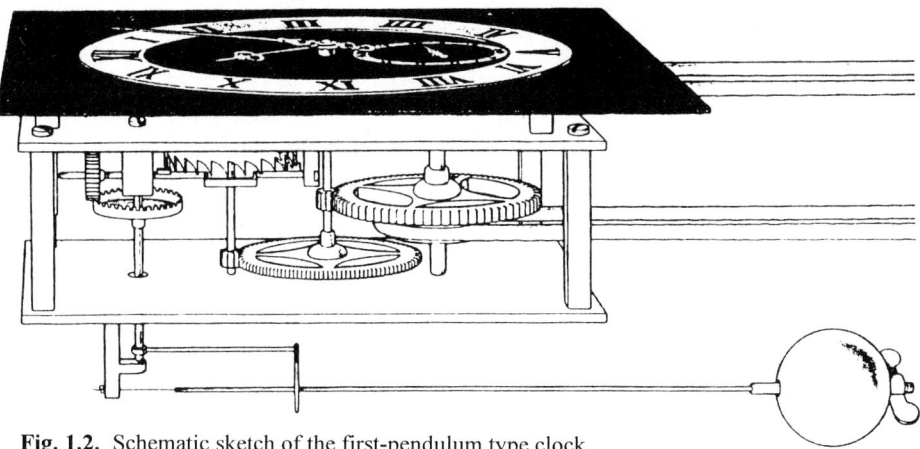

Fig. 1.2. Schematic sketch of the first-pendulum type clock.

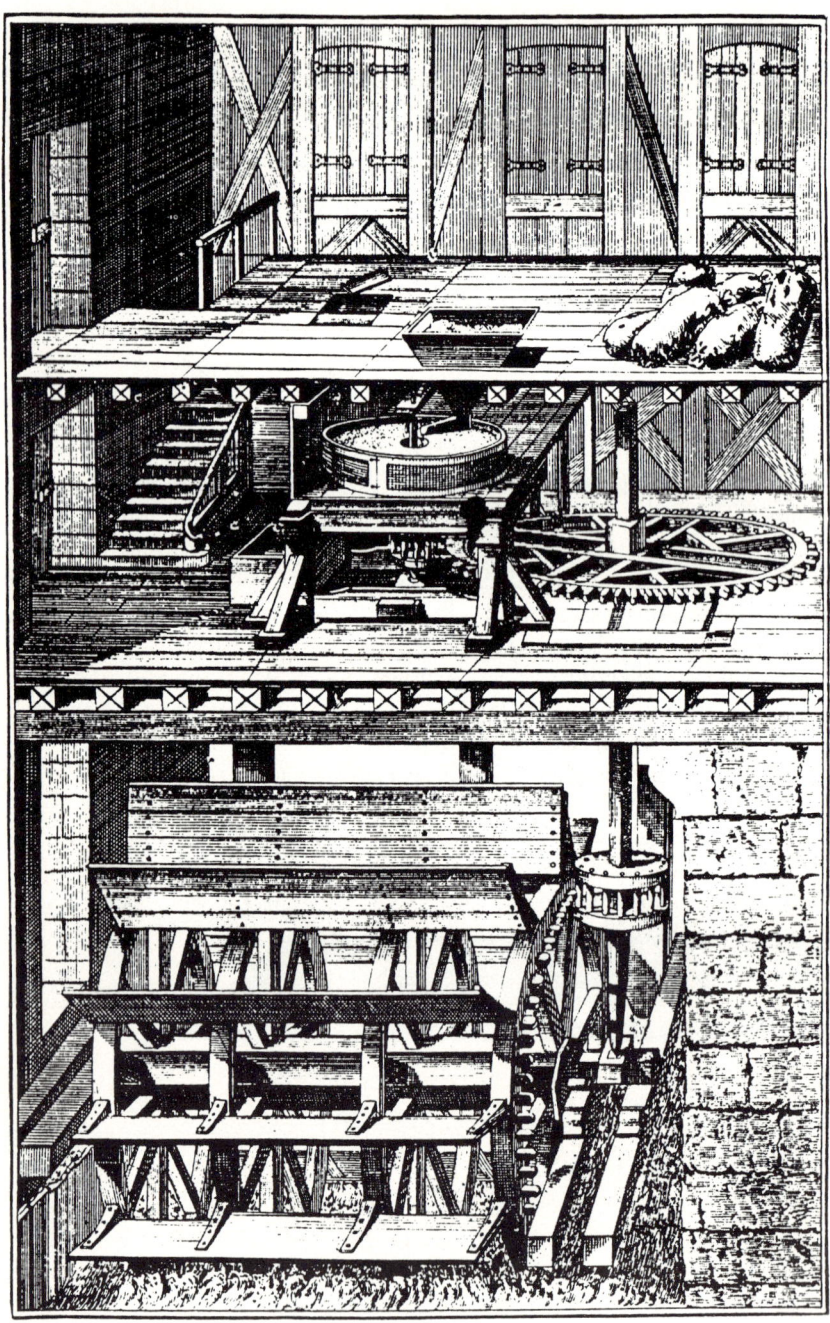

Fig. 1.3. Schematic sketch of a grist mill in service in the United States near the end of the 18th century. Note two stages of wooden gears.

1.2 Development of Gear Technology

We are now almost at the end of the 20th century. It can be said that *over 90 per cent of the gear technology we use has been developed in this century!* To go a step further, the technical advances in the machinery to make gears and the engineering considerations in gear materials, gear rating and gear lubrication that have developed since 1970 make all the technology developed in the first 70 years of this century seem relatively obsolete.

With automation and computer control, the computer-automated numerical control (CNC) machinery now in use to make gears is a most remarkable advancement. Productivity is greatly increased and much more uniform quality is produced.

Developments in technology with regard to material composition make it possible to achieve longer life, run hotter and reduce the tendency to wear. This recent development is significant in both the gears and in the tools that cut, grind or form the gears.

In earlier days, gear rating was based mostly on "safe" bending stresses and tooth contact stresses. Rating of gears, bearings and shafts is now based on probability of failure in a certain desired product life when the gear element is subjected to a spectrum of loading (so much time at several steps of loading from no-load to full load and operation at two or three levels of temperature).

Lubricant viscosity, additive content and basic molecular structure of the oil all enter into how well the lubricant resists wear and the amount of friction loss in the meshing gear or rolling bearing. Much is now known that was not known 25 years ago.

In the aerospace field, aircraft flying at Match 3 are lubricated with an unusual synthetic oil that will stand temperatures up to 315°C (600°F). Vehicles flying in the hard vacuum of outer space was successfully lubricated by very unusual fluids and coatings that have almost no tendency to lose weight by evaporation. In the industrial field, we can now calculate the thickness of the elasto-hydrodynamic (EHD) oil film that tends to separate two gears in contact or a rolling ball in a bearing from the race against which rolls.

1.3 Development of Gear Manufacturing Organisations

In ancient times, artisans with hand tools whittled pins, drilled holes and made a gear somewhat like they would have made a spoked wheel for a cart or a chariot. Metal gears were filed a slot at a time to make a toothed wheel. (Imagine an ancient man's consternation when he had filed 35 notches for a 36 tooth clock gear and found that the space left was only wide enough for one-half of a tooth! A gear with 35½ teeth is absolutely no good.)

In the late 19th century (e.g. 1890), there was an unprecedented need for gearing. The newly developed electric motor could drive all kinds of fast running or slow running things if it would be provided with gears. The ships at sea could go fast even with no wind for sail with steam engine power. In general the proper propeller speed was different from the steam engine speed, so gears were needed for ships.

Power plants began to make electricity and the light bulb replaced the candle. Plants making electricity often needed gears so that the electric generator's speed was right for alternating cycle electricity. In the early part of the 20th century, fast running steam turbines came into being. Much more accurate gearing was needed for steam turbines than had been needed for the slow running, piston type steam engines.

Perhaps the biggest usage of gears came when motor cars and trucks came into being at the start of the 20th century (gears were not needed on the horse-drawn buggies of the 19th century).

The large electric machinery companies and the large vehicle building companies found that there was no established gear manufacturing industry capable of supplying them with the quantity and quality of gearing they needed. In general there was a strong trend for these companies to establish their own gear making capability.

An exception to this was speciality companies that made tools to cut gears, or manufactured the machines that cut gears and used tools like hobs, shaper cutters, broaches, etc. Many of the world famous tool making companies and gear cutting machine companies were established around 100 years ago.

Independent companies, devoted solely to making gears or complete gear boxes with stages of gears in them, came into being, in some cases, about a century ago. In general, though, the development of highly efficient, specialised gear making companies has been a development in the last half of the 20th century. Bonfiglioli Riduttori is a good example of this trend. This company was founded in 1956.

1.4 Developments at Bonfiglioli Riduttori

The Bonfiglioli Company started out with five important types of gear units. These were coaxial gear boxes, right angle gear boxes, parallel shaft gear boxes, epicyclic gear boxes and worm gear boxes. This basic product line has steadily grown over the last 40 years. Several additional gear box configurations have been added.

The present product line has almost any type of gear unit needed for attached electric motors. In addition, drives are made for earth moving machinery, marine vessels, cement mixers. These units may be track drives, slewing drives, wheel drives, rotary mixer drives or propeller drives. The builders of industrial machinery or vehicles can find a wide array of geared designs at Bonfiglioli that already proven in service, tooled for production and available for relatively quick delivery and at low cost.

In addition to gear boxes, Bonfiglioli has expanded in the last 20 years to the manufacture of several components that go along with a gear unit. These include torque limiters, electromagnetic brakes, clutches and permanent magnetic electric motors. It is very helpful to those involved in the design of geared machinery to be able to get the right kind of components from a company knowledgeable in all the characteristics of gear boxes and a power system based on gearing.

In the field of gearing, the overall *quality* of the gearing is of much importance. "Organi di Trasmissione" (an Italian technical magazine devoted to gearing) had a 25th anniversary issue for July/August 1994. Several worldwide leaders in the gear field were interviewed and their answers to certain questions were published. One of these was Clementino Bonfiglioli, the founder and President of Bonfiglioli Riduttori.

Mr. Bonfiglioli gave this statement in regard to a question about Quality.

"Only and always Bonfiglioli has made much of its efforts and paid the best attention towards quality. The high reliability in production, together with the remarkable technological content, have never accepted any improvisation. For this reason, to come to certification of the Company's quality system must be considered as natural."

The Bonfiglioli gear operation is certified under ISO 9001 by Det Norske Veritas through Sincert.

The general concept of gear quality needs some discussion to clarify the meaning to the general public. Essentially, "quality" involves three different things.

Geometric quality. This is the accuracy of gear tooth profiles, tooth alignment with axis of rotation, tooth spacing, concentricity, and the right tooth thickness. It also involves the accuracy of gear casings in regard to bore sizes, correct center distances, flatness of joint surfaces, and appropriate doweling and bolting.

Metallurgical quality. This involves the metal in the gear teeth, the metal in the gear casings, the material in shafts, in seals, and the material in the bolts and dowels. Case hardened gear teeth need to have the right case depth and case pattern around the contour of the tooth. The hardness of the case must be up to specified hardness at the rubbing surface and the change in hardness going across the case must be appropriate for established good practice in carburizing, nitriding or induction hardening.

The gear casing material and shaft material must be clean in regard to dirt, sand, etc., and free from inclusions, voids or cracks. Seal materials are, of course, very critical in regard to wear which may result in oil leakage. Even nuts and bolts are critical; it is necessary to have the right tensile strength so that bolted joints do not loosen over long time periods.

Design suitability. This is an area of quality often overlooked. To begin with, the size of gear unit supplied needs to be large enough to have a length of life in service that meets customer needs. (No one is happy with a unit too small that fails prematurely in service).

Besides size, the suitability of the gear unit depends on the right type of unit for the power package.

Gear types are divided into in-line drives and right angle drives. In-line gears, whether spur or helical, may have input and output shafts concentric or offset.

In some cases the concern is to have enough power in a small package. Epicyclic designs are very useful where concentrated power is needed. The vehicle and the aerospace gear fields make much use of epicyclic gears.

Another consideration is the number of gears in a gear package. Just two elements in a worm gear package can handle reduction ratios going from 7 to 1 to over 100 to 1. Spur and helical gears typically work best in ratios of 1 to 1 to 9 to 1. If a 70 to 1 ratio is desired, helical gears would require at least four toothed parts in two stages or might use six toothed parts in three stages.

There are, of course, many practical considerations that enter into the choosing of a gear type for an application and in sizing the gear unit right for the specifications of the application.

Bonfiglioli is meeting the "quality design" requirements by an extensive array of available gear types and sizes. Their catalogues and their Technical Department have been developed to give the customer an excellent quality of service in obtaining a fully suitable gear unit for a specific application.

The handling of geometric quality is a two-step process. First, the automated production lines that cut or grind gear teeth and the automated lines that machine casings or shafts are fully equipped to measure all the dimensions. For instance, the involute profile, the tooth alignment with the axis (helix angle) and the tooth spacing are all measurable on the production line. The quality control procedure is worked out to take readings as a given lot of parts start going through. Periodic checks are then made to make sure all the parts in a lot meet specification. If a period check shows too much error, automatic production is stopped and all parts manufactured since the last acceptable check are either scrapped or sorted by individual checks of each part.

Bore sizes, center distances, joint dimensions, etc., are handled in a similar manner to the checking of gear teeth. They are done directly on the automated production line. Quality control inspectors are on hand on all production lines to monitor quality and to make sure that the resumption of production on a shut-down line is producing parts to a specified accuracy.

In addition to production line inspection, there is a geometric inspection laboratory. This laboratory makes checks of parts coming from production runs to make sure that the production line measuring equipment is working right. In addition, this laboratory checks master gears, measuring blocks, etc., to make sure that tooling and measuring devices in use on production lines are within rigid calibration. As Mr. Bonfiglioli stated, the best attention is being paid to quality control.

In regard to metallurgical quality, the suppliers of row material like castings, forgings, bar stock, etc., are required to check the quality of their material and certify that rigid quality standards have been met. Also those who provide special heat treatment services like carburizing, plating, etc., must inspect and certify the quality of their work.

This primary material quality control is augmented by inspection work done at Bonfiglioli by the metallurgical laboratory. This laboratory checks a wide variety of things like hardness, depth of case, microstructure, etc.

Metal parts are checked for possible cracks, inclusions, voids in structure and corrosion. The laboratory can study microstructure up to 1000 times size. Hardness can be checked by microhardness checkers as well as the more common macro hardness checkers. Chemical composition can be determined. The capability of this laboratory for industrial work is comparable to the capability of laboratories controlling material quality for the aerospace industry. Again, the presence and use of this laboratory is in keeping with the Bonfiglioli objective of paying the best attention to quality control.

1.5 The Content of this Technical Handbook

This handbook is divided into several parts. The third and longest part is devoted to gears and gear boxes. The next is concerned with electric motors used to drive geared speed reducers. Finally, the philosophy of quality is discussed.

Gearing is covered in 15 chapters. Each chapter is planned to give the reader basic information about one aspect of gear technology. The plan for the chapters is to start with basic fundamentals and then proceed to more complex technical considerations. For instance, the first chapter explains gear materials and their physical properties. Chapters 2 to 4 explain and define gears. This is a necessary background to prepare the reader for the calculation of how large the gears need to be for a given application, which is discussed in Chapter 5.

The sizing of gears for an application involves a determination of the pitch diameter of the pinion and its working face width. The pitch diameter of a spur or helical gear automatically becomes the pitch diameter of the pinion multiplied by the ratio. (For worm gears the size of the gear member is not the pitch diameter of the pinion multiplied by the ratio due to the complication of the worm axis being at 90 degrees to the axis of the worm gear.)

A further step in sizing spur and helical gears is the choice of the tooth size. When the tooth size needed (by module or by diametral pitch) has been determined, the appropriate numbers of teeth for the pinion and the mating gear are established.

The sizing just described is carried out by "rating" calculations. These are based on an initial set of specifications defining the gear ratio needed, the kinds of material selected for the pinion and the gear, and the amount of power to be transmitted.

After the rating calculations have been made and the size of the gear set is determined, there are several other things to consider. A gear box is a piece of machinery. There is input and output shafting. There is a bolted together casing; gears and bearings inside the gear box must have adequate lubrication; pitch line speeds need consideration; both very slow and very fast pitch line speeds can be a complication. These just mentioned are covered in Chapters 6 to 10 of this book.

The complete gear box is a piece of machinery in a power system that involves a prime mover and a driven device. In most Bonfiglioli applications the prime mover is an electric motor. The driven device may vary considerably. It may be a pump, a conveyor, a fan, a hoist, a materials mixer, or something else like the old-time grist mill to grind wheat, rice or corn into flour.

In general, the gear box is marketed as a complete unit. The unit has a rating for the whole unit not just the gears inside the box. Sometimes the difficulty in air cooling a splash lubricated gear unit may result in a "thermal" rating that is lower than the mechanical rating of the gears inside the gear box. In this case the rating of the gear box has to be the lower thermal rating.

Gear units may vibrate and cause trouble with connections like coupling, seals, bearings, etc. Gears also make noise. The noise of a gear unit may be troublesome in some sensitive locations. The technology of gearing involves a design and manufacture so that the gear unit, in its power system, has vibration and noise characteristics that are acceptable from the standpoint of other components in the power system. Also those operating or being close to the running gears need to be in an environment where the noise is within established health standards.

The various aspects of the whole gear unit in a power system are covered in Chapters 11 to 13.

Chapter 14 is the final chapter of the text on gearing. This chapter gives information on how gear units are manufactured.

There has been very rapid progress in gear technology in the last 15 years. This is particularly true of the machinery used to make gears. The cutting and grinding of gear teeth can now be done with CNC machines that are considerably more versatile than the hand-operated machines used in the early part of this century.

The CNC machine can do almost anything. Not only are gear teeth and shafts machined, all kinds of operations such as boring, machining of casing joints, drilling of bolt holes and even assembly of parts can be done by the CNC type machine. These machines also carry out inspection operations to accept or reject parts that do not meet precise quality limits.

The CNC type of advanced machinery makes possible automated manufacturing. The Bonfiglioli plants in about 15 years' time have become essentially fully automated. This leads to low cost manufacture, thus the product quality and uniformity is improved. The worker's skill is directed towards developing and managing complex machinery rather than hand and eye skills in reading dials and making delicate settings to remove a tiny amount of metal to meet a close dimension.

The last chapter in the third part gives examples of gear boxes applications.

The part of the book concerned with electric motors is divided into variable speed drives and drives switched directly to the power supply. The variable speed drives are used where the rotative speed of some operation is subject to frequent changes that are necessary to achieve the desired result. For instance, rudder control on a ship may be electrical where the desired steering wheel movements go by electrical means to the geared motor that moves the rudder. (In small boats, mechanical connections transmit muscle power by the person at the steering wheel directly to the rudder there is no electric motor involved). CNC type machinery makes a considerable use of computer controlled geared motors to convert a computer signal into a desired mechanical motion.

The geared motors that are switched directly on and off are often used at a relatively constant speed. For instance, a pump in an irrigation system is turned on to pump water for some time period like 1 hour or 10 hours. During the watering interval the pump runs at a constant speed. Even in CNC machinery there are often drives that run at a constant speed. For instance, a lubrication system for a large machine may provide a constant supply of oil at 25 pounds per square inch pressure to all the main drive gears and bearings. If oil circulation is needed before start-up and after shut down, the oiling system needs a separate electric motor driven pump to provide oil under pressure.

The basic technical items for variable speed drives are covered in nine chapters. The technical items for motors in relatively constant speed drives are in the last chapters of the part on motors.

Bonfiglioli Riduttori S.p.A.
Head Quarter - Lippo di Calderara - Bologna - Italy

Bonfiglioli Riduttori S.p.A.
Trasmital Division
Forlì - Italy

Translator

16   Darle W. Dudley, Part I

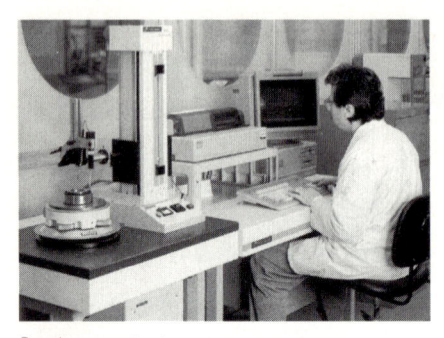

Roundness controll using Taylor-Hobson Talyrond 250 roundometer.

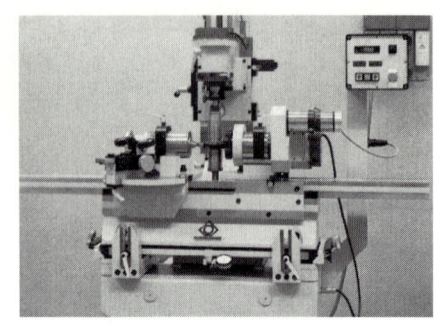

Single-flank control for worm gear pairs using Klingelnberg PSR 500.

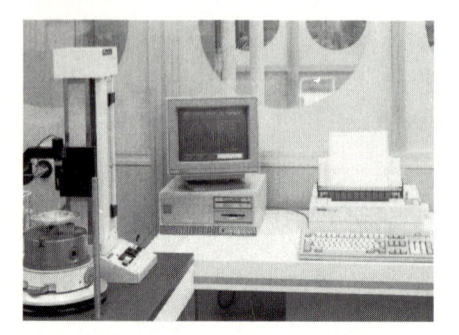

Geometrical tolerance controls using Taylor-Hobson Talyrond 250 computerized system with graphic resolution and with roundometer.

Automatic lathe

1 Introduction   17

Chemical analysis of the product using
Hilger Analytical Polivac E 960 spectrometer.

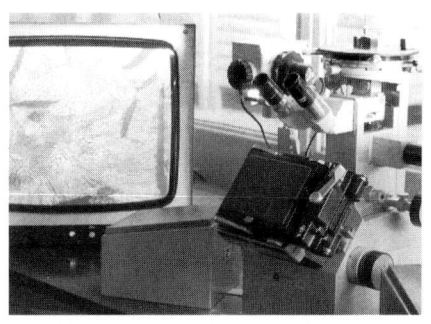

Metallographic analysis using Reichert Microscope.

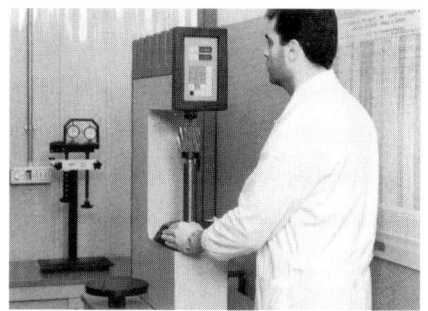

Hardness tests using Galileo D 141 - D.G. 501hardometer.

Dimensional control of hobs using Klingelnberg PNC 40 VA.

Graphic processing of the rolling deviations on single flank
of the finished component.

In-process controls.

Profile controls using Baty R 202.

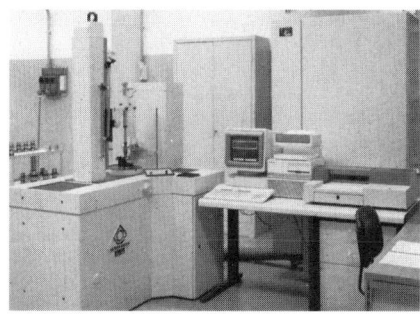

Controls of worm gear pairs, hobs and shaving tools using
Klingelnberg PNC 40 VA.

Part II

The Dynamics of Solids

and

Strength of Materials

Jacques Sprengers
Chairman ISO/TC 60

1 The Dynamics of Solids

1.1 Introduction

Mechanical machines are subject to the laws of general mechanics.

General mechanics include statics (the study of the speeds and the accelerations of solids under the action of forces) and kinematics (the study of particular forces produced by the movement and the mechanical energies induced).

This first section, which has to be considered as an introduction for gearboxes, includes only the study of the dynamics of machines. We will study the specific forces engaged in the machines (engines, gearboxes and driven machines) as well as the work of these forces and the necessary power to carry out this work under the given speed conditions.

1.2 Gravity Centre of Solids

If we exert on a solid body a field of force, i.e. whole of forces parallel to each other, applied to all the elementary masses of the solid and proportional to these masses, a resultant of these forces will be derived; a resultant parallel to these forces and applied to a particular point of the solid. Gravity is an example of a field of forces.

In reality, if we consider only a field of forces, we can only define a line of action of the resulting force, a line that passes for the point of application of this resultant. In order to fully establish the gravity centre exactly, three fields of forces not included in the same plane should be considered. The three resultants will then pass through the same point that is the gravity centre.

Let us now consider an elementary mass of a finished solid. The solid is made up of many of these elementary masses that, added together, give the mass of the solid. Let us now apply to all these masses an acceleration a_x in the direction of Ox, an acceleration a_y in the direction of Oy and an acceleration a_z in the direction of Oz, being Ox, Oy and Oz three orthogonal axes.

Every mass will be affected by a force proportional to its acceleration. Let us now consider the moment of the forces parallel to Ox with regard to the plane OxOz. By following the definition of the resultant, we will have:

$$F_y X = \sum dm\, a_x\, x_x. \qquad (1.001)$$

Or, since the resulting force F_y is equal to $m\, a_x$,

$$X = \frac{\sum dm\, x_i}{m}. \qquad (1.002)$$

By making a similar reasoning, taking into account the three fields and the three planes, we would have:

$$Y = \frac{\sum dm\, y_i}{m} \qquad (1.003)$$

and

$$Z = \frac{\sum dm\, z_i}{m}. \qquad (1.004)$$

These three equations form the definition of the position of the gravity centre with respect to the three chosen axes.

We can thus infer that the gravity centre is situated on the axes of symmetry of the solids.

If the solids are homogeneous, the axes of symmetry will be the geometric symmetry axes. A sphere's gravity centre is situated in the centre of the sphere; that of cylinder is positioned on the axis of the cylinder at mid-height. The gravity centre of a parallepiped is halfway between the three dimensions.

If a solid is made up of several elementary solids whose mass and gravity centre have a clear position, it will be possible to find its gravity centre through the application of equations (1.001), (1.002) and (1.003) by replacing the masses dm with the masses of the elementary solids and the co-ordinates of these masses with those of the elementary solids known gravity centres.

1.3 Movements of Solids

Solids can be driven by two types of simple movements: translation and rotation.

The movement of rotation is characterized by a rotation of all the masses *dm* forming a solid around an axis passing through a gravity centre that keeps still.

The movement of translation is characterized by a movement of all the masses *dm* that follow a trajectory parallel to the trajectory of the gravity centre. The most complicated movement is the superimposition of many of these simple movements.

1.4 Rectilinear Translating Movement

If the gravity centre draws a straight line during its movement, the latter will be called rectilinear. If this movement has a constant speed, it will be called uniform; otherwise, it will be called varied. If the varied movement acceleration is constant, it will be called uniformly varied.

1.5 Movement of Rotation

The main characteristic of the movement of rotation is that the angular speed and the angular acceleration are the same for all the points of the solid. The linear speeds and accelerations change from one point to another apart from the points situated at a same distance from the axis of rotation.

1.6 Mechanical Energy, Work and Power

1.6.1 Introduction

The different energies change from one to another. They are the electric energy, heat energy, mechanical energy, etc. Mechanical energy is the form of energy linked with movement. Work uses mechanical energy directly and in traslating movement is expressed by $dW = dF\ x$. The energy that is utilized or that can be utilized in a unit of time is power.

1.6.2 Potential Energy

The potential energy is a mechanical energy accumulated in a solid that can turn into work. This energy is the energy accumulated into a body with a mass *m* subjected to gravity with an acceleration g and situated at a height h. It can be expressed by:

$$E_p = mgh. \qquad (1.005)$$

If the mass m falls from the height h, the released work is equivalent to this potential energy. If the mass falls from h to h_0, the work saved will be:

$$W = mg(h - h_0). \qquad (1.006)$$

Potential energy can also be the energy accumulated into a spring. If the elastic constant of a spring is defined by the ratio of the force applied to this spring divided by its deformation, and if this constant is represented by c, the deformed spring's energy will be given by:

$$E_p = \frac{1}{2} c \lambda^2 \qquad (1.007)$$

λ being the valve's deformation. If the valve extends from λ to λ_0, the mechanical energy saved will be:

$$W = \frac{1}{2} c (\lambda^2 - \lambda_0^2). \qquad (1.008)$$

The elastic constant of a spring stretched by a movement of rotation is given by the relation between the spring's angular deformation and the torque producing it.
The potential energy will be given by:

$$E_p = \frac{1}{2} c_t \theta^2 \qquad (1.009)$$

if the constant is expressed by c_t and the deformation by θ. If the deformation changes from θ to θ_0, the work saved will be:

$$W = \frac{1}{2} c_t (\theta^2 - \theta_0^2). \qquad (1.010)$$

1.6.3 Kinetic Energy

Kinetic energy is a mechanical energy linked with speed. Let us consider a mass m falling from a height h under the action of gravity. The mass m will be subjected to an acceleration g, the height h will be covered in a time t so that:

$$t = \sqrt{\frac{2h}{g}} \qquad (1.011)$$

and the speed reached will be:

$$v = gt = \sqrt{2gh} \, . \qquad (1.012)$$

It follows that the height h expressed as a function of speed will be:

$$h = \frac{v^2}{2g} \, . \qquad (1.013)$$

The potential energy will be then given by:

$$E_p = mgh = \frac{mv^2}{2} \, . \qquad (1.014)$$

This energy, expressed as a function of the speed, is the kinetic energy equivalent to the developed potential energy.
So:

$$E_c = \frac{1}{2} mv^2 \, . \qquad (1.015)$$

For a movement of rotation, every elementary mass of the solid turning around the rotating axis at an angular speed ω, is driven by a linear speed $v = \omega\rho$. The whole of the elementary masses will then have an energy resulting from the sum of the energy of each mass, i.e.

$$E_c = \frac{1}{2} \iiint_V dm \, \omega^2 \rho^2 \, . \qquad (1.016)$$

The angular speed, being the same for all the points of the solid, will give:

$$E_c = \frac{1}{2} \omega^2 \iiint_V dm \, \rho^2 = \frac{1}{2} J_\Delta \omega^2 \qquad (1.017)$$

where J_Δ is the moment of inertia (see equation 1.039)

1.6.4 Work in the Uniform Rectilinear Movement

The uniform rectilinear movement is achieved by an active force opposite to a passive or resistant force. The different forms of passive forces will be analyzed later.

The active force originates a positive work opposite to the negative work of the passive forces. Being the identical forces and having an opposite direction, the solid is driven by a uniform rectilinear movement. The work of passive forces is equal to the product of the force by its shift in the direction of the force. The work of passive forces will be given by the same expression. If the force is inclined with respect to the shift, only the component of the force in the direction of the movement will be taken into consideration.

1.6.5 Work in the Uniform Movement of Rotation

If a solid is driven by a movement of rotation around an axis, an elementary mass dm is subjected to a force dF and its shift x is equal to the angular shift θ multiplied by the distance ρ from the centre of rotation ($x = \theta \rho$).

The elementary work of this mass is:

$$dW = dF \theta \rho \quad . \tag{1.018}$$

This elementary force gives a torque with respect to the centre of rotation with a movement equal to:

$$dT = dF \rho \quad . \tag{1.019}$$

If dF is taken out of this equation and if all the masses of the solid are taken into account, we will have:

$$W = \iiint_V dT \theta \quad . \tag{1.020}$$

Being the same angular shift for all the points of the solid, we will finally have:

$$W = \theta \iiint_V dT = T \theta \quad . \tag{1.021}$$

1.6.6 Power in the Uniform Rectilinear Movement

As a definition, power is the work comprised in a unit of time. If the work has been carried out in time t and if the distance covered at the uniform speed v is x, as a definition, this distance is equal to the speed multiplied by the time. We can then write:

$$P = \frac{F v t}{t} = F v \quad . \tag{1.022}$$

In the uniform rectilinear movement, power is equal to the product of force by speed.

1.6.7 Power in the Movement of Rotation

In the movement of rotation, the angular distance is the product of the angular speed by time ($\theta = \omega t$). If ω is the angular speed, the power will be:

$$P = \frac{W}{t} = \frac{T \theta}{t} = T \omega \quad . \tag{1.023}$$

Let us consider a force acting on a circumference with a diameter d, and provoking the rotation of a solid at the rotating speed ω. The force will be expressed in Newtons, the diameter in mm and the rotating speed in min^{-1}. The torque, expressed in N.m is equivalent to the product of the force by the radius, i.e. the half of the diameter expressed in mm:

$$T = \frac{1}{2000} F d \quad . \tag{1.024}$$

It follows that (being $\omega = 2\pi n$) the power is equal to:

$$P = \frac{1}{2000} F d \frac{2 \pi n}{60} \frac{1}{1000} \tag{1.025}$$

or:

$$P = \frac{F d n}{1.91 . 10^7} = \frac{T n}{9540} \quad . \tag{1.026}$$

1.7 Inertia

1.7.1 Introduction

The natural movement of a solid is the uniform rectilinear movement. Only an acceleration can modify this movement. Any kind of acceleration can be obtained only by applying to the solid a force proportional to its mass and to the desired acceleration. This force should overrule the natural tendency of the solid to keep a uniform movement. This natural tendency is the inertial force which is opposed to the acceleration that is about to be provoked.

1.7.2 Inertia of Solids During Translation

For a translation of a solid following a line, the direction of acceleration is given by following this line. All the elementary masses *dm* comprised in the solid are subject to the same acceleration. So, the inertial forces of these masses *dm* form a field of forces. If the acceleration is *a*, the inertial force F_i of the solid, applied to its gravity centre, will be equal to:

$$F_i = -m\,a \quad . \tag{1.027}$$

So, during a translation, the solid is reduced to a mass *m* concentrated at the gravity centre of the solid.

If the acceleration is the acceleration of gravity, the resulting force is the weight of the solid.

For a translation following a circumference, the movement will be a circular one. Let us take into account the uniform circular movement, i.e. the movement whose angular speed around the centre of the circumference is constant and then let us reduce the solid to a mass *m* concentrated in its gravity centre. The linear speed of the gravity centre is constant to its extent, but changes in direction. There is then a vector acceleration.

Let us consider (Fig. 1.1) the speed at time *t* and at time *t+dt*. The two speeds have the same extent, but they form an angle $d\theta$ that is equal to the angular shift of the considered mass. If *R* is the distance of the mass (i.e. of the gravity centre) from the centre of rotation and if ω is the constant angular speed, we will have:

$$\frac{v}{d\theta} = \omega\,\frac{R}{dt} \quad . \tag{1.028}$$

The acceleration is vector $\overline{AB}$:

$$\overline{AB} = \frac{v\,d\theta}{dt} \tag{1.029}$$

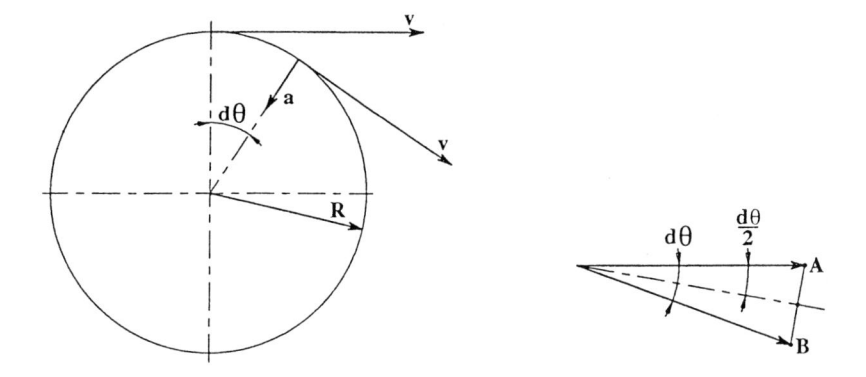

Fig. 1.1. Circular translation

divided by the time dt, as:

$$v = R\omega \qquad (1.030)$$

we will have acceleration a, expressed by the following equation:

$$a = R\omega \frac{d\theta}{dt} \qquad (1.031)$$

or:

$$a = \omega^2 R . \qquad (1.032)$$

This acceleration is directed towards the centre of rotation. It is called centripetal acceleration. The force to be gained is in the same direction, but has the opposite sense of this acceleration. It is called centrifugal force. It is equal to:

$$F_c = -m\omega^2 R . \qquad (1.033)$$

The work of the inertial forces during translation is the product of the inertial force by its shift. The power is the product of the inertial force by the speed reached at the end of the movement divided by 2 if the acceleration is constant. If the acceleration is not constant, the inertial force and speed are variable. The power can thus be obtained by an integral of the product of the two functions expressed according to time; an integral calculated according time.

1.7.3 Balance of Rotors. Imbalances

A rotor is a solid turning around an axis. If the rotation takes place at a constant angular speed, the movement can be considered a circular translation of the gravity centre around the axis considered. If the gravity centre is situated on this axis, the centrifugal force is null since the distance R is null. If the gravity centre is not situated on the axis of rotation, a centrifugal force given by equation (1.033) arises. The rotating solid is then subject to this force that turns with it. If the rotor is held up by supports, as well as supporting its normal functioning and the rotor's weight, these supports also maintain the effects of this force. The resulting reactions then move with the rotor and sometimes cause additional energies or vibrations. These energies and vibrations are stronger as the speed of rotation becomes higher. In the expression of the centrifugal force we can on the hand consider the influence of speed that is dependent solely on the rotor in use and the product of the mass exterted by the rotor's lack of centring or, on the other hand the product mR that depends only on the rotor's geometry. This product is called statical imbalance.

But, if we consider any point of the rotor's axis, the centrifugal force produces, at this point, a moment given by $\omega^2 m R x$. The product $m R x$ is called dynamic imbalance in regard to the point considered. For any point, this dynamic imbalance depends on the position of the centrifugal force in the rotor. Generally, and mainly at high speeds of rotation, these imbalances should be avoided. If the rotor turns around a symmetrical axis and if the gravity centre is situated exactly on this axis, the imbalances will be null. However, heterogeneities and which may be found in the manufacturing of rotors could put the real gravity centre out of the axis of rotation, as does happen. In order to ensure a correct movement, the imbalances should be balanced. If a static imbalance, equal and opposite to the existing balance, is placed in any of the rotor's points that does not exactly correspond to the point where the static imbalance of itself is placed, a supplementary moment will be introduced thus aggravating the imbalance. In order to balance the rotor, the compensating imbalance should be placed just on the plane of the existing imbalance. Often this is not possible, so the solution is to place two imbalances into the two planes in order to achieve this balance. The rotor's balancing is made by placing some static compensating balances onto two arbitrarily chosen planes. The value of these imbalances can be obtained by measurement on adequate machines.

1.7.4 Inertia During Accelerated Rotation. Moments of Inertia Solids

Let us consider (Fig. 1.2) a solid balanced with reference to an axis and turning around this axis with an angular acceleration B defined as follows:

$$B = \frac{d^2\theta}{dt^2} \quad . \tag{1.034}$$

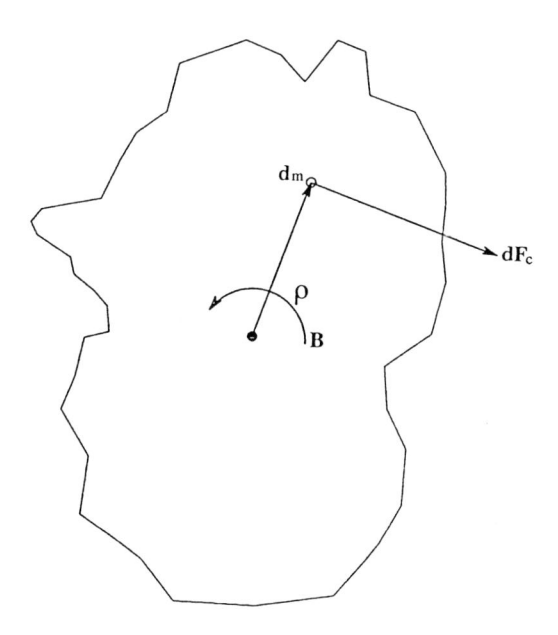

Fig. 1.2. Moment of inertia

Inside this solid we consider an elementary mass dm situated at the distance ρ from the axis of rotation. Its linear acceleration is equivalent to:

$$a = B\rho \quad . \tag{1.035}$$

The elementary mass then has an inertial force given by:

$$dF_c = dm\, a = B\rho\, dm \quad . \tag{1.036}$$

With regard to the axis of rotation, this force creates a moment given by:

$$dT_c = B\rho^2\, dm \quad . \tag{1.037}$$

The whole of the elementary masses that constitute the solid considered, have a moment given by:

$$T_c = \iiint_V B \rho^2 \, dm = B \iiint_V \rho^2 \, dm \quad . \quad (1.038)$$

The triple integral, when extended to the volume of the solid, depends solely on the geometrical elements and on the density of the solid, similar to the mass. It is called the "Mass-moment of inertia" of the solid as in the solid under consideration. It is represented by J_Δ.

$$J_\Delta = \iiint_V \rho^2 \, dm \quad (1.039)$$

The product of the moment of inertia using angular acceleration is the inertial torque of the solid around the axis considered. This is similar to the movement of translation since inertia is explained by the product of an acceleration by a characteristic of the solid. In the case of translation, the mass is multiplied by the linear acceleration and in the case of rotation the moment of inertia is multiplied by the angular acceleration. In the first case the result is a force, in the second a couple's moment. Let us also remember the fundamental difference in this sense that, if the acceleration is the same, the inertial force during translation will always be the same with any kind of movement; while, in the movement of rotation the moment of inertia and also the torque's movement depend on the axis of rotation. Consequently, there is an inertial torque for every axis of rotation considered.

For the prismatic solids with a length L, the mass dm can turn into the product of a section ds, with a constant length L and with a density δ, if the axis of rotation is also the axis of the prismatic solid. It follows that:

$$J_\Delta = L \delta \iint_S \rho^2 \, ds \quad . \quad (1.041)$$

The double integral is the polar moment of inertia of the section of the prismatic solid with regard to the opening point of the solid's axis in this section, i.e. as to the centre of this section. So, for a cylinder with a diameter d and a length L, the moment of inertia versus the cylinder's axis is equal to:

$$J_D = L \frac{\pi d^4}{32} \delta = \frac{\pi d^2}{4} L \delta \frac{d^2}{8} \quad . \quad (1.042)$$

If we observe that the cylinder's mass is :

$$m = L \delta \frac{\pi d^2}{4} \quad (1.043)$$

and if we define a inertial radius R_G given by:

$$R_G = \frac{d}{\sqrt{8}} \qquad (1.044)$$

we will have:

$$J_D = m R_G^2 . \qquad (1.045)$$

For a parallelepipedon with dimensions a, b and c with respect to its axis parallel to c, we will have:

$$m = a b c \delta \qquad (1.046)$$

$$R_G = \sqrt{\frac{a^2 + b^2}{12}} \qquad (1.047)$$

and equation (1.023) is still valid.

If a solid turns around an axis different from the one passing by its gravity centre without turning around this latter, the movement is a circular translation movement. In order to accelerate the solid a torque should be applied capable of overcoming inertia. This inertia is calculated by:

$$J_D = m R^2 \qquad (1.048)$$

where m is the mass of the solid and if R is the distance from its gravity centre to the axis considered.

If the solid turns around the axis passing by its gravity centre and around the axis parallel to this one, with the same we will have:

$$J_\Delta = J_{\Delta 0} + m R^2 . \qquad (1.049)$$

The moment of inertia with an index 0 is that which is calculated around the axis passing by the gravity centre of the solid.

If we consider the equation defining the moment of inertia, we will notice that this moment is as high as the mass is larger, but also that it is as high for a given mass the faster is this mass from the centre of rotation. It follows that if we want to produce a great moment of inertia with the weakest mass, we should move this mass from the centre of rotation. A rim with a large diameter will have, for a given moment of inertia, a weaker mass than that of a full disk. By contrast, if we want to obtain the smallest moment of inertia for a given mass, the solid should have the most compact shape around the rotation. It is also clear that if some masses have to be frequently accelerated, they also have to be as small as possible during translation (with the lightest of materials) or as small and as concentrated as possible around the axis of rotation in the case of a movement of rotation.

1.8 Friction

1.8.1 Fundamental Equation

If a solid stands on another solid with a normal force F_n and if there is the tendency to shift the first solid onto the other one without losing the force, a movement will be produced. It is an opposition that will turn into a force F_f which is proportional to the normal force and tangential to the two surfaces. The coefficient of proportionality is called the coefficient of friction of the two surfaces (one on the other) μ :

$$F_f = \mu \, F_n \, . \tag{1.050}$$

This coefficient of friction depends on the two materials of the two surfaces and also on the materials that, if necessary, could be put between the two surfaces. Here we are going to analyze the immediate friction, i.e. the friction between the two surfaces with nothing between them.

1.8.2 Angular Value of the Coefficient of Friction

Let us consider (Fig. 1.3) a solid on an inclined plane with a variable inclination α.

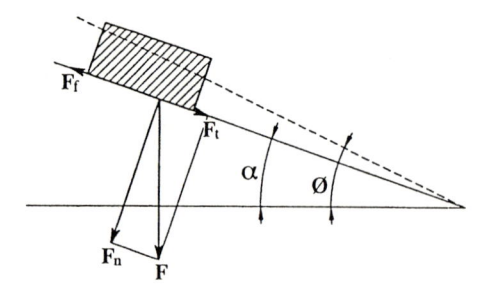

Fig. 1.3. Inclined plane

1 The Dynamics of Solids   35

The solid's weight F can be decomposed into a force F_n normal to the inclined plane and a force F_t tangential to this plane.
We have:

$$F_n = F \cos \alpha \qquad (1.051)$$

$$F_t = F \sin \alpha \qquad (1.052)$$

Under the action of the tangential force the solid tends to decompose. Under the action of the normal force, a tangential force standing against the shift is created: the force of friction. For a value ϕ of α, the two forces are the same. We can then write:

$$F \sin \phi = \mu \, F \cos \phi \qquad (1.053)$$

It follows that:

$$\mu = \tan \phi \qquad (1.054)$$

The angle ϕ is called angle of friction.

1.8.3 The Wedge Effect

Let us consider a corner with an angle α situated under a solid with which this corner is in contact on its inclined plane (Fig 1.4).

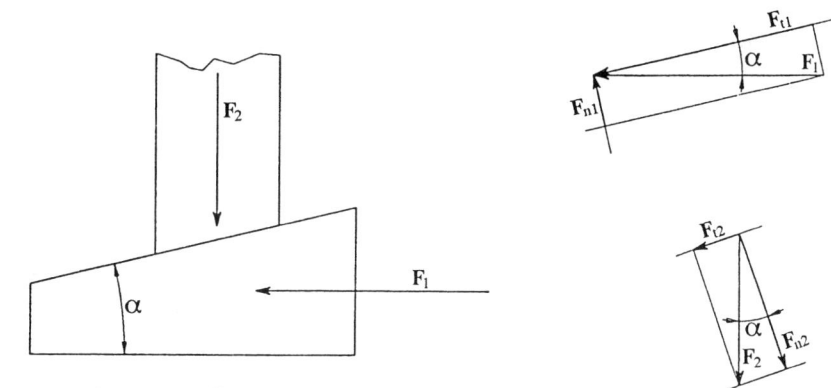

Fig. 1.4. Wedge effect

Let us call F_1 the force acting on the corner and F_2 the force acting perpendicularly on the corner on the second solid.

These forces can be decomposed into a normal force and a tangential force as to the connection plane of the two solids:

$$F_{t1} = F_1 \cos \alpha \qquad (1.055)$$

$$F_{n1} = F_1 \sin \alpha \qquad (1.056)$$

$$F_{t2} = F_2 \sin \alpha \qquad (1.057)$$

$$F_{n2} = F_2 \cos \alpha \ . \qquad (1.058)$$

The normal forces create a force of friction standing against the movement:

$$F_{f1} = F_{n1} \tan \phi = F_1 \sin \alpha \ \tan \phi \qquad (1.059)$$

$$F_{f2} = F_{n2} \tan \phi = F_2 \cos \alpha \ \tan \phi \ . \qquad (1.060)$$

If F_1 is active and F_2 passive, for a shift x along the inclined plane, the work of the tangential force F_{t1} is balanced by the passive works of the other tangential forces:

$$F_{t1} x = F_{t2} x + F_{n1} x + F_{n2} x \qquad (1.061)$$

or:

$$F_{t1} = F_{t2} + F_{n1} + F_{n2} \ . \qquad (1.062)$$

By replacing the forces with their values following equations (1.055), (1.056), (1.059) and (1.060), it will result that:

$$F_2 = F_1 \tan(\alpha + \phi) \ . \qquad (1.063)$$

If F_2 is active and F_1 is passive, it will follow that:

$$F_1 = F_2 \tan(\alpha - \phi) \ . \qquad (1.064)$$

We will notice that, if a force in the direction of F_1 has to be obtained, an infinite force should be applied (if the inclination of the plane is equal or inferior to the angle of friction). It is said that there is irreversibility, i.e. that this action is not possible.

Without friction, the relationship between the two forces would be:

$$F_1 = F_2 \tan \alpha \ . \qquad (1.065)$$

If F_1 is active, efficiency will be given by:

$$\eta = \frac{\tan\alpha}{\tan(\alpha + \phi)} \quad . \tag{1.066}$$

Otherwise, it will result:

$$\eta = \frac{\tan(\alpha - \phi)}{\tan\alpha} \quad . \tag{1.067}$$

This wedge effect is frequently used in mechanics: dovetailing, bolt and nut, worm and worm wheel, etc.

1.8.4 Friction of a Roller on its Axis

A roller whose diameter is D charged with a force F and whose axis has a diameter d is shown in figure 1.5. Under the action of the force F, the friction of the axis in its support gives rise to an effort of friction tangential to the axis $F_f = \mu F$. In order to make the roller turn by moving forward, an effort F_t should be applied at the centre of the roller to create a torque $F_t\, D/2$ capable of overcoming the resistant torque created by the force of friction, a torque equal to $F_f\, d/2$.

We have the relationship:

$$F_t = F\mu \frac{d}{D} \tag{1.068}$$

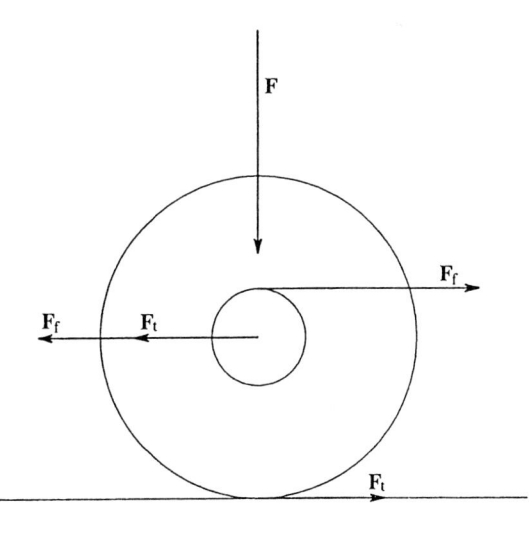

Fig. 1.5. Roller

1.8.5 Friction of a Belt Around its Pulley

Let us consider (Fig. 1.6) a belt on a pulley with a diameter d.

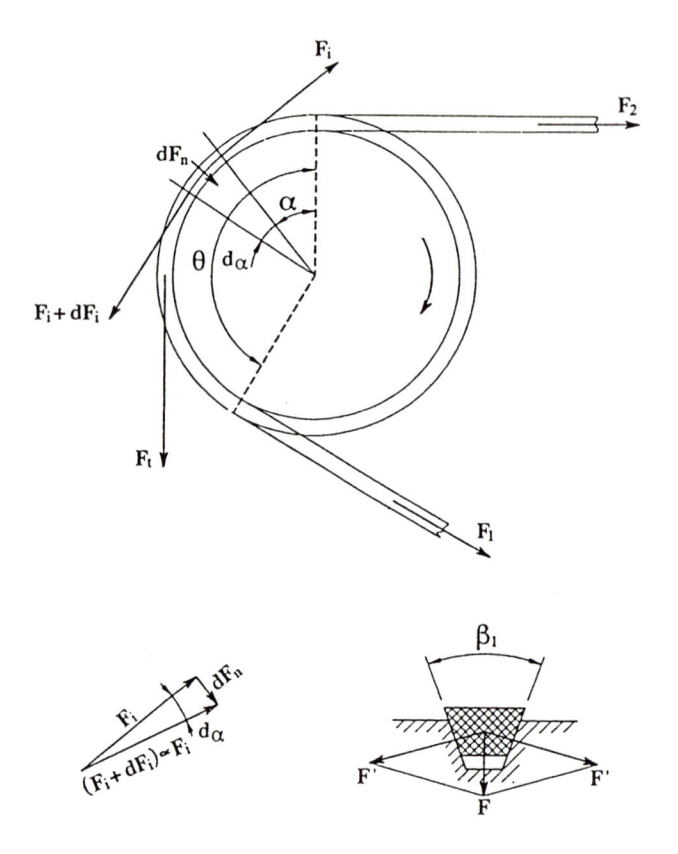

Fig. 1.6. Pulley and belt

Let us call F_1 the force in the upper branch and F_2 the force in the lower branch. A very small element of the pulley can be considered at any of the points forming an angle θ with the radius passing by the tangential point of the force F_1 to the pulley.

The two radiuses delimiting this stumb form an angle $d\alpha$. On one side the force is F_i and, if F_2 is smaller than F_1, the force on the other side of the stumb will be $F_i + dF_i$. These two compound forces give a force dF_n whose value is:

$$dF_n = F_i \, d\alpha \quad . \tag{1.069}$$

1 The Dynamics of Solids 39

This force creates a force of friction $\mu\, dF_n$ that is equal to dF_i. It will then result that:

$$\frac{dF_i}{F_i} = \mu\, d\alpha \qquad (1.070)$$

and by integrating the whole length of the belt in contact with the pulley:

$$\int_{F2}^{F1} \frac{dF_i}{F_i} = \int_0^\theta \mu\, d\alpha \qquad (1.071)$$

By solving this equation it consequently follows that :

$$\ln \frac{F_1}{F_2} = \mu\, \theta \qquad (1.072)$$

$$F_1 = F_2\, e^{\mu\theta} \qquad (1.073)$$

The force that can be transmitted to the pulley by the belt, or the force that can be transmitted by the pulley to the belt, is the force F_t given by:

$$F_t = F_1 - F_2 = F_1 \frac{e^{\mu\theta} - 1}{e^{\mu\theta}} \qquad (1.074)$$

The couple that can be transmitted is:

$$T = \frac{F_t\, d}{2} = \frac{1}{2} F_1\, d\, \frac{e^{\mu\theta} - 1}{e^{\mu\theta}} \qquad (1.075)$$

The above-mentioned theory is valid for plate belts. It can also be applied to trapeizoidal belts, only if the forces on the surfaces of contact with the pulley are taken into account (these surfaces being inclined with regard to the surface of plate belts). The forces to be considered are equal to:

$$F' = \frac{F}{\sin \beta} \qquad (1.076)$$

So the trapeizoidal belts can transmit a couple bigger than that of the plate belts.

1.8.6 Adhesion

When a roller has to shift on ist sliding plane, it has to be equipped with a torque capable of overcoming any resistances. This torque is termed T. The force of traction F_t, d is the roller's diameter, is equal to:

$$F_t = \frac{2T}{d}. \qquad (1.077)$$

The force applied to the roller during its sliding plane is termed F_n. The contact between the roller and the sliding plane creates an effort of friction μF_n creating a torque T_m given by:

$$T_m = \mu F_n \frac{d}{2}. \qquad (1.078)$$

If the torque resulting from friction is inferior to the torque to be transmitted, the roller will skate on its sliding plane. The transmission will be possible only if:

$$F_n > \frac{F_t}{\mu}. \qquad (1.079)$$

See Fig. 1.7

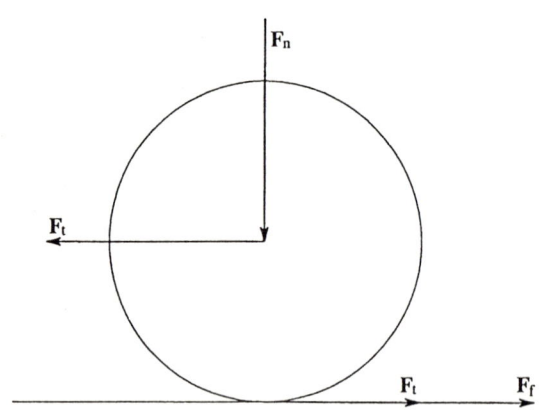

Fig. 1.7. Adhesion

This condition takes place for any kind of vehicle. It can be extended to the lateral adhesion of the road vehicles. If the resistant torque is created by a normal charge to the sliding plane F and the necessary force of traction is equivalent to K

times to this charge, being the number of motor rollers n'. The force of traction for roller will be:

$$F_t = K \frac{F}{n'} \quad . \tag{1.080}$$

If n is the number of motor rollers, the adhesion will be:

$$F_a = \mu \frac{F}{n} \quad . \tag{1.081}$$

As the adhesion F_a is greater than the force of traction for motor rollers, we should have:

$$n' > \frac{K}{\mu} n \quad . \tag{1.082}$$

1.8.7 Friction on Disks (Fig. 1.8)

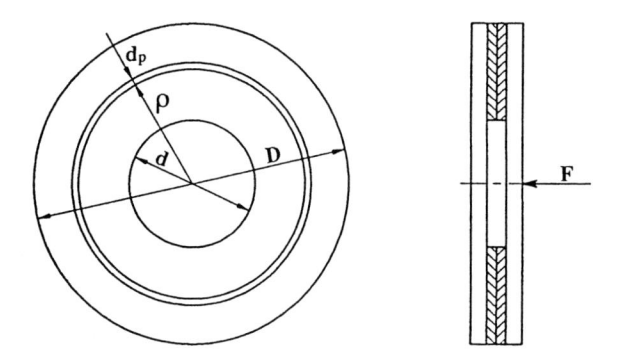

Fig. 1.8. Disks

Let us consider two disks with an outer diameter D and an inner diameter d, standing one on the other with a force F. The force F creates a pressure p on the disks, given by:

$$p = \frac{F}{S} = \frac{4F}{\pi (D^2 - d^2)} \quad . \tag{1.083}$$

This pressure creates a force on a rim with a width $d\rho$ situated on a radius ρ given by:

$$dF_t = 2\mu p \pi \rho \, d\rho \qquad (1.084)$$

and consequently, a couple:

$$dT = dF_t \, \rho = 2\pi \mu p \rho^2 \, d\rho \qquad (1.085)$$

For all the surfaces we will have:

$$T = \int_{d/2}^{D/2} 2\pi p \mu \rho^2 \, d\rho \qquad (1.086)$$

or:

$$T = \pi p \mu \frac{D^3 - d^3}{12} = \frac{1}{3} F\mu \frac{D^3 - d^3}{D^2 - d^2} \qquad (1.087)$$

If the disks are cones, the force to be considered, with a conical angle equal to γ, is:

$$F' = \frac{F}{\sin\frac{\gamma}{2}} \qquad (1.088)$$

1.8.8 Values of some Coefficients of Friction

Steel on steel	0.3 to 0.4
Steel on grey pig iron	0.12 to 0.2
Steel on rubber	0.25 to 0.35
Rubber on rubber	0.35 to 0.6
Steel on wood	0.45 to 0.6
Steel on asbestos	0.3 to 0.4
Steel on lubricated steel	0.06 to 0.12

1.9 Rolling Resistance (Fig. 1. 9)

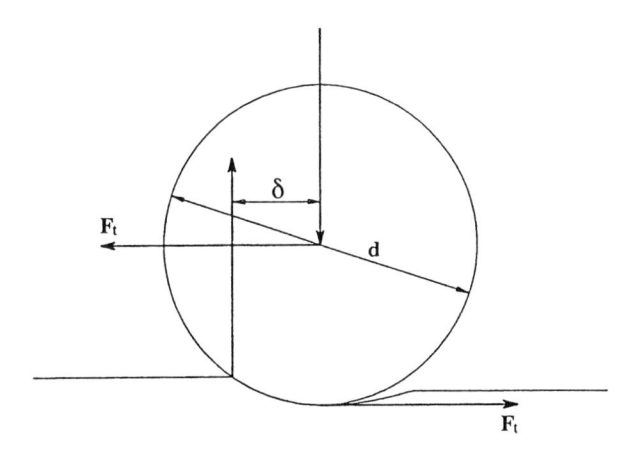

Fig. 1.9. Rolling

When a roller stands on its sliding plane with a force F, the roller and the sliding plane are subject to a deformation. This deformation depends on the effort and the material. Of course, this deformation is higher for a tyre than for a steel wheel: this places the reaction of the roller on the sliding plane advanced of δ in the direction of the movement. The force F and its equal reaction create a torque whose moment T is given by:

$$T = F\delta . \qquad (1.089)$$

If the roller is shifted, a force of traction F_t corresponding to a torque $T_m = F_t\, d/2$ has to be exerted. This torque has to overcome the resisting torque created by the deformation of the sliding plane.

It will follow that:

$$F_t = 2F\frac{\delta}{d} . \qquad (1.090)$$

The rolling resistance is defined through the extent without dimension $\delta_0 = \delta/d$, so the rolling resistance takes the form of a friction resistance:

$$F_t = F\delta_0 . \qquad (1.091)$$

The values of this coefficient of friction are changeable. For steel on steel (wheel on rails) the value can be estimated around 0.01.

1.10 Couple Necessary for a Roller of Translation

Let us consider a translation roller with a diameter D and whose axis has a diameter d. The load on the roller is F. The force of traction to be gained is given by:

$$F_t = F\left(\mu \frac{d}{D} + \delta_0\right) . \tag{1.092}$$

It follows that the couple to be applied on the roller's axis is equal to:

$$T = \frac{1}{2000} FD\left(\mu \frac{d}{D} + \delta_0\right) . \tag{1.093}$$

If the number of carrier-rollers is n and the number of driving rollers is n', F being the total load, the couple on every motor will be given by:

$$F' = \frac{F}{n}\frac{n}{n'}\left(\mu \frac{d}{D} + \delta_0\right) \tag{1.094}$$

or:

$$F' = \frac{F}{n'}\left(\mu \frac{d}{D} + \delta_0\right) . \tag{1.095}$$

This force has to be lower than the adhesion of the driving rollers. The torque will be then:

$$T = \frac{1}{2000}\frac{F'}{n'} D\left(\mu \frac{d}{D} + \delta_0\right) . \tag{1.096}$$

1.11 Cranking Torque

The cranking torque of a mechanical body is constituted by a couple necessary for a constant-speed movement increased by the couple caused by the inertial forces. In some applications, several masses are used to balance useful masses statically, such as, for example, in the case of elevators where a balance weight balances the mass of the cabin, or in a vertical elevator where all the descending elements except the charge are in use balanced by the rising elements. These masses create several opposite couples in the constant-speed movement and so are not taken back by the driving motor. However, when starting, any kind of mass has to be accelerated in the same direction. The masses then take part in the cranking torque. The balancing masses add inertia to starting.

1.12 Kinematic Chain

1.12.1 Definition

A kinematic chain is a collection of rotating axes that mutually carry each other by some transmission units. Consequently, the axes constituting this chain have mutually-dependent speeds.

1.12.2 Torques

Let us call $n_1,....,n_i,....n_j,....n_n$ the speeds of rotation of the axes $1,....,i...j,....n$ respectively. A power that can be considered constant, if we ignore efficiency, passes through the kinematic chain. Let us call this power P. The torque on every shaft will be:

$$T_i = 9550 \frac{P}{n_i} . \qquad (1.097)$$

It follows that the relationship of the torques on two separate axes is given by:

$$\frac{T_i}{T_j} = \frac{n_j}{n_i} . \qquad (1.098)$$

If we consider the efficiency, the driving shaft and the driven shaft have to be defined. Let us call i the leading shaft and j the driven shaft. The torque on the shaft j will be lower than that calculated on the hypothesis of an efficiency equal to 1. It will follow that:

$$\frac{T_i}{T_j} = \frac{n_j}{n_i} \frac{1}{\eta} . \qquad (1.099)$$

1.12.3 Inertial Moments

Let us call J_j the inertial moment on the axis j. As all the axes are tied together, on the kinematic chain it is possible to consider an equivalent inertial moment on a a shaft as for the starting of the whole. The axes' accelerations are linked to each other by the same relationship linking speeds. Let us call B_i the angular acceleration on the axis i and B_j that on the axis j. We can say that:

$$\frac{B_i}{B_j} = \frac{n_i}{n_j} . \tag{1.100}$$

The inertial torque on the axis j is:

$$T_j = J_j B_j . \tag{1.101}$$

Let us call J_{ji} the moment of inertia on the shaft i equivalent to the moment of inertia J_j on the axis j. The corresponding torque on the axis i will be:

$$T_{ji} = J_{ji} B_i . \tag{1.102}$$

It should be equal to the one given by equation (1.101) transferred on the axis by virtue of the law concerning the torque. By ignoring the efficiency, it will follow that:

$$J_{ji} B_i = J_j B_j \frac{n_j}{n_i} \tag{1.104}$$

and finally :

$$J_{ji} = J_j \left(\frac{n_j}{n_i}\right)^2 . \tag{1.105}$$

1.12.4 Elastic Constants of Axes

The elastic constant is the relationship between the couple applied and the deformation it produces. These constants can also be shifted from one axis to another to obtain equivalent constants to be used in the calculation. It will follow that:

$$c_{ji} = c_j \left(\frac{n_j}{n_i}\right)^2 . \tag{1.106}$$

1.12.5 Application of a Lifting Winch on a Bridge Crane

The lifting of a load by a bridge crane is made through the winding of a cable on a drum. Let us call T_t the torque necessary for the drum. The speed of rotation of the drum should be obtained from the motor by a speed reduction gear whose total reduction ratio is i. Consequently, the torque necessary to the motor at a constant lifting speed will be, if η is the reduction gear efficiency:

$$T_m = \frac{T_t}{i} \frac{1}{\eta} \ . \tag{1.107}$$

Upon starting, together with the load acceleration imposed, we know the angular acceleration of the drum. The engine's angular acceleration will be i times higher. Let us call J_t the moment of inertia of the drum and J_c the moment of inertia equivalent to that of the load transformed to the drum, on the driving shaft. So the equivalent moment of inertia will be given by:

$$J_1 = (J_t + J_c) \frac{1}{i^2} \ . \tag{1.108}$$

J_m being the moment of inertia of the driving shaft and B_m the angular acceleration of the driving shaft, the torque necessary for starting will be:

$$T_d = (J_m + J_1) B_m + T_m \ . \tag{1.109}$$

The rated capacity of the engine, if n_m is its speed of rotation, will be:

$$P \geq \frac{T_m \ n_m}{9540} \ . \tag{1.110}$$

So we could chose the motor and verify whether its braking torque is satisfied. During descent, as for the brake supported by the driving shaft, the drum will have to be overcome; being T_t the torque equal to that of lifting, but here it is a driving torque and efficiency that acts in the opposite direction. It will follow that:

$$T_f = \frac{T_t}{i} \eta \ . \tag{1.111}$$

The braking torque will be the driving torque multiplied by the square of the efficiency.

2 Strength of Materials

2.1 Single Traction or Compression

This simple deformation takes place if the resultant of the outer forces on one side of the section becomes a unique force perpendicular to the section and passing by its centre of elasticity.

If this resultant is called F_n, the balance of the outer and the inner forces is given by:

$$\sigma_n = -\frac{F_n}{A} \leq \sigma_{Pn} \qquad (2.001)$$

σ_{Pn} being the allowable stress for traction or compression and A being the area of the considered section.

The deformation is the expansion following the longitudinal axis of the solid. For an initial length dx, the expansion $d\lambda$ will be:

$$d\lambda = \frac{F_n}{EA} dx \qquad (2.002)$$

E being the coefficient of elasticity of the material.

For a length ℓ, the deformation will be:

$$\lambda = \int_0^\ell \frac{F_n}{EA} dx \qquad (2.003)$$

For a prismatic solid (A = constant) and a constant force of traction, we will have:

$$\lambda = \frac{F_n}{EA} \ell \qquad (2.004)$$

The work of deformation is expressed by:

$$W_n = \frac{1}{2} \int_0^\ell F_n \, d\lambda \quad . \tag{2.005}$$

This leads to the two following expressions:

$$W_n = \frac{1}{2} \int_0^\ell \frac{F_n^2}{EA} \, dx \tag{2.006}$$

$$W_n = \frac{1}{2} \int_0^\ell \frac{A \sigma_n^2}{E} \, dx \quad . \tag{2.007}$$

For a prismatic solid subject to a constant force of traction, we will have:

$$W_n = \frac{1}{2} \frac{F_n^2 \ell}{EA} \tag{2.008}$$

$$W_n = \frac{1}{2} EA \frac{\lambda^2}{\ell} \quad . \tag{2.009}$$

2.2 Shearing

If the resultant of the forces situated on one side of the section is a force passing through the centre of elasticity of the section and is situated in the plane of this section, the deformation is called shearing. As for traction, we have the following equations:

$$\sigma_c = \frac{F_c}{A} \tag{2.010}$$

$$\gamma = \frac{d\lambda}{d\ell} = \frac{\sigma_c}{G} \tag{2.011}$$

where G is the transverse coefficient of elasticity of the material, σ_{Pc} is the max. shear allowable, F_c is the strenghts resultant and γ is the shearing factor.

2.3 Simple Bending

If the outer forces situated on one side of the section become a couple whose moment is on the section's plane, the deformation is called bending. The moment of the couple is called the "bending moment" and is represented by M_b. If the bending moment (sometimes called moment of flexure) is confused with one of the section's main inertial axes (or confused with a symmetry axis, if there is one), the bending is called plane bending. Otherwise it is called left bending.

2.3.1 Plane Bending

The section's main inertial axes should be such that the geometrical inertial moment is at its maximum if it follows one of these axes and at its minimum if it follows the other. On the other hand, M_b has to be the bending moment confused with the axis that, if followed by the geometrical inertial moment, makes this latter become maximum. In a section parallel to this axis and situated at a distance y from this axis, the result will be a stress given by:

$$\sigma_y = \frac{M_b \, y}{I_{max}} \qquad (2.012)$$

Following the axis ($y = 0$) the stress is null. This axis is called neutral fibre. On one side of the section, by following the sign of the moment, the stress is positive while on the other it is negative. As these stresses are perpendicular to the section, they are traction stresses or compression stresses, respectively. So one part of the section is compressed and the other is stretched. If the bending moment was confused with the other axis, similar equations (but with I_{min} instead of I_{max}) would result. The maximum compression stress is achieved for the maximum distance $v-$ of y in the compressed area and the maximum traction stress is achieved at the highest distance $v+$ of y in the stretched area. If the material's resistance to traction and compression are different, it should be verified that each calculated stress is inferior to the corresponding permissible stress. If the two materials have the same resistance to compression and traction (as for steels), the maximum traction and compression stresses could be different if the corresponding distances v are not the same. This will be checked with the furthest point of the neutral fibre. If the section is symmetric to the neutral fibre, we will have:

$$\sigma_y = \frac{M_b}{I/v} \qquad (2.013)$$

where I is the geometrical inertial moment versus the neutral fibre. I/v, that is a peculiarity of the section, is called the bending factor.

For a circular section:

$$I = \frac{\pi d^4}{64} \qquad (2.014)$$

and

$$I/v = \frac{\pi d^3}{32} \qquad (2.015)$$

For a rectangular section:

$$I = \frac{b a^3}{12} \qquad (2.016)$$

and

$$I/v = \frac{b a^2}{6} \,. \qquad (2.017)$$

with b being the dimension parallel to the neutral fibre.

The section, under the moment's action, turns around the neutral fibre and with respect to a section at a distance dx, it turns around an angle given by:

$$d\phi = \frac{M_b}{EI} dx \,. \qquad (2.018)$$

The moment varies along the solid's axis, the section and the coefficient of elasticity. For all the points of the solid we will have:

$$\phi = \int \frac{M_b}{EI} dx \,. \qquad (2.019)$$

The angle ϕ is the section's rotation at the point x.

The solid's axis deformation is a shift onto the plan perpendicular to the moment (and so the neutral fibre) designated by an arrow. This shift varies along the axis and its equation is given by:

$$y = \int \phi \, dx \,. \qquad (2.020)$$

If the solid is prismatic and homogeneous, E and I can be taken out of the integral.

2.3.2 Left Bending

If the bending moment is not aligned with one of the main vertical axes, the bending is called left bending. Let us call ψ the angle of the moment with the inertial axis as to which the inertial moment is at its maximum. The neutral fibre is not aligned with the moment, but it forms an angle β with the same main inertial axis given by:

$$\tan\beta = \tan\psi \, \frac{I_{max}}{I_{min}} . \qquad (2.021)$$

The maximum stress is the maximum of the value given by:

$$\sigma_b = \frac{M_b \cos(\beta - \psi)}{I_{FN}/v} \qquad (2.022)$$

with:

$$I_{FN} = I_{max} \cos^2\beta + I_{min} \sin^2\beta \qquad (2.023)$$

and

$$v = y \cos\beta - z \sin\beta \qquad (2.024)$$

y and z being the co-ordinates of a point on the section's outline with regard to the inertial axis with the greatest inertial moment and to the axis with the smallest inertial moment, respectively.

We notice the following:

(1) In a circular section all the diameters have symmetrical axes and the inertial moments are all the same. We cannot speak of left bending in the strict sense of the word.

(2) The left bending of a rectangular section can easily be calculated by projecting the bending moment onto the two symmetry axes of the section. Every projection creates a plane bending to which the equations outlined in the paragraph of plane bending can be applied. In this way we can obtain a maximum bending stress by following one side of the rectangle and a minimum bending stress by following the other. Let us call σ_y and σ_z these two stresses. The resulting stress is

given by:

$$\sigma_b = \sigma_y + \sigma_z \quad . \tag{2.025}$$

The deformation is calculated for every simple bending. We obtain a deformation y inside one plane and a deformation z in the other. The resulting arrow is:

$$\omega = \sqrt{z^2 + y^2} \quad . \tag{2.026}$$

The resulting deformation is not always situated in the same plane, which is why it is called left bending.

2.4 Torsion

There is a torsion when the outer forces situated on one side of the section become a torque whose moment is perpendicular to the section. The torsion's mathematical theory can be correctly applied only to circular sections.

The stress at the section's edge, T being the torsional moment, is given by:

$$\tau_{t\ max} = \frac{T}{I_0/r} \tag{2.027}$$

I_0 being the polar inertial moment and I_0/r being the section's coefficient of torsion. For a circular section with a diameter d, it will result that:

$$I_0 = \frac{\pi\ d^4}{32} \tag{2.028}$$

and

$$I_0/r = \frac{\pi\ d^3}{16} \quad . \tag{2.029}$$

For non-circular sections, the calculation is empirical. The deformations provoked by torsion are given by:

$$\theta = \int \frac{T}{G I_0} dx \quad . \tag{2.030}$$

For a prismatic and homogeneous stump with a constant torsional couple, we have:

$$\theta = \frac{T\ell}{GI_0} \qquad (2.031)$$

ℓ being the stump's length.

2.5 Buckling

The combined compressing and bending stress is a lateral deformation under the action of a compressive force. The risk of buckling depends on the rusting coefficient that is equal to:

$$\alpha = \frac{\ell_f}{i} \qquad (2.032)$$

with:
ℓ_f = free length of combined compressing and bending stress, and i = radius of gyration of the section

$$i = \sqrt{\frac{I}{A}} \qquad (2.033)$$

being I the minimum moment of inertia and being A the area of the section.
The material is carachterized by the extent:

$$\beta = \pi^2 \frac{E}{\sigma_E} \qquad (2.034)$$

E being the coefficient of elasticity and σ_E the elastic limit.
If:

$$\alpha \geq \sqrt{\beta} \qquad (2.035)$$

the Euler analytic formula can be applied:

$$\sigma_{crit} = \frac{\beta \, \sigma_E}{\alpha^2} \qquad (2.036)$$

Otherwise, an empirical formula is applied. The most common one is the Tetmayer formula:

$$\sigma_{crit} = \sigma_E + \gamma \left[\sqrt{\beta} - \alpha\right] \quad . \tag{2.037}$$

For steels, the coefficient of elasticity is equal to 206000 MPa and $\gamma = 1.14$.

The defined critical stress is used as a compressive stress. However, the permissible stress is weaker than a pure compressive stress because of the very high risk provoked by the buckling.

If buckling is combined with bending, an equivalent bending stress may be calculated to be added to the bending stress and given by the Rankine formula:

$$\sigma_f = \frac{F_n}{A}\left[1 + \frac{1}{\beta}\alpha^2\right] \quad . \tag{2.038}$$

The useful length of the buckling depends on the supports. This length is shown in Fig. 2.1

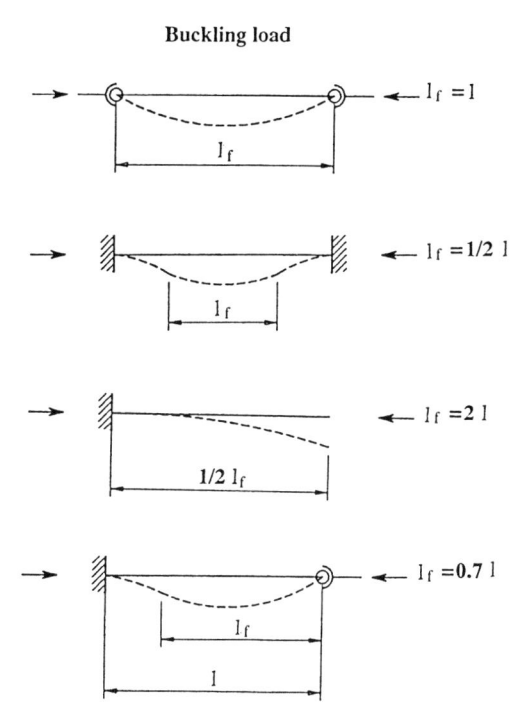

Buckling load

$l_f = l$

$l_f = 1/2\ l$

$l_f = 2\ l$

$l_f = 0.7\ l$

Fig. 2.1. Buckling

2.6 Composition of Simple Stresses

If some simple stresses are applied contemporaneously, they should be combined together. If these stresses have the same direction, they can be added to each other. If they have a different direction, they must be combined by following an empirical criterion. The most widely used criterion is the Von Mieses and Hencky. If the stresses are expressed by a normal effort and a tangential effort, the criterion is expressed by:

$$\sigma_{equ} = \sqrt{\sigma^2 + 3\tau^2} \quad . \tag{2.039}$$

For a spatial distribution (three axes) of the stresses, if σ_I, σ_{II} and σ_{III} are the main stresses in three orthogonal directions, the stress of comparison will be:

$$\sigma_{equ} = \sqrt{\frac{1}{2}\left[(\sigma_I - \sigma_{II})^2 + (\sigma_{II} - \sigma_{III})^2 + (\sigma_{III} - \sigma_I)^2\right]} \quad . \tag{2.040}$$

Main stresses are normal stresses applied in section where the tangential stress is equal to zero. They are orthogonal.

In a plane defined by the axes O_x and O_y, if we have normal stresses σ_x and σ_y and a shearing stress τ_{xy} the main stresses are defined by:

$$\sigma_I = \frac{1}{2}(\sigma_x + \sigma_y) + \frac{1}{2}\sqrt{(\sigma_x - \sigma_y)^2 + 4\tau_{xy}^2} \tag{2.041}$$

$$\sigma_{II} = \frac{1}{2}(\sigma_x + \sigma_y) - \frac{1}{2}\sqrt{(\sigma_x - \sigma_y)^2 + 4\tau_{xy}^2} \quad . \tag{2.042}$$

For a spatial distribution of stresses we have three orthogonal main stresses, σ_I, σ_{II} and σ_{III}.

In the case of a plane distribution with $\sigma = \sigma_x$, $\sigma_y = 0$ and $\tau_{xy} = \tau$, we have

$$\sigma_I = \frac{1}{2}\sigma + \frac{1}{2}\sqrt{\sigma^2 + 4\tau^2}$$

$$\sigma_{II} = \frac{1}{2}\sigma - \frac{1}{2}\sqrt{\sigma^2 + 4\tau^2}$$

$$\sigma_{III} = 0$$

and equation (2.040) becomes equation (2.039).

For a distribution with σ_x, σ_y and τ we have

$$\sigma_{equ} = \sqrt{\sigma_x^2 + \sigma_y^2 - \sigma_x \sigma_y + 3\tau^2} \qquad (2.043)$$

Part III

Gear Reduction Units

Jacques Sprengers
Chairman ISO/TC 60

1 Stresses and Materials

1.1 Introduction

Gearboxes are made up of mechanical components, which comply with a series of common rules as far as operational reliability is concerned. They are also made of materials which appear very similar even if completely different. Generally, gear reduction units are composed of metallic materials showing similar features that need to be accurately defined. These components are subject to torque stresses resulting from power transmission and from the rotating speed of the gearing parts. The relation existing between torque stresses, power and the rotation speed is then particularly important as to the load capacity of gearboxes.

In this chapter we will generally analyse the main subjects of this work. We will study them more accurately when we examine unit calculations.

1.2 Stresses

Generally, a constant power (efficiency excluded) runs through gearboxes while the gearing units run at different speed: the transmission shaft has to support a torque that is directly proportional to power and inversely proportional to rotation speed. Power is expressed in kW (kilowatt), while the rotation speed is expressed in revolutions per minute (min^{-1}). The torque is calculated in Newton metres (Nm) and is equal to:

$$T = 9550 \frac{P}{n} \qquad (1.001)$$

where P represents the power and n the rotation speed.

Transmission units, in particular gears, transform the torque into forces which are applied to the different components of gears.

Torques and forces produce bending stress and/or traction-compression stress that can cause damage to the different parts of gears. If the value of such stresses is higher than that allowed, the units can break down.

Stresses comply with a series of rules concerning the strength of materials already known. Their application is shown further on when we examine the various gear components.

1.3 Materials: Properties and Features

1.3.1 Permissible Static Stress

Calculation of the stress allowed is based on the normalized tensile test on a given material. During this test, a test bar must bear and contemporarily measure a stress, the intensity of which is progressive and characterized by a slow increasing speed.

The test bar is cylindrical in shape and shows a smooth unchanging section (without notches) the dimensions and tolerances of which are normalized. At the same time it is possible to measure the test bar elongation under the effects of the stress. They report on the axis of the abscissa, the test bar elongation and on the axis of ordinate, the corresponding stress. Looking at the graph in figure we notice that:

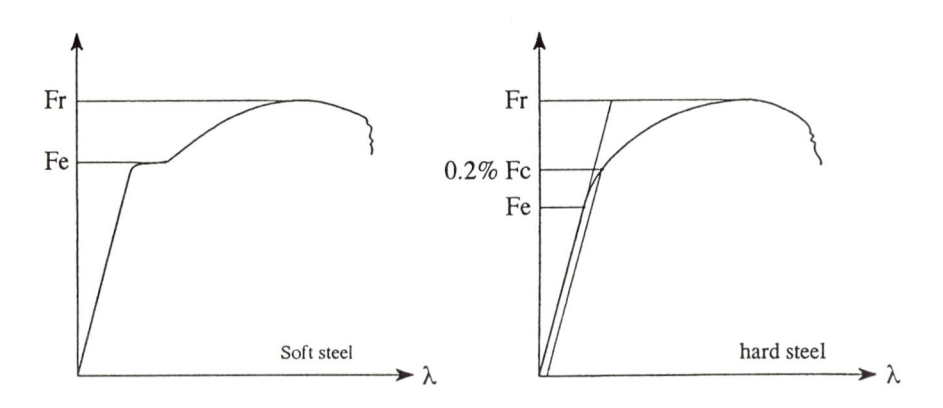

Fig. 1.1 Traction-Graphs

First of all the curve reaches a maximum value after which the stress intensity diminishes. Such reduction seems to depend on the increase in elongation. In reality the section of the test bar diminishes and such decrease ends with the breaking down of the test bar.

The maximum value can easily be calculated: the division of the corresponding stress by the original section of the test bar enables experts to determine the breaking stress of the tested material.

Then we notice that at the beginning of the graph, the stress and the elongation are proportional for a determined interval of time: if during such lapse of time the stress stops, the test bar will regain its original shape.

At a certain point, things change: experts can obtain the elastic limit of the material considered during the tensile test through the division of the quotient of tensile stress at this point by the original section of the test bar. It stands to reason that such a point must not be depassed otherwise elastic failure will occur. The resulting elastic stress represents the maximum allowable stress. The values relating to such stress do not always clearly emerge from the graph, so experts prefer to multiply the breaking stress by a factor, which is lower than the unit, in order to find the standard limit of elasticity. However, such stress does not represent the stress allowed for a material forming a given unit.

First of all any industrial working in proximity to the considered section does not enable workers to obtain a surface-layer equal to that of the test bar. A different superficial layer can give rise to breaks and to a limit of elasticity, which is different from those of the test. Therefore experts have to modify the normalized value on the basis of a roughness factor.

The unit does not always have a constant section and during some experiments, experts noticed that, supplementary stresses would be produced, if variations in section took place. This phenomenon is called stress concentration. It can be prevented by applying a modification factor.

The stresses following operation are not all tensile stresses; a different factor taking such diversities into consideration must be applied. The shape and the dimension of the mechanical unit do not correspond to those of the normalized test bar, therefore experts must also consider such aspects as far as the application of shape and dimension factors are concerned. Often additional factors are deduced empirically.

The stress obtained after such modifications is known as permitted stress and is expressed as follows: σ_{statP}

1.3.2 Fatigue. Endurance Limit

Periodical Stresses. In a mechanical component stresses under the action of loads can show periodic variations in value (maximum value and minimum value).

$$\sigma_{max} = \sigma_m + \sigma_a \qquad (1.002)$$

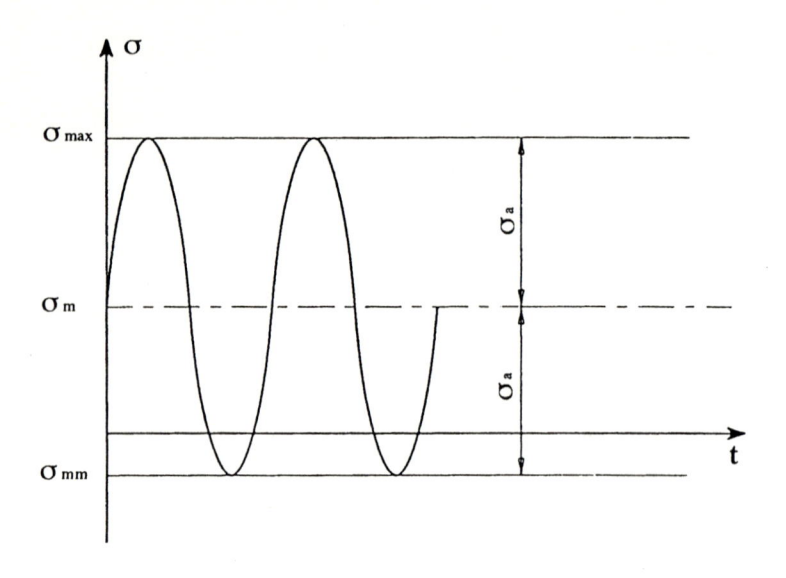

Fig. 1.2 Periodical stresses

$$\sigma_{min} = \sigma_m - \sigma_a \qquad (1.003)$$

$$\sigma_m = \frac{1}{2}(\sigma_{max} + \sigma_{min}) . \qquad (1.004)$$

Periodical stresses can be characterized by the value of the middle stress σ_m and by the ratio expressed by the letter κ representing the value defined as follows:

$$\kappa = \frac{\sigma_{min}}{\sigma_{max}} = \frac{\sigma_m - \sigma_a}{\sigma_m + \sigma_a} . \qquad (1.005)$$

The Fig. 1.3 represents the particular values of periodical vibrations, where:

κ= -1 : Alternate stresses

$$\sigma_m = 0$$
$$\sigma_{max} = \sigma_a$$
$$\sigma_{min} = -\sigma_a$$

κ= 0 : Repeated stresses (pulsating stresses)

$$\sigma_m = \sigma_a$$
$$\sigma_{max} = 2\sigma_a$$
$$\sigma_{min} = 0$$

$\kappa = +1$: Constant effort

$$\sigma_a = 0$$
$$\sigma_{min} = \sigma_{max} = \sigma_m$$

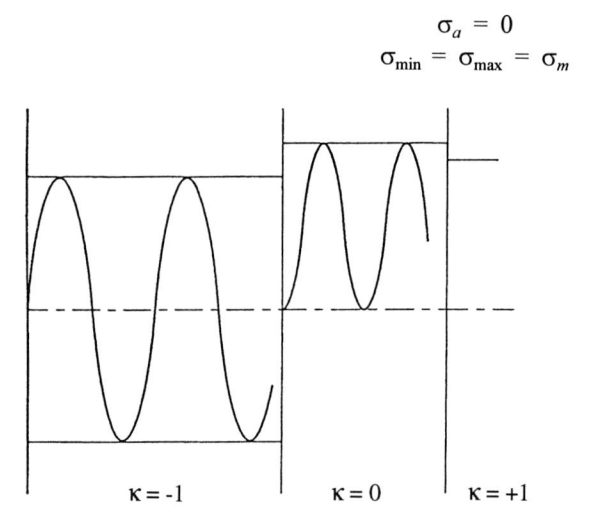

$\kappa = -1$ $\kappa = 0$ $\kappa = +1$

Fig. 1.3. Some periodical vibrations

Endurance Limit. Let us consider a number of polished test bars without notches made of a given material, which are subject to a periodic stress of κ with an index. After the application of many load cycles, a fixed number of test bars breaks down. The application of a further stress applied of the same number of load cycles will have a different break percentage. A curve can be traced. It indicates the probable breaking percentage for this material in accordance with the number of load cycles applied. In statistics, this curve is called Galton's curve.

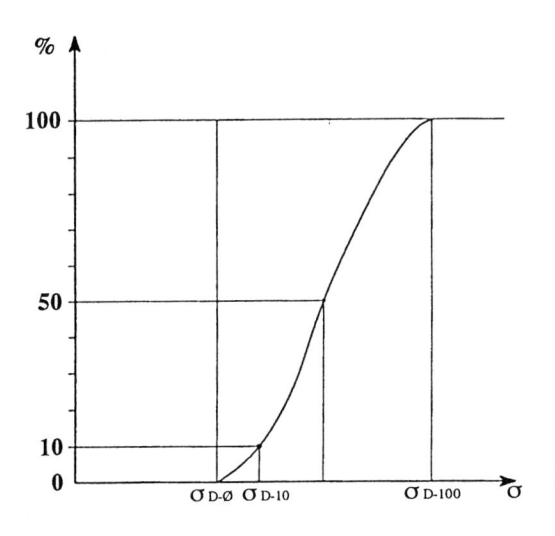

Fig. 1.4. Galton's curve

A breaking percentage of 10% will probably correspond to a breaking risk of 10% marked by σ_{D-10} if we decide to choose such value on Galton's curve as an example.

Such risk is defined as follows σ_{D-10} and is known as the endurance limit of the material to the applied stress (breaking risk of 10% and consequent reliability of 0.9). In order to generalize the following explanations, such a limit will be indicated as follows σ_D as its reliability is always implicit.

For a unit made of the same material, which has to bear the same stress and shows the same degree of reliability, the real endurance limit is given by the a.m. value modified on the basis of factors similar to those considered for the breaking point and corrected in accordance with the following factors: roughness, stress concentration and reduction of the elastic limit, whenever it exists.

Such stress is called allowable breaking stress and is indicated as follows $\sigma_{D\!P}$.

Woehler's Curve (σ- N curve). The value of the permitted endurance strength is lower than that of the permitted stress. Starting from a relatively small number of load cycles, with such value depending on the material (e.g. 10^3), we notice that the permissible stress diminishes to reach asymptotically the value of the endurance strength, if the number of load cycles increases. Starting from a high number of load cycles $N_{D\!P}$ (e.g. 3 10^6), the difference between the real curve and the asymptote is negligible with the allowable stress reaching the value of the endurance strength, being equal to this. We obtain a graph that is divided into three parts: the first one is constant up to the value N_{stat} and equal to the allowable stress σ_{statP} ; the second develops as an hyperbole and the third joins the second part in the point corresponding to N_D. It is a straight line showing a constant value σ_D.
The part developing as an hyperbole is expressed as follows:

$$\frac{\sigma_1}{\sigma_2} = \left(\frac{N_2}{N_1}\right)^{\frac{1}{p}} \qquad (1.006)$$

$$p = -\frac{\log\left(\dfrac{N_D}{N_{stat}}\right)}{\log\left(\dfrac{\sigma_{statP}}{\sigma_{DP}}\right)} \qquad (1.007)$$

The a.m. curve is σ-N curve and is always represented under the form of a logarithm (see Fig. 1.5); the different sections are straight lines.

This curve represents the values of the allowable stress as a function of the number of load cycles taken by the unit. It is valid only for a stress of κ index, for a given material with given roughness, stress concentration and reliability.

Fig. 1.5. σ - N curve

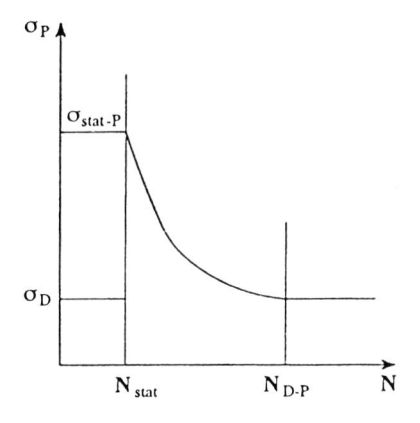

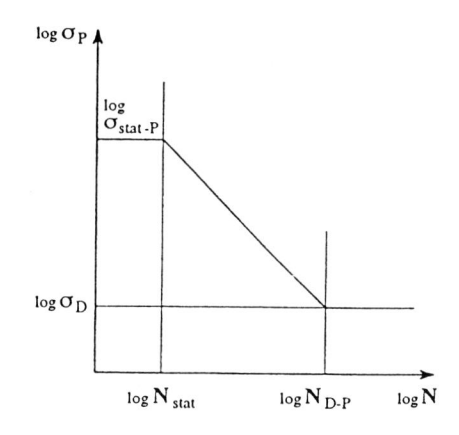

It is possible to trace a σ - N curve corresponding to the breaking points of the polished test bar and to apply modification factors subsequently.

The calculated stress is always proportional to the torque taken by the machine or at least to a torque power expressed as follows:

$$\sigma = K T^r . \qquad (1.008)$$

The substitution of this value in equation (1.006) enables us to obtain:

$$\frac{T_1}{T_2} = \left(\frac{N_2}{N_1}\right)^{\frac{r}{p}} . \qquad (1.009)$$

Mostly, the value of r corresponds to the unit even where exceptions exist (see Hertz's theory concerning punctiform normal pressures or pressures on a generatrix).

Stability Under a Single Load. If the momentary load shows a unique maximum value during the unit's life, stability will be assured with a proper degree of reliability if the point representing the maximum value of stress shown on the curve, such value being extended by the end of the unit's life, stays below or upon such curve .

In this case, the value of the unit's life is equal to the duration of the stress shown on σ - N curve. If ΔN corresponds to the unit's life and N corresponds to the duration of the stress on σ - N curve, stability will be assured if:

$$\frac{\Delta N}{N} \leq 1 . \qquad (1.010)$$

If inequality is assured, the value of the real stress σ will be lower than that given by the a.m. curve for a fixed duration ΔN and will be equal to σ_P. The ratio of σ_P to σ is the safety factor.

Stability Under a Load Spectrum. Cumulative damage or Palmgren-Miner's rule.

The maximum value of the momentary load is not necessarily constant. During a unit's life different maximum values can be generated. Loads can be classified following a decreasing order. Loads having the same intensity can be grouped by the same duration that represents the sum of each load duration. The loads classified following a decreasing order form a load spectrum. It is clear that each element of the load spectrum contributes to cause damage to the part that must support its destroying action. In the case of a single load, the rule expressed in equation (1.010) is applied.

The rule of Palmgren-Miner assumes that, for the lapse of time not greater than N_D, each load acts proportionally to its duration.

If for each stress σ_i resulting from the application of an elementary load of the spectrum (or for each torque T_i that forms the spectrum) the duration of the application is ΔN_i and the corresponding duration on σ - N curve is N_i, stability will be assured if:

$$\sum \frac{\Delta N_i}{N_i} \leq 1 \quad . \tag{1.011}$$

See Fig. 1.6

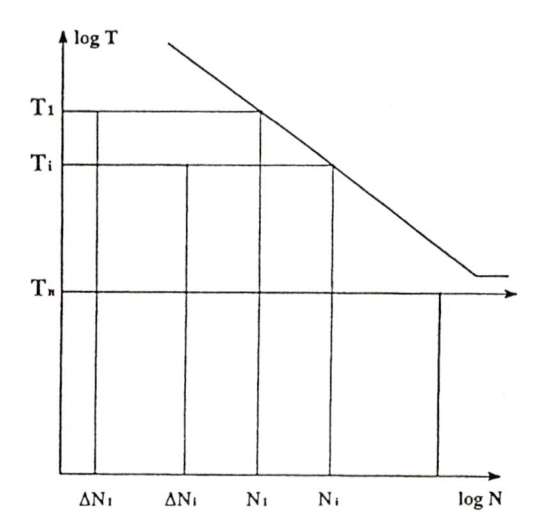

Fig. 1.6. Miner's rule

Equivalent Load. Equation (1.006) can be applied to a given load spectrum; any value can be chosen for N_2. It will be expressed by N_{equ} to which the equivalent torque corresponds T_{equ}.

For each level of the spectrum:

$$\frac{T_i}{T_{equ}} = \left(\frac{N_{equ}}{N_i}\right)^{\frac{r}{p}} . \qquad (1.012)$$

Applying equation (1.011) with N_i resulting from this relation:

$$\sum \frac{\Delta N_i}{N_i} = \left[\sum \Delta N_i\, T_i^{\frac{p}{r}}\right] \frac{1}{N_{equ}\, T_{equ}^{\frac{p}{r}}} = 1 \qquad (1.013)$$

and

$$T_{equ} = \left[\sum \frac{\Delta N_i}{N_{equ}} T_i^{\frac{p}{r}}\right]^{\frac{r}{p}} . \qquad (1.014)$$

The equivalent load is related to equivalent number of load cycles chosen arbitrarily. It forms a single load replacing the spectrum on condition that the number of cycles associated with the spectrum is taken into consideration.

1.4 Materials

1.4.1 General Features

The materials used in the production of gearboxes are generally metal materials apart from polymers used to make joints or rings and elastic materials used in the production of transmission devices.

The main metal materials are ferrous materials and bronze.

1.4.2 Ferrous Materials

Ferrous materials are made of iron and carbon alloys. The percentage of carbon affects the production: indeed it may be steel (containing less than 1.2% carbon) or cast-iron (containing more than 2.5% carbon).

Carbon Steel. Steel is made of ferrite crystals and of an amalgam of small crystals of ferrite and cementite called pearlite. Ferrite is iron ore while cementite is an iron carbide. Steels show different properties that vary according to their carbon content.

Resistance to traction and hardness increase proportionally to the carbon content, while resistance to shocks and the following elongations are inversely proportional to the inner quantity of carbon in the steel (resistance will diminish if the percentage of carbon increases).

Special Steels. Often ordinary steels are enriched with metals or various substances such as Manganese (Mn), Chrome (Cr), Nickel (Ni) and Molybden (Mo) in order to improve their chemical and mechanical properties. The products resulting from such addition are known as special steels.

The important addition of Cr and Ni (18% and 8%), for example, results in stainless steel. Even a smaller quantity of the a.m. substances adds to a general increase in the mechanical properties of steel. Such improvement does not affect ordinary steels free from this addition which have a similar carbon content.

Cast-iron. After fusion the cooling rate affects materials, in particular cast iron and its features. If cooling is fast, it means that the cast iron is made of ferrite and cementite (an iron carbide); the resulting product is white cast-iron, which is very hard, very fragile and not easily malleable. Through slow cooling this cast iron can be obtained without free cementite. Such iron is made of a metallic matrix which is similar to that of hard steel (high carbon content) containing flakes of free graphite resulting from cementite decomposition. The product showing such features and resulting from this slow cooling is known as grey cast iron. Grey cast iron has good mechanical properties, which are similar to those of hard steels. However, such properties are not as good as those of hard steels because grey cast iron can break. Such risk is due to the presence of graphite flakes.

If graphite flakes are melted, under the action of addition, they change into small bodies (globules) from which originates SG iron. The mechanical properties of SG iron are better than those showed by grey cast-iron, but SG iron production is much more expensive.

Thermic Treatment of Steel. Temperature affects the structure of steels and their properties. Such an effect is known as thermic treatment.

- Annealing. Above 910°C iron transforms into another crystallographic variety which can dissolve carbon (below such temperature carbon cannot be dissolved). In an alloy containing 0.9% of carbon, the a.m. temperature will diminish to 720°C if the content of carbon increases. When steel is subject to a higher temperature, the crystalline structure turns into small crystals of an homogeneous solution of carbon mixed with iron, from which austenite originates. In the presence of increasing temperatures, all the unwanted particles existing in steel at ambient temperature tend to disappear. Such unwanted particles result from distortions caused by different stresses or by a faulty crystallization during cooling. At the same time the dimensions of austenite grains increase. If cooling is slow, the transformation of austenite into ferrite and pearlite takes place again and traces of past structures disappear. The dimensions of grains again increase. Such thermic treatment is called homogenizing diffusion annealing.

If a structure characterized by large grains undergoes such treatment again at a temperature above the transformation range, grains of austenite will be formed again and the structure ferrite-pearlite resulting from a slower cooling will show fine grains which are known to develop the good mechanical properties. This latest treatment is known as normalizing annealing.

- Hardening. As stated before, iron is heated above the transformation temperature in order to obtain austenite which is immediately water or oil cooled. During this process iron undergoes the a.m. crystallographic transformation, even when the cooling rate prevents ferrite and pearlite from separating. The resulting structure is called martensite, which is very hard and very resistant to traction. This kind of treatment goes under the name of hardening (or quenching) and it is made easier by the carbon content. Steels with a high carbon content (from 0.4% to 1.0%) show a high degree of hardenability, while steels poor in carbon cannot be hardened. The addition of metal (Cr, Ni, Mn) as well as that of carbon helps hardening.

Steels containing a carbon percentage that goes from 0.35% to 0.42% with additions like Cr (from 1% to 1.5%) and Ni can be hardened without difficulty and the results are satisfactory. Steels that give such satisfactory results are called hardening steel. These steels are used in the production of shafts and gears. The hardening capacity is defined by the Jominy test, which consists of heating a bar of the tested material which is then cooled by means of water or oil. The final hardness varies depending on the basis of the J distance from the hardened extremity. Some empirical formulae enable one to make provisions about the degree of hardness at the J distance on the basis of steel compounds. These formulae are approximate but they demonstrate the effect of any additions. According to Just, for steels containing a percentage of carbon lower than 0.25% we find that:

$$HRC_J = 74\sqrt{C} + 14Cr + 5,4Ni + 29Mo + 16Mn - 16,8\sqrt{J} + 1,386J + 7 \quad . \quad (1.015)$$

While for steels containing a percentage of carbon higher than 0.25% (to 0.60%) :

$$HRC_J = 102\sqrt{C} + 22Cr + 21Mn + 7Ni + 33Mo - 15,47\sqrt{J} + 1,102J - 16 \quad . \quad (1.016)$$

The chemical symbols represent the percentage of the different elements. For a steel containing 0.18% of C, 1.5% of Ni, 1.5% of Cr and 0.2% of Mo, with a Jominy distance=12 the value found is: HRC12= 31.7.

The same steel without Cr, Ni and Mo has a "negative" hardness calculation and will show it to have a low degree of hardenability.

Hardening can be executed by means of water or oil at ambient temperature (water or oil quenching) or it can be performed in baths at a higher temperature. The resulting structure is called bainite which is quite different from martensite. Bainite is a very hard and extremely resilient material and results from martempering (using a bath at a constant temperature) or from austempering (following two baths at a decreasing temperature).

- Surface hardening (induction hardening or flame hardening). The hardest steels, which are also the most fragile with regard to shocks, are those showing the highest degree of hardenability.

If a high degree of hardness and resilience is required, it is sufficient to harden the surface of steels. The surface hardening consists of heating the exterior part of a piece of steel down to a fixed depth and in hardening the hot surface. The exterior part gets harder while the inner part, the core, remains unchanged (it shows the same degree of resilience). The exterior section can be heated following two different methods: a pipe may be used that results in an immediate rise in the temperature of the exterior surface on which oil or water may be poured in order to make the cooling faster to prevent the core from heating. It is also possible to wind a solenoid round the piece, through which runs a flow of high frequency current. Induction, due to the electric field, creates a flow of high frequency induced current on the surface of the piece. As a consequence of Joule's effect and of the skin effect such a flow does not penetrate into the core that is created. To harden the exterior section, it should be bathed it with water or oil. Steels which are subject to shell hardening and those fully tempered are of the same type. These processes are used in gear production.

- Carburizing. Surface-hardened steels have a low degree of resilience as surface-hardening is performed on hard steels which are not very resilient.

To obtain a high degree of resilience and hardness it is necessary to choose a mild steel which is not particularly hard; its peripheral area must be turned into hard steel which shows a good degree of hardenability. By leaving an ordinary or specially mild steel in an atmosphere rich in carbon at a temperature which is higher than that of the transformation point (1000°C), an increase in the carbon content occurs, starting from the peripheral area of the section. Austenite on the surface dissolves carbon that penetrates slowly inwards. When the quantity of carbon penetrated is satisfactory (attention should be to avoid any excess of carbon in the peripheral area) the operation can stop. After this treatment, that lasts several hours, the piece is taken out of the oven and subject to oil or water quenching. The layer enriched with carbon (case-hardened layer) is tempered while the core itself does not undergo any treatment. The material is now resilient and has a high degree of shell hardness that is deep. Steels chosen to undergo such treatment are poor in carbon (from 0.15% to 0.20%). Cr-Ni- or Cr-Mn are added in quantities from 1.5% to 2%. Such steel is known as case-hardening steel. As this treatment necessitates a long lasting heating, distortions may occur. The case-hardened layer creates compressive stress in the mass which have beneficial effects in some technical applications.

- Nitriding. When metal is heated at a temperature of about 500-530°C in an atmosphere which is rich in Nitrogen, Nitrogen forms together with iron into, extremely hard metal nitrides. Such nitrides penetrate slowly into the grains and moves towards the core. By leaving a piece of metal under the a.m. conditions for any length of time (about 100 hours) penetration can reach 1 millimetre. In this way the degree of metal hardness without modifying its original resistance to shocks can increase. This treatment gives the same results as carburizing (case-hardening). Different from case-hardening the thickness of the hardened layer is smaller and the duration of the treatment is longer. Therefore it is much more expensive than case-hardening even if the value of distortions is lower as the treating temperature is lower.

This treatment is addressed to pieces that may be infrequently worked. Nitriding can be performed in a gaseous atmosphere rich in atomic Nitrogen, coming for example from ammonia decomposition. Nitriding can also be executed in a bath of melted salts (melted cyanides). In this case nitriding is accompanied by carburization. The resulting treatment goes under the name of nitrocarburizing. Steels chosen for such treatment are the same as those used for case-hardening. Aluminium makes Nitrogen penetration easier but does not make the core resistant. It is for this reason that aluminium steels are not often used.

- Tempering. Hardened steels often hide residual stresses that can be harmful to a mechanical unit when it is working. If the hardened unit is heated once again at a temperature below 700° C stresses will prevail (at a lower temperature, 200°C, it is the only effect obtained).

At a higher temperature the hardened steel becomes milder and hardness and resistance diminish. Such treatment is known as tempering and creates homogeneous pieces free from residual stresses which have a correct degree of hardness.

In general hardening and tempering are given to crude materials which will be worked without being subject to any further treatment.

1.4.3 Bronzes

Bronzes are alloys of copper and tin, zinc, aluminium or lead with a percentage of copper higher than 50%. Bronzes are generally used to build components which have to withstand strong friction such as sliding bearings and wormwheels. Often bronzes are cast on another support before being worked.

2 Geometry and Gears

2.1 General Features

Gears are made of two or more wheels with teeth around their edges that fit into the teeth of another wheel. They have been designed to transfer power or movement by means of shafts.

Gears can be classified on the basis of the position of the axes, these may be:

- parallel gears: axes are on the same plane; one axis is parallel to the other one
- gear pair with intersecting axes: axes are on the same plane but they share a theoretical common point
- gear pair with nonparallel axes : axes are not on the same plane.

Wheels can be classified according to their general shape; these may be:

- cylindrical gears: teeth are placed on a cylinder;
- bevel gears: teeth are placed on a cone.

Cylindrical gears are used in parallel gears while bevel gears are used in gear pairs with intersecting axes. Gear pairs with nonparallel axes are provided with bevel gears (hypoid wheels) or cylindrical gears (worm and wormwheel).

Wheels are made of even surfaces characterized by straight lines (in cylindrical gears, straight lines are parallel to the axis, in bevel gears, straight lines start from the vertex of the cone). Teeth can be classified on the basis of the line they follow as :

- spur gear teeth,
- spiral teeth,
- helical teeth.

They also vary with their position as to the wheels; these are:

- internal teeth, if they are internally positioned,
- external teeth, if they are externally positioned.

Two external wheels can be mated in order to form an external gear.

An external wheel and an internal wheel can also be mated to form an internal gear while internal wheels cannot be mated.

Teeth must fit perfectly so that the gear can engage correctly. On the other hand teeth must ensure the regular movement of the driven gear which depends on the correct movement of the driving gear (i.e. without shocks or discontinuity). Involute teeth meet such requirements. Teeth having different shapes gave satisfactory results too, but their production is rather complex; it is for this reason that technology has focused on the study and production of wheels whose profiles result from involutes to a circles.

2.2 Involute to a Circle

The involute to a circle, from now onwards simply called the involute, is a geometric curve that can be defined as follows: first of all it can be considered as a curve drawn by a pencil placed on the extremity of a rope which is rolled up around a disk; during the unrolling operation the rope is kept taut.

Put geometrically, the involute is the locus of the point fixed on a straight line that rolls without sliding on a circle (Fig. 2.1). The circle or the disk is referred to as the base circle.

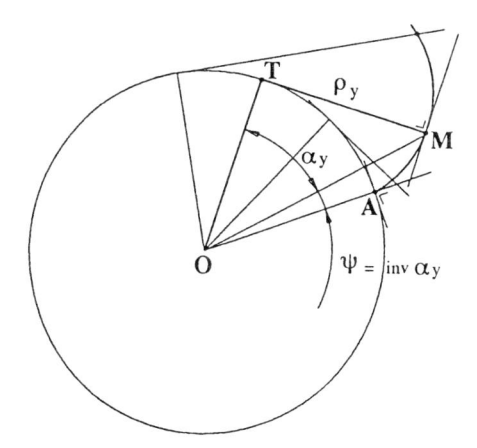

Fig. 2.1. Definition of involute

The involute has many series of properties as follows:

(1) The involute is orthogonal to the base circle with diameter d_b. This means that the tangent at the base circle and the tangent at the involute are perpendicular or that the tangent at the involute on this point is the radius of the base circle.

(2) The arc of the base circle (contained), between the point on the base circle and any point of the involute, is equal to the length of a segment of the tangent at the base circle, created by the same point (included) between this point and the tangent point:

$$arc\ TA = segment\ TM$$

(3) The tangent to the base circle is orthogonal to the involute; in other words the tangent at any point M is perpendicular to the tangent TM at the base circle.

(4) The length of the tangent is the curvature radius of the involute at the point taken into consideration; that is

$$curvature\ radius\ in\ M\ (\rho_y) = TM$$

(5) There are no involutes inside a base circle.

(6) The involute of a straight line (circle of infinite radius) is a straight line. These definitions enable us to draw the polar co-ordinates of the involute. If we chose the angle α_y as variable, we notice that:

$$TM = \frac{1}{2} d_b \tan\alpha_y \tag{2.001}$$

$$TM = arc\ TA = \frac{1}{2} d_b (\alpha_y + \psi) \tag{2.002}$$

$$\psi = \tan\alpha_y - \alpha_y \tag{2.003}$$

and

$$R = OM = \frac{d_b}{2\cos\alpha_y} \tag{2.004}$$

The function ψ is known as the involute function of α_y and is represented by "inv". So we have:

$$inv\alpha_y = \tan\alpha_y - \alpha_y \tag{2.005}$$

as the value of the angle α_y is expressed in radians.

This angle is called the angle of incidence. If this angle corresponds to a diameter d_y, the following relation will be valid:

$$\cos \alpha_y = \frac{d_b}{d_y}. \qquad (2.006)$$

2.3 Elementary Geometric Wheel. Meshing Theory

Let us consider two circles and their diameters d_{b1} and d_{b2}. Circles are placed at a distance of the centres O_1 and O_2 which is longer than the half the sum of the diameters (Fig. 2.2).

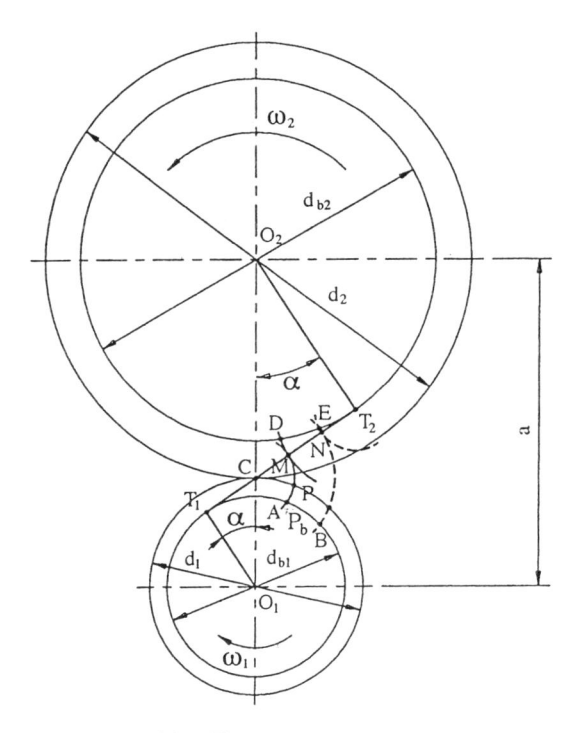

Fig. 2.2. Meshing Theory

We draw the common and internal tangent T_1T_2 at the two circles (circle 1 and circle 2). Passing through a point M, we draw two involutes to a circles (one for each circle). These involutes have the point M in common and they are at a tangent. In fact they are both perpendicular to the straight line T_1T_2. If we consider any point N and we do the same as before, the two involutes passing through N are both at a tangent.

If wheel 1 turns at the speed of ω_1 engaging its involute in M that in turn engages the involute of circle 2 and the circle 2 itself, the involutes will still be tangent and after a period of time t they will be in N. Circle 2 will have turned at a speed of ω_2 that can easily be calculated. In effect the arc AB of circle 1 will have a length corresponding to $0.5\, d_{b1}\, \omega_1\, t$ and the arc DE of circle 2 will have a length corresponding to $0.5\, d_{b2}\, \omega_2\, t$. These two arcs are equal to the segment MN. Such equality results from the involutes.

The two arcs are equal and:

$$\frac{\omega_1}{\omega_2} = \frac{d_{b2}}{d_{b1}}. \tag{2.007}$$

This equation is the base of the meshing theory. The common tangent meets the center line at the point C. Passing through this point we draw a circle concentric with the first base circle 1 and a circle concentric with base circle 2. The diameters of these new circles are d_1 and d_2. They are tangent at the point C and are called reference circles as they are used to define all the features of wheels and gears from a geometrical point of view. We notice that the radii O_1T_1 and O_1C form an angle α having the same features as the angle formed by the radii O_2T_2 and O_2C. We remark the existence of the following relations:

$$d_{b1} = d_1 \cos \alpha \tag{2.008}$$

and

$$d_{b2} = d_2 \cos \alpha . \tag{2.009}$$

The angle α is called the pressure angle and the two wheels have it in common. As there are only two involutes (one for each circle) the contact will occur in T_1 and will cease at T_2. In order to ensure the continuity as regard to movement, it is necessary to multiply involutes, leaving among them a constant and equal space whose value is lower than T_1T_2.

Given the number of involutes z_1 and z_2, the space among involutes will be:

$$p_b = \pi \frac{d_{b1}}{z_1} = \pi \frac{d_{b2}}{z_2} \tag{2.010}$$

$$p_b = \pi \frac{d_1}{z_1} \cos \alpha = \pi \frac{d_2}{z_1} \cos \alpha \tag{2.011}$$

$$p = \pi \frac{d_1}{z_1} = \pi \frac{d_2}{z_2} . \tag{2.012}$$

The quantities p_b and p are the base pitch and the reference pitch of the two wheels, last relation being:

$$\frac{d_1}{z_1} = \frac{d_2}{z_2} = m \quad . \tag{2.013}$$

The quantity represented by the letter m is called a module.

Involute teeth can engage correctly if the pressure angle as well as the module of the two wheels have the same values and the value of the base pitch is lower than the length of the tangent the wheels share in common.

Contact takes place constantly along the tangent which is common to the base circles. Such a tangent is called the action line and represents the locus where contact takes place.

Wheels can be defined in such a way only if the figure is closed, suggesting the section of a finite solid. Involutes opposite to those drawn before (non-corresponding involutes) and the definition of a conventional distance of the reference circle that creates the tooth thickness will complete the geometrical drawing of wheels. The space existing between the first involutes drawn will be divided into two sections and the new base involutes to a circle will pass through the points resulting from such a division. A tip circle and a root circle, whose diameters are d_a and d_f will limit the tooth depth. The radial depths above and below the reference circle will be expressed by h_a and h_f.

Equations will be then the following:

$$d_a = d + 2\, h_a \tag{2.014}$$

$$d_f = d - 2\, h_f \quad . \tag{2.015}$$

Equation (2.013) allows us to write:

$$d = m\, z \tag{2.016}$$

$$d_a = m\, (z + 2\, h_a^*) \tag{2.017}$$

and:

$$d_f = m\, (z - 2\, h_f^*) \tag{2.018}$$

The marked value * is reduced to a module which is equal to 1.

2.4 Standard Basic Rack Tooth Profile

Given a pressure angle α, an addendum reduced to the module h_a^* and a dedendum reduced to the module h_f^*, the wheels that have the same number of teeth will be very much alike and the similar ratio will be the ratio of modules. There is a wheel whose diameter is infinite. A circle having an infinite diameter is a straight line and an involute to a straight line is a straight line. The resulting wheel, shown on Fig. 2.3, is made up of straight lines.

Such a profile is called reference profile and characterizes a family of wheels, indipendentely on the teeth number.

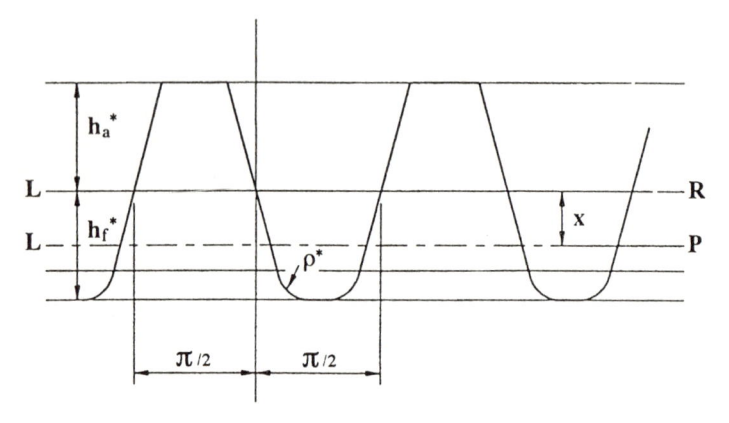

Fig. 2.3. Standard basic rack tooth profile

Dimensions are defined in relation to a line (LR), called the reference line.

2.5 Geometrical Wheel - Basic Rack Profile

A wheel whose reference diameter is given by the number of teeth (module =1) results from the envelope outline of the profile when its datum line turns without sliding on the reference circle of the wheel, as this movement is similar to that of any wheel fitting with the wheel representing the standard rack profile. We can also obtain a wheel by letting the standard basic rack profile roll on the reference circle following a line parallel to its reference line (which is at a distance x from the wheel) instead of its standard basic rack profile.

The quantity expressed by x is the rack shift factor of the resulting wheel. Such a factor will be positive if the profile is distanced from the centre of the wheel.

Failing this, it will be negative. Such a wheel will have the following dimensions:

$$d^* = z \qquad (2.019)$$

$$d_a^* = z + 2(h_a^* + x) \qquad (2.020)$$

$$d_f^* = z - 2(h_f^* - x) \quad . \qquad (2.021)$$

The line of the basic rack profile that rolls on the reference circle of the wheel (LP) is called the pitch line. The action line passes through a point of the reference line that is tangent to the reference circle of the wheel. Any point forming the action line is at the same time on a line parallel to the reference line and on a circle concentric with the reference circle. It is also possible to define the relation existing between each point forming the wheel tooth profile and each point forming the standard basic rack profile.

The following dimensions are referred to a wheel with a module equal to 1. Given a module m, the resulting wheel has the same geometrical features of the wheel mentioned above. It is sufficient to multiply the dimensions by m.

$$\text{Reference pitch: } p = \pi\, m \qquad (2.022)$$

$$\text{Base pitch: } p_b = \pi\, m \cos \alpha \qquad (2.023)$$

$$\text{Base diameter: } d_b = z\, m \cos \alpha \qquad (2.024)$$

$$\text{Reference diameter: } d = z\, m \qquad (2.025)$$

$$\text{Tip diameter: } d_a = m\left[z + 2(h_a^* + x)\right] \qquad (2.026)$$

$$\text{Root diameter: } d_f = m\left[z - 2(h_f^* - x)\right] \quad . \qquad (2.027)$$

2.6 Tooth Thickness

The tooth thickness s_y is the length of the arc of a circle with a diameter d_y which is contained between the two involutes that limit the tooth (Fig. 2.4).

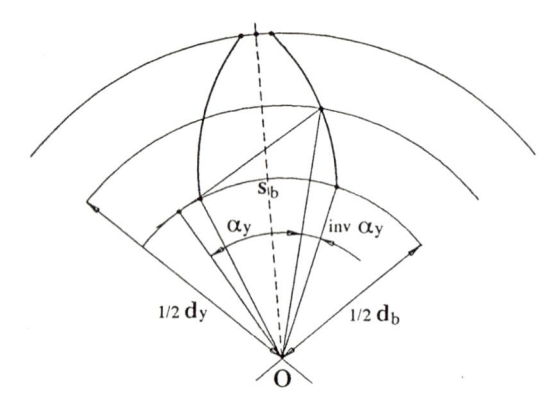

Fig. 2.4. Tooth thickness

s_b represents such distance. The value of the half angle in the centre is expressed by $\psi_b = s_b/d_b$.

Looking at the figure we notice that the half angle in the centre is equal to $\psi_y = \psi_b - \text{inv } \alpha_y$.

Thickness can be obtained by multiplying the half angle by the diameter:

$$s_y = d_y \left(\frac{s_b}{d_b} - \text{inv} \alpha_y \right) . \qquad (2.028)$$

Between the value of thickness of a circle with diameter d_{y1} and that of a circle with diameter d_{y2} the following general relation exists:

$$\frac{s_{y1}}{d_{y1}} + \text{inv} \alpha_{y1} = \frac{s_{y2}}{d_{y2}} + \text{inv} \alpha_{y2} . \qquad (2.029)$$

We notice that (Fig. 2.5) on the standard basic rack profile the reference line is $\pi/2\ m$ and that the thickness on the pitch line is $(\pi/2 + 2 \times \tan \alpha)m$. The arc of the reference circle has the same length and we can affirm that:

$$s = \left(\frac{\pi}{2} + 2 \times \tan\alpha \right) m . \qquad (2.030)$$

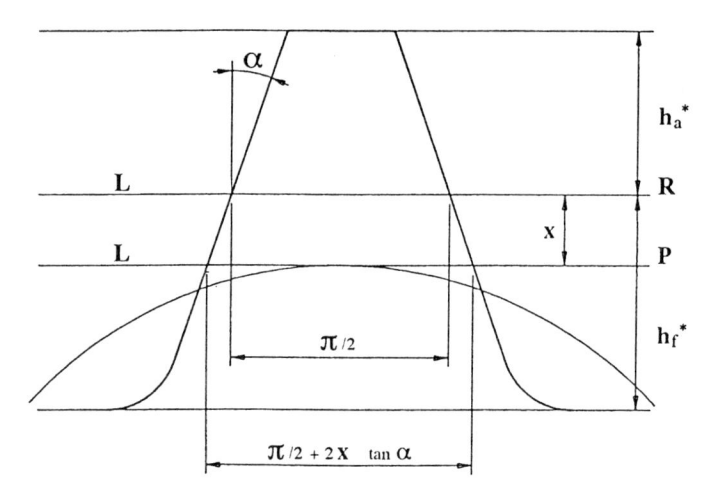

Fig. 2.5. Original (Standard - Basic) thickness.

The value of the thickness of a circle with any diameter d_y is:

$$s_y = d_y \left(\frac{\frac{\pi}{2} + 2x\tan\alpha}{z} + \text{inv}\alpha - \text{inv}\alpha_y \right) . \quad (2.031)$$

In particular the value of the tip thickness is:

$$s_a = d_a \left(\frac{\frac{\pi}{2} + 2x\tan\alpha}{z} + \text{inv}\alpha - \text{inv}\alpha_a \right) \quad (2.032)$$

with:

$$\cos\alpha_a = \frac{d_b}{d_a} = \frac{z\cos\alpha}{z + 2(h_a^* + x)} \quad (2.033)$$

and the value of the base thickness is:

$$S_b = m z \cos\alpha \left(\frac{\frac{\pi}{2} + 2x \tan\alpha}{z} + \text{inv}\alpha \right) \qquad (2.034)$$

or

$$S_b = m \left(\frac{\pi}{2} \cos\alpha + 2x \sin\alpha + z \, \text{inv}\alpha \, \cos\alpha \right). \qquad (2.035)$$

2.7 Influence of the Rack Shift Factor. Undercut and Meshing Interference

The rack shift factor affects the shape of teeth. Such an effect can clearly be seen from the a.m. equations concerning the tooth thickness.

If the value of the rack shift factor increases, the value of the tip thickness will diminish and the factor can assume a value that makes the thickness value null. The resulting tooth is tipped and its dimensions are still the same.

Given a diameter lower than the tip diameter, the value of thickness will be null if the value of the factor is subject to a further increase.

On the other hand, the value of the tip diameter, corresponding to a null thickness, diminishes. The resulting tooth is tipped and much shorter. A limit lower than the value of the rack shift factor could also be defined.

Let us consider the standard basic rack tooth profile in comparison with the base circle. When the action line intersects with the base of the tooth profile on the left of the point T_1 (Fig. 2.6), the tooth profile moves inwards towards the base circle. As stated previously, it is not possible to have involutes inside a base circle.

During the generation of the tooth profile, the base of such a profile creates a curve called trochoid. There is an inverse curvature of such line in regard to that of the involute. The resulting profile does not have the shape of an involute and the tooth root is not as thick as was first thought: such phenomenon is called the undercut. If we mate a wheel with the wheel obtained following the a.m. procedure and if contact is made between them by means of an involute, wheels will not engage correctly causing damage to the unit and jeopardizing its workings. Such inconvenience is called Meshing Interference and may be avoided if a limit lower than the value of the rack shift factor is chosen. This limit makes

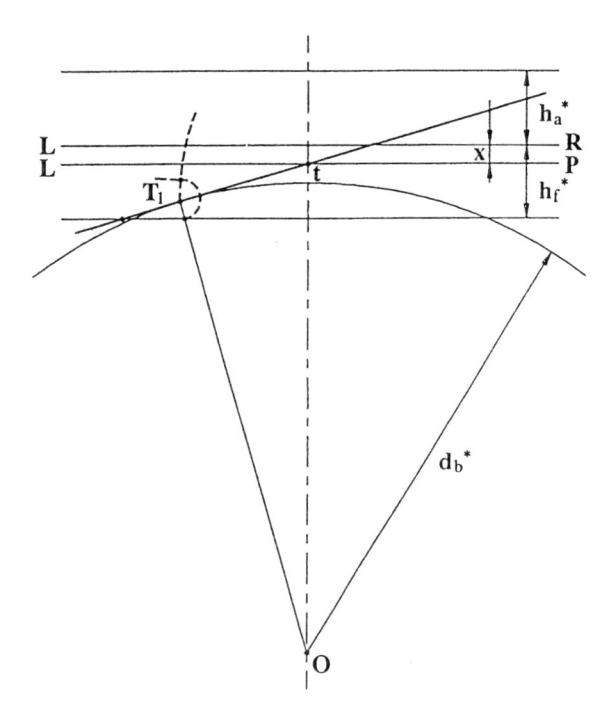

Fig. 2.6. Undercut

the line below the standard basic rack tooth profile intersect with the action line at the point in T_1 (Fig. 2.7).
With:
$$T_1 C = 0.5 \, d \sin a$$

and
$$T_1 C = (h_f - mx)/\sin a$$

we have:
$$x > h_f^* - \frac{1}{2} z \sin^2 \alpha \quad . \tag{2.036}$$

Given a rack shift factor x and a defined standard basic rack tooth profile, interference may be limited by taking the number of teeth to the following value:

$$z \geq \frac{2 \, (h_f^* - x)}{\sin^2 \alpha} \quad . \tag{2.037}$$

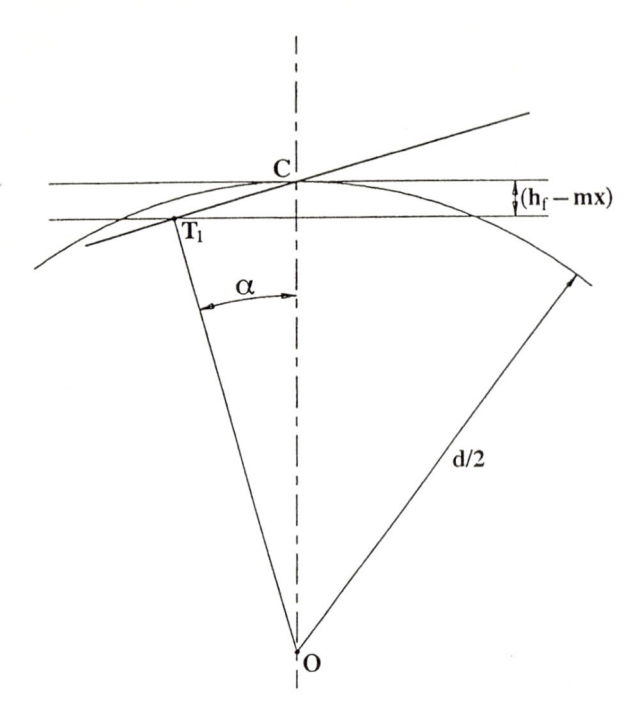

Fig. 2.7. Limit of the rack shift factor

2.8 Reference Centre Distance. Modified Centre Distance

The two wheels defined in paragraph 2.5 are mated so as to ensure correct meshing. The validity of meshing rules is confirmed by the definitions given previously. Wheels are governed with their tangential reference diameters. In this case wheels are at a reference centre distance and the sum of rackshift factors for the two wheels is null:

$$\Sigma x = x_1 + x_2 = 0. \tag{2.038}$$

The value of the reference centre distance is:

$$a = \frac{1}{2}(d_1 + d_2) = \frac{1}{2}m(z_1 + z_2). \tag{2.039}$$

If the distance between the centres of the wheels increases (Fig. 2.8) the common tangent will cut the line of the centres across a point C; the circles

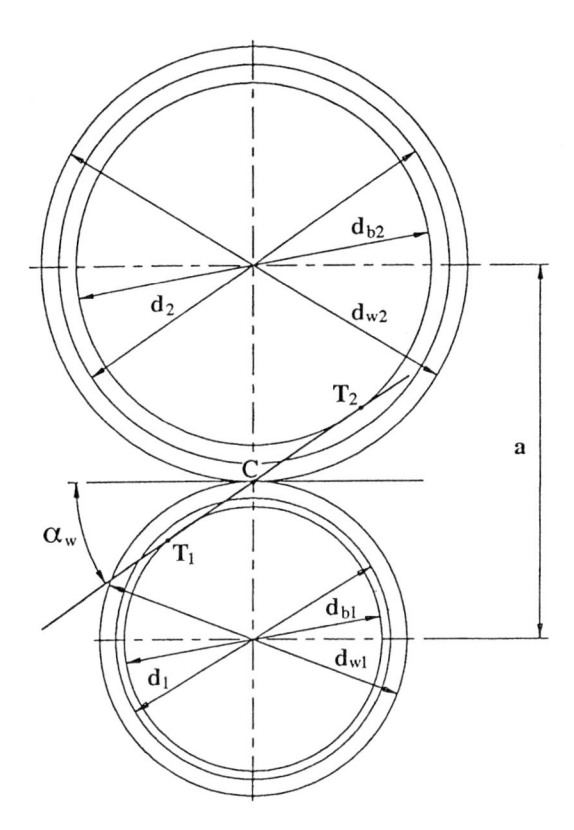

Fig. 2.8. Reference centre distance. Modified centre distance

concentric with the reference circles will pass through such a point. Those circles are the pitch circles and their diameters are the pitch diameters.

During meshing, the two circles roll one on the other without sliding. Reference circles do the same during working at a reference centre distance. The above mentioned arrangement is called the modified centre distance. Such an arrangement is produced when the sum of the modification coefficients is different from 0. Given the centre distance a and half the sum d_m of the reference diameters working will be correct if sum of the rackshift factor follows the relation mentioned below:

$$\Sigma x = \frac{z_1 + z_2}{2 \tan\alpha} (\text{inv}\alpha_w - \text{inv}\,\alpha) \qquad (2.040)$$

88 Jacques Sprengers, Part III

where α_w is the working pressure angle expressed by:

$$a = \frac{d_{b1} + d_{b2}}{2 \cos\alpha_w} \qquad (2.041)$$

$$a \cos\alpha_w = d_m \cos\alpha \qquad (2.042)$$

where d_m is given by:

$$d_m = \frac{d_1 + d_2}{2} = m \frac{z_1 + z_2}{2} \qquad (2.043)$$

2.9 Contact Ratio

When a wheel engages with another one by means of teeth, teeth are in contact on the line of action (Fig. 2.9).

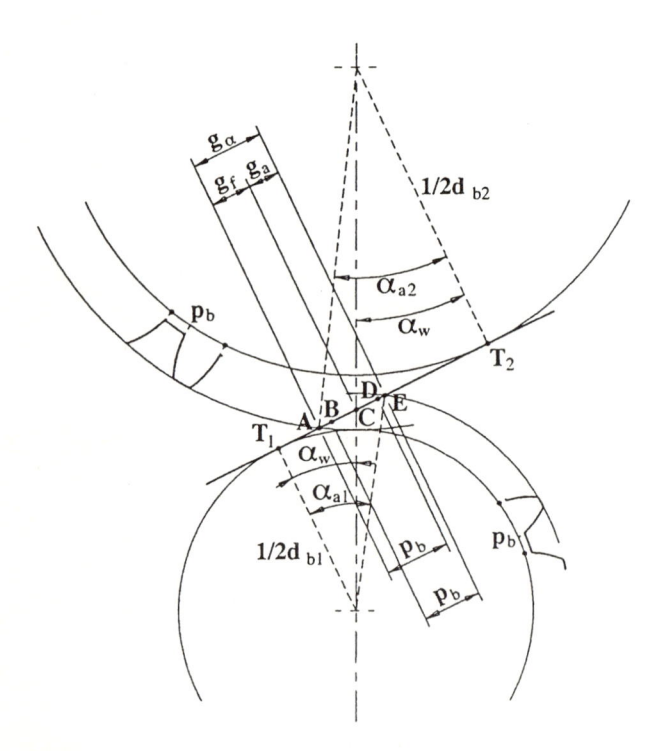

Fig. 2.9. Contact ratio

The driving gear tip circle cuts the line of action across the point E, while the tip circle of the driven gear cuts this line across the point A. Before A, contact between the two teeth cannot take place as the driven gear does not exist. After the point E, the same thing occurs: the driven gear does not exist anymore and contact between teeth cannot take place. The straight line segment AE is called the path of contact. The segment AC is called the approach path while segment CE is called the recess path. g_α represents the path of contact length, g_f represents the approach path length, while g_a represents the recess path length.

We have:

$$g_\alpha = g_f + g_a . \tag{2.044}$$

The equation concerning such segments can easily be deduced from the Fig. 2.9:

$$g_a = \frac{d_{b1}}{2} (\tan\alpha_{a1} - \tan\alpha_w) \tag{2.045}$$

and

$$g_f = \frac{d_{b2}}{2} (\tan\alpha_{a2} - \tan\alpha_w) . \tag{2.046}$$

The distance between two corresponding involutes (i.e. two corresponding flanks) is called the base pitch. If the value of the base pitch is higher than the value of the path of contact, contact between the two following teeth does not occur, when the two mating teeth release, thus continuity in meshing is not guaranted. The correct working of the gear will be assured if the value of the path of contact is higher than that of the base pitch. The ratio given by the following formula :

$$\varepsilon_\alpha = \frac{g_\alpha}{p_b} \tag{2.047}$$

is called the transverse contact ratio.

It also results in the following equations:

$$\varepsilon_\alpha = \frac{g_f + g_a}{\pi\, m\, \cos\alpha} = \varepsilon_1 + \varepsilon_2 \tag{2.048}$$

or

$$\varepsilon_\alpha = \frac{z_1}{2\pi}(\tan\alpha_{a1} - \tan\alpha) + \frac{z_2}{2\pi}(\tan\alpha_{a2} - \tan\alpha_w) . \tag{2.049}$$

Let us consider a segment AD starting from A which is drawn on the action line and equal to the base pitch and a segment EB starting from E equal to the base pitch. The path of contact is divided into three parts. When the contact takes place in A, a contact between two other pairs of teeth that fit in D occurs; if contact continues two other pairs of teeth will come into contact. In B, as the other pairs of teeth are at a base pitch distance, the teeth shown at in E, the contact between the couple of teeth stops, and from B to D there is only a couple of teeth in contact. Starting from A the contact between A and B takes place once again. The sections AB and DE of the path of contact are in double contact while the section BD is in single contact. The position of the point C is not defined as it depends on the rackshift factor. The point B is called the "lower point of single contact" while the point D is known as the "upper point of single contact". The curvature radius of the tooth of the driven gear as well as that of the driving gear can easily be calculated.

$T_1A = T_2T_1 - T_2A$ represents the curvature radius in A, i.e.

$$\rho_{A1} = a \sin\alpha_w - \frac{1}{2} d_{b2} \tan\alpha_{a2} \qquad (2.050)$$

T_1E represents the curvature radius in E, i.e.

$$\rho_{E1} = \frac{1}{2} d_{b1} \sin\alpha_{a1} \qquad (2.051)$$

at the point B the curvature radius is $T_1E - p_b$ i.e.

$$\rho_{B1} = \rho_{E1} - \pi\, m \cos\alpha \qquad (2.052)$$

while at point D the curvature radius is $T_1A + p_b$ i.e.

$$\rho_{D1} = \rho_{A1} + \pi\, m \cos\alpha \qquad (2.053)$$

For the driven gear we have:

$$\rho_{E2} = a \sin\alpha_w - \frac{1}{2} d_{b1} \tan\alpha_{a1} \qquad (2.054)$$

$$\rho_{A2} = \frac{1}{2} d_{b2} \tan\alpha_{a2} \qquad (2.055)$$

$$\rho_{D2} = \rho_{A2} - \pi\, m \cos\alpha \qquad (2.056)$$

$$\rho_{B2} = \rho_{E2} + \pi\, m \cos\alpha \qquad (2.057)$$

2.10 Calculation of an Angle Starting from its Involute

If we know the value of the involute inv α we calculate:

$$q = (\text{inv } \alpha)^{0.66667} .$$

With the following constant:

$$\begin{aligned}
C1 &= 1.040042 \\
C2 &= 0.324506 \\
C3 &= -0.003209 \\
C4 &= 0.0088336 \\
C5 &= 0.0031898 \\
C6 &= 0.0004772
\end{aligned}$$

and it will result that:

$$p = 1 + q(C_1 + q(C_2 + q(C_3 + q(C_4 + q(C_5 + q(C_6 + q))))))$$

and

$$\cos \alpha = 1/p .$$

3. Mechanical Gears

3.1 General Spur Features

Mechanical gears are made of real wheels that must comply with the meshing geometrical rules. As outlined previously, there are cylindrical, bevel and parallel gears, and gears with intersecting or non intersecting axes, while toothing can be external or internal. We will take also into consideration spur gears, spur parallel gears, external and internal cylindrical gear pairs and parallel gears made of such wheels. Then we will examine in details gears made of bevel gears and cylindrical worm gears.

3.2 Cylindrical Wheels

3.2.1 Definition

Cylindrical wheels are in the form of a cylinder whose section is perpendicular to the circular axis. Spur teeth follow a generatrix of the cylinder. A wheel with an infinite diameter has a straight section which is a straight line and a surface that is a plane. Such definition corresponds to a rack.

3.2.2 Geometry

The straight section of a wheel is a geometric wheel having all the features that characterize the wheel described in the above paragraphs.
   All the rules quoted before can be applied integrally. The point of reference is still the reference profile which applied to the real wheel becomes the basic rack.
   The reference circle of the geometric wheel becomes then the reference cylinder, the tip circle becomes the tip cylinder and the root circle becomes the cylinder. Such cylinders have the same diameters of their respective

circumferences. When we examine the properties of wheels we will also consider the straight sections which all are equal. These circumferences are simply called circles.

One important aspect must be emphasize: wheels are mechanical components that are subject to stresses, therefore any sudden change in section must be avoided; the connection between the tooth and the root cylinder will be made gradually by means of a fillet. The reference profile includes a circular section of the radius ρ_F whose envelope (curve) when generating the profile, forms the fillet profile.

3.2.3 Standardized Basic Profile and Modules

Gear teeth are defined by the basic profile and by module. It is interesting to note that the recourse is limited to a standard basic rack tooth profile together with modules in order to limit the use of tools, as to economic reasons which are defined by the same quantities characterizing gear teeth. The ISO has defined a standardized basic rack tooth profile (ISO 53) and normalized modules (ISO 54).

The profile defined by standard ISO 53 is shown on Fig. 3.1.

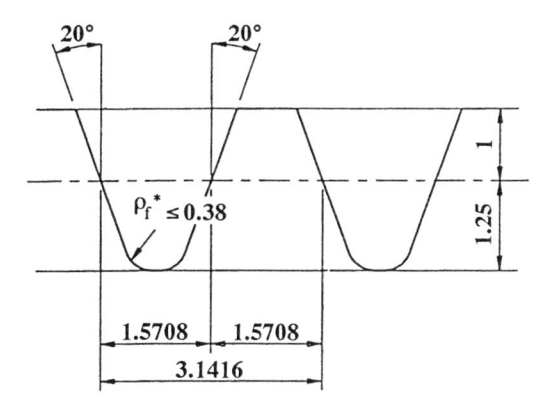

Fig. 3.1. ISO 53 standardized basic rack profile.

The fillet radius cannot have a too high value otherwise the straight section of the profile generating the involute will diminish. However, such a value must be sufficiently high to avoid stress concentration. Given a profile addendum equal to 1 the active dedendum will also be equal to 1. The junction radius ρ_F of the profile will be limited to:

$$\rho_F \leq \frac{h_f^* - 1}{1 - \sin\alpha} . \qquad (3.001)$$

Given an angle α of 20° and a dedendum of 1.25, the junction radius of the tooth will be lower to 0.38.

Additional profiles have been studied to allow for lump cutting. Such profiles will have a standardized dedendum equal to 1.40. Modules are arranged in two lists: a reference list and a reported list providing further information.

3.2.4 Facewidth

The length of the generatrix of the reference cylinder (and of the other cylinders concentric with this one) is called facewidth. Its value ranges from 0.5 to 1.8 times the value of the reference diameter. Such variation depends on applications.

3.2.5 Undercut

In the above paragraph we notice that if a section of the reference tooth profile is below the base circle, the resulting profile will perturbe the curvature which is opposite to that of the involute. In a real wheel, such a profile penetrates into the base of the tooth causing a remarkable decrease in the tooth root thickness and in the mechanical resistence of the tooth. Mating two wheels, the involute of the mating gear will make in contact with the perturbing profile, causing serious problems such as noise, additional stresses and so on. Such phenomenon must be absolutely avoied; the problem can be eliminated and will not occur if we increase the number of teeth according to a given modification coefficient or if we increase the modification coefficient of the tooth belonging to the considered wheel.

3.2.6 Tooth Thickness

The equations concerning the thickness of the teeth on circles with different diameters are the same as those reported in the second section. We notice that the value of head width must be higher than 0 as a tipped tooth is not strongly resistant. The value of the head width in teeth which have not undergone surface hardening is 0.2. The width head value of case-hardened steel teeth is 0.3 to avoid the destruction of the case-hardened layer during operation. On the tip of the tooth, the shearing section is perpendicular to the involute. The value of this section is as low as the value of the tip thickness and as high as the value of the tip angle. The value of the tip thickness of teeth on a wheel with many teeth must be higher than that of a pinion with few teeth.

3.2.7 Internal Toothing

Internal toothing is a complementary of the external toothing: the shape of the tooth corresponds to that of the dedendum of external teeth and the shape of the dedendum of the internal teeth corresponds to that characterizing the tooth belonging to external toothing. It is possible to define the equations of diameters and the tooth thickness by following the rules stated before. A useful method is to change the signs of the diameters of the number of teeth as well as rack shift factor and apply equations for the external toothing and after that to reach positive dimensions. To give an example, we consider the reference thickness of the internal toothing (Figure 3.2).

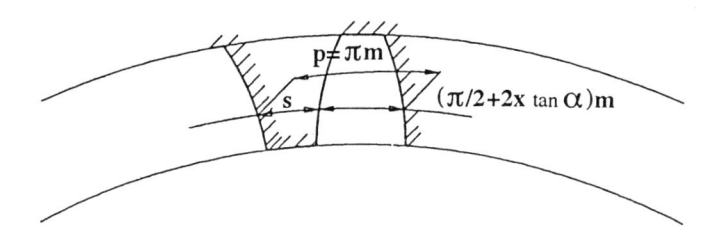

Fig. 3.2. Reference thickness of internal teeth.

We notice that the value of the reference thickness s of internal gear wheel is equal to the value of the reference pitch minus the value of the reference thickness of external gear wheel.

$$s = \pi\, m - m\left(\frac{\pi}{2} + 2x\tan\alpha\right) \qquad (3.002)$$

$$s = m\left(\frac{\pi}{2} - 2x\tan\alpha\right). \qquad (3.003)$$

If we had applied the a.m. method, the result would have been the same. Internal gear wheel will be given by the following equations:

$$d = m\,z \qquad (3.004)$$

$$d_b = m\,z\cos\alpha \qquad (3.005)$$

$$d_a = m\,z - 2(h_a^* - x)\,m = m\left[z - 2(h_a^* - x)\right] \qquad (3.006)$$

$$d_f = m\left[z + 2(h_f^* + x)\right] \qquad (3.007)$$

$$p = \pi\, m \qquad (3.008)$$

$$p_b = \pi\, m\, \cos\alpha \qquad (3.009)$$

$$S_b = m\left(\frac{\pi}{2}\cos\alpha - 2x\sin\alpha - z\,\mathrm{inv}\alpha\,\cos\alpha\right) \qquad (3.010)$$

$$S_a = d_a\left(\frac{\frac{\pi}{2} - 2x\tan\alpha}{z} - \mathrm{inv}\alpha + \mathrm{inv}\alpha_a\right). \qquad (3.011)$$

The surfaces limiting the teeth that intersect with the transverse plane following involute profiles are called flanks. The different cylinders (the reference cylinder, the tip cylinder and the base cylinder) intersect with flanks following straight lines parallel to the axis.

3.3 Parallel Spur Gears

In order to obtain a parallel spur gear, it is necessary to mate two spur cylindrical wheels of the type defined in the above paragraphs. The two axes of the cylinders are parallel while the two reference surfaces or the two pitch surfaces depending on whether they are working at reference centre distance or at modified centre distance, are tangent. Given the value of the sum of the rack shift factors, which is equal to 0, the working at the reference centre at a distance is considered to be a particular case deriving from the more general case of the modified centre distance working. In this case the pitch diameters are the same as the reference diameters. If two external wheels are mated the resulting unit will be an external gear pair, while if an external wheel (one necessarily having a smaller diameter) is mated with an internal wheel, the resulting unit will be an internal gear pair. Meshing will be assured on condition that the value of the module and of the pressure angle are the same for both wheels. The following equations characterize the external gear pair:

$$a = \frac{1}{2}(d_{w1} + d_{w2}) \qquad (3.012)$$

$$d_{w1,2} = d_{1,2}\frac{\cos \alpha}{\cos \alpha_w} \qquad (3.013)$$

$$a = \frac{1}{2}(d_1 + d_2)\frac{\cos \alpha}{\cos \alpha_w} \qquad (3.014)$$

$$a = \frac{m}{2}(z_1 + z_2)\frac{\cos \alpha}{\cos \alpha_w} \qquad (3.015)$$

$$\Sigma x = \frac{z_1 + z_2}{2 \tan \alpha}(\text{inv } \alpha_w - \text{inv } \alpha) \ . \qquad (3.016)$$

Given the value of the centre distance, the number of teeth and the basic profile, it is possible to calculate α_w and the sum of the rack shift factors that must be shared between the two wheels.

Given the number of teeth, the value of the module and the sum of the rack shift factors, it is possible to calculate inv α_w, α_w and a for an internal gear pair.

$$a = \frac{1}{2}(d_{w2} - d_{w1}) = \frac{m}{2}(z_2 - z_1)\frac{\cos \alpha}{\cos \alpha_w} \ . \qquad (3.017)$$

It results in a difference of rack shift factors and in a subtraction instead of a sum.

$$\Delta x = x_2 - x_1 = \frac{z_2 - z_1}{2 \tan \alpha}(\text{inv } \alpha_w - \text{inv } \alpha) \ . \qquad (3.018)$$

The value of the two rack shift factors will be shared between the two wheels.

The contact between flanks takes place following a straight line which is parallel to axes. This line moves along the flanks in contact and is called the line of contact. It represents the tooth flank line for each wheel.

3.4 Helical Gears. Involute Helicoid

Let us consider a base cylinder with a diameter d_b and a plane tangent at such a cylinder. On this plane we consider a straight line inclined at an angle β_b with respect to the generatrix of the cylinder.

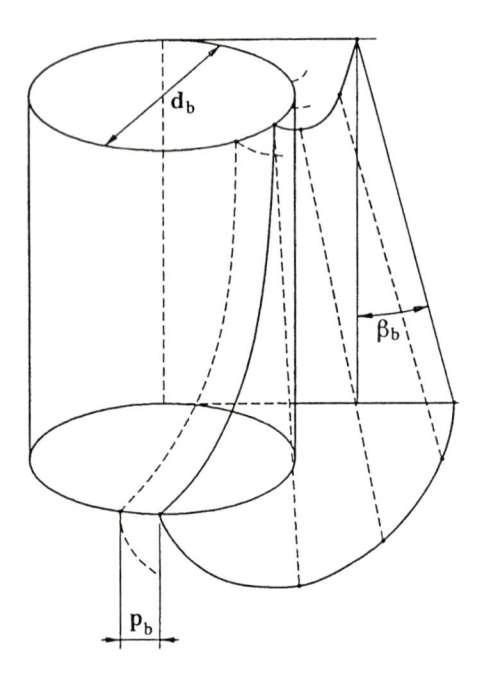

Fig. 3.3. Involute helicoid.

If this plane rolls on the cylinder without sliding, the straight line will originate a surface called involute helicoid.

Each point of the straight line generates an involute. Any straight section of the cylinder is called the transverse surface.

The intersection of the involute helicoid with the transversal section creates an involute while the intersection of the involute helicoid with the base cylinder is an helix whose tangent is parallel to the generatrix: this means that the base helix angle is β_b.

In the base plane, the equidistant base helixes can be drawn on the a.m. plane as the distance between them is the base pitch. The base pitch has the following value:

$$p_{bt} = \pi \frac{d_b}{z}. \quad (3.019)$$

Let us consider another cylinder with a diameter d:

$$d_b = d \cos \alpha_t. \quad (3.021)$$

We will call α_t the transverse pressure angle, see equation (3.035). On this cylinder the helix angle can be reached taking into consideration that the helix axial pitch is the same on all cylinders. Figure 3.4 shows that:

$$\frac{\tan \beta}{\tan \beta_b} = \frac{d}{d_b} = \frac{1}{\cos \alpha_t}. \quad (3.022)$$

The angle β is the reference helix angle. Given a cylinder with any diameter d_y:

$$\frac{\tan \beta_y}{\tan \beta_b} = \frac{1}{\cos \alpha_y} \quad (3.023)$$

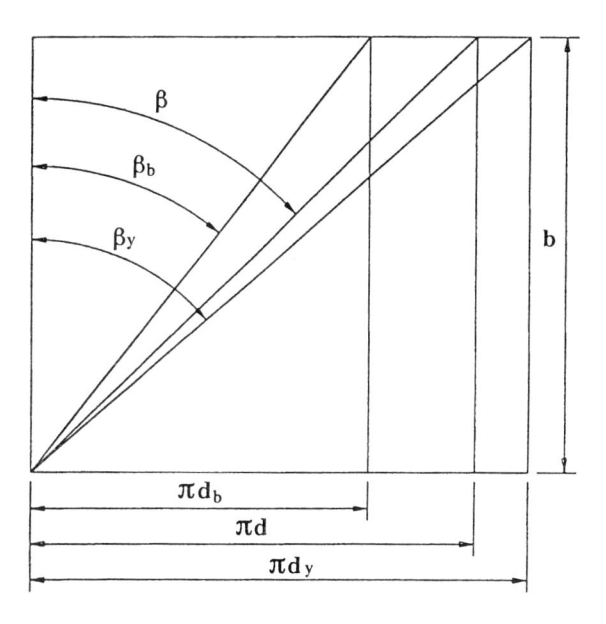

Fig. 3.4 Helix angle

and

$$\tan \beta_y = \tan \beta \, \frac{\cos \alpha_t}{\cos \alpha_y} \qquad (3.024)$$

with

$$\cos \alpha_y = \frac{d_b}{d_y} \, . \qquad (3.025)$$

3.4.1 Helical Wheel

Definition. By forming a closed solid by means of z non corresponding involute helixes (the involute having originated on the transverse plane and are corresponding involutes); we cut the reference cylinder across a point at a distance equal to half the value of the pitch or module of the first helixes drawn and we consider two cylinders concentric with the reference cylinders known as the tip cylinder and the root cylinder: we are defining an helical gear.

On the transverse plane we draw an elementary geometric wheel having a module m_t which is defined by the following equation:

$$m_t = \frac{d}{z} \, . \qquad (3.026)$$

This module is called the transverse module.

Reference Rack and Diameters. The wheel characterized by an infinite diameter is a rack whose transverse section is the transverse profile. The rack is characterized by edges delimiting its profile that form together with the transverse plane an angle β (Fig. 3.5).

The plane perpendicular to such edges is called the normal plane. It determines the reference normalized tooth profile of the module m which is also normalized. We notice that there is a relation between the pitch of the transverse profile p_t and the pitch of the real profile:

$$p_t = p/\cos \beta \qquad (3.027)$$

dividing by π we obtain:

$$m_t = \frac{m}{\cos \beta} \, . \qquad (3.028)$$

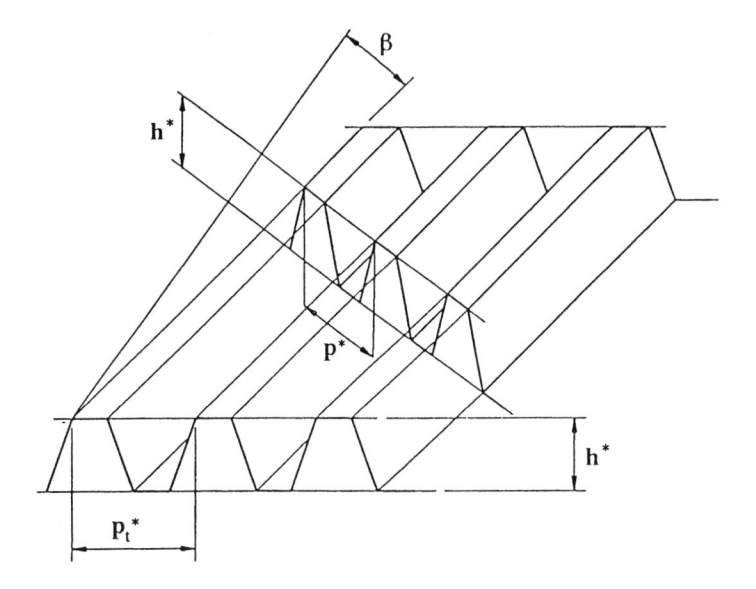

Fig. 3.5. Basic rack.

The head height of the reference normalized tooth profile is $h_a^* m$ while the foot height of the base is $h_f^* m$.

On the basis of the a.m. rules defined by the ISO (ISO 701), the dimensions of the helical wheel are the following:

$$d = \frac{m\,z}{\cos\beta} \qquad (3.029)$$

$$d_b = \frac{m\,z}{\cos\beta} \cos\alpha_t \qquad (3.030)$$

$$d_a = m\left[\frac{z}{\cos\beta} + 2(h_a^* + x)\right] \qquad (3.031)$$

$$d_f = m\left[\frac{z}{\cos\beta} - 2(h_f^* - x)\right] . \qquad (3.032)$$

The following equations are for an internal gear wheel:

$$d_a = m\left[\frac{z}{\cos\beta} - 2(h_a^* - x)\right] \qquad (3.033)$$

and

$$d_f = m\left[\frac{z}{\cos\beta} + 2(h_f^* + x)\right] . \qquad (3.034)$$

Angular Relations. Starting from the following equation:

$$\tan\alpha_t = \frac{\tan\alpha}{\cos\beta} \qquad (3.035)$$

and

$$\tan\beta_b = \tan\beta \; \cos\alpha_t \qquad (3.036)$$

it is possible to calculate the following angular relations:

$$\cos\beta_b = \sqrt{\cos^2\alpha \; \cos^2\beta + \sin^2\alpha} \qquad (3.037)$$

$$\cos\alpha_t \; \cos\beta_b = \cos\alpha \; \cos\beta \qquad (3.038)$$

and

$$\cos\beta_b = \frac{\sin\alpha}{\sin\alpha_t} . \qquad (3.039)$$

Undercut. The application of the equations for the interference in spur wheels allows equation for the helical gear to be defined taking into consideration the transverse plane. It is then to be obtained:

$$z \geq \frac{2(h_f^* - x)\cos\beta}{\sin^2\alpha_t} . \qquad (3.040)$$

Given a determinate rack shift factor, such a relation enables us to calculate the maximum number of teeth in order to avoid interference and, given the number of teeth, it enables us to define the maximum allowable value of the rack shift factor to avoid cut interference. The maximum allowable number of teeth is lower than that of teeth in spur wheels, while the other conditions are the same (also the value of the rack shift factor is lower for a given number of teeth). The undercut resulting in drilling the tooth root dimishing the degree of resistance to stresses and producing a harmful contact between mating flanks (diminishing the degree of resistance to stress) must be avoided.

Tooth Thickness. The equations for spur gear teeth are also valid on a transverse plane.
Transverse reference thickness:

$$S_t = \left(\frac{\pi}{2} m_t \pm 2x \tan\alpha_t\right) \tag{3.041}$$

$$S_t = m_t\left(\frac{\pi}{2} \pm 2x \tan\alpha\right) \tag{3.042}$$

Transverse normal thickness:

$$S_n = S_t \cos\beta = m\left(\frac{\pi}{2} \pm 2x \tan\alpha\right) . \tag{3.043}$$

It is the value found for spur wheel teeth as the section of the tooth is that of a spur wheel teeth.
Transverse base thickness:

$$S_{bt} = d_b\left(\frac{S_t}{d} \pm \text{inv}\alpha_t\right) \tag{3.044}$$

Normal base thickness:

$$S_{bn} = S_{bt} \cos\beta_b = m\left(\frac{\pi}{2}\cos\alpha \pm 2x \sin\alpha \pm z\,\text{inv}\alpha_t\,\cos\alpha\right) . \tag{3.045}$$

Transverse tip thickness:

$$S_{at} = d_a\left(\frac{\frac{\pi}{2} \pm 2x \tan\alpha}{z} \pm \text{inv}\alpha_t \mp \text{inv}\alpha_a\right) \tag{3.046}$$

Normal tip thickness:

$$S_{an} = S_{at} \cos \beta_a \qquad (3.047)$$

with:

$$\tan \beta_a = \tan \beta \frac{d_a}{d} = \tan \beta \frac{\cos \alpha_t}{\cos \alpha_a} . \qquad (3.048)$$

Virtual Spur Gear - Virtual Number of Teeth. If the normal plane (which is perpendicular to the reference helix) cuts across a section of the cylindrical gear, the teeth in the meshing area belong to a standard spur gear called the spur gear to which corresponds a fixed number of teeth called the virtual number of teeth and is expressed by z_n. In order to determine the value of z_n we must consider the bending radius of the ellipsis resulting from the intersection of the reference cylinder with a plane which is not perpendicular to its axis. Such value can also be obtained by considering the cut interference limit.

- If we equated the bending radius with the reference radius of the virtual gear, the value of the reference diameter of such a gear will be:

$$d_n = z_n m = \frac{z m}{\cos^3 \beta} \qquad (3.049)$$

from which:

$$z_n = \frac{z}{\cos^3 \beta} . \qquad (3.050)$$

- Considering the base thickness, the result will be:

$$S_{bn} = m \left(\frac{\pi}{2} \cos \alpha \pm 2 x \sin \alpha \pm z \, \text{inv} \alpha_t \cos \alpha \right) \qquad (3.051)$$

while considering spur gear teeth z_n the result will be:

$$S_{bn} = m \left(\frac{\pi}{2} \cos \alpha \pm 2 x \sin \alpha \pm z_n \, \text{inv} \alpha \cos \alpha \right) . \qquad (3.052)$$

We notice that:

$$z_n = z \frac{\text{inv} \alpha_t}{\text{inv} \alpha} . \qquad (3.053)$$

- Considering the undercut limit, the result will be:

$$z = \frac{2(1-x)\cos\beta}{\sin^2\alpha_t} \qquad (3.054)$$

and

$$z_n = \frac{2(1-x)}{\sin^2\alpha} \qquad (3.055)$$

It results that:

$$z_n = \frac{z}{\cos\beta}\frac{\sin^2\alpha}{\sin^2\alpha_t} = \frac{z}{\cos^2\beta_b \cos\beta} \qquad (3.056)$$

The resulting values vary according to the factors considered.

Such differences are due to the fact that the section in the normal plane is equated with an involute which is not an involute. If we calculate the three values, they will not be so different from the other. Therefore it is possible to adopt them without distinction and with sufficient approximation. In particular, equations (3.050) and (3.056) are very similar and can be adopted in different ways.

3.4.2 Helical Parallel Gear Pairs

Definition. Let us consider two different base cylinders and let us draw the common tangent plane. On such a plane we draw a line at an angle β_b. If the plane rolls on the two cylinders two helicoids will be generated. Such helicoids are in contact following the straight line. Starting from these cylinders and from such helicoids, it is possible to define two gear wheels whose number of teeth is z_1 and z_2, respectively. The module (identical) is m_t for both wheels. If a wheel turns, meshing its teeth together with those belonging to the other wheel and if one of the teeth is in contact with a tooth of the first wheel mentioned, it will start the wheel to which it belongs. Considering the elementary geometric wheel, things were exactly the same. Meshing of the wheels takes place correctly as the mating teeth are in contact following a line that is oblique at the flanks and which moves along the respective flanks. Such an oblique line is called a line of contact. When it moves along the flanks it is called the path of contact, as in the elementary geometric gears.

However one important aspect must be emphasized: the helix angles of the two wheels have the same value but a different direction.

Reference Centre Distance - Modified Centre Distance.

Two tangentially positioned wheels can be mated correctly either following or not following their reference cylinders.

In the first case the working is at reference centre distance with a sum of the rack shift factors (or a difference if we consider an internal teeth) equal to 0. In the second case, the sum (or the difference) is different from 0 and the working is said to be at a modified centre distance.

The resulting equations are the same as those defined for spur toothing but they are taken in from the transverse plane.

$$a \cos\alpha_{wt} = \frac{d_1 + d_2}{2} \cos\alpha_t = m \frac{z_1 + z_2}{2 \cos\beta} \cos\alpha_t \qquad (3.057)$$

$$\Sigma x = \frac{z_1 + z_2}{2 \tan\alpha}(\text{inv }\alpha_{wt} - \text{inv }\alpha_t) \qquad (3.059)$$

or

$$\Delta x = x_2 - x_1 = \frac{z_2 - z_1}{2 \tan\alpha}(\text{inv}\alpha_{wt} - \text{inv}\alpha_t) . \qquad (3.060)$$

Transverse Contact Ratio - Overlap Ratio.

Considering the features of the transverse plane it is possible to define on it a total contact ratio, which is equal to that defined for spur toothing. This equation can be written for an internal toothing (minus sign) or for an external toothing (plus sign).

$$\varepsilon_\alpha = \frac{z_1}{2\pi}(\tan\alpha_{a1} - \tan\alpha_{wt}) \pm \frac{z_2}{2\pi}(\tan\alpha_{a2} - \tan\alpha_{wt}) . \qquad (3.061)$$

The value of the normal contact ratio is:

$$\varepsilon_{an} = \varepsilon_{at} \cos\beta_b . \qquad (3.062)$$

The value of the path of contact is:

$$g_\alpha = \varepsilon_\alpha \, p_{bt} . \qquad (3.063)$$

The value of the length of the approach path is:

$$g_f = \pm \frac{z_2 \, m \cos\alpha_t}{2 \cos\beta}(\tan\alpha_{a2} - \tan\alpha_{wt}) \qquad (3.064)$$

and the length of the recess path will be:

$$g_a = \frac{z_1 \, m \, \cos\alpha_t}{2 \cos\beta}(\tan\alpha_{a1} - \tan\alpha_{wt}) \quad . \tag{3.065}$$

When the contact ceases on a straight section, it continues on the transverse plane that follows, as the generatrice line is oblique at the flank.

If the contact ceases first on an extreme plane of the wheel, the wheel must turn until an arc of the base circle is achieved that is equal to the tooth width b (length of the reference cylinder generatrix) multiplied by the tangent of the base helix.

The overlap ratio is given by the ratio of the arc to the base pitch.

$$\varepsilon_\beta = \frac{b \, \tan\beta_b \, \cos\beta}{\pi \, m \, \cos\alpha_t} \tag{3.066}$$

as well as:

$$\tan\beta_b = \tan\beta \, \cos\alpha_t \tag{3.067}$$

it will result that:

$$\varepsilon_\beta = \frac{b \, \sin\beta}{\pi \, m} \quad . \tag{3.068}$$

The total contact ratio is given by the sum of the transverse contact ratio and the overlap ratio.

$$\varepsilon_\gamma = \varepsilon_\alpha + \varepsilon_\beta \quad . \tag{3.069}$$

Meshing Interference. The limiting point of the involute in proximity to the root of the tooth of the driving gear wheel is at a distance $(1 - x_1) \, m$ from the reference circle (Fig. 3.6).

The value of the diameter of the driving gear wheel corresponding to this point is:

$$d_{A1} = 2\sqrt{\frac{1}{2}d_{b1}^2 - \left(\frac{1}{2}d_{b1}\tan\alpha_t - \frac{(1-x)m}{\sin\alpha_t}\right)^2} \quad . \tag{3.070}$$

During working at a modified centre distance, the diameter corresponding to the first point of contact is given by:

$$d_{B1} = 2\sqrt{\frac{1}{2}d_{b1}^2 - \left(\frac{1}{2}d_{b1}\tan\alpha_{wt} - g_f\right)^2} \quad . \tag{3.071}$$

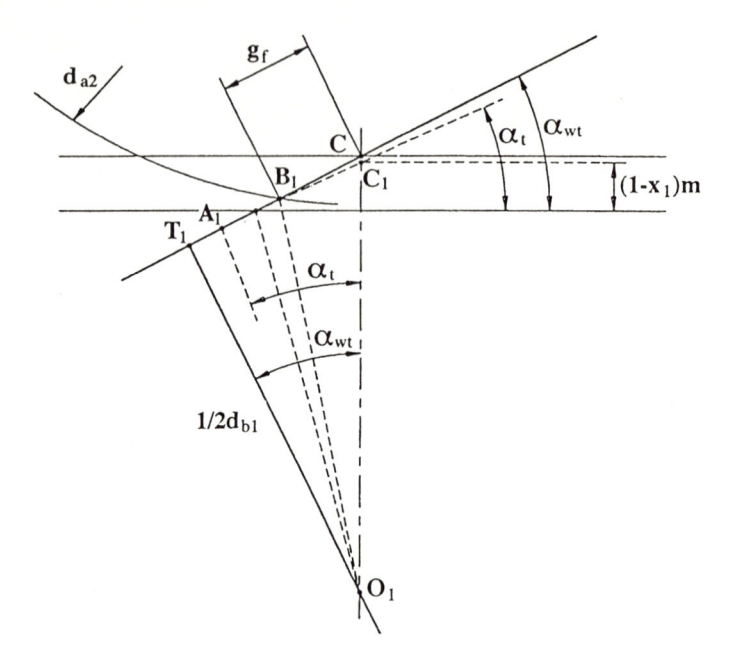

Fig. 3.6 Meshing interference

Contact between the involute of the driven gear wheel and the profile of the fillet will not take place, if d_{A1} is lower than d_{B1} leading to:

$$\frac{1}{2} d_{b1} \tan\alpha_t - \frac{(1-x)m}{\sin\alpha_t} > \frac{1}{2} d_{b1} \tan\alpha_{wt} - g_f \qquad (3.072)$$

i.e.

$$\frac{2(1-x_1)\cos\beta}{z_2 \sin\alpha_t \cos\alpha_t} > \pm(\tan\alpha_{a2} - \tan\alpha_{wt}) - \frac{z_1}{z_2}(\tan\alpha_{wt} - \tan\alpha_t) \qquad (3.073)$$

$$\frac{2(1-x_2)\cos\beta}{z_1 \sin\alpha_t \cos\alpha_t} > \tan\alpha_{wt} - \tan\alpha_{a1} - \frac{z_1}{z_2}(\tan\alpha_{wt} - \tan\alpha_t) \ . \qquad (3.074)$$

In theory, during reference centre distance working, center distance interference does not occur (Fig. 3.7).

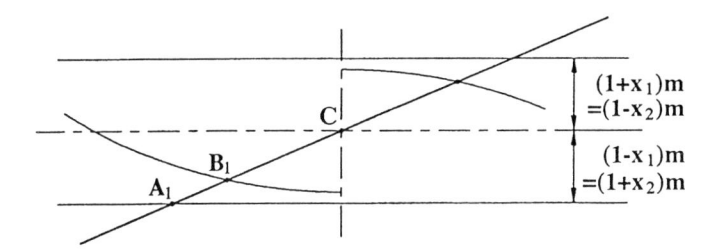

Fig. 3.7 Meshing interference during working at a reference centre distance

However, if negative tolerances of the centre distance together with positive tolerances of the tip diameter are present, the a.m. risk can exist. Therefore tolerances must be carefully chosen. In internal gears such a risk is always present. In order to prevent it, a tooth shortening is executed on the tooth of the internal gear: the value of the addendum is lowered by the following quantity $k\,m$:

$$k = \frac{8\,(1 - x_2)^2}{z_n} \qquad (3.075)$$

Workers often resort to reliefs in all tooth types. Further information will be provided in the following paragraphs. The equations for tip diameters are the following:

$$d_{a1,2} = m\left[\frac{z_{1,2}}{\cos\beta} \pm 2\,(h^*_{a1,2} - k_{1,2} \pm x_{1,2})\right] . \qquad (3.076)$$

In internal gear pairs a secondary meshing interference due to the contact between the tip of the teeth of the wheels outside the path of contact can be produced. Such interference occurs when the number of the teeth of the wheel is not greatly different from the number of teeth of the pinion.

Through the following calculations it is possible to determine the existence of such an interference.

$$\Delta_1 = \text{arc cos}\left[\frac{a}{d_{a1}}\left(\frac{d^2_{a2} - d^2_{a1}}{4\,a^2} - 1\right)\right] \qquad (3.077)$$

$$\Delta_2 = \text{arc cos}\left[\frac{a}{d_{a2}}\left(\frac{d^2_{a2} - d^2_{a1}}{4\,a^2} + 1\right)\right] \qquad (3.078)$$

$$\Psi_1 = \text{inv}\,\alpha_{a1} - \text{inv}\,\alpha_{wt} + \Delta_1 \qquad (3.079)$$

$$\Psi_2 = \text{inv}\,\alpha_{a2} - \text{inv}\,\alpha_{wt} + \Delta_2 \qquad (3.080)$$

$$\Delta = \frac{1}{2}(d_{b1}\,\Psi_1 + d_{b2}\,\Psi_2) \,. \qquad (3.081)$$

There is no interference if:

$$\Delta > 0 \,. \qquad (3.082)$$

The value of such a function varies with the centre distance value. When required, it is necessary to check the maximum and the minimun values taken by the centre distance with regard to tolerances (see paragraph on "Gear precision").

Tip and root clearance. During operation the clearance left between the tip of a tooth belonging to a wheel and the root of the tooth belonging to the mating gear must not be null. Failing this working reliability and lubrication are not assured.
The tip and root clearance is expressed by the following formula:

$$c_{1,2} = a - \frac{1}{2}(d_{a2,1} - d_{f1,2}) \,. \qquad (3.083)$$

We notice the importance of the role played by tooth shortening in order to assure the correct working of the whole unit.

3.4.3 Sliding

Theory. Rolling speed and Sliding speed. On the transverse plane we consider a point M belonging to the line of action. Such a point is common to the pinion and to the wheels but during movement it moves on the line of action; the two wheels continue to be in contact. This point also turns around the centre O_1 as it belongs to the pinion and around the centre O_2 as it belongs to the wheel. Whether this point belongs to the wheel or to the pinion, it moves at the same speed on the line of action (see Fig. 3.8 for the expression of such speed).

If n_1 is the rotation speed of the pinion and n_2 the rotation speed of the wheel, the basic rotation speed of the wheel ($n_2 = n_1 u$) will be :

$$v = \pi\,\frac{n_1 d_{w1}}{60000} = \pi\,\frac{n_2 d_{w2}}{60000} \,. \qquad (3.084)$$

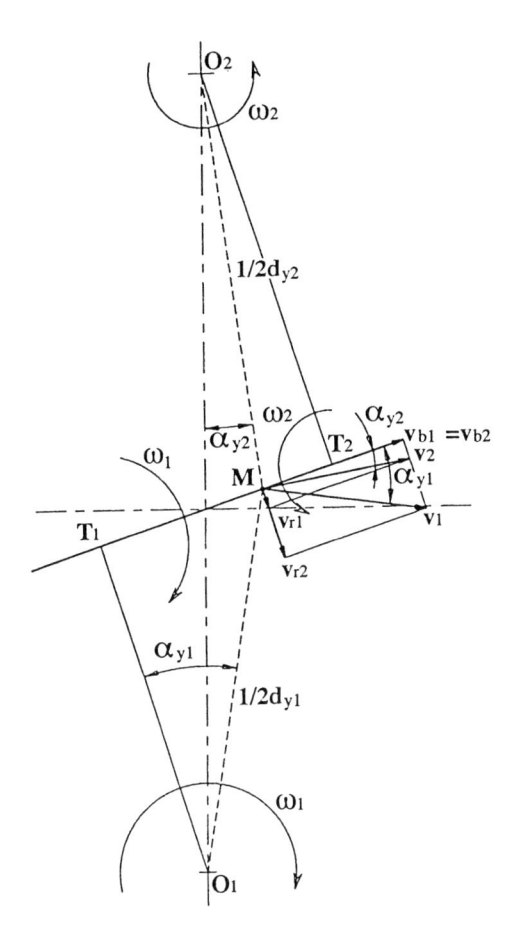

Fig. 3.8 Sliding theory

If the value of the rotation angular speed of the pinion is ω_1 the value of the rotation speed around O_1 will be: $v_1 = 0.0005\, d_{y1}\, \omega_1$ which is perpendicular to O_1M. The tangential speed around O_2 will be: $v_2 = 0.0005\, d_{y2}\, \omega_2$ which is perpendicular to O_2M.

Let us project such speed on a action line and on a line perpendicular to it. The following formula will result from such a projection:

$$v_{r1} = v_1 \sin \alpha_{y1} = 0{,}0005\, d_{y1}\, w_1 \sin \alpha_{y1}$$
$$v_{r2} = v_2 \sin \alpha_{y2} = 0{,}0005\, d_{y2}\, w_2 \sin \alpha_{y2}$$
$$v_{b1} = v_1 \cos \alpha_{y1} = 0{,}0005\, d_{y1}\, w_1 \cos \alpha_{y1}$$
$$v_{b2} = v_2 \cos \alpha_{y2} = 0{,}0005\, d_{y2}\, w_2 \cos \alpha_{y2}$$

as
$$d_{y1} \cos \alpha_{y1} = d_{b1}$$

and
$$d_{y2} \cos \alpha_{y2} = d_{b2}$$

on the basis of the gear definition, we can also write that:

$$v_{b1} = v_{b2} \ .$$

It results that:

$$\frac{\omega_1}{\omega_2} = \frac{d_{b2}}{d_{b1}} = \frac{z_2}{z_1} \ . \qquad (3.085)$$

On the other hand, we get:

$$0.5 \, d_{y1} \sin \alpha_{y1} = T_1 M = \rho_{M1} \qquad (3.086)$$

and

$$0.5 \, d_{y2} \sin \alpha_{y2} = T_2 M = \rho_{M2} \qquad (3.087)$$

and also

$$v_{r1} = \rho_{M1} \, \omega_1 \, /1000 \qquad (3.088)$$

and

$$v_{r2} = \rho_{M2} \, \omega_2 \, /1000 \ . \qquad (3.089)$$

We notice that the instantaneous meshing is formed by a speed of displacement on the line of action and a rotation tangential speed on each tooth around the point T_1 et T_2 respectively. These two different rotation speeds have the same direction, which is perpendicular to the line of action, but they are different.

A rotating speed exists on each point of contact. It is given by:

$$v_g = |v_{r1} - v_{r2}| \ . \qquad (3.090)$$

Figure 3.9 shows the trend of the rolling and sliding speeds for the external toothing.

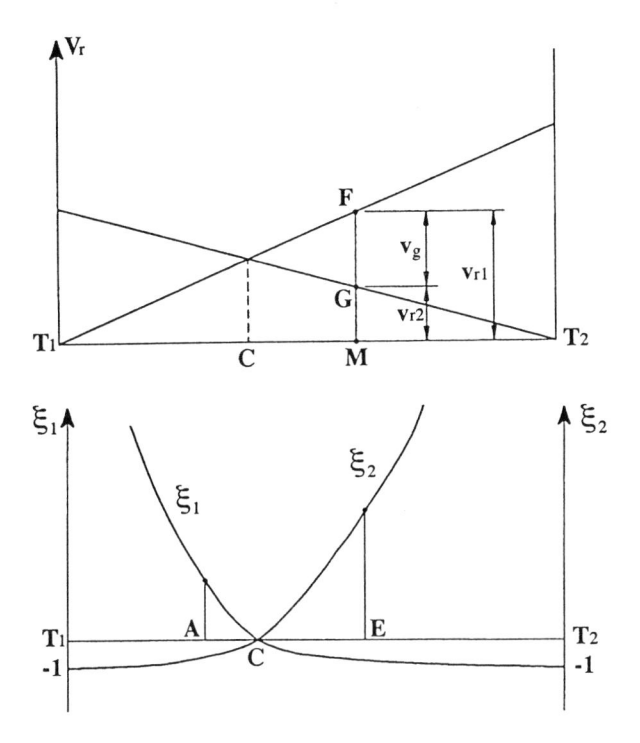

Fig. 3.9 Rolling and sliding speed. External specific sliding

Figure 3.10 provides the same speed for internal toothing.

We notice that on the approch path the driven gear wheel slides on the driving gear and that the driving gear slides on the driven gear wheel on the recess path. On the pitch point the sliding value is null as the two rotation tangential speeds are equal.

A gear can work due to the fact that mating teeth slide and roll.

On the graph representing speeds at a point M the rolling value is MG while the sliding value is GF. Sliding causes wear and overheating, i.e energy loss. Therefore the value of the gear efficiency is lower than the unit.

Specific Sliding. In order to define the importance of sliding it is preferable to consider the relation of the sliding speed to the rolling speed at each point of contact. This relation is called specific sliding. It is positive on one wheel when the mating gear wheel slides on the other or when the wheel undergoes sliding. It is negative when sliding is caused by the considered tooth.

Then:

$$\xi_1 = \frac{\omega_2 \, \rho_{M2} - \omega_1 \, \rho_{M1}}{\omega_1 \, \rho_{M1}} \qquad (3.091)$$

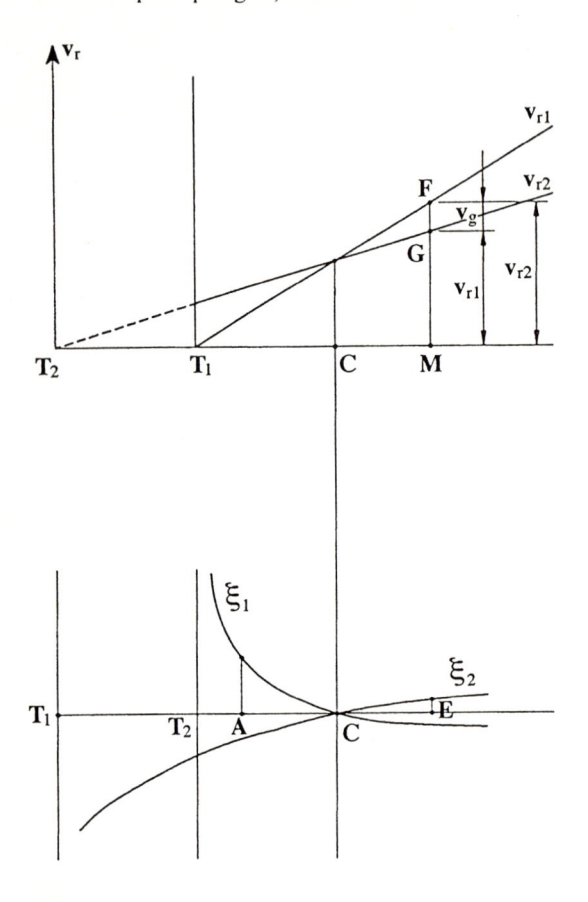

Fig. 3.10 Rolling and sliding speed for internal toothing

and

$$\xi_2 = \frac{\omega_1 \, \rho_{M1} - \omega_2 \, \rho_{M2}}{\omega_2 \, \rho_{M2}} \tag{3.092}$$

or even

$$\xi_1 = \frac{\omega_2 \, \rho_{M2}}{\omega_1 \, \rho_{M1}} - 1 = \frac{z_1 \, \rho_{M2}}{z_2 \, \rho_{M1}} - 1 \tag{3.093}$$

and

$$\xi_2 = \frac{\omega_1 \, \rho_{M1}}{\omega_2 \, \rho_{M2}} - 1 = \frac{z_2 \, \rho_{M1}}{z_1 \, \rho_{M2}} - 1 \tag{3.094}$$

or

$$\xi_1 = \frac{\tan \alpha_{y2}}{\tan \alpha_{y1}} - 1 \qquad (3.095)$$

$$\xi_2 = \frac{\tan \alpha_{y1}}{\tan \alpha_{y2}} - 1 \quad . \qquad (3.096)$$

Figure 3.10 shows the graphs concerning the specific sliding between mating teeth at a point of contact that moves between the two tangent points on the base circles.

If we set points A and E on the line of action representing the beginning and the end of meshing, we notice that at the beginning of meshing the value of the specific sliding reaches the top on the driving gear wheel. When it ends its value reaches the top on the driven gear wheel. Specific sliding causes wear that concentrates particularly on the root at the tooth. All the precautions taken to reduce the value of specific sliding turn out to have strenghtened the tooth resistence to wear.

In the paragraph "rack shift factor" we will deal with the methods aimed at reducing such values.

3.4.4 Forces

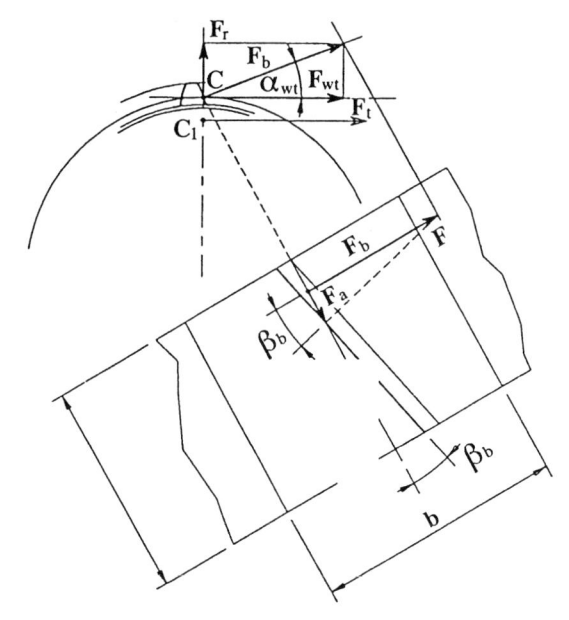

Projected base plane   **Fig. 3.11** Forces

If a torque T_1 is applied to the axis of the driving gear wheel, a force F will be generated between the mating teeth. Such force is positioned on the base plane perpendicular to the generatrix line as such direction is perpendicular to the two surfaces of contact. This force is an action on the driven gear wheel and a reaction on the driving gear wheel. On the base plane it is possible to divide such a force into a force F_b which is positioned on the transverse plane and into another force F_a an axial force, which is parallel to the axis.

The force F_α is given by:

$$F_\alpha = F \sin\beta_b \quad (3.097)$$

while for force F_b we have:

$$F_b = F \cos\beta_b \quad (3.098)$$

Such a force can be also defined starting from the torque T_1

$$F_b = 2000 \frac{T_1}{d_{b1}} \quad (3.099)$$

The force F_b applied to the reference point can also be factorized into two forces: a force F_{wt} which is tangent at the pitch cylinders and a force F_r which is perpendicular to the wheel axes. Such forces are both reactionary and have actions.

The value of the force F_r is:

$$F_r = F_b \sin\alpha_{wt} = F_{wt} \tan\alpha_{wt} \quad (3.100)$$

A reference force F_t tangentially at the reference cylinder can be determined. Its torque is the same as the force F_{wt} that is to say:

$$F_t = F_{wt} \frac{d_{wt}}{d} = F_{wt} \frac{\cos\alpha_t}{\cos\alpha_{wt}} \quad (3.101)$$

If we adopt F_t as the parameter, the following forces will result:
reference tangential force:

$$F_t = 2000 \frac{T_1}{d_1} = 1.908 \; 10^7 \frac{P}{d_1 \, n_1} \quad (3.102)$$

effective tangential force:

$$F_{wt} = F_t \frac{\cos\alpha_{wt}}{\cos\alpha_t} \qquad (3.103)$$

radial force:

$$F_r = F_t \frac{\sin\alpha_{wt}}{\cos\alpha_t} \qquad (3.104)$$

axial force:

$$F_a = F_t \tan\beta \qquad (3.105)$$

and then the force resulting from the sum of the a.m. forces the total force :

$$F = \frac{F_t}{\cos\alpha \, \cos\beta} \qquad (3.106)$$

The force resulting from F_r and F_{wt} is F_b and is positioned on the transverse plane tangentially at the base cylinder. The value of such a force is:

$$F_b = \frac{F_{wt}}{\cos\alpha_{wt}} = \frac{F_t}{\cos\alpha_t} \qquad (3.107)$$

Harmful axial forces will occur in helical gears if the value of the torque is high and the helix angles are wide.

To avoid such phenomenon double helical gears are used: they are made of two helical gears with opposite helix angles, which are carried by the same wheel. In this way axial forces are neutralized.

3.4.5 Gear Ratio - Speed Ratio

The gear ratio is the ratio of the number of teeth of the wheel to the number of teeth of the pinion.

$$u = \frac{number\ of\ theeth\ of\ the\ wheel}{number\ of\ teeth\ of\ the\ pinion} = \frac{z_2}{z_1} \qquad (3.108)$$

The speed ratio is the ratio of the speed at the driven gear wheel and turns to the speed at which the driving gear wheel turns.

$$i = \frac{speed\ of\ the\ driven\ gear\ wheel}{speed\ of\ the\ driving\ gear\ wheel} \qquad (3.109)$$

3.4.6 Normal Pressure

The two flanks are pressed one against the other by the force F; contact takes place following the line of contact. Such a situation corresponds to Hertz's theory, concerning the contact between two cylinders. According to this theory the value of the normal pressure following local distortions equal to c, varies from 0 to σ_H at the centre of c which is given by (where b is the length of the contact):

$$\sigma_H = \sqrt{\dfrac{F}{b} \dfrac{\dfrac{1}{\rho_1} + \dfrac{1}{\rho_2}}{\pi \left[\dfrac{1-v_1^2}{E_1} + \dfrac{1-v_2^2}{E_2}\right]}} \qquad (3.110)$$

where:
ρ_1 is the curvature radius of the tooth belonging to the driving gear wheel at the point considered,
ρ_2 is the curvature radius of the tooth belonging to the driven gear wheel at the point considered,
v_1 is Poisson's coefficient of the material forming the driven gear wheel,
v_2 is Poisson's coefficient of the material forming the driving gear wheel,
E_1 is the module of elasticity of the material forming the driving gear wheel,
E_2 is the module of elasticity of the material forming the driven gear wheel.

Deformation occurs on a section whose width is given by:

$$c = \dfrac{4}{\pi} \dfrac{F}{b\sigma_H} \qquad (3.111)$$

Such pressure creates a shearing stress in the material. The value of the shearing stress reaches the top at a depth of $0.39\ c$ below the surface. The maximum value of the shearing stress is:

$$\tau_{max} = 0.304\ \sigma_H . \qquad (3.112)$$

If we analyse the variation of σ_H along the path of contact in a particular gear, it is sufficient to consider the following variation:

$$\sigma_{H\ geom} = K \sqrt{\dfrac{1}{\rho_1} + \dfrac{1}{\rho_2}} \qquad (3.113)$$

which is made of two variables, the sum of which is equal to the segment $T_1 T_2$.

The representative curve starts from the infinite in page T_1, it passes through the centre of $T_1 T_2$ where it has a minimum value and then ends at T_2 where it regains the infinite value.

Figure 3.12 represents the trend of such variation. We set the points A and E corresponding to the length of the path of contact. We also set points B and D that are the lower points of single contact and the higher points of single contact. Between A and B and D and E, the load is distributed on the two teeth in contact and the value of the contact pressure is lower than the length of the line represented. Between B and D there is a single contact and the load is applied integrally. The contact pressure follows then the curve trend. If the reference point C is between B and D, the value of the pressure contact reaches the top at the lower point of single contact.

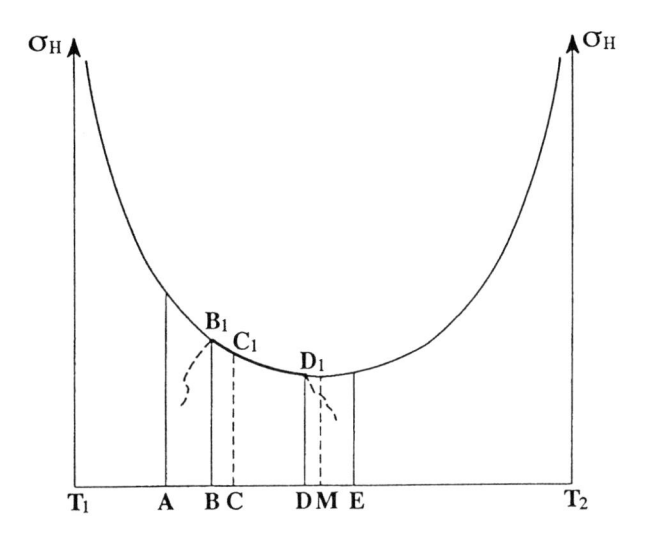

Fig 3.12. Normal pressure

3.4.7 Rack Shift Factor

In the paragraphs above the rack shift factor enables working at the modified centre distance. However such a coefficient affects the gear and its working in many other ways.

(1) The rack shift factor, if it is wisely chosen, is useful to prevent undercut.

(2) If the value of the rack shift factor is too high, the tooth will be too tipped.

(3) The value of rack shift factor is proportional to the value of the tooth thickness in proximity of the root of the tooth. If the value of the factor increases, the thickness too will augment, increasing the tooth resistence to breaking point. During working at the reference centre distance, an increase in the value of the

rack shift factor applied to a wheel decreases the value of the rack shift factor applied to the other wheel. A solution that has turned out to be advantageous for one wheel cannot be suitable for the other one. Different from what happens during working at the reference centre distance, when working at modified centre distance a positive sum of the rack shift factors increases the value of the rack shift factor applied to the wheels. An increase in the value of rack shift factor sum has turned out to improve the tooth resistence to bending.

(4) An increase in the value of the toothing rack shift factor applied to the driving gear wheel makes the path of contact move on the thrust line and provokes a decrease in the value of the overlap ratio.

If we analyse the effect generated by the displacement of the path of contact towards the contact line, we notice that the specific sliding and the normal pressure on the tooth of the driven gear wheel diminish. To a certain extent, such an increase is beneficial to working.

There is a optimal rack shift factor value which does not necessarily provide equality of the rack shift factors. In effect the value of the wear of the pinion is higher under equal conditions than that of the wheel whose teeth are not so often involved in meshing.

3.4.8 Gears that Reduce Speed/Gears that Increase Speed

When the smaller wheel (the pinion) meshes with the bigger one (wheel) the speed of the wheel is lower than that of the pinion (i is lower than the unit). In this case gears are called reducing gear pairs. Otherwise ($i > 1$) gears are known as increasing gears. When the teeth of the wheels come into contact with A, there will be two teeth in contact with D, which are the only ones transmitting the torque. The driving tooth bends backwards while the driven gear bends forwards. The pitch of the driving gear diminishes, while that of the driven gear increases. In A the driving tooth is in advance while the driven gear is behind. When contact takes place, an interference between the teeth is produced, causing an excessive stress.

If we have not taken special measures involving the tooth profile, we notice that at the beginning of meshing the value of the applied stress increases. At the same time the value of the specific sliding reaches the top. In addition the shifting of the point of contact is contrary to the sliding thereby strenghtening the harmful effects.

It has been necessary to increase the value of the rack shift factor applied to the teeth belonging to a driving gear.

In speed reducing gears the driving wheel is the pinion. Through the increase in the value of the rack shift factor applied to the pinion, it is possible to reduce interference and to improve working. The following decrease concerning the wheel is not remarkable as the number of the teeth of the wheel is higher and the risk deriving from undercut is lower.

In speed increasing gears, it is necessary to increase the rack shift factor applied to the pinion to avoid interference and its harmuful effects. At the same time it is also necessary to increase the value of the rack shift factor applied to the wheel and to diminish it on the pinion in order to neutralize the perturbing effect.

In speed increasing gears it is necessary to adopt a compromise solution complying with the two trends. Such a compromise is not required in speed reducing gears. In fact the solution suitable for speed reducing gears is not the same as that suggested for speed increasing gears.

In a gear designed to reduce speed, an increase in speed must be achieved with caution. If necessary the load applied must be diminished.

3.4.9 Tooth Modifications

Transverse modifications. Teeth that are in contact at the higher point of single contact are subject to distortions causing an increase in the value of the pitch of the driven gear wheel and a decrease in the pitch value of the driving wheel.

To soften movement and to avoid an increase in the intensity of the load at the point corresponding to the beginning of meshing, it is possible to diminish the pitch of the smoothing of the driving gear thereby relieving the tip of the tooth. Smoothing consists of decreasing the value of the tip thickness by C_a and reaching the theoretical profile progressively at the theoretical depth of h_c. (Fig. 3.13).

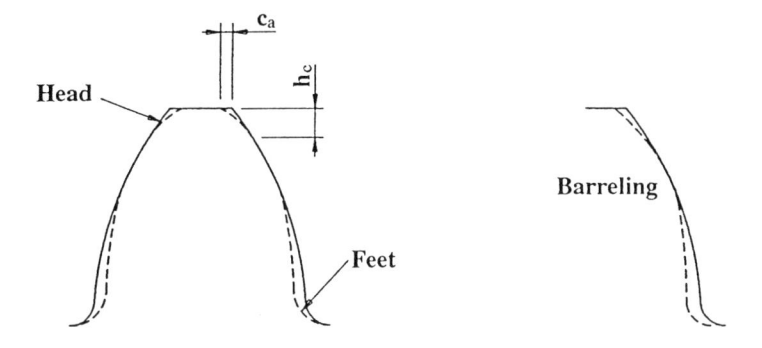

Fig. 3.13. Tooth modifications

It is also possible to relieve the root of the tooth or to relieve the root and the tip jointly. The result is a progressive application of the stress that accompanies contact at the beginning and a progressive decrease in the application of the stress that accompanies the contact at the end. If both the root and the tip are relieved the profile will be called "barelled".

If we calculate the elastic constant of two teeth on the basis of an analytical system on condition that the load is absolutely constant, it will be possible to calculate the tooth distortion at the higher point of single contact and the value of the relief (given by the sum of the tip relief and the root relief) will be equal to such a value.

If the tip of the tooth has been relieved, the value of the tip relief will be equal to that of distortion. In many gearboxes, the load is not constant and in gearboxes having normal classfications the load capacity is unknown until the unit is set in operation. Therefore, it is not possible to calculate the value of the relief. In this case we can attribute a medium empirical value to the tip relief, that will probably vary with load. However, such practice has turned out to be advantageous for the tooth even if the relief is not perfect.

Longitudinal modifications. The shaft supporting wheels and wheels that distort are subject to bending and torsion stresses.
If theoretical contact takes place following a straight line along the tooth flank, the real contact will take place differently because of such distortion. The resulting contact will not be correct and will take place on a smaller surface causing an increase in the local stress on each tooth.

To have theoretical contact after distortion, the tooth must not have the shape that it has after manufacturing: it must have the shape that becomes the theoretical shape after distortion. So during manufacturing it is possible to modify (to correct) the theoretical shape of the helix.

Such modifications are called longitudinal correction, as follows.

- Helix modification: distortion due to torsion is an helical distortion.

If we wish to correct such distortion we must modify the helix angle by considering an angle equivalent to the torsional angle opposite to the distortion. Such modification is called helix correction. It can be analytically determined only if the torque applied to the gear is constant.

If the torque varies or is unknown, it will be necessary to recourse to an average empirical correction.

- Profile crowning: distortion due to bending is an arc that adds to the distortion caused by torsion.

It is possible to correct the tooth transversally, giving - with regard to its theoretical shape - a modification equal but in an opposite direction to the combined deformation.

The resulting profile is called crowned (Fig. 3.14). Such correction can be calculated only if the torque applied to the gear is constant. Failing this, the crowning calculated together with a given torque is no longer valid for the others. However, an empirical crowning resulting from the experience of the designer always has a positive effect on the loading capacity of the gear.

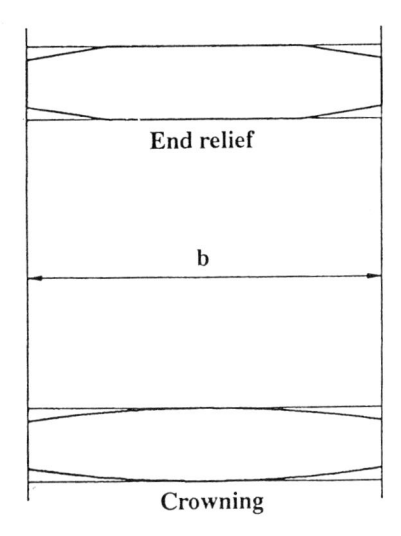

Fig. 3.14. Profile Crowning and end relief

Instead of defining a profile crowning that is a longitudinal relief all over the width of the tooth, it is possible to limit the end relief to a fixed length at each extremity.

Such end relief joins up together progressively with the non-corrected tooth. This longitudinal correction is not perfect but it is an advantage for the loading capacity. The value of this correction is empirical and based on a designer's experience.

3.4.10 Mechanical Losses in Gears

Friction between teeth under the action of the applied load creates mechanical losses that produce overheating.

The lost power is given by the product of the force multiplied by the sliding speed. The force and the sliding speed are variable on the place of contact in the same way. The result is an integral to the product. The lost power is equal to the product of the transmitted power, of the average coefficient of friction between teeth and factors depending exclusively on the geometry of the gear.

$$P_f = P \, \mu_m \, H_g \, . \tag{3.114}$$

If we approximate the force distribution between the teeth in double contact we will find that:

$$H_g = \frac{\pi \, (u + 1) \, \cos^2\beta}{u \, z_1} \frac{\varepsilon_1^2 + \varepsilon_2^2}{\varepsilon_\alpha} \tag{3.115}$$

with

$$\varepsilon_1 = \frac{g_f}{p_{bt}} \qquad (3.116)$$

$$\varepsilon_2 = \frac{g_a}{p_{bt}}. \qquad (3.117)$$

Gears efficiency will be developed in the chapter on the thermal power of gears.

3.5 Bevel Gear Pair

3.5.1 Definition

The simple bevel gear pair is made of two wheels whose axes intersect at S and whose reference surfaces (the tip surface and the root surface) are in the shape of a cone with the vertex at S.

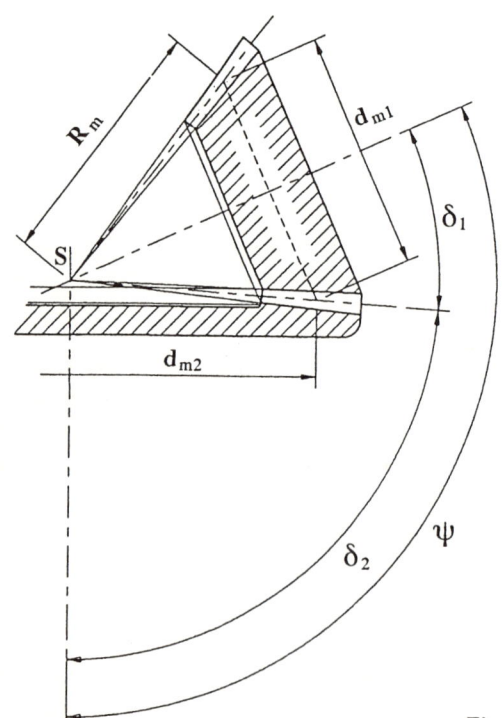

Fig. 3.15a. Bevel gear pair

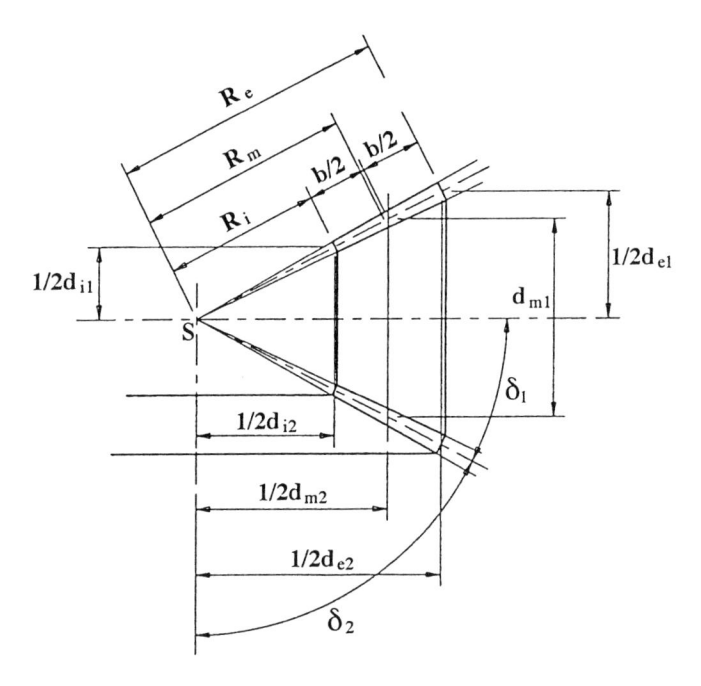

Fig. 3.15b Bevel gear pair

Theeth of the wheels that form the gear do not have the same depth. If M is the median point of the tooth width, if δ_1 is the half angle of the driving gear, if δ_2 is the half angle of the driven gear and ψ is the angle formed by the two axes, it will result that:

$$\psi = \delta_1 + \delta_2 \quad . \tag{3.118}$$

Let us define the gear characterized by medium wide teeth referring to the mean point of the band width, i.e. at the point M R_m is the distance SM d_{m1} which expresses the diameter of the reference cone of the driving gear in M while d_{m2} expresses the diameter of the reference cone of the driven gear in M.

We notice that:

$$R_m \sin\delta_1 = \frac{1}{2} d_{m1} \tag{3.119}$$

and

$$R_m \sin\delta_2 = \frac{1}{2} d_{m2} \quad . \tag{3.120}$$

It results that:

$$\frac{\sin\delta_1}{\sin\delta_2} = \frac{d_{m1}}{d_{m2}} \qquad (3.121)$$

As the two wheels have z_1 and z_2 number of teeth respectively, we can define a module m_{mt} given by the quotient of the reference diameter divided by the number of teeth. It will result that:

$$\frac{\sin\delta_1}{\sin\delta_2} = \frac{z_1}{z_2} = \frac{1}{u} \qquad (3.122)$$

For two axes positioned at 90°:

$$\delta_2 = \frac{\pi}{2} - \delta_1 \qquad (3.123)$$

$$\tan\delta_1 = \frac{z_1}{z_2} = \frac{1}{u} \qquad (3.124)$$

We notice that:

$$R_m = \frac{z_1\, m_{mt}}{2\sin\delta_1} = \frac{z_2\, m_{mt}}{2\sin\delta_2} \qquad (3.125)$$

$$R_i = R_m - \frac{b}{2} \qquad (3.126)$$

and

$$R_e = R_m + \frac{b}{2} \qquad (3.127)$$

It will result that:

$$m_{ti} = m_{mt}\frac{R_i}{R_m} = m_{mt} - \frac{b}{z_1}\sin\delta_1 = m_{mt} - \frac{b}{z_2}\sin\delta_2 \qquad (3.128)$$

and

$$m_{te} = m_{mt}\frac{R_e}{R_m} = m_{mt} + \frac{b}{z_1}\sin\delta_1 = m_{mt} + \frac{b}{z_2}\sin\delta_2 \qquad (3.129)$$

The minimum value of diameters is:

$$d_{i1} = z_1\, m_{ti} \tag{3.130}$$

$$d_{i2} = z_2\, m_{ti}\,. \tag{3.131}$$

The maximum value of diameters is:

$$d_{e1} = z_1\, m_{te} \tag{3.132}$$

$$d_{e2} = z_2\, m_{te}\,. \tag{3.133}$$

3.5.2 Equivalent Gears. Trestgold's Hypothesis

We cut the bevel gear pairs by means of a plane characterized by two axes. Let us take a plane perpendicular to the cone that intersects with the plane characterized by the axes.

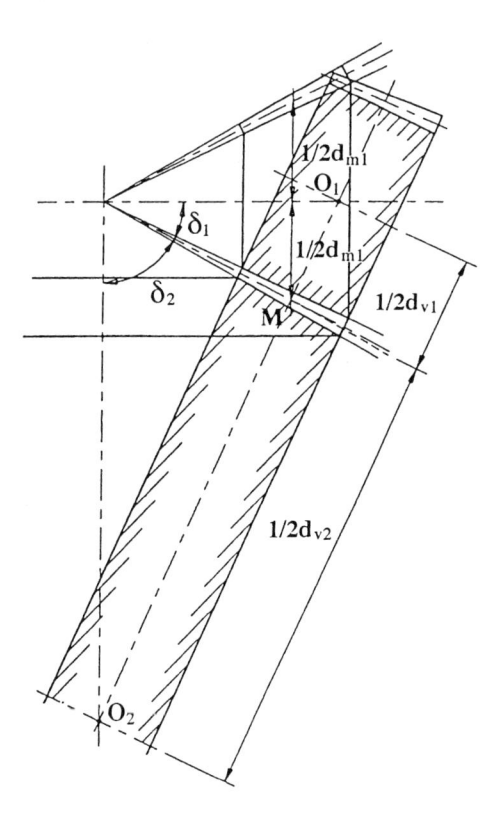

Fig. 3.16. Virtual wheels

O_1 is the point of intersection of the perpendicular plane with the axis of the driving gear wheel while the point O_2 is the same point on the axis of the driven gear wheel. It is possible to consider these two points as the centres of the geometric plane wheels representing the considered bevel gear pair considered. We will have:

$$O_1 M = \frac{1}{2} d_{v1} = \frac{d_{m1}}{2 \cos \delta_1} \qquad (3.134)$$

or

$$d_{v1} = \frac{d_{m1}}{\cos \delta_1} \qquad (3.135)$$

and

$$d_{v2} = \frac{d_{m2}}{\cos \delta_2} \qquad (3.136)$$

Now we consider the parallel axes; the only one to be considered from now on is:

$$d_{v1} = \frac{d_{m1}}{\cos \delta_1} \qquad (3.137)$$

$$d_{v2} = \frac{d_{m2}}{\sin \delta_1} \qquad (3.138)$$

$$z_{v1} = \frac{z_1}{\cos \delta_1} \qquad (3.139)$$

$$z_{v2} = \frac{z_2}{\sin \delta_1} \qquad (3.139 \text{ bis})$$

$$\cos \delta_1 = \frac{z_1}{z_{w1}} \qquad (3.140)$$

As well as:

$$\tan \delta_1 = \frac{1}{u} = \frac{\sin \delta_1}{\cos \delta_1} \qquad (3.141)$$

For trigonometrical development we find that:

$$z_{v1} = z_1 \frac{\sqrt{u^2 + 1}}{u} \qquad (3.142)$$

$$z_{v2} = z_2 \sqrt{u^2 + 1} \qquad (3.143)$$

and

$$u_v = \frac{z_{v2}}{z_{v1}} = u^2 \qquad (3.145)$$

If we adopt as the virtual number of teeth the numbers defined above and if we consider the module m_{mt}, the resulting gear is a parallel cylindrical gear. Such a gear is equivalent to the bevel gear pairs and will replace it in the following studies.

3.5.3 Constant Tooth Depth Bevel Gear

We can draw teeth having constant depth teeth on the reference cone (Fig. 3.17)

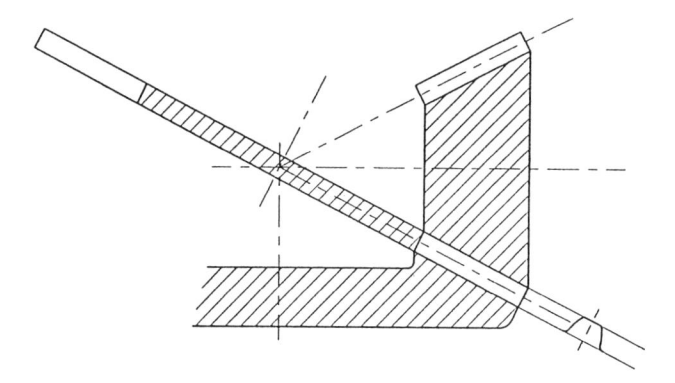

Fig. 3.17. Constant depth of tooth

We lengthen the point and the root of the tooth to achieve exact similarity with respect to the cones' apex. The resulting wheel is a crown wheel with a medium diameter $2\,R_m$. Such wheel meshes correctly both with the driving gear wheel and the driven gear wheel.

If we replace the tool used to shape the tooth profile, the two wheels could be milled by means of the same tool. Such a tool will allow the inside of the crown wheel arcs of the circle and curves to be described that will create mating curves on the bevel gear wheels which can mesh correctly.

The two wheels can also be worked by the crown wheel, which is now characterized by non-intersecting axes. There are a number of combination possibilities and it is for this reason that engineers look favourably at this type of gear. The angle formed by the tangent at the profile (at the point corresponding to half the width of the profile) and the radius of the wheel is equal to an helix angle. The wheel obtained by means of Trestgold's hypothesis is a helical gear that can turn into a spur gear wheel.

The virtual number of teeth that allow bevel gears to be located as spur gears is:

$$Z_{vn1} = z_1 \frac{\sqrt{1 + u^2}}{u \cos^3 \beta_m} \qquad (3.146)$$

and

$$Z_{vn2} = z_2 \frac{\sqrt{1 + u^2}}{\cos^3 \beta_m}. \qquad (3.147)$$

The theory concerning spur gears can be applied to constant tooth height or variable tooth depth bevel gears only if we consider wheels whose number of teeth corresponds to the equivalent number.

3.5.4 Constant Tooth Depth Gear

Gears can also be classified according to the shape of the curves generated into the gear wheel and on the basis of tool and machine availability.

These consist of the following:

- Gleason: curves are arcs of a circle. The dimensions of the tooth are not constant and diminish from the extremity to the vertex of the cone. The spiral angle ranges from 0° to 35° but generally it measures 35°. It is possible to draw spirals on the right and on the left.
- Klingelnberg ("Palloide"): curves are involutes to a circle. The inclination angle ranges from 35° to 38°. The tooth depth is constant.
- Klingelnberg ("Ciclo-palloide"): curves are epicycloids. The tooth depth is constant and the value of the inclination angle ranges from 0° to about 45°.
- Module - Kurvex: curves are arcs of a circle. The tooth depth is constant and the value of the inclination angle varies from 25° to 45°. Mating gears can be milled by means of the same tool.
- Oerlikon ("Spiromatic"): curves are epicycloids. The tooth depth is constant.

Two types of toothing can be distinguished: N toothing and G toothing. In the first type of teeth the module reaches its maximum value at the centre of the tooth and diminishes in proximity of the two extremities. The value of the inclination angle varies from 30° to 50°. In the second type of toothing the value of the a.m. angle ranges from 0° to about 50°.

3.5.5 Forces on Bevel Gear Pairs

Let us consider two tangent wheels for their two reference cones. The tangential force at half the length of the two cones is given by:

$$F_{mt} = 2000 \frac{T_1}{d_{m1}} = 2000 \frac{T_2}{d_{m2}} \qquad (3.148)$$

An axial force F_1 (involving the virtual cylindrical gear pair) and a force F_2 due to the tooth inclination if the wheel does not stand in a straight position to the virtual parallel gear correspond to the tangential force considered on the virtual gear (Fig. 3.18).

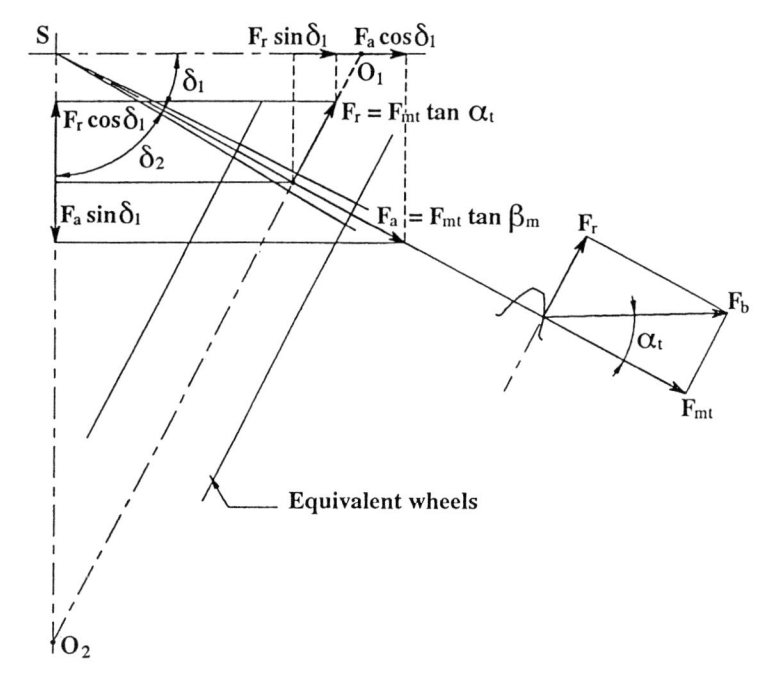

Fig. 3.18. Forces on bevel gear pairs

$$F_1 = F_{mt} \frac{\tan \alpha_t}{\cos \beta_m} \qquad (3.149)$$

and

$$F_2 = F_{mt} \tan \beta_m . \qquad (3.150)$$

The forces F_1 and F_2 considered on the bevel gears create axial and radial components being respectively as the radial component of one of the wheels is the axial component of the other wheel. It results that:

$$F_{r1} = F_{a2} = F_{mt}(\tan \alpha_t \frac{\cos \delta_1}{\cos \beta_m} \mp \tan \beta_m \sin \delta_1) \qquad (3.151)$$

and

$$F_{r2} = F_{a1} = F_{mt}(\tan \alpha_t \frac{\sin \delta_1}{\cos \beta_1} \pm \tan \beta_m \cos \delta_1) . \qquad (3.152)$$

In spur gear the second term is null.

A curved profile diminishes the axial stress on a wheel, which increases on the other wheel. The radial stress increases on a wheel while it diminishes on the other one.

The direction of the tooth curvature helps us to understand which of the wheels is touched by the increase or by the decrease mentioned above.

3.5.6 Dimensions of the Blanks Forming the Gears

The following equations can be deduced from Fig. 3.19.
We have already defined the following equations:

$$\tan \delta_1 = \frac{z_1}{z_2} = \frac{1}{u} \qquad (3.153)$$

$$d_{m1} = m_{mt} z_1 \qquad (3.154)$$

and

$$d_{m2} = m_{mt} z_2 . \qquad (3.155)$$

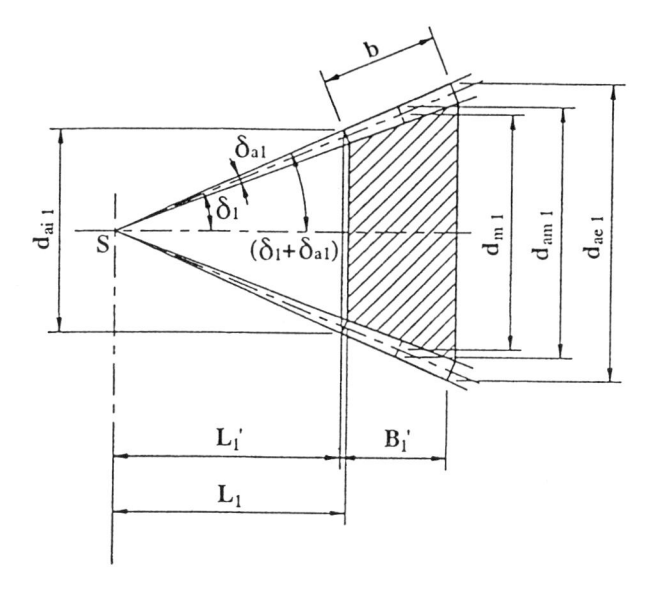

Fig. 3.19. Dimensions of the blanks forming the gears

We also have:

$$R_m = \frac{1}{2}\frac{d_{m2}}{\cos\delta_1} \quad . \tag{3.156}$$

So:

$$m_{te} = m_{mt}\frac{d_{m1} + b\cos\delta_1}{d_{m1}} \tag{3.157}$$

$$m_{ti} = m_{mt}\frac{d_{m1} - b\cos\delta_1}{d_{m1}} \tag{3.158}$$

$$d_{am1} = m_{mt}\left[z_1 + 2(h_{a1}^* + x_1)\cos\delta_1\right] \tag{3.159}$$

$$d_{am2} = m_{mt}\left[z_2 + 2(h_{a2}^* + x_2)\sin\delta_1\right] \tag{3.160}$$

$$d_{ae1} = d_{am1} \frac{m_{te}}{m_{mt}} \tag{3.161}$$

$$d_{ai1} = d_{am1} \frac{m_{ti}}{m_{mt}} \tag{3.162}$$

$$d_{ae2} = d_{am2} \frac{m_{te}}{m_{mt}} \tag{3.163}$$

$$d_{ai2} = d_{am2} \frac{m_{ti}}{m_{mt}} \tag{3.164}$$

$$\tan\delta_{a1} = \frac{(\overset{*}{h}_{a1} + x_1)\, m_{mt}}{R_m} \tag{3.165}$$

$$\tan\delta_{a2} = \frac{(\overset{*}{h}_{a2} + x_2)\, m_{mt}}{R_m} \tag{3.166}$$

We obtain the heights of the cone and the mounting dimensions by:

$$L_1' = \frac{d_{am1}}{2\tan(\delta_1 + \delta_{a1})} - \frac{b\cos^2(\delta_1 + \delta_{a1})}{2\cos\delta_1} \tag{3.167}$$

$$L_2' = \frac{d_{am2}}{2\tan(\delta_2 + \delta_{a2})} - \frac{b\cos^2(\delta_2 + \delta_{a2})}{2\sin\delta_1} \tag{3.168}$$

$$B_1' = b\,\frac{\cos^2(\delta_1 + \delta_{a1})}{\cos\delta_1} \tag{3.169}$$

$$B_2' = b\,\frac{\cos^2(\delta_2 + \delta_{a2})}{\sin\delta_1} \tag{3.170}$$

$$L_1 = L_1' + (\overset{*}{h}_{a1} + \overset{*}{h}_{f1})\, m_{ti} \sin\delta_1 \tag{3.171}$$

$$L_2 = L_2' + (\overset{*}{h}_{a2} + \overset{*}{h}_{f2})\, m_{ti} \cos\delta_1 \tag{3.172}$$

3.6 Worms and Wormwheels

3.6.1 Definition

Let us consider a cylinder with a diameter d_{m1}.

We place z_1 teeth on such cylinder following an helical arrangement so that the solids are equal to the spaces and the toothing is divided from each other with regard to the cylinder - equal to a heigth h_a above and h_f below. The cylinder is called the reference cylinder and the diameter of such a cylinder is called the reference diameter.

The resulting figure is a worm. We can associate such a worm with a toroidal gear whose teeth fit perfectly with the worm and have the same dimensions of the teeth of the worm as well as the same inclination z_2 is the number of teeth of such wheel. In the median section of the wheel we find a gear wheel that is the tranverse section of the helical gear. When the worm turns at a speed of n_1, the wheel turns at the speed of n_2. It will result that:

$$n_2 = n_1 \frac{z_1}{z_2} \qquad (3.173)$$

The resulting unit is called the worm and the wormwheel. The axes of rotation are at 90°. As the number of teeth of the worm which are called worm threads is extremely low (it can even be equal to the unit) the value of the speed ratio can be extremely high. Worms and wormwheels make the transmission between the two non intersecting orthogonal axes possible with reduced overall dimensions and high reduction power.

3.6.2 Main Dimensions

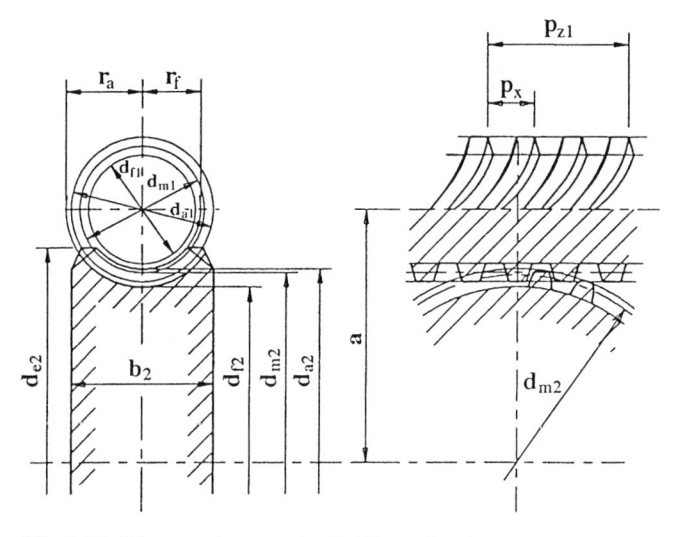

Fig. 3.20. Worm and wormwheels (dimensions).

- Worm: each tooth forms an helix whose inclination angle is γ_m. Such a helix has an axial pitch p_{z1} that is linked to the reference diameter and to the lead angle by means of the following relation:

$$p_{z1} = \pi\, d_{m1} \tan\gamma_m \qquad (3.174)$$

as there are z_1 equidistant teeth the value of the tooth pitch p_x will be:

$$p_x = \frac{p_{z1}}{z_1} = \pi\, \frac{d_{m1} \tan\gamma_m}{z_1} \,. \qquad (3.175)$$

A module m_t is defined as:

$$m_t = \frac{p_x}{\pi} = \frac{d_{m1} \tan\gamma_m}{z_1} \,. \qquad (3.176)$$

The diametric quotient q can be obtained by dividing the reference diameter by the module:

$$q = \frac{d_{m1}}{m_t} \,. \qquad (3.177)$$

We can assume that the value of the tooth addendum is m_t and that the value of the dedendum is $m_t + c_1$ (generally with $c_1 = 0.2\, m_t$) as such quantity c_1 is useful to guarantee a backlash between the head of the tooth and the bottom of the worm tooth. It will result that:

$$d_{a1} = d_{m1} + 2\, m_t \qquad (3.178)$$

$$d_{f1} = d_{m1} - 2(m_t + c_1) \,. \qquad (3.179)$$

It is also possible to define a normal module perpendicular to the helix. The value of such a module is:

$$m_n = m_t \cos\gamma_m \,. \qquad (3.180)$$

The value of the thickness of the tooth placed on the reference cylinder is:

$$s = \frac{p_x}{2} = \pi\, \frac{m_t}{2} \qquad (3.181)$$

while the value of the normal tooth thickness is:

$$s_n = \frac{\pi \, m_t \, \cos\gamma_m}{2} \quad . \tag{3.182}$$

Different shaped teeth will be considered in the following chapters. If a tooth has an involute profile, it is possible to define a base cylinder the diameter of which is given by:

$$d_{b1} = d_{m1} \frac{\tan\gamma_m}{\tan\gamma_b} \tag{3.183}$$

with:

$$\cos\gamma_b = \cos\gamma_m \, \cos\alpha_1 \quad . \tag{3.184}$$

For the width of the worm we fix the value:

$$b_1 = 2 \, m_t \, \sqrt{z_2 + 1} \quad . \tag{3.185}$$

- Wormwheel. We define a geometric wheel belonging to an helical gear with an helix angle γ_m on the median plane of the wormwheel.

The dimensions of such section will be those of the transverse section of a wheel having a module m_{t2}, as the addendum is m_{t2} the dedendum is $m_{t2} + c_2$ and the tooth is defined by involutes having a α_t angle.

The transverse module will be the same as the worm axial module, as the transverse pitch is equal to the worm axial pitch.

$$m_t = m_{x1} = m_{t2} \tag{3.186}$$

c_2 is the backlash at the root of the tooth which is generally equal to 0.2 m_t; if we apply a rack shift factor x to the wheel, it will result that:

$$d_{m2} = z_2 \, m_t \tag{3.187}$$

$$d_{a2} = d_{m2} + 2 \, m_t \, (1 + x) \tag{3.188}$$

$$d_{f2} = d_{m2} - 2 \, (m_t + c_2 - m_t \, x) \quad . \tag{3.189}$$

If a is the centre distance (the smaller distance between the two orthogonal axes), the toroidal shape of the wheel will be defined by two radii whose values are for the point of the tooth and for the root of the tooth.

$$r_a = a - \frac{d_{f2}}{2} \qquad (3.190)$$

and

$$r_f = a - \frac{d_{a2}}{2} \qquad (3.191)$$

The value of the toothing width is close to:

$$b_2 = 2\, m_t\, \sqrt{q+1}\ . \qquad (3.192)$$

The value of the external diameter is generally:

$$d_{e2} = d_{a2} + m_t\ . \qquad (3.193)$$

The value of the reference diameter and that of the pitch diameter are:

$$d_2 = d_{m2} - 2\, x\, m_t \qquad (3.194)$$

$$a = \frac{1}{2}(d_{m1} + d_{m2})\ . \qquad (3.194)\text{bis}$$

3.6.3 Efficiency of Worms and Wormwheels. Reversing

Worm and wormwheel are subject to a higher standard of friction during operation and to larger heat losses with respect to other gears. The calculation of such losses will be shown in the section dealing with the gear thermal capacity. If P_v is the power lost during transmission, P_1 is the power from the worm and P_2 the power from the wheel, two cases must be taken into consideration:

(1) the worm is the driving unit while the wheel is the driven unit

$$\eta_G = \frac{P_2}{P_2 + P_v} = \frac{P_1 - P_v}{P_1} \qquad (3.195)$$

(2) the wheel is the driving unit

$$\eta'_G = \frac{P_1}{P_1 + P_v} = \frac{P_2 - P_v}{P_2}\ . \qquad (3.196)$$

In order to simplify such calculations, we can assume that the working of the unit containing the worm and wormwheel is similar to an inclined plane (Fig. 3.21).

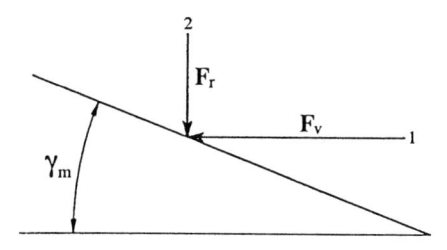

Fig. 3.21. Inclined plane of forces

F_r is the force resulting from the torque applied to the tooth of the wheel and F_v is the force resulting from the torque applied to the worm. These two forces are orthogonal forces. We apply the work theory to them and consider a coefficient of friction $\tan \phi$.

- If F_v is active and F_r is passive (i.e. if the worm is the driving unit) it will result that:

$$F_v = F_r \tan(\gamma_m + \phi) \ . \qquad (3.197)$$

Otherwise it will result that:

$$F_v = F_r \tan(\gamma_m - \phi) \ . \qquad (3.198)$$

In this latest case we notice that for an inclination angle equal to the friction angle, the force F_r must be infinite independent on the value of the force F_v. Failing this, working is not possible. It will happen the same if the value of the inclination angle is lower than that of the friction angle. When the value of the worm helix inclination is lower than the value of the friction angle, the worm does not mesh with the wheel. This phenomenon is called Non-Reversing.

3.6.4 Tooth Profile

Worm profile. We can distinguish between many different worm tooth profiles whose shape varies with working.

(1) *ZA worm*: the section of the worm obtained by the intersection of a plane passing through its axis creates spur profiles. Such worms are manufactured by means of trapezoidal section tools. Profile grinding is made by means of a grinding wheel.

(2) *ZN worm*: the section of the worm that follows the plane perpendicular to the teeth (normal section) creates a straight profile. Such a profile can be shaped by means of an end mill cutter whose cutting edges are in the form of a cone or by a turning tool following the worm inclination. Profile grinding is made by means of a trapezoidal thread milling cutter.

140   Jacques Sprengers, Part III

(3) *ZK* worm: the tool positioned on the normal section is in the shape of a trapezium. The profile can be shaped by means of big milling cutters or generating tools used for turning.

(4) *ZI* worm: the profiles result from involutes. They are obtained by generation (as helical gears) or by turning by a generating tool positioned on the base plane.

(5) *ZH* worm: the concave profile is obtained by means of a wheel whose section has a proper convex profile.

Wheel profile. The median section of the wheel is a geometric gear wheel whose profile results from an involute. Following the normal plane the profile results from an involute generated by a reference profile with a pressure angle α_n.

3.6.5 Forces

T_1 is the torque applied to the worm axis while T_2 is the torque applied to the wheel axis (Fig. 3.22).

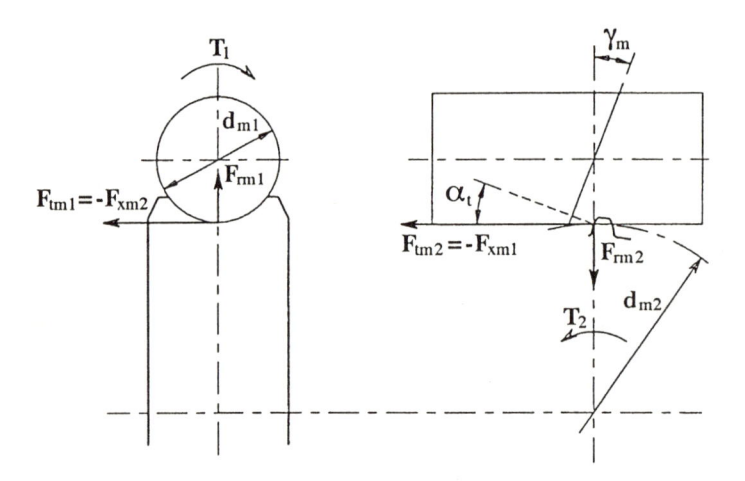

Fig. 3.22. Forces on worm wheels

(1) If the worm is the driving unit and η is the efficiency of the transmission, it will result that:

$$F_{tm1} = 2000 \frac{T_1}{d_{m1}} = 2000 \frac{T_2}{d_{m1} \eta \frac{z_2}{z_1}} \qquad (3.199)$$

and

$$F_{tm2} = 2000 \frac{T_2}{d_{m2}} = 2000 \frac{T_1 \eta \frac{z_2}{z_1}}{d_{m2}} \quad (3.200)$$

(2) If the wheel is the driving unit, it will result that:

$$F_{tm1} = 2000 \frac{T_1}{d_{m1}} = 2000 \frac{T_2 \eta'}{\frac{z_2}{z_1} d_{m1}} \quad (3.201)$$

and

$$F_{tm2} = 2000 \frac{T_2}{d_{m2}} = 2000 \frac{T_1 \frac{z_2}{z_1}}{\eta' d_{m2}} \quad (3.202)$$

However, the a.m. tangential force exerted (i.e. generated) up on one of the units is equal to the axial force exerted on the other component.
Following the definitions given above, it will result that:
(1) In the case where the worm is the driving unit :

$$F_{tm1} = -F_{xm2} = F_{tm2} \tan(\gamma_m + \phi) \quad (3.203)$$

$$F_{tm2} = -F_{xm1} \quad (3.204)$$

(2) If the wheel is the driving unit, it will result that:

$$F_{tm1} = -F_{xm2} = F_{tm2} \tan(\gamma_m - \phi) \quad (3.204 \text{ bis})$$

$$F_{tm2} = -F_{xm1} \quad (3.205)$$

In both cases units are characterized by a radial force for driven wheel:

$$F_{rm1} = F_{rm2} = \frac{F_{tm1} \tan \alpha_n}{\sin(\gamma_m + \phi)} \quad (3.207)$$

and for driving wheel:

$$F_{rm1} = F_{rm2} = \frac{F_{tm2} \tan\alpha_n}{\cos(\gamma_m - \phi)} \qquad (3.208)$$

where α_n is the normal pressure angle of the wheel.

3.7 Single Planetary Gear Train

3.7.1 Description (Fig. 3.23)

(S) represents a wheel called Solars that meshes with n wheels called Planets (P) that in turn fit with an internal gear called the annulus. Planets on the circle, that receive their axes, are equidistant one from another. Planets are fixed to a component that turns concentrically with the solar called the Planet Carrier. The resulting unit is called the planetary gear train.

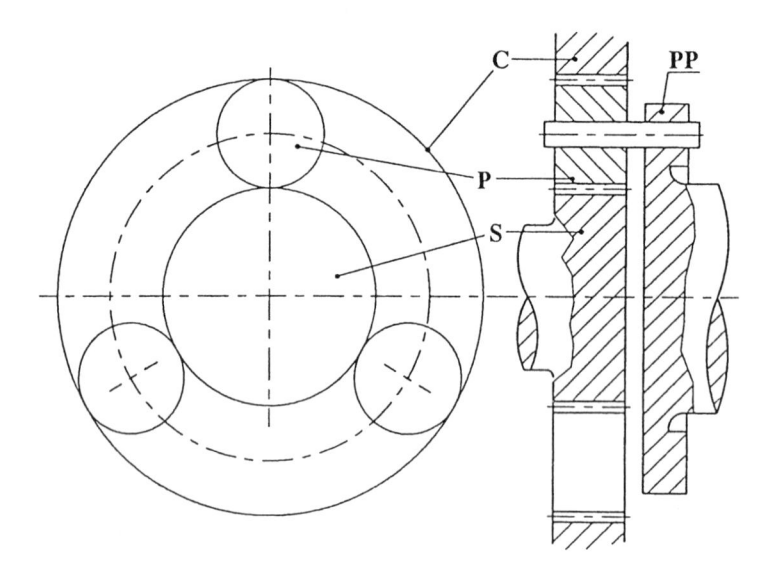

Fig. 3.23. Planetary gear train

The indexes 1, 2, 3 and e are assigned to the solar, to each planet, to the internal gear wheel and to the planet carrier respectively. The meshing wheels are at the reference centre distance or at the modified centre distance. In order to make things easier we will consider working at the modified centre distance. As wheels are then tangents following pitch cylinders it will result that:

$$d_{w3} = d_{w1} + 2\, d_{w2} \qquad (3.209)$$

and as a consequence:

$$z_3 = z_1 + 2\, z_2 \qquad (3.210)$$

if we assume that the pitch cone angles of the couple solar-planet and planet-gear wheel have similar versus they are considered as equal. Mounting can be executed successfully if:

$$z_1 + z_3 = \textit{Multiple of } n \qquad (3.211)$$

where n is the number of planets.

Planets must be positioned on the circle that gathers their centres. Therefore it is necessary that:

$$\frac{\pi\,(d_{w1} + d_{w3})}{n} > d_{w2} \qquad (3.212)$$

or following the same hypothesis:

$$n < \textit{Integral}\left(\pi\, \frac{\frac{z_3}{z_1} + 1}{\frac{z_3}{z_1} - 1}\right). \qquad (3.213)$$

3.7.2 Kinematic Relation

n_1, n_2, n_3 and n_e are the rotating speed of the components around their respective axis. Such speeds are considered according to their sign.

According to an observer positioned on the planet carrier, the speeds are respectively:

$$n_1' = n_1 - n_e \qquad (3.214)$$

$$n_2' = n_2 - n_e \qquad (3.215)$$

and

$$n_3' = n_3 - n_e \qquad (3.216)$$

these kinematic relations have the same characteristics as normal gears:

$$\frac{n_2'}{n_1'} = -\frac{z_1}{z_2} \qquad (3.216 \text{ bis})$$

$$\frac{n_2'}{n_3'} = \frac{z_3}{z_2} \qquad (3.217)$$

from which it derives that:

$$\frac{n_1'}{n_3'} = \frac{n_1 - n_e}{n_3 - n_e} = -\frac{z_3}{z_2}\frac{z_2}{z_1} = -\frac{z_3}{z_1} \qquad (3.219)$$

Such a relation is known as Willis' relation.

In the planetary train gear (type 1) the internal gear wheel is fixed. As $n_3 = 0$ it results that:

$$\frac{n_1 - n_e}{-n_e} = -\frac{z_3}{z_1} = -K \qquad (3.220)$$

or

$$\frac{n_1}{n_e} = K + 1 \qquad (3.221)$$

3.7.3 Forces

Let us consider the forces on the transverse plane. The forces perpendicular to that plane act only as thrusts towards bearings but they do not intervene during working of gears.

Let us consider:

$$F_{wt1} = 2000\,\frac{T_1}{d_1\,n} \qquad (3.222)$$

The forces to be considered are F_{b1} and F_{b2} whose values are respectively:

$$F_{b1} = F_{wt}\,\frac{1}{\cos\alpha_{wt1}} \qquad (3.223)$$

and

$$F_{b2} = F_{wt}\,\frac{1}{\cos\alpha_{wt2}} \qquad (3.224)$$

Such forces will be equal if the pitch pressure angles have the same value. Such forces do not differ much one from the other. On the other hand, on the axes of the planets there are the following forces $2F_{wt}$, corresponding to a torque applied to the planet carrier which is equal to:

$$T_e = 2 F_{wt1} n \frac{a}{1000} \qquad (3.225)$$

where a is the internal and external centre distance of the two gear pairs. If we calculate the relation of T_1 to T_e after a simplification it will result that:

$$\frac{T_e}{T_1} = K + 1 \quad . \qquad (3.226)$$

The forces F_{b1} form a closed polygon on the solar and in this way they neutralize. The solar centers exactly on itself. The forces F_{b2} too neutralize on the annulus. With regards to stresses on the planet shaft, we must emphasize that the centrifugal force on it is far from being negligible. If m_p is the mass, such a force will be expressed by the following formula:

$$F_c = \frac{m_p \pi^2 n_e^2 a}{900} \quad . \qquad (3.227)$$

3.7.4 Advantages Derived from the Use of Planetary Train Gears

The distribution of power among planets enables planetary train to transmit greater power even when they are very compact. Such power transmission also takes place when a particularly high reduction ratio is accompanied by a high degree of compactness.

4 Gear Measuring

4.1 Measurement of the Tooth Thickness

4.1.1 Base Tangent Measurement

The value of the tangent at the base circle on the transverse plane limited by teeth κ (Fig. 4.1) within the proximity of extreme tooth profiles, is:

$$W_k = (k-1)p_{bt} + s_b \qquad (4.001)$$

or if we replace the pitch and the thickness by their respective values:

$$W_{kt} = \frac{m \cos\alpha_t}{\cos\beta}\left(k\pi - \frac{\pi}{2} + 2x\tan\alpha + z\,\text{inv}\alpha_t\right) \qquad (4.002)$$

The tangent on the normal plane is given by the following formulas:

$$W_k = W_{kt}\cos\beta_b \qquad (4.003)$$

or

$$W_k = m\left[(2k-1)\frac{\pi}{2}\cos\alpha + 2x\sin\alpha + z\,\text{inv}\alpha_t\cos\alpha\right]. \qquad (4.004)$$

The number of teeth is chosen so as to limit the length W_k by means of two parallels, positioned at half the depth of the tooth, that are tangent at the extreme

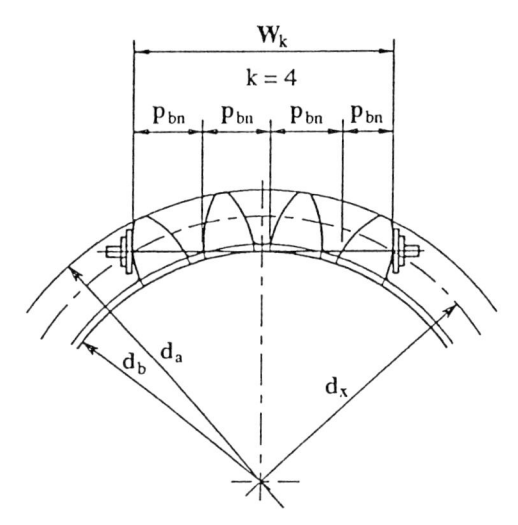

Fig. 4.1. Base tangent length

profiles. Such a value is given by:

$$k = \frac{\left(\sqrt{d_x^2 - d_b^2}\right)\cos\beta_b - S_{bn}}{p\, m\, \cos\alpha_x} + 1 \qquad (4.005)$$

to make a round figure of it we proceed as follows:

$$d_x = d_a - 2m \qquad (4.006)$$

$$\cos\alpha_x = \frac{d_b}{d_a - 2m} \qquad (4.007)$$

and

$$S_{bn} = m\left(\frac{\pi}{2}\cos\alpha + z\,\text{inv}\alpha_t\,\cos\alpha + 2\,x\,\sin\alpha\right). \qquad (4.008)$$

Such formulas are valid for a tooth belonging to an external teeth and for the void in a set of internal teeth. If we know the base tangent length and we want a tooth characterized by a thickness whose value is lower than that of the theoretical thickness, being the deviation Δs_b, the control reference value will be:

$$W_{k\,ref} = W_k - \Delta s_b. \qquad (4.009)$$

If the value of the base thickness is contained between two limiting values (tolerances) it will result that:

$$W_{k\ min} = W_k - \Delta s_{b\ max} \qquad (4.010)$$

and

$$W_{k\ max} = W_k - \Delta s_{b\ min} \qquad (4.011)$$

and

$$W_{k\ min} < W_{k\ mes} < W_{k\ max} \ . \qquad (4.012)$$

The base tangent length measurement is made by means of a micrometer or a sliding caliper which is less precise than the first instrument mentioned. Such a measurement can be executed on a milling machine. Such a measurement has two limits which are linked to the dimensions of the gauging instruments necessary to measure large gears and to the possibility of positioning the two disks of the micrometer on the extreme teeth depending on the tooth width and on the size of the helix angle.

These requirements are met if:

$$W_{kt} < \frac{b}{\sin \beta_b} \ . \qquad (4.013)$$

Internal teeth can be measured only if they belong to a spur gear. Generally internal teeth are measured by means of a gauging ball or a gauging pin.

4.1.2 Normal Chordal Tooth Thickness Measurement

Such a measurement consists in positioning a depth micrometer on the point of the tooth which is regulated to let the probe which is perpendicular to the depth micrometer touch the tooth flanks on the reference circle. At this point the aim is to measure the distance between the two probes. The gauging is executed on the normal plane and the resulting value corresponds to the value of the normal chordal tooth thickness subtending the reference circle between the two profiles corresponding to the tooth (Fig. 4.2).

The first micrometer is placed at a height h_c which is given by:

$$h_c = h_m \pm \frac{m\ z_v}{2} (1 - \cos\psi) \qquad (4.014)$$

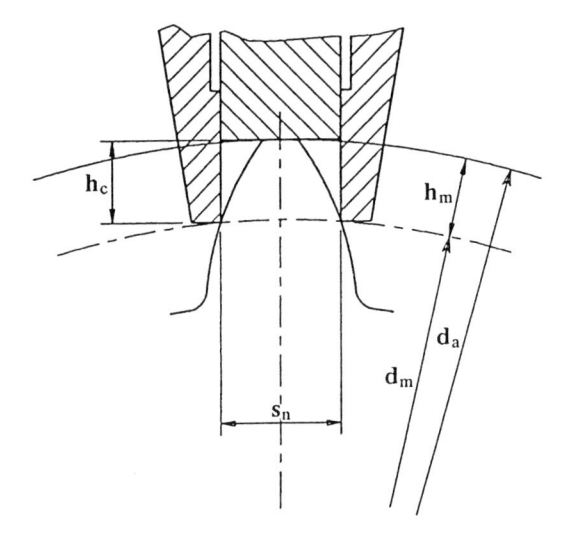

Fig. 4.2. Measurement of the normal chordal tooth thickness

the value of the normal chordal tooth thickness is:

$$\overline{s_n} = m\, z_v\, \sin\psi \qquad (4.015)$$

with:

$$h_m = \left|\frac{d_a - d}{2}\right| = h_a - m(k \pm x) \qquad (4.016)$$

$$z_v = \frac{z}{\cos^3\beta} \qquad (4.017)$$

$$\psi = \frac{s}{z_v} \qquad (4.018)$$

or

$$\psi = \frac{\frac{\pi}{2} \pm 2x\tan\alpha}{z_v}. \qquad (4.019)$$

The plus signs correspond to an external set of gear teeth while the minus signs correspond to an internal set of gear teeth. We notice that the thickness s is measured indirectly and that its expression is not explicit. Such measurement does not provide the deviation existing between the real value and the theoretical value. The measurement is performed on the tip circle whose value is not equal to the theoretical value. Instead of following a tangent, the gauging instrument follows an edge (a point on the profile). Such a measuring technique is not considered as particularly interesting and is used when it is not possible to determine the base tangent length on κ teeth.

4.1.3 Measuring by Balls and Pins

Two balls or two pins (having a diameter of D_M) will be positioned on the same transverse plane in the void of the opposite teeth if there is an odd number of teeth or in the void nearest to this situation if there is an even number of teeth. If α_{kt} is the pressure angle relating to the contact of the ball with the tooth flank and if M_d is the value of the tangent at the ball, it will result that:

$$\cos\alpha_{kt} = \frac{d_b}{M_d \mp D_M} \qquad (4.020)$$

if there is an even number of teeth and

$$\cos\alpha_{kt} = \frac{d_b \cos\frac{\pi}{2z}}{M_d \mp D_M} \qquad (4.021)$$

if there is an odd number of teeth.
It will result that:

$$S_{bn} = z\,m\,\cos\alpha\,\operatorname{inv}\alpha_{kt} \mp D_M \qquad (4.022)$$

The plus sign refers to an external set of gear teeth, while the minus sign refers to an internal set of gear teeth. Such a measurement is a good method for measuring and checking the teeth. However, as the balls must be positioned and kept on the same transverse plane, such a measurement cannot be easily performed in stores and it is for this reason that it cannot be classified among methods used for checking in stores. The wheel to be checked is placed on a check plane in a laboratory where the correct measurement is carried out.

Part of the balls and the pins is external to the tip diameter so that gauging can be carried out. It is for this reason that balls and pins must have particular

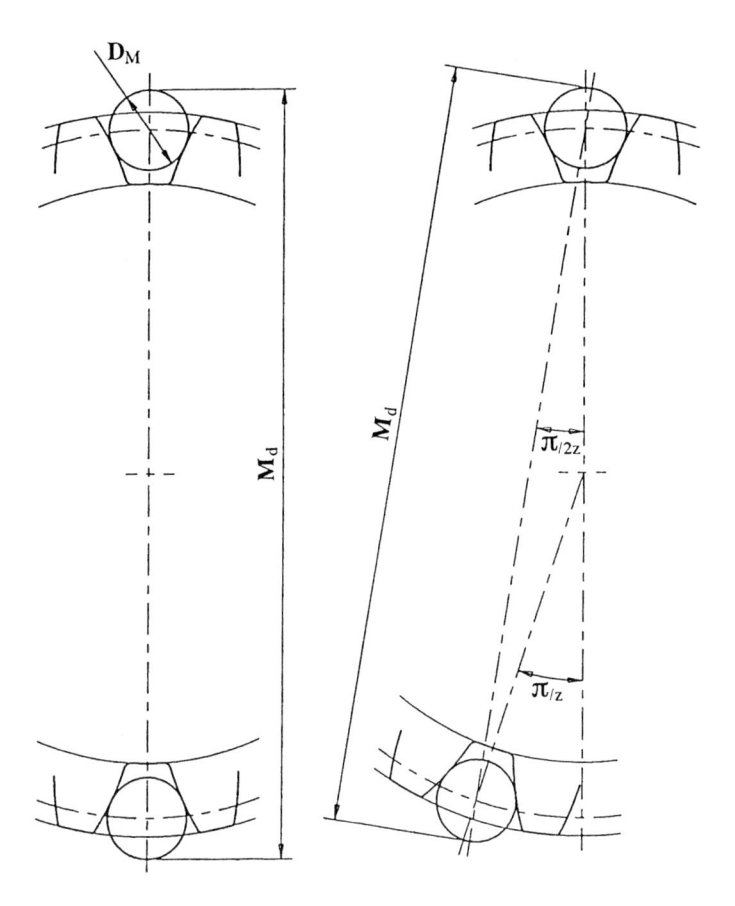

Fig. 4.3. Measurement by balls or pins

diameters. Success in measuring also depends on the geometry of the teeth, on the number of teeth, on the rack shift factor, on the helix angle and so on. For an internal gear teeth we can choose a value which is twice the value of the module of the ball diameter. For an external tooth the value of such a diameter is 2.2 times the module for $z < 20$ and 1.8 times the module for $z > 20$.

4.2 Pitch Measurement

In a theoretical gear, the pitch is divided equally around the wheel. In real gear manufacturing this causes deviations. The pitch deviation is the algebraic difference between the real value of the pitch and the theoretical value of the pitch (Fig. 4.4).

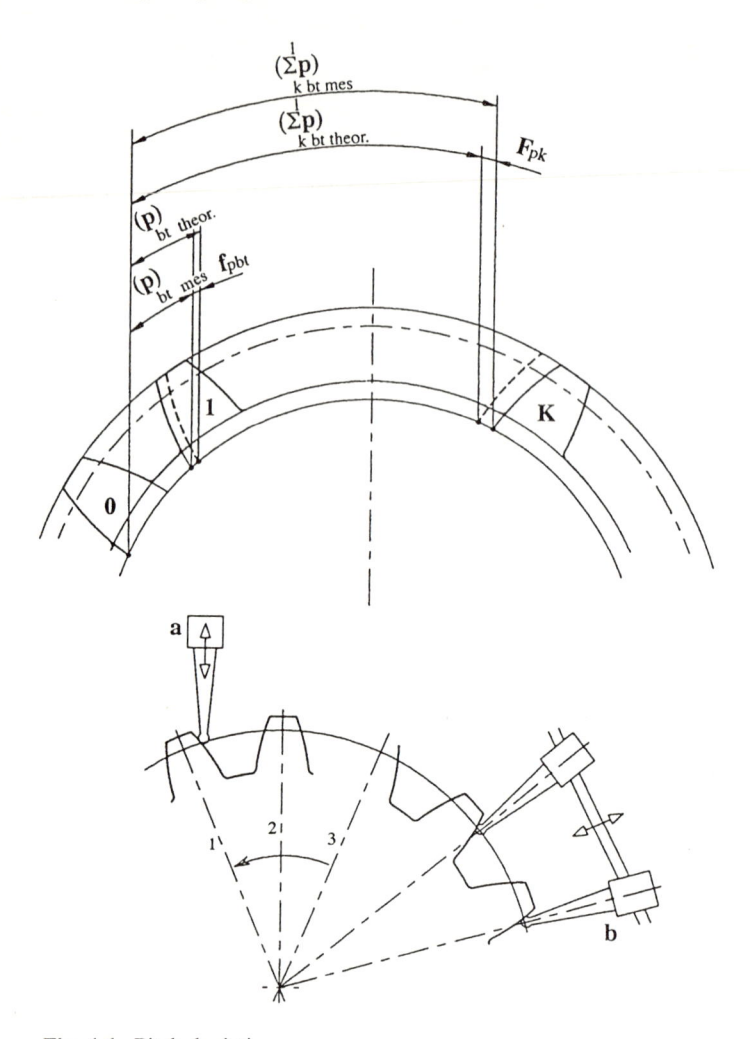

Fig. 4.4. Pitch deviation

The base tangent length on k pitches is the difference between the real arc on such k pitches and the theoretical arc for the same k pitches. It is possible to distinguish between the transverse pitch, the normal pitch and the transverse base pitch.

The transverse pitch can be measured by two different methods.

The first method consists in placing a probe radially to the wheel (fig 4.4 a) so that it can touch the tooth flank reaching the reference circle. We let the wheel turn around its axis forming an angle equivalent to the reference pitch ($2\pi/z$) thus the sensor deviation has to be detected. Such deviation is the simple deviation for the reference pitch. We must be sure that the rotating axis of the wheel during gauging is equal to the rotating axis in operation.

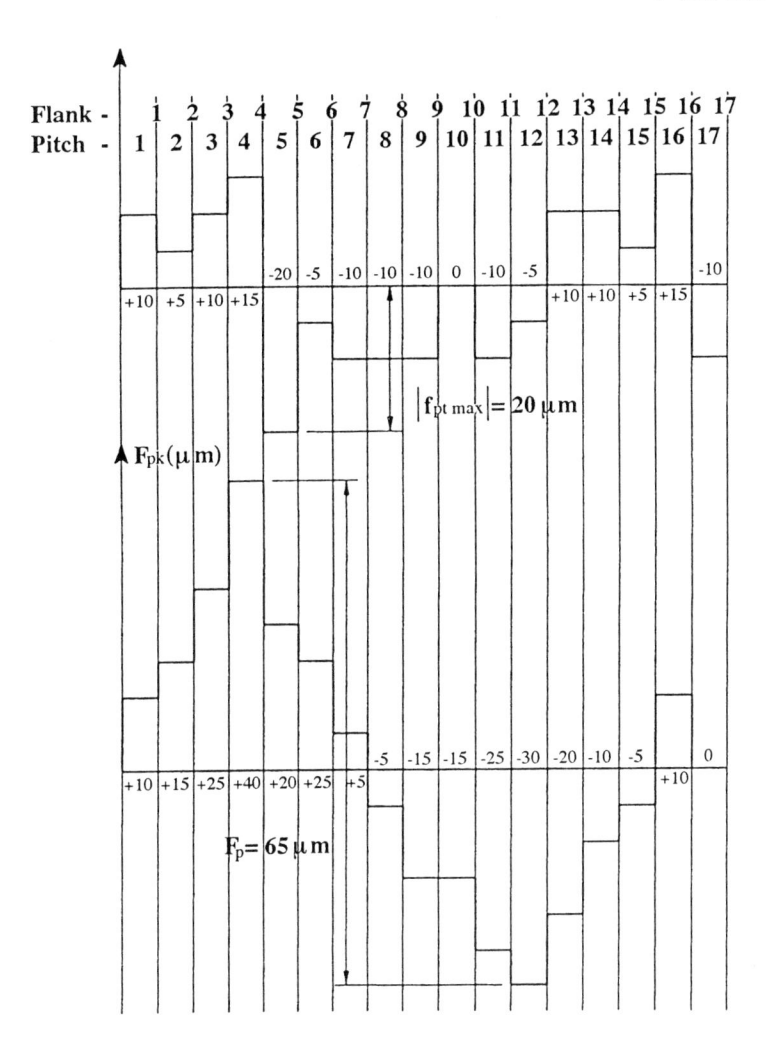

Fig. 4.5. Pitch deviation graphs

The second method consists in using two probes regulated at a distance which is approximately equal to the pitch to be measured. The two probes are placed radially to the wheel. The wheel turns around its axis and at each pitch the probes are made to move forward until they make contact with the wheel, level with the reference circle. The deviation between the probes can be seen easily. After a complete revolution, the sum of the registered deviations is not null because of the coarse regulation of the pitch measured. However, the sum of the deviations around the wheel must be null because of the return to the original flank. The sum of the deviations registered, divided by the number of teeth, provides the quantity by which each value must be corrected. The pitch deviation is given by the algebraic difference between the value resulting from measurement and the average value obtained. In addition it is possible to calculate such values when considering only some sectors of k pitches.

Such measurement is carried out on gears having a high number of teeth (at least 60 teeth) for economical reasons but results are not particularly satisfactory.

If we totalize the deviations we obtain the accumulated pitch deviation. The resulting graphs are equal to those shown on Fig. 4.5.

In this way we obtain the graph representing the single transverse pitch deviations f_{pt} and the graph representing the cumulated transverse deviations F_{pk}. The figure also shows the difference between the deviations F_{pk} which are expressed by F_p. Such a quantity is called the total cumulated deviation. By means of special instruments it is also possible to measure the base pitch deviation. The single pitch deviation affects the distribution of the forces between teeth. We have to make sure that the value of deviation is not higher than those that have already been standardized; such values are called tolerances. Cumulated pitch deviations demonstrate variations of the meshing speed, that prove the existence of dynamic stresses resulting from acceleration due to movement. Tolerances must be attributed to such stresses.

4.3 Profile Measurements

The profile resulting from hobbing or grinding differs from the theoretical involute shape with a determined pressure angle. If dimensions are made larger in order to make the picture clearer, the case shown on Fig. 4.6 will occur.

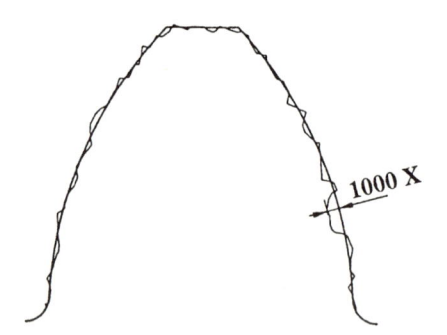

Fig. 4.6. Real shape of the profile

Such a profile can be checked by means of an instrument that makes a probe move along the flanks of the tooth following a pure involute to a circle. If we register the probe deviation under the flank's influence in comparison with its normal path, flank deviations could also be registered. If we draw a graph in which the axis of the abscissa corresponds to a length proportional to the position of the point of measurement on the pressure line and the axis of the ordinates corresponds to the deviations registered (which have been extremely enlarged), the profile of a tooth without deviations will be a horizontal line. Such a graph is shown on Fig. 4.7.

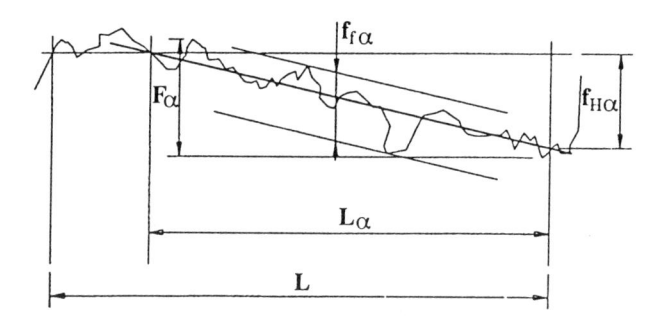

Fig. 4.7. Profile graph

We notice that an involute to a circle having a pressure angle different from the one desired provides a straight line on the graph that slopes at an angle to the horizontal line.

The graph also provides the total deviation of the profile F_α. Such a measurement is carried out between the two parallels enclosing the graph. It is also possible to draw the graph's average straight line. Such a line meets the statistic rule of the minimum square and is represented by an involute. If we enclose the profile graph with two straight lines parallel to the average straight line, we will determine a vertical distance representing the form deviation of the profile expressed by $f_{f\alpha}$. If we draw a parallel to the theorical profile in correspondence to a point placed on one of the extremities of the profile drawing and if we measure the deviation between the parallel line and the average straight line on the other extremity, we will determine a perpendicular distance to the reference straight line, thus giving the profile slope deviation $f_{H\alpha}$. All that has been said for an involute profile is also valid for a modified profile. It is sufficient to replace the straight line corresponding to the profile with a curve representing the modified profile.

The average straight line will become an average curve having a different inclination. The definitions are still the same. The deviation affecting inclination results from the base diameter deviation. The base diameter deviation can be calculated as follows:

$$f_{db} = f_{H\alpha} \frac{d_b}{L_\alpha} \qquad (4.023)$$

where L_α is the length of evaluation of the graph.

The pressure angle deviation is calculated by considering the deviation expressed by the following formula:

$$f_\alpha = - \frac{f_{H\alpha}}{L_\alpha \tan\alpha_t} \qquad (4.024)$$

The instrument used to carry out such a measurement is made of a disk whose diameter corresponds to that of the wheel to be measured that moves a rule (Fig. 4.8).

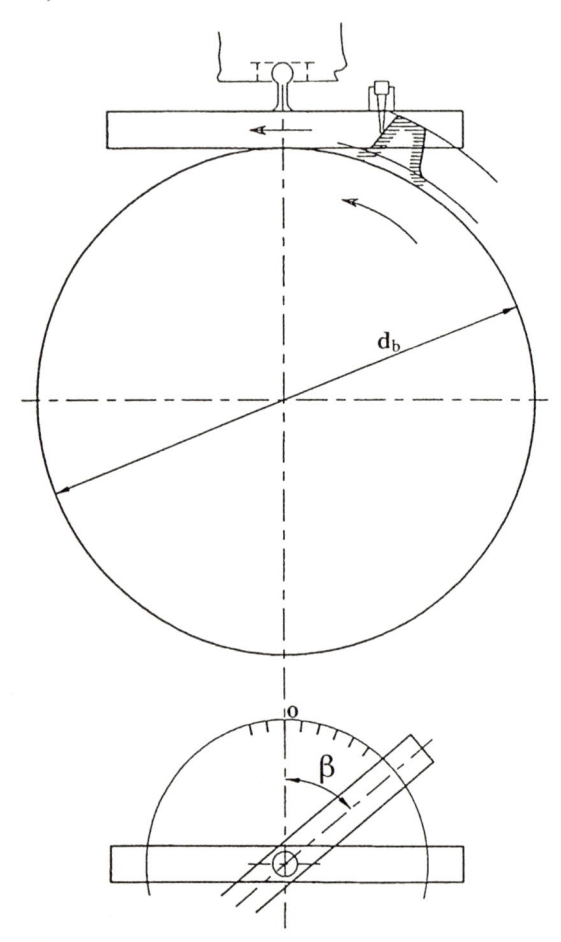

Fig. 4.8. Profile gauging machine

The wheel to be measured turns together with the disk while a probe that is on the same vertical plane as the plane tangent at the disk (plane of friction) moves with the rule. In order to measure the different transverse section the probe can move perpendicularly to the plane of the disk. In helical gears, the probe is set in motion after which an oblique movement pre-regulated at the value of the helix angle takes place. To avoid turning a disk for each wheel, a machine having a base diameter that may be regulated is used (Fig.4.9).

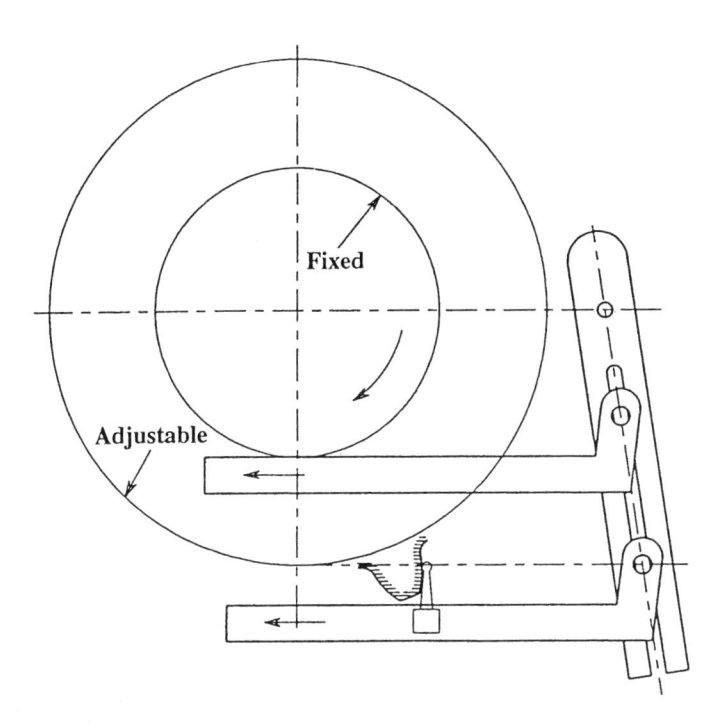

Fig. 4.9. Base diameter machine that may be regulated

4.4 Measurement of the Helix Deviation

By means of the machine described in the above paragraph it is also possible to control the development of the helix angle. If we place the sensor along the tooth flank and if we move it vertically following the angle pre-regulated at the value of the helix angle, it will be possible to register the deviation of the position of the sensor in relation to the theoretical helix. We can obtain a diagram of the deviations on an axis in relation to the length of toothing on a perpendicular axis. If the resulting helix is perfect, the graph will be a horizantal line. Failing this, a graph similar to that shown on Fig. 4.10, will be obtained.

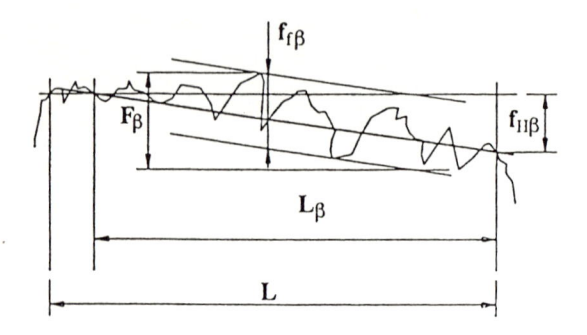

Fig. 4.10. Inclination graph

Following the same rules adopted for the profile we can distinguish between:

- the total helix deviation F_β,
- the helix form deviation $f_{f\beta}$
- the helix slope deviation $f_{H\beta}$.

4.5 Composite Tangential Deviation

If a wheel that must be tested is placed near a master wheel at the reference centre distance and the two wheels mesh correctly one with the other, according to the kinematic relation of meshing, the regular movement of one gear should cause the regular movement of the other gear.

$$\frac{\omega_2}{\omega_1} = \frac{z_1}{z_2} \quad . \tag{4.025}$$

In reality meshing does not occur because deviations exist in the tested gear (as the master gear is considered to be perfect in comparison with the gear under testing).

By means of mechanical units (such as clutch gears whose diameters are equal to the reference operating diameters) or electronic devices (counters and drives working at proportional speed), it is possible to obtain the value of the theorical speed and to compare it with the real speed. In this way a graph equal to those shown on Fig. 4.11 is obtained.

Such a graph represents the composite tangential deviation.

We notice a kind of undulation corresponding to the teeth belonging to the tested gear wheel. It is the composite tangential deviation from tooth to tooth ($f''_i{}_i$).

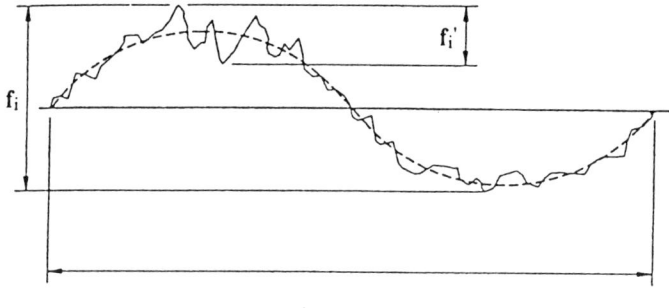

Fig. 4.11. Composite tangential deviation

The graph also shows the distance between the highest peak and the lowest peak. Such distance is known as the composite total deviation (F'_i).

The composite deviation enables us to understand gear working, even if it does not provide useful information about the causes generating deviations.

For example, it cannot be used to define a global quality even if its consultation is very interesting. As the master gear wheel is not completely free from imperfections, the gear wheel to be tested must be of a lower quality than the master gear.

Such condition limits the use of such a form of measurement.

4.6 Composite Radial Deviation (Measurement on the two Flanks)

A gear to be tested and a master gear are placed so that their teeth come into contact. If one of the two axes is mounted on a spring, the centre distance could vary as the two teeth are still in contact along the two flanks. We can register the displacement of the moving axis (variation of the centre distance). If the master gear is considered to be without imperfections, the resulting graph of the deviation in relation to the tested gear lenght of rotation will be a straight line.

Deviations will provide a graph like the one shown on Fig. 4.12.

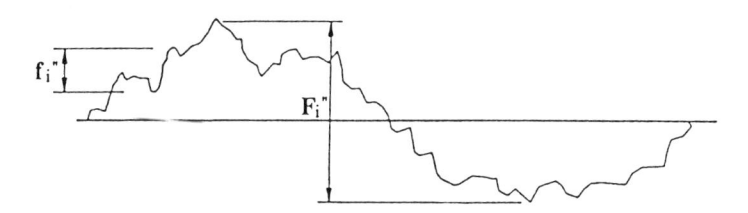

Fig. 4.12. Composite radial deviation

If we assume that the master gear shows deviations having a negligible value when compared with those of the tested gear, we are aware that the composite radial deviation reflects all the deviations existing in the tested gear.

We can also distinguish the composite radial deviation from tooth to tooth f''_i and the composite radial total deviation F''_i. Such a test does not provide useful information on the gear working but can be easily executed and be used to regulate the machine manufacturing the same piece on the basis of the measurements carried out on the master gear. If a real gear with deviations works without clearance (the flanks of the teeth are in contact during working) and if the axes are strongly fixed, stresses will occur when the movement of the shaft is hampered.

Such stresses are harmful to the gear working and can be highly concentrated.

This results that gears are at fixed centre distance so that the value of clearance between teeth cannot be null.

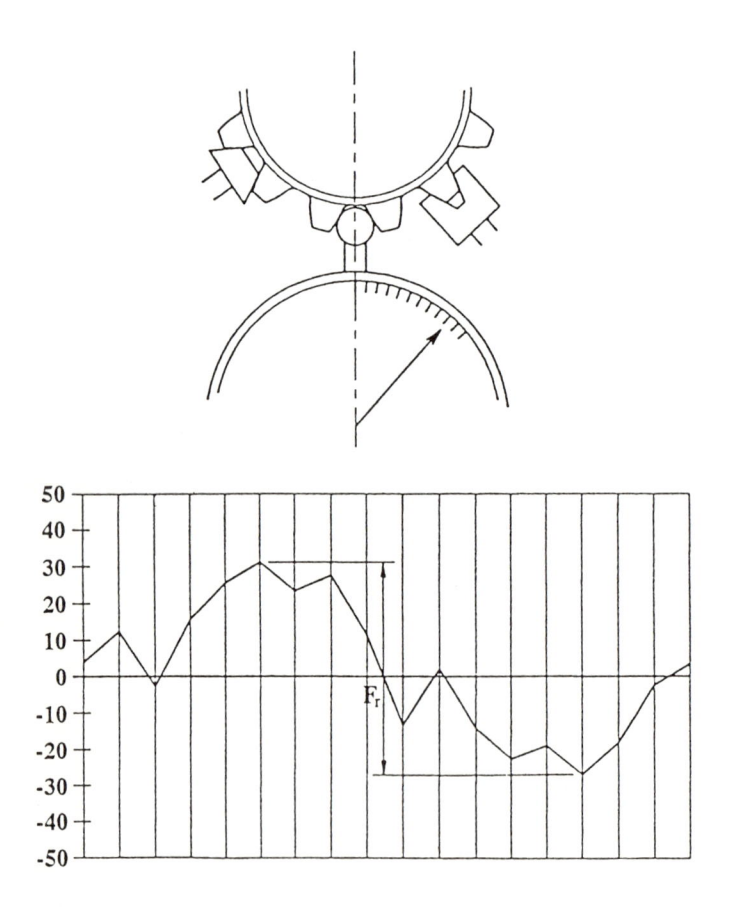

Fig. 4.13. Eccentricity

4.7 Eccentricity

The radial run out resulting from working causes a miscentering of the reference diameter in relation to the gear axis. This eccentricity is expressed by F_r and is measured by means of a probe fitted with a ball which is placed inside the space of the teeth.

Another gauging method consists in placing the probe fitted with a fork on the tip of the tooth. We measure the deviation in each position.

Figure 4.13 demonstrates the principle and shows a graph equal to that obtained during test.

4.8 Precision in Gear Blank

The degree of precision characterizing the wheels and affecting gear working, is strictly connected with the general degree of precision that we would like to obtain. Boring, for example, must be precise enough to ensure the correct centring of the gears. Even when the tip diameter does not play an important role as it does not take part in movement, tolerances having negative values are applied to avoid meshing interference.

The tolerance, following the parameters defined by ISO in the field of coupling, is expressed by h8 or h7.

The boring has a tolerance indicated by H and the quality varies on the basis of the quality required by the teeth. However, if we consider the tip of the tooth as the base of the measurement (see the normal chordal thickness measurement) we also attribute a great importance to the tolerance of the tip circle during turning.

In order to assure the correct working of the gear, mounting must not affect negatively the degree of quality achieved during manufacturing.

To avoid such a perturbating effect, reference surfaces perpendicular to the axis and concentric with it must be adopted during milling, grinding and during control and mounting operations.

The surface perpendicular to the axis has a tolerance regarding its position to assure its perpendicularity. Such a surface is used as a base for fixing in milling machines, grinding machines and control instruments.

The cylindrical surface concentric to the gear axis has a tolerance towards dimensions, concentricity and eccentricity.

It is used as a base for the positioning of gears following their axes on different manufacturing and control machines.

In integral pinions, the areas where bearings are in contact can be used as reference surfaces. Tolerances vary with the degree of precision required by the gear (Fig. 4.14).

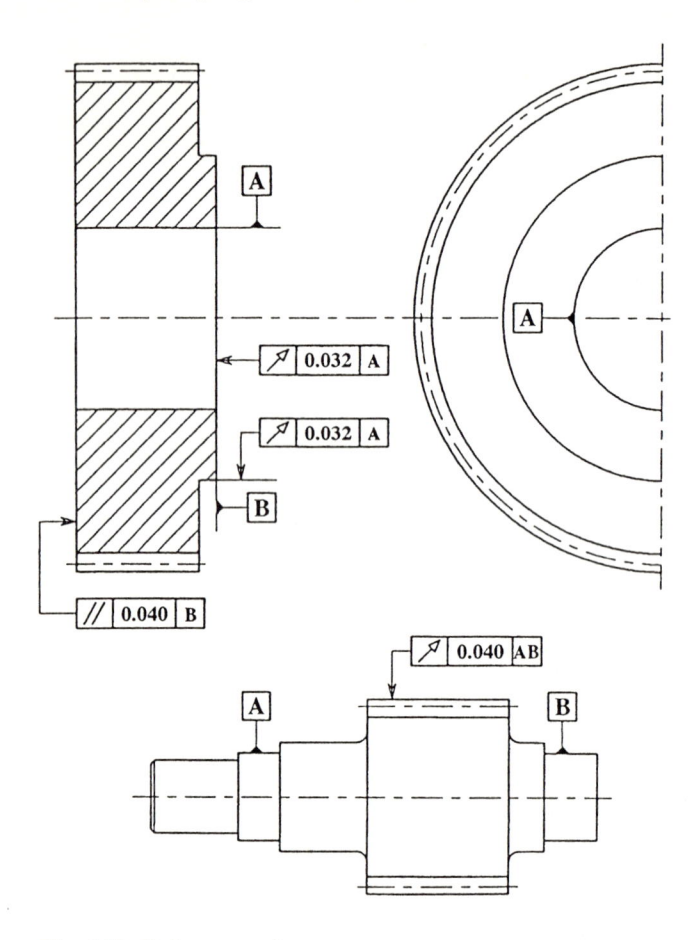

Fig. 4.14. Reference surfaces

4.9 Centre Distance Deviation and Axis Parallelism

The axes of two wheels fixed at their supports can show deviations due to the geometry of the supports (bearings).

Generally we distinguish between two types of deviation: the first one occurs on the plane of the axes ($f_{\Sigma\delta}$) while the second one occurs outside the plane of the axes ($f_{\Sigma\beta}$) (Fig. 4.15).

It is the deviation occurring outside the plane of the axes that affects mostly the gear working. Tolerances are attributed to such deviations.

Such tolerances are expressed in mm per mm, it means that deviations have an angular nature. To determine these deviations we choose one of the axes and a

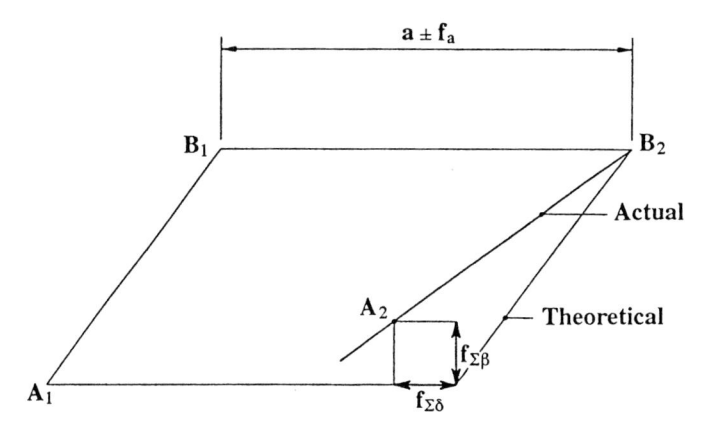

Fig. 4.15. Axis parallelism

point belonging to the other one in order to define a plane. We measure the deviation from another point belonging to the second axis with respect to the plane that has just been defined (measurement is carried out on a plane perpendicular to the plane defined) and the deviation in relation to the reference axis on this plane. The deviation to be considered is the relation of the distance measured on the first plane and on the second one to the distance of the shaft from the point chosen to the point used to determine the reference plane. The distance between the axis chosen and the point determining the reference plane is the real centre distance. It differs from the theoretical centre distance by a value of f_a. Positive and negative tolerances may be attributed to such deviation. To avoid meshing interference the value of the negative tolerance must be lower than that of the tolerance of the gear tip circle.

4.10 Information on Standards

There are some standards defining the tolerances allowed for different qualities. We refer to the standard ISO 1328 underlining also the fact that for domestic use standard DIN 3991 exists. The standard ISO 1328-1 defines 13 classes of quality which are marked with numbers from 0 to 12, the lowest number corresponds to the most precise quality (negligible deviation). Such a standard does not provide instructions about gear manufacturing or working; it simply provides reference parameters. However, it is obvious that sometimes particular qualities are achieved following the standard suggestions and that some qualities are much more suitable for certain applications than others. The qualities characterized by numbers from 1 to 3 are for master gear wheels requiring a high degree of precision; those marked with numbers from 4 to 6 are for gears that have undergone grinding while those marked with numbers from 7 to 10 are for milled gears (in this case the number defining the quality varies with the milling degree of precision). The qualities 11 and 12 define the raw gears that are used in foundry. Such prescriptions offer simple information available for engineers. Wheels requiring a high degree of precision for gears when incorporate the effects of deviations like, for example, the gears used in turbines and those designed for positioning objects; wheels having a lower degree of precision are used in gears having larger dimensions and turning at a low speed.

Standard ISO 1328-1 provides tolerances for the pitch deviation of the profile and distortion in relation to the diameter and the module or to the tooth width (according to the deviations considered) and the module.

The values provided are tolerances for an interval of diameter, module or tooth width; the value given for each interval is calculated for a value of the geometric mean of the interval limits by means of formulas given by the standard. Such formulas define the quality marked by the number 5. The other qualities have tolerances that increase or decrease in proportion to the index with a pitch of $2^{0.5}$ for an index of quality. In this way the quality marked by the number 7 has tolerances with a value that is twice as high the value of the quality 5 and twice low as the quality marked by the number 9. With regard to tolerances that must be applied to the profile, the standard defines a length of evaluation L_α that is equal to 92% of the length of the usable flank as 8% of such length is ignored on the tip of the tooth unless an excess of matter is produced here.

The central part of the standard enables tolerances to be applied to the total deviation F_α while an informative appendix provides the tolerances $f_{f\alpha}$ and $f_{H\alpha}$. The same is for the deviation of distortion for whom the value of the total deviation F_β is standardized.

The deviations $f_{f\beta}$ and $f_{H\beta}$ are considered in the appendix.

Standard ISO 1328-1 also provides tolerances that must be applied to composite tangential deviation and deals with tolerances to be applied to wheels and axes.

4.11 Backlash

When a tooth is in the space of the mating tooth of a theoretical gear, the flank of the first tooth is in contact with the flank of the second. In a gear with backlash, when the working flanks are in contact the others are distant. Such a dividing distance is called the backlash.

Backlash can be classified on the basis of its direction. There may be the radial backlash j_r that follows the radii of the circles forming the gear (i.e. the centre distance), the rotational backlash j_t that follows the pitch circle and the normal backlash j_n which is perpendicular to flanks. There is also the angular backlash j_θ that is the angle by which a wheel can turn when the other one is rotating.

The relation between these different backlashes is as follows:

$$j_t = \frac{j_n}{\cos\alpha_{wt} \cos\beta_b} \qquad (4.026)$$

$$j_r = \frac{j_n}{2\sin\alpha_{wt} \cos\beta_b} \qquad (4.027)$$

The angular backlash is given by the following relation:

$$j_\theta = \frac{j_t}{\frac{1}{2}d_w} \qquad (4.028)$$

Backlash occurs when tooth thickness diminishes (Fig. 4.16).

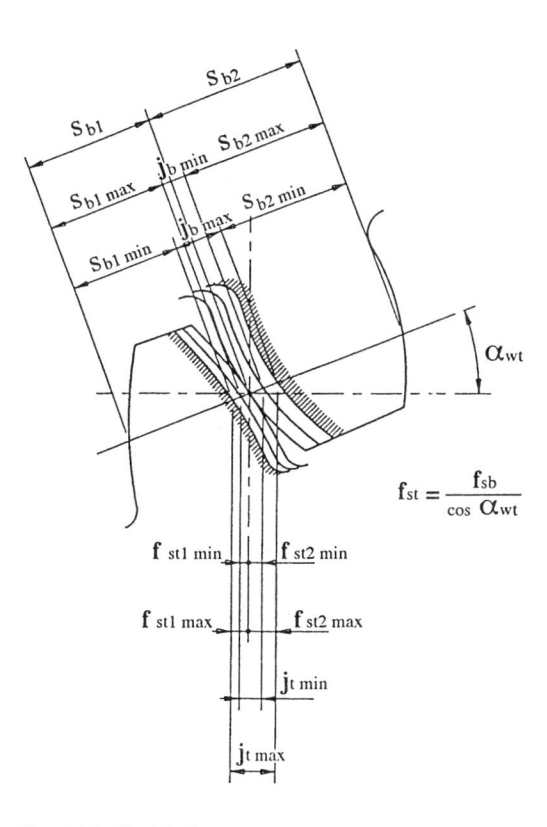

Fig. 4.16. Backlash

If we apply tolerances to this thickness, it is possible to define the minimum thickness and the maximum thickness of a tooth. $s_{b1\ min}$ and $s_{b1\ max}$ indicate the maximum thickness of the pinion and $s_{b2\ min}$ as well as $s_{b2\ max}$ indicate the maximum thickness of the wheel. If the gear is theoretically perfect the minimum backlash will be defined by the following formula:

$$j_{t\ min} = \frac{(s_{b1} - s_{b1\ max}) + (s_{b2} - s_{b2\ max})}{\cos\alpha_w} \qquad (4.029)$$

or

$$j_{t\ min} = \frac{1}{\cos\alpha_{wt}}(f_{sb1\ min} + f_{sb2\ min}) \ . \qquad (4.030)$$

In reality we must consider many other factors like, for example, the centre distance deviations, the tolerances that must be applied to the pitch, to the profile and to distortion. We must also consider eccentricity tolerances and deviations due to the running temperature. The effect of these deviations varies with the tooth quality. The influences that deviations have on the gear in accordance with the tooth quality are symbolically represented by K_q while the influence that expansion due to temperature deviations has on the gear is indicated by K_T.

If f_a is the centre distance negative deviation the minimum backlash will be:

$$j_{t\ min} = f_{sb1\ min} + f_{sb2\ min} - 2 f_a \sin\alpha_{wt} - K_q - K_T \qquad 4.031)$$

and the value of the normal backlash will be

$$j_{n\ min} = j_{t\ min} \cos\beta_b \ . \qquad (4.032)$$

The value of the normal backlash must be higher than 0. Failing this the gear will not work correctly. The theoretical backlash given by the sum of teeth minimum deviations must take the quality of the performance and the temperature deviations into account. For current quality, the backlash could assume high values. It happens the same when high temperature deviations are said to occur. Theoretical backlashes having a module equal to 0.1 are not considered to be exceptional. If we desire a gear without backlash (or with an extremely reduced backlash as gears without backlash are inconceivable) the elements forming the gear must be particularly efficient and the gear dimensions must correspond to the running temperature as much as possible.

4.12 Contact Pattern Measurement

Contact pattern measurements are often carried out on gears. Generally a thin coloured layer is placed between the teeth in order to register and analyse the print produced by idle stroke. The way in which the measurement is carried out and the interpretation of the resulting data are subjective. Such a test has turned out to be useful for engineers in particular fields of application as it can provide interesting data that can be easily compared with previous information. This test and the interpretation of the results must be carried out by an engineer as during working any contact between the teeth takes place differently from what happens during idling.

Furthermore, in the presence of tooth modification, the idle contact cannot take place correctly but can take place perfectly when a load is applied as the principle at the base of the tooth modification is to create teeth that during an idle stroke detach from the theoretical profile and provide such profile again when a load is applied. Non-experts could be misled by all these factors. A loading capacity test to be carried out on the bevel gear pair has turned out to be particularly interesting: two mating bevel gears come into contact on a special appliance and an engineer tries to determine the position providing the best loading capacity during idling (experience and perfect knowledge of distortions due to overload complete the calculation). Such a position is determined by displacing the pinion axially.

The position of the reference surface of the pinion is noted in relation to the vertex of the cone and such value is adopted as the assembly quota of the pinion in relation to the gear wheel.

5 Gear Calculation

5.1 Aspects of Teeth After Operation

After running teeth can assume different aspects that vary according to their way of working and their applications. Standard ISO 12825 ("Wears and damage to teeth - Terminology") provides all the useful information about this matter. The standard also indicates a series of situations that could produce harmful effects.

Teeth can be damaged by many types of wear: abrasive wear, erosion wear, normal wear, cavitation and so on. Normal wear occurs when teeth slide. It can be reduced through the application of an adequate modification coefficient. Other types of wear result from working in prohibitive conditions. There are no calculations enabling engineers to prevent wear in parallel gears and in the bevel gear pairs. Worm gears are not touched by this problem as friction is preponderant. Pitting is a phenomenon that occurs very frequently in gears. It produces small holes damaging the surface of the teeth. In case of initial pitting, which is not harmful, holes tend to disappear with the passing of time. In other cases holes multiply becoming bigger and bigger (destructive pitting). Destructive pitting causes the destruction of the surface of the flanks producing the breaking of the tooth. Such form of pitting exerts a destructive action on the gear and on its working; it is for this reason that destructive pitting must be avoided absolutely. Fatigue or overload cause the teeth to break. Such a defect can jeopardize the working of the gear; therefore it must be prevented. Sometimes, even in presence of a sufficient quantity of lubricant scuffing can be produced. This is very dangerous as it occurs instantaneously and can cause the detachment of matter and the breaking of teeth. Generally such phenomenon takes place in gears that must bear excessive loads and turning at high speed. The load capacity calculation aims at preventing pitting, breaking due to fatigue and sometimes scuffing (see section 8). In worms and gears wear due to friction must be prevented.

5.2 Parallel Gear and Bevel Gear Pair Load Capacity

5.2.1 Prevention of Pitting (Contact Stress)

General Features. Pitting generally appears late in gears history. It results from the application of more and more excessive loads and thermal treatment increasing the gear bending load capacity (such treatments must comply with the breaking point). Niemann was the first to foresee the existence of a relation between contact stress and pitting.

Contact stress (or Hertz's pressure as the study was carried out by Hertz) provokes a shearing stress just below the surface which is proportional to the contact stress, as the depth is about a third the length of the superficial area subject to distortion. The shearing stress can be provoked statically or following fatigue due to the cracking of the sub-layer of the material.

Cracking can reach the surface of the material and under the action of the lubricant pressure (resulting from the viscosity of the lubricant, from the sliding speed and from the roughness of the surface) it can further increase the detachment of material particle. It is obvious that any increase in pitting is linked to an excessive normal stress.

Contact Stress (Hertz). In paragraph 3.4.7 we noticed that a pressure given by equation (3.110) was generated between two teeth in contact following the path of contact line. Such an equation can also be applied to a virtual spur gear of parallel helical gear or of bevel gear pair.

If such an equation is applied to a parallel helical gear, using the equations defined in the paragraphs above, we will find that:

$$\sigma_{H\,theor} = Z_H Z_E \sqrt{\frac{F_t}{d_1 b} \cdot \frac{u+1}{u}} \qquad (5.001)$$

supposing that the force is applied to the pitch point the factors are as follows:

$$Z_E = \frac{1}{\sqrt{\pi \left[\frac{1-v_1^2}{E_1} + \frac{1-v_2^2}{E_2}\right]}} \qquad (5.002)$$

where $v_{1,2}$ is Poisson's coefficient of the material of the pinion and of the wheel while $E_{1,2}$ is the module of elasticity of the materials forming the pinion and the wheel.

And:

$$Z_H = \sqrt{\frac{2 \cos \beta_b \, \cos \alpha_{wt}}{\cos^2 \alpha_t \, \sin \alpha_{wt}}} \qquad (5.003)$$

In a parallel gear the diameter d_1 corresponds to the diameter of the pinion while, in a gear pair with intersecting axes, diameter d_1 is the diameter of the virtual pinion of the bevel gear pair.

The theoretical contact stress must be corrected on the basis of factors that consider:

- the influence of the transverse contact ratio and the influence of the overlap ratio. The distribution of forces between teeth is affected by these relations. A decrease in effective pressure follows an increase in stresses.

- the influence of the angle β. In a helical gear the contact line slopes at an angle to the generatrix of the cylinder, while it is parallel to such generatrix in a spur gear. This means that the force is better divided in a helical gear than in a virtual spur gear.

In spur gear teeth or in other teeth where overlap ratio has a low value, the value of contact stress reaches maximum in correspondence to the lower point of single contact and not in correspondence to the pitch point (i.e. when the pitch point is after the lower point of single contact on the action line). The resulting stress can be compared with the stress allowed resulting from a test on the material or on a gear which is more or less similar to the gear considered. Such stress is also influenced by other factors such as the difference existing between the gear tested and the test gear used to determine the limit of stress (stress allowed).

The stress allowed also depends on S - N curve concerning pitting produced on the material considered on the basis of the load cycles stressing the teeth.

The value of the stress resulting from this superficial pressure must be lower than the value of the stress permissible which is calculated taking into consideration all actual conditions. The value of the stress resulting from the superficial pressure can also be equal to that of the stress allowed.

5.2.2 Stress Applied to the Tooth Root

Materials break when they are subject to overload, to fatigue or to an excessive stress applied to the root of the tooth. Part of the tooth section is subject to bending and compressive stress under the action of the load which is perpendicular to the virtual spur toothing considered, but due to the inclination of the profile at the pitch point which is oblique at the tooth axis.

Let us consider the component load which is perpendicular to the tooth axis at any height (Figure 5.1).

Given a distance x from the point of intersection of the force support with the tooth axis, if σ_x is the stress due to bending, it will result that:

$$\sigma_x = \frac{F\,x}{\dfrac{b\,\overline{s}_x^{\,2}}{6}}\,. \qquad (5.004)$$

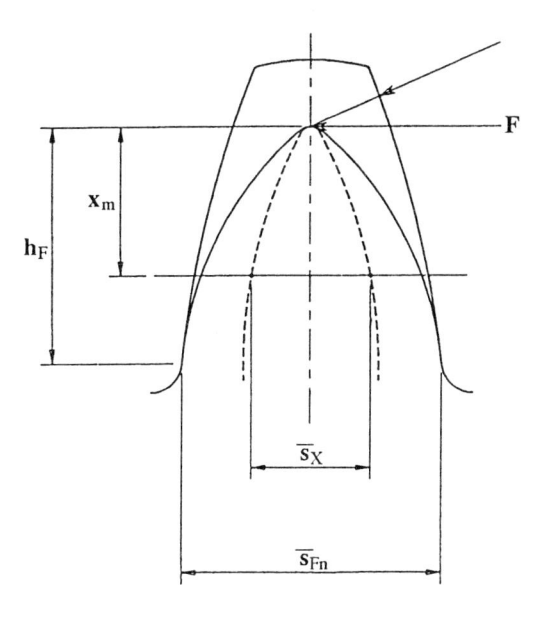

Fig. 5.1. Bending on the tooth root

If we want to keep the bending stress constant, independent on the value of the distance marked x, the profile must be parabolic in shape:

$$\bar{s}_x = \sqrt{\frac{x}{\sigma_x}} \sqrt{\frac{6F}{b}} . \qquad (5.005)$$

The parabola depends on the stress. It is possible to draw a parabola whose vertex is the point where the direction of the force intersects with the axis of the wheel; such a parabola is tangent at the profile. Tangency occurs in correspondence to the fillet profile. In this section, called the critical section, the value of the bending stress is maximum (Lewis):

$$\sigma_{theor} = \frac{6 F h_F}{b \bar{s}_{Fn}^2} . \qquad (5.006)$$

The dimensions of the section and those of the strenght lever are expressed as a function of the module. Therefore it results that:

$$\sigma_{theor} = \frac{F}{b m} \frac{6 h^*_{Fn}}{\bar{s}^{*2}_{Fn}} . \qquad (5.007)$$

As at the higher point of single contact, where the quantities depending on the index e are influencial, the bending stress is maximum as it corresponds to the major bending arm and the whole load, from which it results that:

$$\sigma_{F\,theor} = \frac{F_t}{b\,m} \frac{6\,h^*_{Fe}\,\cos\alpha_{Fne}}{\overline{s}^{*2}_{Fn}\,\cos\alpha_n} \qquad (5.008)$$

or

$$\sigma_{F\,theor} = \frac{F_t}{b\,m}\,Y_F \qquad (5.009)$$

with:

$$Y_F = \frac{6\,h^*_{Fe}\,\cos\alpha_{Fne}}{\overline{s}^{*2}_{Fn}\,\cos\alpha_n} \qquad (5.010)$$

If we consider the compression stress resulting from the component of the force along the axis of the tooth, a further stress will be created:

$$\sigma_{N\,theor} = \frac{F_t}{b\,m} \frac{\sin\alpha_{Fne}}{\overline{s}^*_{Fn}\,\cos\alpha_n} \qquad (5.011)$$

The stress on the tooth is the sum of these two stresses. As the shape of the teeth is variable we must take stress concentration into account and, as for the stress resulting from the contact stress, we must consider factors provided for helical gear, overlap ratio and contact ratio.

The resulting stress is compared to a stress obtained by means of tests during which the stress found is influenced by corrective factors that consider the cycles of solicitation, the roughness of the profile corresponding to the critical section and the different degrees of notch sensitivity of the material to the shearing stress in comparison with the material used during tests.

5.3 Standard ISO 6336 (or DIN 3990)

5.3.1 Introduction

Gear calculation has been formalized by the ISO. The standard regulating such calculations is standard number 6336, parts 1, 2, 3 and 5 (the content of this standard is quite the same as standard DIN 3990 apart from some descriptions quoted below). Such standard is based on a.m. principles and has been defined as being applied to different gears in different applications. It is a complex regulation and the text in its integral form must be consulted for further details. Standards deriving from the general standard are going to be developed for gears for engeneering applications such as high speed gears, gears used in the nautical field and in cars. The different factors that appear in the standard can be calculated following different methods that are classified by the alphabetical letters A,B,C,D and so on in order of simplicity. Such methods are clearly applied with some restrictions. When engineers pass from the method marked with the letter B to that marked with the letter C, the number of precautions taken in terms of safety increases. It is the same for the passage from a method to another marked with a letter that occupies a higher position in the alphabet.

In this section we will examine the methods adopted for the calculation of industrial gears in order to provide a complete overview of such standard.

Differences also appear when the material changes and it is for this reason that we have considered a single type of material: case hardened steel. The method indicated by the letter A is not standardized. It is implicitly an analytical or an experimental method. The analytical method presupposes theoretical knowledge of all the phenomena, while the experimental method presupposes the execution of measurements on the test gear that must have the same feature as the gear calculated. This procedure is rather expensive and recourse to it can only be justified by mass production. Engineers use method C which is widely used, for calculating the different factors in industrial gears. We will comply with this rule even if sometimes the base principle of method B, as the principle characterizing method C is derived from the method B, is quoted. To perform such calculations we must also consider four different aspects:

(1) The calculation of the influencing factors.
(2) The calculation with reference to the contact stress.
(3) The calculation with reference to the bending stress at the root of the tooth.
(4) The study of the material properties.

5.3.2 Influence Factors

At the beginning of this chapter we said that the force applied was as a function of the tangential force at the reference circle. However, the value of the effective force applied to the tooth can be higher than the value of the theoretical force because of internal and external causes. The influencing factors take the increase in the value of force into account. Therefore we can distinguish between an application factor, a dynamic factor and a face load factor and a transverse load factor.

Application Factor (K_A). Such calculations considering first a nominal power and then a tangential force resulting from the normal capacity of the motor considered. However, operating conditions do not ensure that the resulting force keeps a constant value. Driving machines can show dynamic variations (shocks): it is rather different to start a gear by means of an electric motor or a turbine than starting it by a multi-cylinder or single-cylinder internal combustion motor. On the other hand driven machines work in very different dynamic conditions. Let us consider a ball mill and a belt conveyor for package: their working is completely different even if their nominal capacity is equal.

The application factor is represented by K_A. It considers all the above mentioned differences. In the case of an application that incorporates a load spectrum, it is possible to calculate the gear for a nominal power which is different from the power equivalent to the load spectrum following Miner's rule. If we know the load spectrum and we apply Miner's equivalent load, the value of the application factor will be equal to 1. If we know the value of the equivalent load of the spectrum and if we apply a rating load, the application factor will be given by the relation of the equivalent load to the rating load. The application factor is chosen empirically on the basis of the experience gained after having studied a particular machine type.

The application factor must not be confused with service factor. We will deal with the service factor in the section dealing with gearboxes.

Dynamic factor (K_v). Teeth have a certain degree of elasticity while units are characterized by masses. Elasticity and masses are affected by the intermittent working of teeth, the vibrations created (see section 12) and dynamic internal stresses that must be added to the stresses already transmitted. Method B is based on the elastic behaviour of teeth in relation to the masses of wheels. If c_γ is the elastic costant of teeth in contact, which is expressed by the tooth width unit, and if m_{red} is the reduced mass of wheels, the study of vibrations will compare the frequency of the system defined above with the excitation frequency ($z_1\ n_1$). The resulting ratio is without dimensions.

$$N = \frac{\pi\ n_1\ z_1}{30000} \sqrt{\frac{m_{red}}{c_\gamma}} \cdot \qquad (5.012)$$

In case of resonance, the value of such a ratio is equal to 1.

Four ranges are defined: a sub-critical range, a main critical range, an intermediate critical range and a higher critical range. The reduced mass is obtained by considering the moment of inertia of wheels (J_1 and J_2) which are divided by the width b:

$$m_{red} = \frac{J_1^*\ J_2^*}{J_1^*\ r_{b2}^2 + J_2^*\ r_{b1}^2} \cdot \qquad (5.013)$$

Method B provides some formulas that enable engineers to calculate K_v in each range. Such a calculation is influenced by the total contact ratio, by pitch and profile deviations as well as by any profile modifications in relation to ISO qualities classified by numbers lower than 5. Method C is derived from method B but it is simpler. This method considers the industrial gears for engineering applications in the sub-critical range.

The factor K_v is calculated as follows:

$$K_v = 1 + \left[\frac{K_1}{K_A \dfrac{F_t}{b}} + K_2 \right] \frac{z_1 \, v}{100} \sqrt{\frac{u^2}{1 + u^2}} \qquad (5.014)$$

where:

$K_2 = 0.0193$ for spur gear teeth
$ = 0.0087$ for helical gears
$K_1 = 14.9$ for spur gear teeth having an ISO quality marked by 6
$ = 26.8$ for spur gear teeth having an ISO quality marked by 7
$ = 39.1$ for spur gear teeth having an ISO quality marked by 8
$ = 13.3$ for helical gears having an ISO quality marked by 6
$ = 23.9$ for helical gears having an ISO quality marked by 7
$ = 34.8$ for helical gears having an ISO quality marked by 8

v is the reference tangential speed, while K_A is the application factor.

For teeth whose overlap ratio is between 0 and 1, we interpolate linearly the results obtained for spur gear teeth and for helical gears as a function of the overlap.

Face load factors ($K_{H\beta}, K_{F\beta}$). There is a proper factor that enables to calculate the contact stress ($K_{H\beta}$) and the tooth bending strength ($K_{F\beta}$). Such factors are linked one to the other. Following distortion deviation as well as shaft and wheel distortions, teeth that are theoretically in contact following a line of contact along their flanks, come into contact at a single point. Following tooth distortion, such contact stretches for a given length that depends on tooth elasticity and on the application force that differs from the theoretical width. If the tooth is rigid, the contact will take place at a point where the force by unitary length is maximal. The relation of such maximum force to the force resulting from the equal distribution of load on the flank represents the longitudinal distribution factor $K_{H\beta}$. As the tooth is rigid, $F_{\beta y}$ is the deviation between the two flanks, before tooth distortion at the end opposite to the point where contact between teeth takes place (Fig. 5.2).

The tooth distortion can be either a partial contact or a complete contact.
Partial contact occurs when the force is weak and the deviation has a high value.

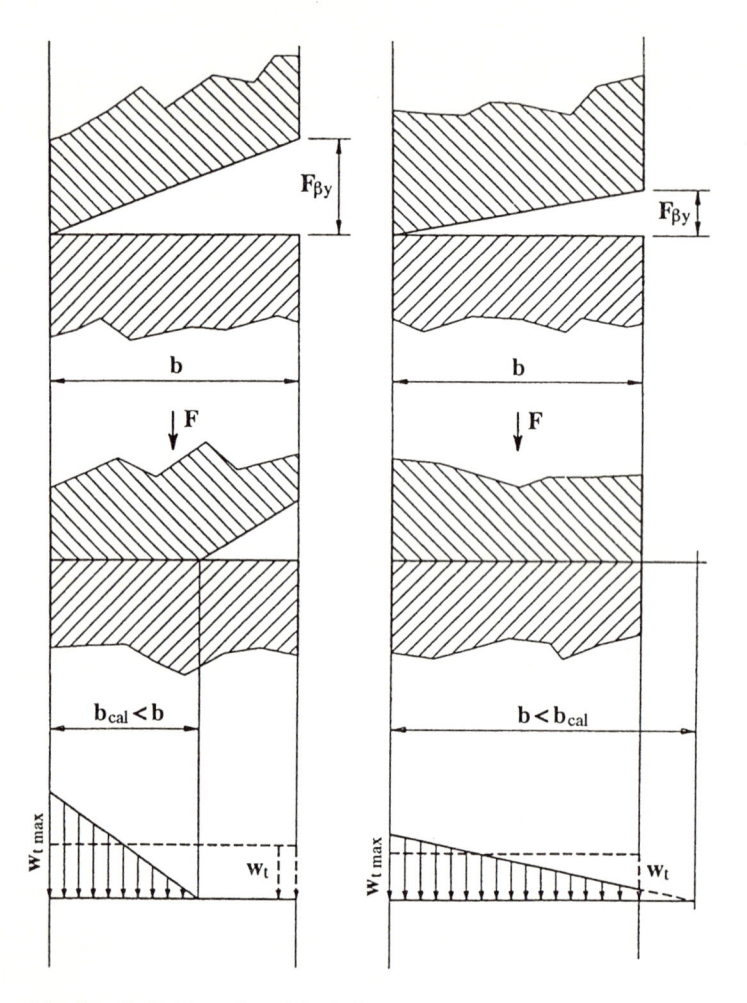

Fig. 5.2. Definition of total deviation

On the other hand, partial contact is given by the value of the variable:

$$\frac{F_{\beta y} \, c_\gamma}{2 \, \dfrac{F_t \, K_A \, K_v}{b}} \geq 1 \quad . \tag{5.015}$$

In this case:

$$K_{H\beta} = \sqrt{2 \, \frac{F_{\beta y} \, c_\gamma}{\dfrac{K_A \, K_v \, F_t}{b}}} \quad . \tag{5.016}$$

Otherwise:

$$\frac{F_{\beta y} \; c_\gamma}{2 \, K_A \, K_v \, F_t} < 1 \qquad (5.017)$$
$$b$$

$$K_{H\beta} = 1 + \frac{F_{\beta y} \; c_\gamma}{2 \, K_A \, K_v \, F_t} \cdot \qquad (5.018)$$
$$b$$

The elastic constant of teeth can be equal to 20 N/ (mm µm).
The deviation $F_{\beta y}$ results from distortions affecting the position assumed by teeth, which are corrected by running in.
The elements that contribute to the misalignment of wheels are the following:

- shafts distortion
- bearings distortion
- housings distortion
- wrong deviations of wheels
- misalignment of bearings.

For such calculation we consider that the shaft and pinion distortions are equivalent to the helix slope tolerances of the wheel characterized by a lower degree of precision.
The shaft distortion is represented by f_{sh}, the deviation due to manufacturing is represented by f_{ma}, while the influence of running in is represented by y_β. It will result that:

$$F_{\beta y} = 1.33 \, f_{sh} + f_{ma} - y_\beta \; . \qquad (5.019)$$

For a tooth without helix modifications $f_{ma} = f_{H\beta}$
For a tooth with helix modification $f_{ma} = 0.5 \, f_{H\beta}$
For a tooth with end relief $f_{ma} = 0.7 \, f_{H\beta}$.
For case hardened steels:

$$y_\beta = 0.15 \, (1.33 \, f_{sh} + f_{ma}) \; . \qquad (5.020)$$

The method used to determine f_{sh} varies according to the position of the pinion. If the pinion is centred in relation to bearings method C1 will be adopted, otherwise method C2 will be adopted.
(a) Method C1. Such method is analytical. We consider distortion due to bending and torsion. The standard includes cases providing many examples (for instance those which are helix modifications or the case of planets in planetary

train gears). We will just consider gears for normal transmission without helix modifications. The calculation of f_{sh} is directly integrated in $K_{H\beta}$.

For spur gears having single helix:

$$K_{H\beta} = 1 + \frac{4000}{3\pi} \chi_\beta \frac{c_\gamma}{E}\left(\frac{b}{d_1}\right)^2\left[5.12+\left(\frac{b}{d_1}\right)^2\left(\frac{l}{b}-\frac{7}{12}\right)\right] + \frac{\chi_\beta c_\gamma f_{ma}}{K_A K_v F_t/b}. \quad (5.021)$$

For gears having a double helix:

$$K_{H\beta} = 1 + \frac{4000}{3\pi} \chi_\beta \frac{c_\gamma}{E}\left[3.2\left(\frac{2b_B}{d_1}\right)^2+\left(\frac{B}{d_1}\right)^4\left(\frac{l}{B}-\frac{7}{12}\right)\right] + \frac{\chi_\beta c_\gamma f_{ma}}{K_A K_v F_t/b_B} \quad (5.022)$$

where:

χ_β = 0.85 in case hardened steels
E = 206000 N/mm² in steels
b_B = the width of one of the half helical gear of the double helix gear
B = the total width of the double helical gear
l = the length of the shaft between bearings.
Such a method cannot be applied to case hardened steel if:

$$\frac{K_{H\beta} - 1}{\chi_\beta\left(\frac{c_\gamma/2}{K_A K_v F_t/b}\right)} 0.15 > 6. \quad (5.023)$$

In this case we will apply method C2 explained below.

(b) Method C2: this method is used when the pinion is not centred in relation to the bearings. s is the distance of the pinion median plane from the centre of the distance between the bearings. By means of the following formula we can calculate a factor γ based on the dimensions of the pinion and on its position on the shaft:

$$\gamma = \left[|1 + K'\frac{l\,s}{d_1^2}\left(\frac{d_1}{d_{sh}}\right)^4 - 0.3| + 0.3\right]\left(\frac{b}{d_1}\right)^2 \quad (5.024)$$

where
l = distance between the centre of bearings
d_{sh} = average diameter of the shaft
K' = a constant depending on the position of the shafts of the pinion and on the position of the wheel.
Figure 5.3 provides the values of K'.

Factor K' with Stiffening	Factor K' without Stiffening	Fig.		
0.48	0.8	a)	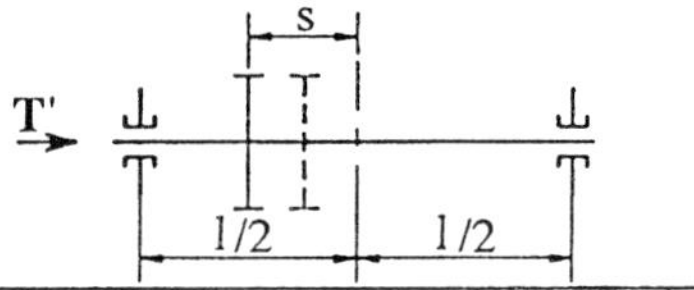	with s/l <0.3
-0.48	-0.8	b)	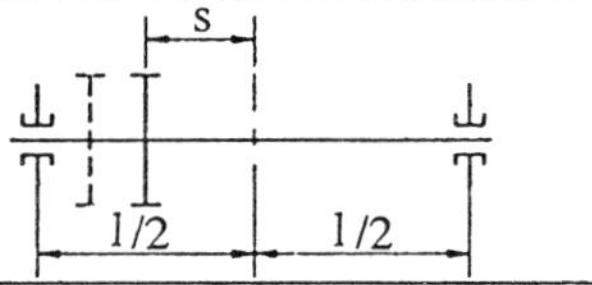	with s/l <0.3
1.33	1.33	c)	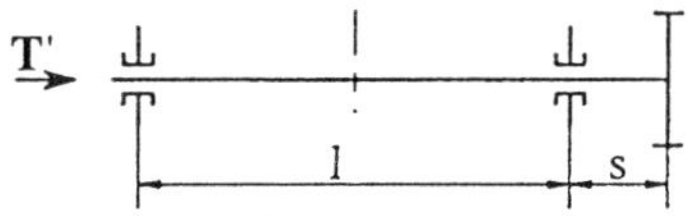	with s/l <0.5
-0.36	-0.6	d)	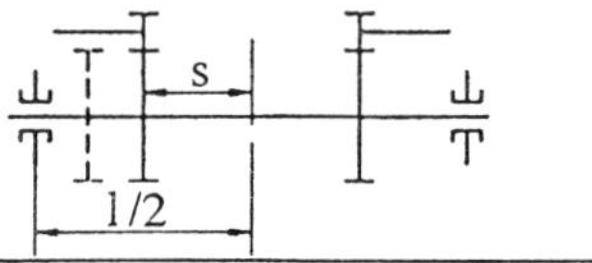	with s/l <0.3
-0.6	-1.0	e)	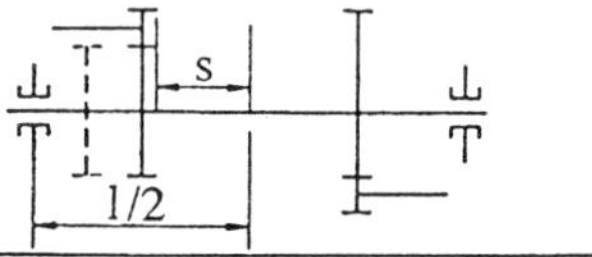	with s/l <0.3

Fig. 5.3. Values of K'

Therefore it is possible to calculate the value of the distortion for an unitary value of the specific load in wheels without longitudinal modifications

$$f_{sh0} = 0.023 \, \gamma \quad . \tag{5.025}$$

For gears with profile crowning we can take half the value while for gears with end relief we can assume the two-third of the value.

The value of the total empirically calculated elastic distortion of shafts is:

$$f_{sh} = \frac{K_A \, K_v \, F_t}{b} \, f_{sh0} \tag{5.026}$$

$K_{H\beta}$ is calculated as above.

If the limits 1/s expressed in Figure 5.3 are bypassed, such method could not be applied.

To calculate f_{sh} we should adopt an analytical method and follow the a.m. procedure.

The factor $K_{f\beta}$ can be calculated as follows:

$$K_{F\beta} = K_{H\beta}^{NF} \tag{5.027}$$

with:

$$NF = \frac{(b/h)^2}{1 + b/h + (b/h)^2} \tag{5.028}$$

where b is the tooth width and h is the total depth of tooth.

If the teeth of the wheel pinion and those of the wheel have different depths we will chose the smallest ratio b/h.

Transverse load distribution factor ($K_{H\alpha}$, $H_{F\alpha}$). Pitch deviations can produce changes in the way in which contact between teeth takes place as well as variation of the transmitted stress.

In order to evaluate whether the stress reaches a maximum value during meshing, we consider the factor given by the relation of the maximum value to the nominal value. Such a factor is the transverse load distribution factor.

Method C is a simplified method which is based on method B and which provides values that are completely independent from the variation of the base pitch.

The following values refer to case hardened steels and helical gears:

- Quality ISO 5: $K_{H\alpha} = K_{F\alpha} = 1.0$
- Quality ISO 6 $K_{H\alpha} = K_{F\alpha} = 1.0$
- Quality ISO 7 $K_{H\alpha} = K_{F\alpha} = 1.1$

The standard provides a complete table of values.

5.3.3 Contact Stress Calculation

Stresses. The contact stress is calculated on the basis of Hertz's theory above. Equation (5.001) provides the value of the theoretical contact stress. As before, the total contact ratio and the helix angle must be taken into account during the calculation of the contact stress.

The equation of the nominal contact stress corresponding to the pitch point is the following:

$$\sigma_{H0} = Z_H Z_E Z_\varepsilon Z_\beta \sqrt{\frac{F_t}{b\, d_1} \cdot \frac{u \pm 1}{u}} \tag{5.029}$$

where:
  Z_H is the zone factor: see equation (5.003)
  Z_E is the elasticity factor: see equation (5.002) and the following information
  Z_ε is the contact ratio factor
  Z_β is the helix angle factor

We must also consider the fact that the maximum stress is not exerted on the pitch point but on the lower point of single contact.
The following formulae refer to the pinion and the wheel respectively:

$$\sigma_H = Z_B \sqrt{K_A \ K_v \ K_{H\beta} \ K_{H\alpha}} \ \sigma_{H0} \qquad (5.030)$$

$$\sigma_H = Z_D \sqrt{K_A \ K_v \ K_{H\beta} \ K_{H\alpha}} \ \sigma_{H0} \ . \qquad (5.031)$$

Factors K were defined in the previous paragraph.
Factors Z_B et Z_D are the factors of single pair tooth contact factor for the pinion and the wheel.
The value of the stress resulting from contact stress must be lower than the stress permissible given by:

$$\sigma_{HP} = \sigma_{H\lim} \ Z_{NT} \ (Z_L \ Z_v \ Z_R) \ Z_W \ Z_X \qquad (5.032)$$

where:
  $\sigma_{H\lim}$: allowable stress number
  Z_{NT} : life factor
  $(Z_L \ Z_v \ Z_R)$: lubrication factor
  Z_W: : work hardening factor
  Z_X : size factor

The following relation must exist:

$$\frac{\sigma_{HP}}{\sigma_H} \geq S_{H\lim} \qquad (5.033)$$

where $S_{H\lim}$ is the minimum working safety factor allowed.

Contact Pressure Stress Factors. - Zone factor (Z_H): see equation (5.003). - Elasticity factor (Z_E) : see equation (5.002).
Theoretical factors result from the elastic features of materials forming gears. For a steel wheel, the value of Poisson's coefficient is 0.3 while the values of the modules of elasticity are 206000.
For steels it results that $Z_E = 189.8 \ (N/mm^2)^{0.5}$.

- Contact ratio factor (Z_ε): this factor considers the total contact ratio which indicates the distribution of load on teeth. Such a factor is calculated as follows:

$$Z_\varepsilon = \sqrt{\frac{4 - \varepsilon_\alpha}{3}(1 - \varepsilon_\beta) + \frac{\varepsilon_\beta}{\varepsilon_\alpha}} \quad . \tag{5.034}$$

In the case of spur gear teeth $\varepsilon_\beta = 0$; if $\varepsilon_\beta > 1$, use $\varepsilon_\beta = 1$.

- Helix angle factor (Z_β): such a factor considers the fact that the line of contact along the tooth flank of a helical gear is not parallel to the axis of the wheel. The maximum stress due to contact stress decreases while the line of contact extends.

Such a factor is expressed by:

$$Z_\beta = \sqrt{\cos \beta} \quad . \tag{5.035}$$

- Single pair tooth contact factors (Z_B/Z_D): due to these factors the contact stress is transposed from the pitch point to the lower point of single contact of each wheel. In Hertz's equation, the curvature radii belonging to the pitch point define the exact position. So it is sufficient to replace the curvature radii belonging to the pitch point with the curvature radii of teeth on the lower point of single contact. We also note that the sum of the curvature radii at any point is constant and is equivalent to the distance between the two tangent points at the base circles that belong to the action line. The following equations will explain the values of the a.m. curvature radii.

$$Z_B = \sqrt{\frac{\rho_{C1} \, \rho_{C2}}{\rho_{B1} \, \rho_{B2}}} \tag{5.036}$$

and

$$Z_D = \sqrt{\frac{\rho_{C1} \, \rho_{C2}}{\rho_{D1} \, \rho_{D2}}} \quad . \tag{5.037}$$

If the value of such factors is higher than 1, they are integrally applied; otherwise they are ignored. We note that the maximum value of the contact stress is the same for the pinion and the wheel.

Permissible Contact Stress Factors . - Allowable stress number ($\sigma_{H\,lim}$) : such number is taken into consideration in the part 5 of the standard. We notice that this value corresponds to a precise value of the number of load cycles borne by the tooth. For case hardened steels such value will be determined out of 5.10^7 cycles if pitting is not tolerated or out of 10^9 if a ligth pitting is tolerated. The value given is related to a degree of reliability equal to 0.99, i.e to a breaking risk of 1%.

- Life factor (Z_{NT}): such factor is related to S - N curve for the materials considered (Fig. 5.4).

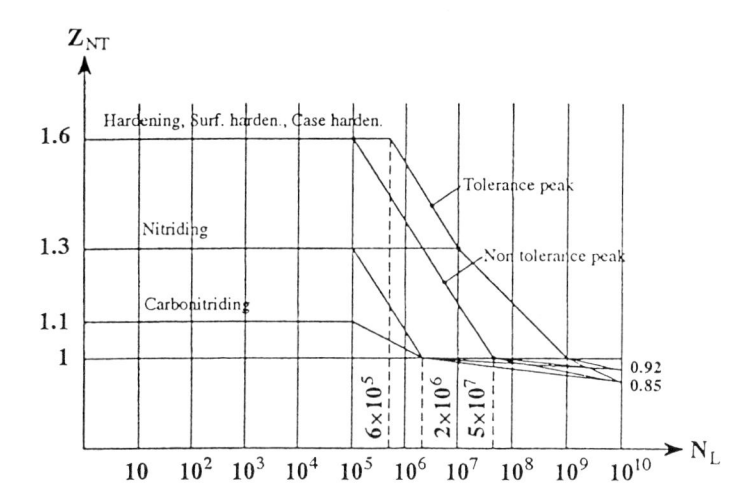

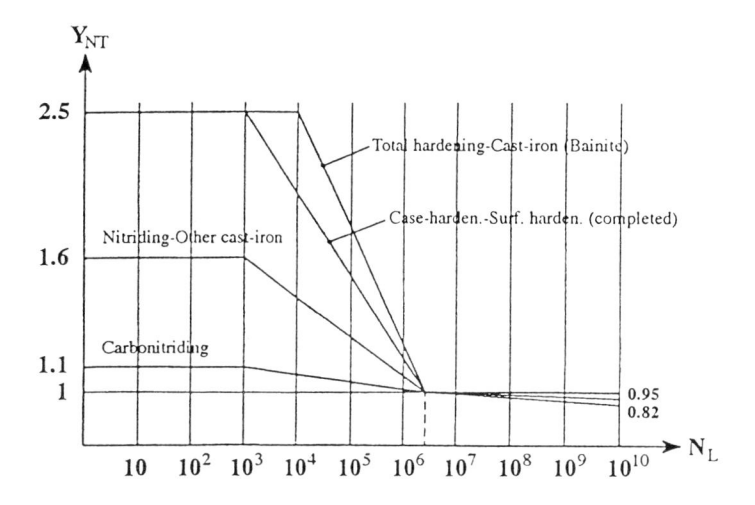

Fig. 5.4. Life factor

The curves relate to and are represented in structural steels, through hardened steel, perlitic grey cast iron, case hardened steels and surface hardened steels (such materials all have the same S - N curve); we also define a curve where pitting is tolerated (higher curve) and not tolerated (lower curve). The values depend on the number of cycles estimated for the pinion and for the wheel. The number of cycles can be calculated as follows:

$$N_{L\,1.2} = 60\, h\, n_{1,2} \qquad (5.038)$$

where h is the number of hours of total life duration of gear.
The following relations are drawn:
(a) Pitting is tolerated

- for $N_L < 6 \cdot 10^5$ $Z_{NT} = 1.6$

- for $6 \cdot 10^5 < N_L < 10^7$ $Z_{NT} = 4.3739(N_L)^{-0.0756}$

- for $10^7 < N_L < 10^9$ $Z_{NT} = 3.2584(N_L)^{-0.0570}$

- for $10^9 < N_L < 10^{10}$ $Z_{NT} = N_L^{-0.004}$

b) Pitting is not tolerated:

- for $N_L < 10^5$ $Z_{NT} = 1.6$

- for $10^5 < N_L < 5 \cdot 10^7$ $Z_{NT} = 3.8198(N_L)^{-0.0756}$

- for $5 \cdot 10^7 < N_L < 10^{10}$ $Z_{NT} = (N_L)^{-0.0047}$

- Lubrication factors (Z_L, Z_R, Z_v): pitting is influenced by the elastic-hydrodynamic pressure of the lubricant (see section 8) that in turn is influenced by the viscosity of lubricants, by the roughness of the material and by the sliding speed. Method B enables engineers make a separate and empirical calculation of these three factors (viscosity, roughness and speed). Method C is more severe than B. It provides a value of the product of the three factors with regard to the static situation and endurance. In this way we obtain a S - N curve representing such a factor. The values are the following:
(a) A light pitting is tolerated:

- for $N_L < 6 \cdot 10^5$ $Z_L\, Z_R\, Z_v = 1$

- for $6 \cdot 10^5 < N_L < 10^9$ $Z_L\, Z_R\, Z_v = (N_L/6 \cdot 10^5)^{3.104 \log Z_{lub}}$

- for $10^9 < N_L$ $Z_L\, Z_R\, Z_v = Z_{lub}$

b) Pitting is not tolerated:

- for $N_L < 10^5$ $Z_L\, Z_R\, Z_v = 1$

- for $10^5 < N_L < 5 \cdot 10^7$ $\quad Z_L Z_R Z_v = (N_L/10^5)^{0.3705 \log Z_{lub}}$

- for $5 \cdot 10^7 < N_L$ $\quad Z_L Z_R Z_v = Z_{lub}$

The value of Z_{lub} depends on the working of teeth.
For a grinded toothing $\quad Z_{lub} = 1 \quad$ if $R_{z\,10} < 4$ µm
$\quad\quad\quad\quad\quad\quad\quad\quad\quad\quad Z_{lub} = 0.92$ if $R_{z\,10} \geq 4$ µm
for a milled toothing $\quad Z_{lub} = 0.85$
The value of $R_{z\,10}$ will be given, if R_{z1} is the roughness of the pinion and R_{z2} is the roughness of the wheel:

$$R_{z10} = 0.5(R_{z1} + R_{z2}) \sqrt[3]{\frac{10}{\rho_{red}}} \quad\quad (5.039)$$

with:

$$\frac{1}{\rho_{red}} = \frac{1}{\rho_1} + \frac{1}{\rho_2}. \quad\quad (5.040)$$

- Work hardening factor (Z_W): this factor considers the degree of hardness of the tooth of the wheel during working. Its value depends on whether the degree of hardness of the wheel is lower than that of the pinion. Let us take the case of a case hardened pinion that meshes with a steel wheel obtained by annealing as an example: if the two wheels have the same degree of hardness such factor will be equal to 1. If the wheel has a lower degree of hardness than the pinion, such factor will be considered in the calculation of the permissible contact stress. Generally two methods are used: method B that includes an empirical calculation and method C that does not take this factor into account (or considers it as equal to 1).
- Size factor: Standard ISO 6336 considers such value equal to 1 while standard DIN 3990 provides empirical values depending on the module size.

5.3.4 Calculation of the Tooth Bending Strength

Stresses. Standard ISO 6336, part 3 deals with the stresses due to bending at the root of the tooth while it ignores the compressive stresses which are said to be included in the calculation of the factors adopted.

The equation for theoretical stress has been quoted in paragraph 5.2.2 - equation (5.009). Such theoretical value must be corrected on the basis of a stress concentration factor that considers any changes in the section (value of relief in the critical section). The stress reaches the maximum value allowed when load applied corresponds to the higher point of single contact. However, it is possible to calculate both the form factor and the stress concentration factor when the load is applied, at the higher point of single contact (the value is correct according to the stress) or on the tip of the tooth (in this case it is necessary to apply an empirical correction factor). The empirical factor that takes the factor calculated for the load applied to the tip of the tooth to the value defined for the load on the higher point of single contact, depends on the contact ratio and it is for this reason that it is called the contact ratio factor. The value of the load is influenced by the factors defined above. As the gears are helical gears and not spur gears the line of contact is inclined at an angle to the flanks and is not parallel to the axis. Therefore an helix angle factor must be applied.

For bending stress:
for the force applied to the higher point of single contact:

$$\sigma_F = K_A \, K_v \, K_{F\beta} \, K_{F\alpha} \frac{F_t}{b \, m_n} Y_F \, Y_S \, Y_\beta \quad . \quad (5.041)$$

For the force applied to the tip of the tooth:

$$\sigma_F = K_A \, K_v \, K_{F\beta} \, K_{F\alpha} \frac{F_t}{b \, m_n} Y_{Fa} \, Y_{Sa} \, Y_\varepsilon \, Y_\beta \quad . \quad (5.042)$$

The second method is explained in the following paragraphs.
The value of such stress must be lower than the permissible stress given by:

$$\sigma_{FP} = \sigma_{F\,lim} \, Y_{ST} \, Y_{NT} \, Y_{\delta\,rel\,T} \, Y_{R\,rel\,T} \, Y_X \, . \quad (5.043)$$

The factors will be explained below.

$$\frac{\sigma_{FP}}{\sigma_F} \geq S_{F\,min} \quad . \quad (5.044)$$

Definition of Factors. Critical section. We said that the critical section is determined by the tangent point of the parabola having the same degree of resistance as the fillet profile. According to the ISO standards, such points can result from the tangent at the profile that forms a 30° angle together with the axis of the tooth. The value of the tooth thickness corresponding to the critical section is the same as that for application of the load at the tip of the tooth or as that for the load at the higher point of single contact. For internal toothing, the tooth is that belonging to the basic reference profile.

For an external tooth, the following equations will result (Fig. 5.5):

$$E = \frac{p}{4} \, m_n \, - \, h_{fP} \, \tan \alpha_n \, + \, \frac{s}{\cos \alpha_n} \, - \, (1 \, - \, \sin \alpha_n) \frac{\rho_{fP}}{\cos \alpha_n} \quad (5.045)$$

where s indicates the value of the protuberance that occurs when the tooth is in its final state.

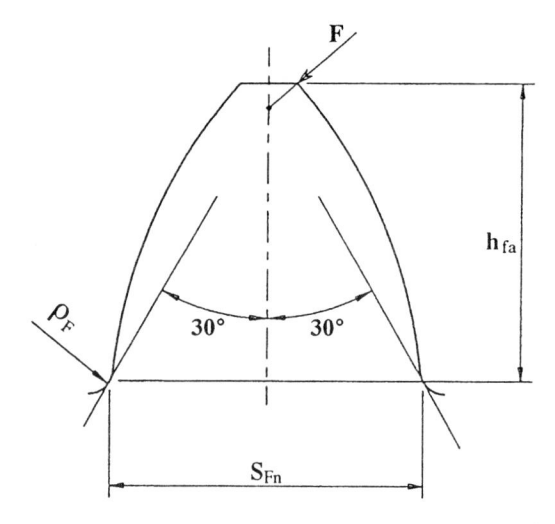

Fig. 5.5. Critical section

h_{fP} is the addendum of the tool, while ρ_{fP} indicates the tip radius of the tool.
These two values can be assimilated to the values defined for the standard profile of the wheel.

$$G = \frac{\rho_{fP}}{m_n} - \frac{h_{fP}}{m_n} + x \tag{5.046}$$

$$H = \frac{2}{z_n}\left(\frac{\pi}{2} - \frac{E}{m_n}\right) - \frac{\pi}{3} \tag{5.047}$$

$$\theta = \frac{2G}{z_n} \tan\theta - H . \tag{5.048}$$

This latest equation is solved by iteration putting $\theta = \pi/6$ and θ is replaced with the value found until the equation is verified (2 or 3 iterations are sufficient). The resulting value will be introduced in the equation by the critical thickness:

$$\frac{S_{Fn}}{m_n} = z_n \sin\left(\frac{\pi}{3} - \theta\right) + \sqrt{3}\left(\frac{G}{\cos\theta} - \frac{\rho_{fP}}{m_n}\right) . \tag{5.049}$$

The curvature radius in the critical section is given by:

$$\frac{\rho_F}{m_n} = \frac{\rho_{fP}}{m_n} + \frac{2G^2}{\cos\theta\,(z_n \cos^2\theta - 2G)} . \tag{5.050}$$

For the load on the tip of the tooth the height of the lever arm for the bending moment is:

$$\frac{h_{Fa}}{m_n} = 0.5\ z_n \left[\frac{\cos \alpha_n}{\cos \alpha_{Fan}} - \cos\left(\frac{\pi}{3} - \theta\right) \right] + 0.5 \left(\frac{P_{fP}}{m_n} - \frac{G}{\cos \theta} \right) \quad (5.051)$$

with:

$$\alpha_{an} = \arccos \left[\frac{\cos \alpha_n}{1 + \frac{(d_a - d)}{m_n\ z_n}} \right] \quad (5.052)$$

and

$$\alpha_{Fan} = \tan \alpha_{an} - \operatorname{inv} \alpha_n - \frac{0.5\ \pi + 2\ x \tan\alpha_n}{z_n} . \quad (5.053)$$

- Tooth form factor (Y_{Fa}): following equation (5.010) the value of the stress factor for the force applied on the head of the tooth is:

$$Y_{Fa} = \frac{6\ \dfrac{h_{Fa}}{m_n} \cos \alpha_{Fan}}{\left(\dfrac{S_{Fn}}{m_n}\right)^2 \cos \alpha_n} . \quad (5.054)$$

- Stress correction factor (Y_{Sa}): the stress correction factor considers the differences between the real stress and the stress calculated following an analytical method. It also considers the stress concentration and the fact that stress results not only from bending but also from compression. Its value is given by the parameter q_s and by the parameter L_a that depend on the dimensions of the critical section. It will result then:

$$q_s = \frac{S_{Fn}}{2\ r_F} \quad (5.055)$$

$$L_a = \frac{S_{Fn}}{h_{Fa}} \quad (5.056)$$

$$Y_{Sa} = (1.2 + 0.13\ L_a)\ q_s^{[1/(1.21 + 2.3/L_a)]} . \quad (5.057)$$

5 Gear Calculation

- Contact ratio factor (y_ε): such a factor empirically transposes the value resulting from the application of the force to the tip of the tooth into an approximative value (in terms of reliability) for the force applied to the higher point of single contact. It only depends on the normal contact ratio which is drawn from the transverse contact ratio by means of the following equation:

$$\varepsilon_{an} = \frac{\varepsilon_a}{\cos^2 \beta_b} \qquad (5.058)$$

It will result that:

$$Y_e = 0.25 + \frac{0.75}{\varepsilon_{an}} \qquad (5.059)$$

- Helix angle factor (Y_β): for bending strength the helix angle factor has the same meaning that it has for the contact pressure. It considers the degree of inclination of the lines of contact along the tooth flank and is expressed by the following formula:

$$Y_\beta = 1 - \varepsilon_\beta \frac{\beta}{120°} \qquad (5.060)$$

In this equation overlap ratio will be limited to the unit, if it depasses such value, while the helix angle which is expressed in degrees, is limited to a maximum of 30°.

- Bending stress number ($\sigma_{F\ lim}$): such stress provides a breaking risk equal to 1% for a given number of cycles that for most materials correspond to 3.10^6 cycles. It is determined by means of measurements on a test gear and it is calculated following the equation defined by the standard starting from the load measured. Bending stress is considered while compressive stress is completely ignored. The value for the different materials are indicated in the 5th part of the standard.

- Life factor (Y_{NT}): such a factor considers the variation of the stress value on the bases of the number of cycles following S - N curve that varies with the material. As example we consider the values concerning case hardened steels. The variation is the same as for surface hardened steels and corresponds to three zones of values linked to the number of cycles desired on the basis of the total life duration of the gear (N_L). With regard to wheel and pinion, things change because materials are different and because the number of cycles for the same life time (expressed in hours) can result in differences:

for $N_L < 10^3$ $Y_{NT} = 2.3$

for $10^3 < N_L < 3.10^6$ $Y_{NT} = 5.5118\ (N_L)^{-0.1144}$

for $3.10^6 < N_L < 10^{10}$ $Y_{NT} = 1.166\ (N_L)^{-0.0103}$

- Notch sensitivity factor ($Y_{\delta\ rel\ T}$): such a factor considers the different sensibilities of the material to the effects of notches. Method B provides a complete empirical calculation. Method C is derived from method B and is simpler. It can be applied when the stress factor has been calculated by the application of the force on the tip of the tooth. A value for the zone pertaining to endurance and also unlimited endurance is provided.

For static load conditions (lower than 1000 cycles) it is possible to assume value equal to 1. For limited life we will have (for hardened steels):

$$Y_{\delta\ rel\ T\ stat} = 0.52\ Y_{Sa} + 0.20 \quad . \tag{5.061}$$

To obtain the value for limited life (between 10^3 and $3 \cdot 10^6$) we interpolate the following:

$$Y_{\delta\ rel\ T} = \left(\frac{3 \cdot 10^6}{N_L}\right)^{\log\frac{Y_{\delta\ rel\ T\ stat}}{3.4771}} \quad . \tag{5.062}$$

- Surface factor ($Y_{R\ rel\ T}$): this considers the difference in roughness existing between the root of the tooth belonging to the wheel calculated and the root of the tooth belonging to the master wheel. In most cases in gearboxes for general applications it is equal to 1.

- Size factor: (Y_X): the module for the wheel and the test gear are not necessarily the same. The influence of such a difference on the stress allowed which takes into consideration the dimension factor is taken into account. It is calculated following the method adopted for shearing stress calculation. For static load conditions we assume the value to be equal to 1, while for the period of unlimited life the following values are adopted:

for $5 < m_n < 25$

$$Y_{X\infty} = 1.05 - 0.01\ m_n \tag{5.063}$$

for $m_n < 5$

$$Y_{X\infty} = 1.0 \tag{5.064}$$

and for $m_n > 25$

$$Y_{X\infty} = 0.8 \quad . \tag{5.065}$$

For limited endurance ($10^3 < N_L < 3 \cdot 10^6$) they are:

$$Y_X = \left(\frac{N_L}{10^3}\right)^{\frac{\log Y_{X\infty}}{3.4771}} \quad . \tag{5.066}$$

- Test gear factor (Y_{ST}) : such a factor represents the value of the test wheel stress correction factor and is equal to 2.

5.3.5 Allowable Stress Number

Part 5 of the standard deals with allowable stress number. It provides tables showing such stresses as a function of the type of material, of the treatment and of the presumed quality of materials.

The standard distinguishes between three levels of quality: a low level of quality indicated by ML for metallic materials, a medium level of quality MQ, which is the most current, and a high level of quality which is marked by ME and corresponds to materials which are specially treated and controlled.

Generally such materials are used in expensive products where the high quality is reflected in the price.

For each level there is a corresponding table that indicates the metallographic and the chemical requirements as well as the necessary tests.

For further details it is suggested consulting the standard in its integral form. For case hardened steels the following values are provided:

for the quality ML $\sigma_{H\,lim}$ = 1300 MPa $\sigma_{F\,lim}$ = 325 MPa

for the quality MQ $\sigma_{H\,lim}$ = 1500 MPa

The value of $\sigma_{F\,lim}$ depends on the features of the materials used for the core. If the hardness in the core of the material is equal to 30 HRC or if it depasses this value, $\sigma_{F\,lim}$ = 500 MPa will be assumed. If the hardness degree corresponding to the core of the material is lower than the a.m. value but higher than 25 HRC, one can distinguish steels with a Jominy hardness degree at 12 mm higher than 28 HRC (in this case the value adopted is $\sigma_{F\,lim}$= 460 MPa) and steels with a Jominy hardness at 12 mm which is lower than 25 HRC (in this case the adopted value is $\sigma_{F\,lim}$ = 425 Mpa).

-For quality ME $\sigma_{H\,lim}$ = 1650 MP
$\sigma_{F\,lim}$ = 500 MP

5.4 Standard ISO 10300 (Bevel Gear Pairs)

5.4.1 Introduction

Standard ISO 10300 is generally used to calculate the load capacity of bevel gears. It complies with the same principles of standard ISO 6336. The calculation is done by considering the virtual spur gear to the equivalent to the bevel gear pair. Some factors will be different while others will take the difference between spur gears and bevel gear pairs into account. These differences have not yet been considered by the factors examined so far.

5.4.2 Influence Factors

- Dynamic factor (K_v) : such a factor is calculated following method C which is simplified. The resulting index of quality depends on pitch tolerances, the normal module and the number of teeth. This index is given by:

$$C = -0.5048 \ln(z) - 1.144 \ln(m_n) + 2.852 \ln(f_p) + 3.32 \qquad (5.069)$$

where "ln" indicates the natural logarithm which is also known as the Neperian logarithm. For very precise gears (where C is lower than 6) it is possible to chose a factor K_v between 1 and 1.1.

For other types of gear the following calculations must be done:

$$A = 50 + 56(1.0 - B) \qquad (5.070)$$

$$B = 0.25(C - 5.0)^{0.667} \qquad (5.071)$$

$$K_v = \left(\frac{A}{A + \sqrt{200 \ v_t}}\right)^{-B} \qquad (5.072)$$

where v_t is the reference speed given by:

$$v_t = \frac{\pi \ d_{m1} \ n_1}{60000} \qquad (5.073)$$

The speed is limited to the area where resonance takes place, which is given by:

$$v_{t \ max} = \frac{[A + (14 - C)]^2}{200} \qquad (5.074)$$

- Face load factor ($K_{H\beta}$): the method enabling us to calculate such a factor is based on the load capacity of the surface between the teeth. We measure the width of such surface (b_e) and we compare it with the tooth width (b).
If $b_e > 0.85.b$,

$$K_{H\beta} = 1.5 \ K_{H\beta} \ b_e \ .$$

Otherwise:

$$K_{H\beta} = 1.5 \, K_{H\beta be} \, \frac{0.85}{b_e/b} \, . \qquad (5.075)$$

The values of $K_{H\beta} \, b_e$ are given in Table 5.1:
- Curving length factor (K_{FO}): such factor is specifically designed for conical gears. For spur teeth its value is equal to 1.00. For spiral teeth, if r_{eo} is the cutting tool radius, we will reach

$$K_{FO} = 0.211 \left(\frac{r_{eo}}{R_m}\right)^q + 0.789 \qquad (5.076)$$

with

$$q = \frac{0.279}{\log_{10}(\sin \beta_m)} \qquad (5.077)$$

R_m is the mean distance from the cone
β_m is the mean spiral angle.

- Transverse load factor ($K_{H\alpha}$, $K_{F\alpha}$): this is the same as found in cylindrical wheels.

Table 5.1 Values of $K_{H\beta} \, b_e$

Performance quality	Mounting conditions		
	No wheel in overhanging load	One wheel in overhanging load	Two wheels in overhanging load
High quality	1	1.1	1.25
Standard quality	1.2	1.32	1.5

5.4.3 Contact Area

Bevel gear pairs often have a crowning and a barelling. Such arrangement of the surface which is made by the tooth width and the length of contact of the virtual gear, provides an elliptical profile as shown in Figure 5.6.

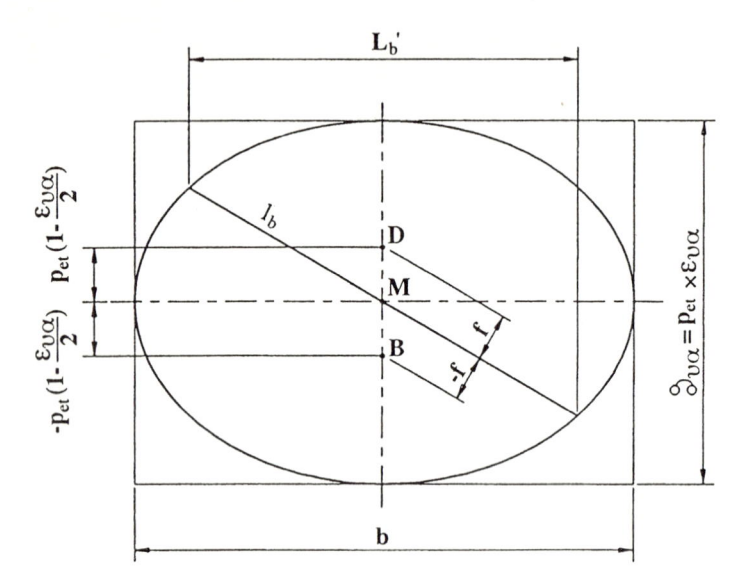

Fig. 5.6. Ellipse of contact

Inside this profile, it is possible to set a higher point (D) and a lower point (B) of single contact, as well as the point that is at the centre of B and D (M).

In a spur gear, with regard to material resistence to contact stress, the critical point is at point B but with regards to tooth bending strength the critical point is D.

In a spiral gear for both the cases the critical point is M. If β_{vb} is the value of the helix angle the distance of the line of contact passing through any value M is given by this value and by the overlap ratio of the virtual gear whose value must be ε_{vb} at a distance f which is calculated for:
- The contact stress:
  the critical point

$$f = -P_{et} \cos \beta_{vb} (1 - 0.5\, \varepsilon_{v\alpha})(1 - \varepsilon_{v\beta}) \qquad (5.078)$$

the tip of the tooth belonging to the driving gear is:

$$f = -P_{et} \cos \beta_{vb} [(1 - 0.5\, \varepsilon_{v\alpha})(1 - \varepsilon_{v\beta}) - 1] \qquad (5.079)$$

the root of the tooth of the driven gear:

$$f = -P_{et} \cos \beta_{vb} [(1 - 0.5\, \varepsilon_{v\alpha})(1 - \varepsilon_{v\beta}) + 1] \ . \qquad (5.080)$$

- The stress due to bending on the root of the tooth is:
the critical point is:

$$f = P_{et} \cos \beta_{vb}(1 - 0.5\, \varepsilon_{v\alpha})(1 - \varepsilon_{v\beta}) \qquad (5.081)$$

the tip of the tooth is:

$$f = P_{et} \cos \beta_{vb}[(1 - 0.5\, \varepsilon_{v\alpha})(1 - \varepsilon_{v\beta}) + 1] \qquad (5.082)$$

the root of the tooth is:

$$f = P_{et} \cos \beta_{vb}[(1 - 0.5\, \varepsilon_{v\alpha})(1 - \varepsilon_{v\beta}) - 1] \, . \qquad (5.083)$$

P_{et} is the transverse base pitch of the equivalent cylindrical gear.
In all these cases, if the overlap ratio $\varepsilon_{v\beta}$ is higher than 1, the value 1 will be chosen. With these values it is possible to calculate the length of the line of contact at each point. The values are those of the virtual gear of the bevel gear pair (v index).

$$l_b = b\, g_{v\alpha} \frac{\sqrt{g_{v\alpha}^2 \cos^2\beta_{vb} + b^2 \sin^2\beta_{vb} - 4 f^2}}{g_{v\alpha}^2 \cos^2\beta_{vb} + b^2 \sin^2\beta_{vb}} \qquad (5.084)$$

where β_{vb} is the angle given by:

$$\sin \beta_{vb} = \sin \beta_m \cos \alpha_n \, . \qquad (5.085)$$

The length of the contact line projected parallel to the generatrix of the equivalent cylinders is given by:

$$l_b' = l_b \cos \beta_{vb} \, . \qquad (5.086)$$

5.4.4 Contact Stress Calculation

General equations. The contact stress is that of the virtual gear which must bear the force F_{mt}.
Therefore it will result that:

$$\sigma_{H0} = Z_H\, Z_E\, Z_{LS}\, Z_\beta\, Z_K\, Z_{M\text{-}B} \sqrt{\frac{F_{mt}}{d_{v1}\, l_b} \frac{\sqrt{u+1}}{u}} \qquad (5.087)$$

and

$$\sigma_H = \sigma_{H0}\sqrt{K_A K_v K_{H\beta} K_{H\alpha}} \quad . \qquad (5.088)$$

Such stress will be compared to the permissible stress:

$$\sigma_{HP} = \sigma_{H\lim} Z_{NT} Z_X (Z_L Z_R Z_v) Z_W \quad . \qquad (5.089)$$

The safety factor will be:

$$S_H = \frac{\sigma_{HP}}{\sigma_H} \quad . \qquad (5.090)$$

Factor determination. Factors are similar or equal to those defined for cylindrical gears. They are:
- Zone factor (Z_H):

$$Z_H = 2\sqrt{\frac{\cos \beta_{vb}}{\sin 2\alpha_{vt}}} \quad . \qquad (5.091)$$

- Elasticity factor (Z_E): such a factor is equal to that defined for parallel gears and for a couple of steel wheels it is equal to 189.8.
- Helix angle factor (Z_β): this is a factor similar to those defined for parallel gears:

$$Z_\beta = \sqrt{\cos \beta_m} \quad . \qquad (5.092)$$

- Bevel gear pair factor (Z_K): such a factor refers to the bevel gear pair. It is an empirical factor that considers the differences between parallel gears and bevel gear pairs and is generally equal to 0.8.
- Dimension factor (Z_X): according to the standard ISO such a factor is equal to 1.
- Distribution load factor (Z_{LS}): this factor also refers to the bevel gear pair and replaces the contact ratio factor referring to parallel gears. It is calculated as follows: for each line of contact at the root of the driving tooth and at the critical point as well that at the tip of the driving tooth, we determine the value of f as said in paragraph 5.4.3. For each path of contact:

$$p^* = 1 - \left(\frac{|f|}{f_{max}}\right)^{1.5} \qquad (5.093)$$

where the value of f_{max} is given by:

$$f_{max} = 0.5 \, \varepsilon_{v\gamma} \, P_{et} \cos \beta_{vb} \qquad (5.094)$$

with:

$$\varepsilon_{v\gamma} = \sqrt{\varepsilon_{v\alpha}^2 + \varepsilon_{v\beta}^2} \quad . \tag{5.095}$$

If:

$$p^* \geq 0 \tag{5.096}$$

A^*_i is calculated by means of the following formulas:

$$A^*_i = 0.25 \, \pi \, p^*_i \, l_b \tag{5.097}$$

where the value of the index i is 1 for the line of contact at the tip, 2 for the medium contact and 3 for the contact occurring at the root of the tooth. Such numbers represent surfaces if p^*_i is positive or null. At the end it will result that:

$$Z_{LS} = \sqrt{\frac{A^*_2}{A^*_1 + A^*_2 + A^*_3}} \geq \sqrt{0.7} \quad . \tag{5.098}$$

- Middle zone factor (Z_{M-B}): such factor transforms stress on the reference point in a stress in the critical point and it is equivalent to:

$$Z_{M-B} = \frac{\tan \alpha_t}{\sqrt{\left(\left(\frac{d_{va1}}{d_{vb1}}\right)^2 - 1 - F_1 \frac{\pi}{z_{v1}}\right)\left(\left(\frac{d_{va2}}{d_{vb2}}\right)^2 - 1 - F_2 \frac{\pi}{z_{v2}}\right)}} \tag{5.099}$$

with the following values:

$$F_1 = 2 + (\varepsilon_{v\alpha} - 2)\varepsilon_{v\beta} \tag{5.100}$$

and

$$F_2 = 2 \, \varepsilon_{v\alpha} - 2 + (2 - \varepsilon_{v\alpha}) \, \varepsilon_{vb} \quad . \tag{5.101}$$

The value of ε_{vb} is limited to 1.
- Lubrication factor (Z_v, Z_R, Z_L): this factor is equal to that defined for parallel gears.
- Work hardening factor (Z_W): this factor too is equal to that defined for parallel gears.
- Life factor (Z_{NT}): this factor too is equal to that defined for parallel gears. It is linked to the value of the stress allowed that is also identical to that provided for parallel gears (see paragraph 5.4.6).

5.4.5 Calculation of Tooth Bending Strength

General equations. The stress at the root of the tooth is calculated as follows:

$$\sigma_F = \sigma_{F0} \, K_A \, K_v \, K_{F\beta} \, K_{F\alpha} \quad (5.102)$$

and

$$\sigma_{F0} = \frac{F_{mt}}{b \, m_{mn}} \, Y_{Fa} \, Y_{Sa} \, Y_\varepsilon \, Y_K \, Y_{LS} \quad (5.103)$$

The stress allowed is given by:

$$\sigma_{FP} = \sigma_{F\,lim} \, Y_{ST} \, Y_{NT} \, Y_{\delta\,rel\,T} \, Y_{R\,rel\,T} \, Y_X \quad (5.104)$$

The value of the safety factor is:

$$S_F = \frac{\sigma_{FP}}{\sigma_F} \quad (5.105)$$

Factor determination. - Tooth form factor (Y_{Fa}): the shape factor is calculated following the same method adopted to calculate parallel gears even for formulae applied to two cylindrical gears virtual to the bevel gears pair.
  - Stress correction factor (Y_{Sa}): is calculated following the same method. In this case formulae are applied to the virtual cylindrical gears (ISO 6336).
  - Contact ratio factor (Y_ε): for the bevel gear pair, this factor depends on the contact ratio of the equivalent parallel gear. It is calculated as follows:

$$Y_\varepsilon = 0.25 + \frac{0.75}{\varepsilon_{v\alpha}} - \varepsilon_{v\beta} \left(\frac{0.75}{\varepsilon_{v\alpha}} - 0.375 \right) \geq 0.625 \quad (5.106)$$

where the overlap ε_{VB} is limited to 1 if its value is higher than 1.
  - Bevel gear pair factor (Y_K) : this factor has a similar function to (Z_k) it is calculated on the basis of the indications provided below:

$$Y_K = 0.25 \left(1 + \frac{l_b'}{b} \right)^2 \frac{b}{l_b} \quad (5.107)$$

The value l'_b is calculaded in pharagraph 5.4.3 .
  - Load distribution factor (Y_{LS}): this factor has the same function as factor Z_{LS} for the calculation of normal pressure. The factors A^*_i are calculated using the values f

for bending at the root of the tooth. This will result as:

$$Y_{LS} = \frac{A_2^*}{A_1^* + A_2^* + A_3^*} \geq 0.7 \quad . \tag{5.108}$$

- Notch sensitivity factor ($Y_{\delta\,rel\,T}$): in which it is calculated:

$$q_s = \frac{S_{Fn}}{2\,\rho_f} \quad . \tag{5.109}$$

If q_s is higher than 1 or equal to 1, $Y_{\delta\,rel\,T} = 1.0$
If q_s is lower than 1 $Y_{\delta\,rel\,T} = 0.95$
- Surface conditions factor ($Y_{R\,rel\,T}$): if the roughness R_z is lower than 1μm, the value of such a factor will be:

$$Y_{R\,rel\,T} = 1.12 \tag{5.110}$$

for case hardened steels and through hardened steel.
If the roughness R_z ranges from 1μm to 40 μm it will result that:

$$Y_{R\,rel\,T} = 1.674 - 0.529\,(R_z + 1)^{0.1} \tag{5.111}$$

for the same types of steel.
- Size factor (Y_X) used for through hardened steel has the formula

$$Y_X = 1.03 - 0.006\,m_{mn} \tag{5.112}$$

where the value of Y_X is included between 0.85 and 1.00.
For surface hardened steels this is:

$$Y_X = 1.05 - 0.01\,m_{mn} \tag{5.113}$$

being Y_X limited between 0.80 and 1.00.
- Life factor (Y_{NT}): this factor is calculated following the method used to calculate parallel gears (ISO 6336).

5.4.6 Materials

The 5th part of standard ISO 6336 is applied integrally even when the values are the same. This is referred to as ISO 6336-5.

5.5 Calculation with a Load Spectrum

5.5.1 Complete Calculation

Contact stress or bending stress depend on the load factor and consequently on the load. Therefore there are distribution factors for each load forming the spectrum. Each stress is considered independently from the others (wheel and pinion contact stress, pinion and wheel bending stress). Now we choose a safety factor and calculate the stress allowed. We determine the number of cycles for this stress that corresponds to the stress permissible borne by each load level. For each load level there is a fixed number of cycles. This number is divided by the number calculated relevant to the stress allowed. The ratios obtained for each load level are then added up. The total must be equal to the unit. If it is higher than the unit the safety factor must be lessened, but if it is lower the safety factor must be increased. After some iterations the right safety factor can be found. This calculation takes some time and can only be performed by special program.

5.5.2 Approximate Calculation

For contact stress and bending stress the value of the equivalent load can be determined following Miner's rule.
The equation is given by the following formula:

$$F_{t\,equi} = \left[\frac{1}{N_{L\,equi}} \Sigma(F_{ti}^{p}\, \Delta N_{Li})\right]^{\frac{1}{p}} \tag{5.114}$$

where $p = 6.61$ for the contact stress and $p = 8.69$ for bending stress.

In order to execute such calculation loads must be classified in a decreasing order. We eliminate all loads whose partial duration depasses the point corresponding to the limit of endurance except the first one. The duration adopted for each calculation is the equivalent duration.

The classic calculation can be executed using the value corresponding to the equivalent load (4 calculations must be done involving contact stress on the wheel and on the pinion and bending stress on the pinion and on the wheel). It is also possible to execute the calculation using any nominal load with an application factor given by the following equation:

$$K_A = \frac{F_{t\,equi}}{F_{t\,nom}} \tag{5.115}$$

5.6 Worm and Worm Wheel Calculation

5.6.1 General Features

Worm and worm wheel are subject to contact stress root breaking risk and remarkably wear which strikes worm and worm wheel much more frequently than parallel gears and bevel gears pairs. In addition, the material used for worm and worm wheels is liable to sudden changes in temperature due to friction which also causes wear.

This increase in temperature will be studied in the section on the thermal capacity of gearboxes as it concerns not only gears but all also gearboxes components.

5.6.2 Load Capacity for Contact Stress

T_2 is the torque to be transmitted to the driven gear. The stress resulting from normal pressure can be defined as follows:

$$\sigma_H = Z_E \, Z_\rho \sqrt{\frac{1000 \, T_2 \, K_A}{a^3}} \qquad (5.116)$$

The permissible stress for contact stress is:

$$\sigma_{HP} = \sigma_{H\,lim} \, Z_h \, Z_n \qquad (5.117)$$

while the safety factor is:

$$S_H = \frac{\sigma_{HP}}{\sigma_H} \qquad (5.118)$$

Factors are explained as follows:
- Elasticity factor (Z_E): this is characterized by the elastic properties of the two materials usually bronze for the wheel and case hardened steel for the worm with the following result:

$$Z_E = \frac{1}{\sqrt{\pi \left[\frac{1 - v_1^2}{E_1} + \frac{1 - v_2^2}{E_2} \right]}} \qquad (5.119)$$

For bronzes the module of elasticity is about 90000 MPa (it can vary according to the composition of bronzes) while for steel it is equal to 206000 MPa. Poisson's coefficient for metals is 0.3. For the bronze-steel couple it is:

$$Z_E = 150 \text{ MPa}^{\frac{1}{2}} \quad . \tag{5.120}$$

- Contact factor (Z_p): this is an empirical factor that varies with the worm profile. Such value is given as a function of the ratio of the average diameter of the worm to the centre distance (d_{m1}/a).

For a profile ZI it will reach 3.4 if the ratio is equal to 0.25 and up to 2.6 if the ratio is 0.55.

For a profile ZH, in presence of the same ratios, such a factor will have the following values: 3.1 and 2.35.

- Life factor (Z_h): this is S - N factor. It is calculated as a function of a number of operating hours expressing the operating time of the unit bearing constant and uniform load.

$$Z_h = (25000/h)^{\frac{1}{6}} \quad . \tag{5.121}$$

If the load is not constant, it is possible to calculate an equivalent duration following Miner's rule, as the exponent is equal to 3.

$$h = \frac{(h_1 \, F_{t21}^3 + h_2 \, F_{t22}^3 + \ldots\ldots)}{F_{t\,equi}^3} \quad . \tag{5.122}$$

In this case the calculation is executed using the equivalent load corresponding to duration. In all the cases the value of this factor is always lower than 1.6 or equal to 1.6.

- Load variation factor (Z_n): such factor is given for a constant load as a function of the rotation speed:

$$Z_n = \left[\frac{1}{\frac{n_2}{8} + 1} \right]^{\frac{1}{8}} \quad . \tag{5.123}$$

For a constant load at variable speed, it is possible to calculate the value of Z_{ni} on the basis of equation (5.123) for each speed and the value to be adopted for the equation:

$$Z_n = \sqrt{\frac{\Sigma \, (Z_{ni}^2 \, t_i)}{\Sigma \, t_i}} \quad . \tag{5.124}$$

- Allowable stress number ($\sigma_{H\ lim}$): such stress is found in the wheel. The allowable stress number for bronze varies according to its composition. For bronze G-CuSn 12 the value corresponding to 265 MPa can be admitted.

5.6.3 Wear Prevention

We calculate the stress allowed due to wear which is compared with the stress caused by contact pressure. The stress allowed following wear is given by:

$$\sigma_{WP} = \sigma_{w\ lim}\ W_P\ W_R\ W_v \qquad (5.125)$$

which is characterized by the following factor:
- Wear stress allowed ($\sigma_{W\ lim}$): this depends on the wear borne by the unit during working. N_L is the life of the unit (in number of cycles per wheel) and Δm_{lim} is the wear tolerated in Kg, resulting in:

$$\sigma_{W\ lim} = \left(\frac{2.6 \cdot 10^6\ \Delta m_{lim}}{L_W} \right)^{\frac{1}{4}} \qquad (5.126)$$

- Material wear factor (W_P): such a factor depends on the materials forming the unit and on the lubricant used. For a worm made of case hardened steel 16 CrNi 4 and a bronze G-CuSn12 lubricated with a mineral oil (VG 320) such factor is equal to 1.

For a worm made of steel through hardened (42 CrMo 4) such value is equal to 0.63. If synthetic oils are used the value can be 1.71 and 1.56 respectively.

We notice that the higher such values are the more resistant the unit will be.
- Roughness factor concerning wear (W_R): this factor considers the influence of roughness on wear. If R_z is the roughness of wheels (this value is 6 times the average roughness) the following formula is adopted:

$$W_R = 4\sqrt{\frac{R_z}{3}} \qquad (5.127)$$

- Speed factor concerning wear (W_v): this factor considers the sliding speed between wheels. The factor of wear of materials, the roughness factor and the speed factor are elasto-hydro-dynamic factors that affect the coefficient of friction causing overheating and wear. The speed factor can be determined as follows:

$$W_v = 4\sqrt{\frac{n_1(v_{go} + v_{gm}^{1.5})}{u\ v_{gm}}} \qquad (5.128)$$

The value of v_{go} is defined in relation to materials and lubricants. For hardened steel and wheels made of bronze CuSn12 lubricated with a mineral oil the value of such factor is equal to 0.1. For through hardened steel this value is equal to 0.65. If the unit is lubricated with a synthetic oil the values will be 0.10 and 0.85 respectively.

The speed v_{gm} is determined as follows:

$$v_{gm} = v_m \frac{v_{m1}}{\cos \gamma_m} \tag{5.129}$$

$$v_{m1} = \frac{\pi \, d_{m1} \, n_1}{60000} \tag{5.130}$$

The safety factor related to wear is:

$$S_W = \frac{\sigma_{WP}}{\sigma_H} \tag{5.131}$$

We remember that the worm diameter must be sufficiently large so that the worm shaft could resist to the stresses generated and the worm deflection must be limited (see chapter 6 - shafts).

5.6.4 Tooth Strength

The factor preventing teeth from breaking in the presence of shearing is calculated approximately. We calculate a factor U given by:

$$U = \frac{F_{tm1} \, K_A}{m \, b_2} \tag{5.132}$$

for bronze G-CuSn 12 we have $U_{\lim} = 115$
This safety factor is given by:

$$S_F = \frac{U_{\lim}}{U} \tag{5.133}$$

We remember that the worm diameter must be sufficiently large so that the worm shaft is able to resist the stresses generated (we will deal with this problem when we will study shafts).

5.6.5 Standard Information

The calculation shown in the paragraph above is the most current method standardized on a national scale. However, there are other standards that differ from this one. They do not consider contact pressure as one of the causes generating pitting, but as one of the causes generating wear. Such standards do not permit engineers to define the degree of wear that the units must bear. There is still no international standard dealing with this matter. An ISO working group of TC 60 is carrying out such a study. The method proposed by the working group is a kind of thesis. However, conclusions cannot yet be drawn as discussions are still at the beginning. The members forming this team have agreed on a subject: they will consider the resistance to pitting, to wear and material strength at the root of the tooth. The existing standards also do not make any difference between toroidal wheels and cylindrical gears. Cylindrical gears are less resistant to pitting and to wear than toroidal wheels as the surface of contact in cylindrical gears is smaller. The new ISO standard will also deal with this difference taking into consideration the existence of contact curves and their length.

5.6.6 Reversing and Efficiency

On the basis of the helix angle, the unit worm-worm wheel can be more or less reversible. Static reversing can easily be determined as it only depends on the angle.

Dynamic reversing depends on many factors: speed, efficiency and the vibrations generated by the load. Non-reversing ends with the abrupt stopping of the unit when the worms stops working, causing overloads if the torque limiter is not applied. We can assume the following limits for reversing as a function of the helix inclination angle of the worm:

> $25°$ Total reversing
from $12°$ to $25°$ Statically reversible
                            Dynamically reversible
from $8°$ to $12°$ Variable static non-reversing
                            Dynamically reversing
from $5°$ to $8°$ Statically non-reversible
                            Bad dynamic reversing
from $3°$ to $5°$ Statically non-reversible
                            Quite null dynamic reversing
from $1°$ to $3°$ Statically non-reversible
                            Quite null dynamic reversing

The efficiency of transmission by means of worms and worm wheels can be deduced from the equation for the sloping plane.

For the driving worm, if the tangential force at the worm resulting from the applied torque is F_v, the force on the wheel providing the output torque will be:

$$F_r = \frac{F_v}{\tan(\gamma_m + \phi)} \qquad (5.134)$$

if the friction is null, this force will be given by:

$$F'_r = \frac{F_v}{\tan \gamma_m} \qquad (5.135)$$

The efficiency is the relation of the real force with friction to the theoretical force without friction:

$$\eta = \frac{F'_r}{F_r} = \frac{\tan \gamma_m}{\tan(\gamma_m + \phi)} \qquad (5.136)$$

for the driven worm:

$$F_v = F_r \tan(\gamma_m - \phi) \qquad (5.137)$$

$$F'_v = F_r \tan \gamma_m \qquad (5.138)$$

$$\eta = \frac{F'_v}{F_v} = \frac{\tan(\gamma_m - \phi)}{\tan \gamma_m} \qquad (5.139)$$

6 Shafts

6.1 Definition and Function

Shafts are the materialization of the rotating axes of gears; they are generally made by cylindrical elements or truncated cones and they are used in order to position wheels. They transmit the stresses to the supports where reactions are created and they transmit the torques to or starting from gears.

6.2 Shafts and Forces in Gears

We said that teeth generate stresses that are tangential, radial or axial in relation to wheels. The tangential and radial stresses are radial to the shaft and they create stresses due to bending.

Axial stresses are axial to shaft but they are out of centre of half the diameter of the driven gear on the shaft. Therefore bending moments are applied to the shaft. Tangential stresses generate a torque equal to the torque transmitted. Axial stresses have a reaction on the axis of the shaft, where they produce compressive stresses and tensile stresses having the same direction as bending stresses. Sometimes shafts are fixed to elements by pressure (see shrinking). Consequently stresses perpendicular to those generated by bending are produced in shafts. Such stresses result from compression. Therefore shafts must bear bending stress, torsional stress ,compressive stress or tensile stress having the same direction of bending or compression which is perpendicular to bending.

6.3 Study of the Theoretical Stresses

6.3.1 Bending

Plane bending. Plane bending occurs when all the forces applied to the shaft are on the same plane which is axial to the shaft. We can distinguish: a force concentrated between the two supports (Fig. 6.1):

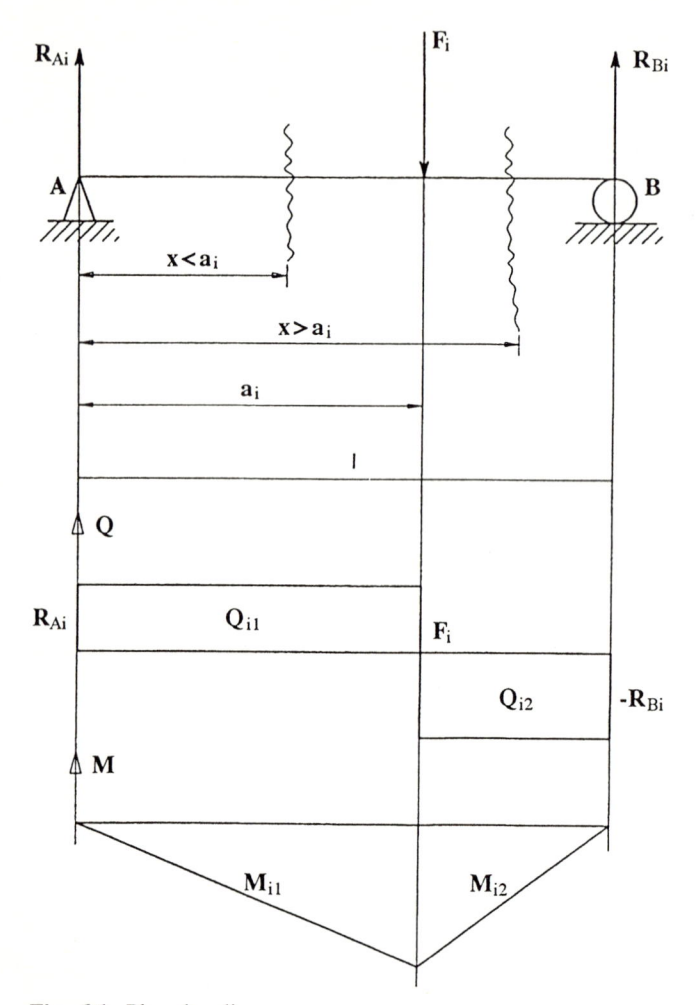

Fig. 6.1. Plane bending

Let us consider a force placed at a distance a_i from the left supports. Such force creates a reaction in A which is given by:

$$R_{Ai} = -F_i \frac{\ell - a_i}{\ell} \tag{6.001}$$

On the basis of the position of the section considered two equations relevant to bending torques are provided (x is the position of the section in relation to the bearing marked with the letter A):

a) If $x < a_i$:

$$M_{i1} = R_{Ai}\, x \ . \tag{6.002}$$

b) If $x > a_i$:

$$M_{i2} = R_{Ai} x + F_i (x - a_i) . \quad (6.003)$$

The shearing stress will be expressed on the basis of the same conditions:
If $x < a_i$:

$$Q_{i1} = R_{Ai} \quad (6.004)$$

and if $x > a_i$:

$$Q_{i2} = R_{Ai} + F_i . \quad (6.005)$$

- Multiple concentrated forces. If multiple forces are applied to the shaft, shaft will be influenced by all these forces $F_1, F_2,...F_n$ are the forces and $a_1, a_2, ...a_n$ the respective distances; the bearing reaction will be given by:

$$R_A = \sum_1^n R_{ai} . \quad (6.006)$$

In a section distant from x, the bending moment is given by the sum of the force moments placed on the left side of the section.
F_1 to F_j are the forces so as the distance a_i are lower than x; it will result that:

$$M = R_A x + \sum_1^j F_i (x - a_i) = \sum_1^j M_{i2} + \sum_{j+1}^n M_{i1} . \quad (6.007)$$

In the same section the value of the shearing load is:

$$Q = R_A + \sum_1^j F_i . \quad (6.008)$$

- Distributed force among supports (Fig. 6.2).
A distributed force is characterized by a force per unit of length q_i on a distance that goes from the abscisse a_i to the abscisse b_i. The left support reaction is the same as that generated by a concentrated force $F_i = q_i (b_i - a_i)$. The bending torque is given by three equations:

a) $x < a_i$ as for a concentrated force F_i which is positioned at half the distance from a_i and b_i
b) $a_i < x < b_i$:

$$M_{i2} = R_A x + q_i \frac{(x - a_i)^2}{2} . \quad (6.009)$$

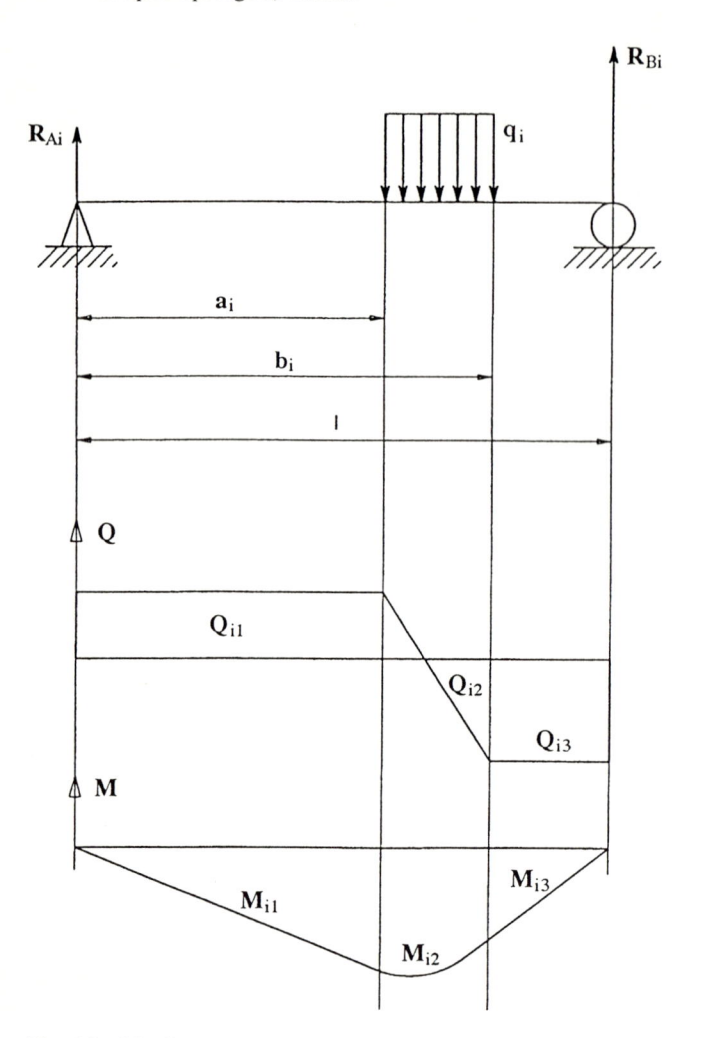

Fig. 6.2. Distributed forces

c) $b_i < x$ as for a concentrated force positioned at half the distance from the two extremities of the concentrated force.

Shearing stresses are also defined by three equations, where the two extremes are equal to those defined for a concentrated load as the median line is given by:

$$Q_{i2} = R_{Ai} + q_i (x - a_i) \ . \qquad (6.010)$$

- Multiple distributed forces. The equation relevant to multiple distributed forces are those resulting from the superposition of the equations defined for a unique distributed force considering the right application of equations of the shearing moments and stresses.

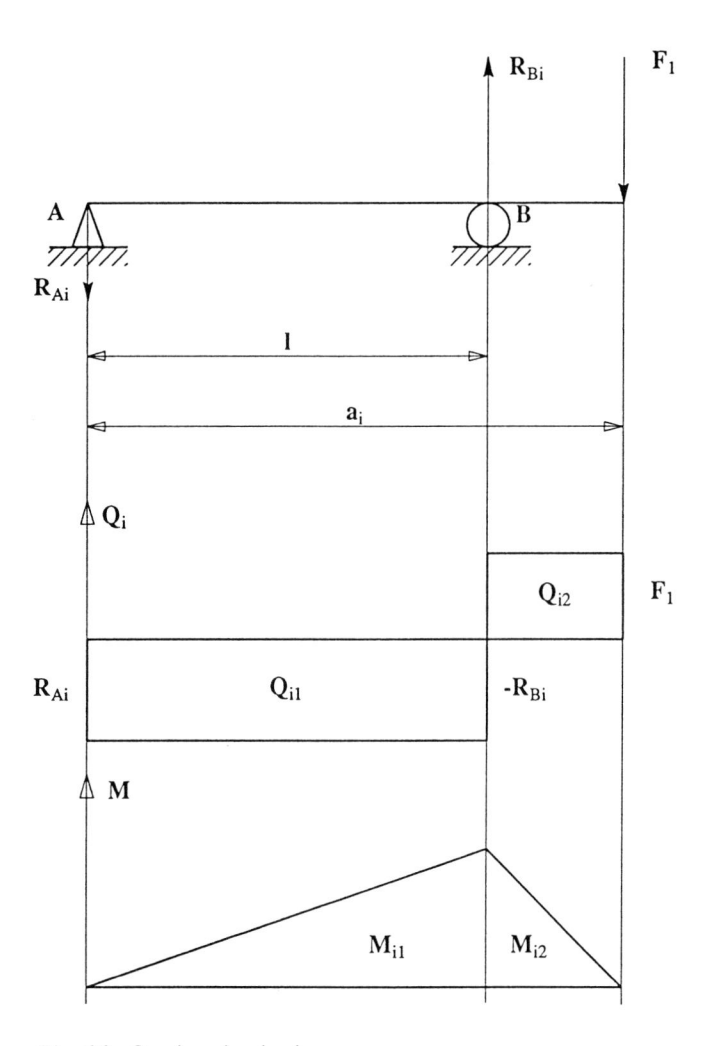

Fig. 6.3. Overhanging load

If a distributed force falls outside the section, concentrated forces occur, as they represent the resultant of each distributed force applied to the centre of the length of the distributed force. If a distributed force whose limits are a_j and b_j falls inside the section, the sum of the moments of the resulting concentrated forces (apart from the first one which is increased by M_{j2}) will represent the value of the torque (calculated by (6.009) with j instead of i).

- Overhanging load (Fig. 6.3)

Let us consider overhanging load at the right side of a support B. The following reactions will take place:

$$R_{Bi} = -F_i \frac{a_i}{l} \qquad (6.011)$$

$$R_{Ai} = -(R_{Bi} + F_i) \quad . \qquad (6.012)$$

The value of the bending moment between the two supports is:

$$M_{i1} = R_{Ai} \, x \qquad (6.013)$$

and

$$M_{i2} = R_{Ai} \, x + R_{Bi} \, (x - l) \quad . \qquad (6.014)$$

The shearing loads are the support reaction in A and the sum of the two support reactions respectively.

The bending moments resulting from the combination of the different cases are given by the sum of the partial torques. As these torques are determined in any section, a stress will occur in the section having a diameter d, whose value is:

$$\sigma_{b0} = \frac{32 \, M}{p \, d^3} \approx \frac{M}{0.1 \, d^3} \qquad (6.015)$$

the stress calculated is the maximum allowable stress that the section can bear.

Such stress is positive in a fibre and negative in another one and it corresponds to compression in a fibre and to traction in the other one. Such fibers are located at the periphery of the shaft (near the two extremities of a diameter). The stress diminishes when we penetrate below the tauter fiber to become null on the perpendicular diameter. If the shaft turns in relation to the load, the same point belonging to the shaft passes alternatively from traction to compression and vice versa.

Deviated Bending. Deviated bending occurs when the forces that are applied to the beam are not on the same plane. We project forces on arbitrary perpendicular planes and we consider the plane bending characterizing each plane together with its forces. Two bending moments M_y et M_z are obtained in the section. They are perpendicular one to the other. The resulting torque is given by:

$$M = \sqrt{M_y^2 + M_z^2} \quad . \qquad (6.016)$$

The stress will be calculated following the equation (6.015).

6.3.2 Torsion

If T is the torque on a section of the shaft (torque transmitted by the shaft) and d is the diameter of the section, it will result that:

$$\tau_0 = \frac{16 \, T}{p \, d^3} \approx \frac{T}{0.2 \, d^3} \quad . \qquad (6.017)$$

6.3.3 Normal Stress

If N is the stress perpendicular to the section having a diameter d, it will result that:

$$\sigma_{n0} = \frac{4N}{p\,d^2} \approx \frac{N}{0.8\,d^2} \quad . \tag{6.018}$$

6.3.4 External Compression

The compressive stress outside the shaft is equal to the pressure exerted by a fitted solid by interference.

6.4 Real Stresses (Stress Concentration Factor - Form Factor)

Stresses given by the resistance of material (they are theoretical stresses) do not represent the real values characterizing the a.m. section. Indeed such stresses are influenced for each stress by the solids shape. Therefore we are obliged to define a form factor Y_{Fk} where k represents the various types of stresses (b represents bending, t represents torsion while n stands for normal stress and d for compression). Shafts do not have a constant section. This feature depends on their construction.

For each variation of section supplementary stresses develop. They are known as stress concentration. It is possible to define a stress concentration factor Y_S that is given by the relation of the real maximum allowable stress in a given section to the calculated stress.

We will study the values of the factor Y_S in another paragraph. The real stress is given by:

$$\sigma_k = \sigma_{k0}\, Y_{Sk}\, Y_{Fk} \quad . \tag{6.019}$$

The index k represents bending ($k=b$), the normal stress ($k=n$) or the compressive stress ($k=d$).

For torsion:

$$\tau = \tau_0\, Y_{Ft}\, Y_{St} \quad . \tag{6.020}$$

6.5 Static Allowable Stress

6.5.1 Static Breaking Stress due to Traction

Tensile tests can be easily carried out on laboratory machinery. In the first chapter we examined the procedure to be followed to determine the breaking stress of a given material. Steels can be classified following three categories: structural steel, treatment steel and case hardening steel.

Structural steels (which are not frequently used for the construction of gearboxes) included carbon steels B60 m (0.6% of C) whose breaking point ranges from 590 MPa to 710 MPa.

To obtain treatment steel generally it is used to be alloy steels to which Chrome-Nickel are added (the 34CrNIMo 6 contains 0.38% of C, 1% of Cr and Ni as well as 0.2% of Mo). This steel has a breaking point that ranges from 980 to 1180 MPa. Among case hardening steels there is the 16 MnCr 5 (0.16% of C, 1.15% of Mn and 0.95% of Cr) whose breaking point varies from 790 to 1080 MPa and the 18 NiCrMo 6 (0.18% of C, 1.55% of Ni and Cr as well as 0.2% of Mo) whose breaking point ranges from 1080 to 1280 MPa.

This stress is represented by σ_R.

For shearing (Torsion) the breaking stress is linked to the tensile stress on the basis of the following relation:

$$\tau_R = \frac{\sigma_R}{\sqrt{3}} \quad . \tag{6.021}$$

6.5.2 Elastic Limit (Factor Y_E)

The elastic limit of steel is important to the resistance of the material. The elastic limit is linked to the breaking stress by the following relation:

$$\sigma_E = \sigma_R \, Y_E \quad . \tag{6.022}$$

The factor Y_E is the elastic limit factor and it is provided by the table 6.1, here below:

Table 6.1. Values of Y_E

Type of steel	Traction
Structural steel	0.6
Treatment steel	0.7
Case hardening steel	0.7

6.5.3 Allowable Static Stress

The allowable static stress is the elastic limit beyond that permanent distortions occur in material which are in contrast with the mechanical functions of the gear.
The allowable static stress is expressed by:

$$\sigma_{P\,stat} = \sigma_R \, Y_E \quad . \tag{6.023}$$

6.6 Allowable Stress for an Unlimited Duration

6.6.1 Endurance (Endurance Factor Y_C)

If a cylindrical polished testbar with a constant or variable section undergoes a periodical high number of cycles of tensile-compressive stress, breaking occurs when the value of the stress is lower than that of the static breaking stress. In reality the result relevant to a great number of tests carried out on many testbar shows a sort of statistical dispersion and it is possible to determine a value providing a given breaking percentage. If we choose 10% as breaking risk value we will find a value σ_{D-1} corresponding to a probable breaking. We could define another value corresponding to a different breaking percentage.

We choose a value related to a probable breaking risk of 10% and we indicate it with σ_D. Such value is linked to the type of the periodic stress, i.e. to the value k which has already been defined in the first chapter. The value of such stress can be defined by a fraction of the static breaking stress. Such fraction too depends on the type of stress then on k. The resulting stress is called endurance stress and it is expressed as follows:

$$\sigma_D = \sigma_R \, Y_C \tag{6.024}$$

where Y_C (the endurance factor) is given by the following equations and by the table 6.2:

Table 6.2. Endurance factor

Type of steel	K1	K2	K3
Structural steel	0.7	0.45	0.675
Treatment steel	0.432	0.41	0.697
Case hardening Steel	0.6	0.4	0.64

For $-1 < k < 0$
$$Y_C = K_3 (1 + k) - K_2 k \quad . \quad (6.025)$$

For $0 < k < 1$
$$Y_C = K_3 \left(1 - \frac{k}{K_1}\right) + \frac{k}{K_1} \quad . \quad (6.026)$$

The value of Y_C is limited to 1.

6.6.2 Factor Modifying Endurance

Reliability factor (Y_ϕ). We said that endurance is defined by a probable breaking risk of 10% (i.e. a reliability R equal to 0.90). Given a different degree of reliability it is necessary to correct the endurance limit on the basis of a factor Y_ϕ given by:

$$Y_\phi = 1.25 (1 - R)^{0.1} \quad (6.027)$$

where R is the desidered reliability factor.

Roughness factor (Y_R). Endurance is calculated on a polished testbar. If the shaft roughness is different from the testbar roughness, endurance too will be different. A modification factor Y_R must be adopted. It is given by:

$$Y_R = A - B \sigma_R \quad . \quad (6.028)$$

In case roughness is equal to $R_t > 1 \mu m$.

$$A = 1.025 - 1.252 \cdot 10^{-3} R_t \quad (6.029)$$

$$B = (0.8472 + 3.82 \cdot 10^{-2} R_t) 10^{-4} \quad . \quad (6.030)$$

In case roughness is equal to $R_t < 1$ μm we have

$A = 1$
$B = 0$.

If roughness is unknown, it is possible to use the values shown on the table 6.3 that are provided according to the type of working, as the values listed in the last column represent the average value of roughness R_t.

Table 6.3. Constant elements for the factor Y_R

Type of working	A	B	Rt
Fine lapping	1	0	<1
Fine grinding	1,024	0,0000885	1
Grinding	1,022	0,0000943	2,5
Fine working	1,017	0,0001088	6,3
Fine roughing	1,005	0,0001458	16
Roughing	0,946	0,0003254	63
Accurate forging and lamination	0,915	0,0004208	80
Normal forging or rolling or fine sands casting	0,904	0,0004553	100

Dimension factor (Y_X). The dimension of the testbar differs from that of the shaft calculated. Dimensions affect resistance which is taken into account by the dimension factor Y_X. This factor depends on the diameter (mm) of the shaft in the section calculated as follows:

$$Y_X = 1 - 0.2 \log \frac{d}{10} \ . \qquad (6.031)$$

6.6.3 Allowable Stress for Unlimited Life Duration $\sigma_{P\lim}$

On the basis of what we said the allowable stress for an unlimited duration can be expressed as follows:

$$\sigma_{P\lim} = \sigma_R \ Y_E \ Y_C \ Y_\phi \ Y_R \ Y_X \ . \qquad (6.032)$$

This value is valid for the application of a high number of stress cycles.

6.7 Allowable Stress (σ_P)

The static allowable stress is valid for a number of cycles going to $5 \cdot 10^3$.
The allowable stress for an unlimited life duration is valid starting from $5 \cdot 10^6$.
The allowable stress varies in function of the number of stress cycles.
The resulting curve is S-N curve (Fig. 6.4).
The value of the allowable stress is:

For $N_L < 5 \cdot 10^3$ $\sigma_P = \sigma_{P\,stat}$ See the equation (6.023)

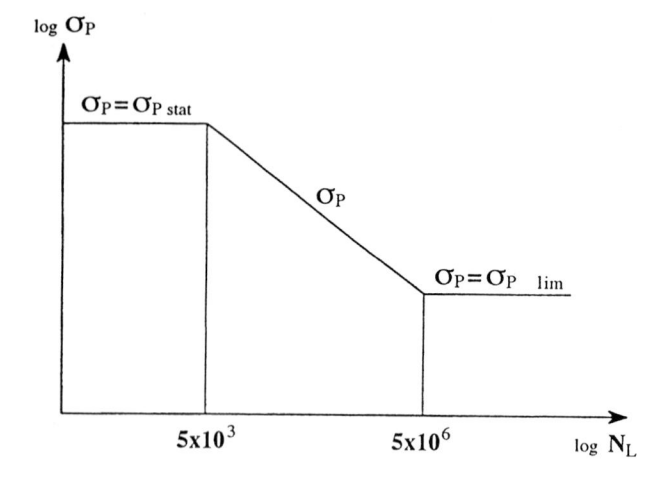

Fig. 6.4. Woehler's curve for shafts

For $5 \cdot 10^3 < N_L < 5 \cdot 10^6$

$$\sigma_P = \sigma_{P\,stat} \left(\frac{N_L}{5 \cdot 10^3} \right)^{\frac{1}{p}} \qquad (6.033)$$

with:

$$p = - \frac{3}{\log \dfrac{\sigma_{P\,stat}}{\sigma_{P\,lim}}} \qquad (6.034)$$

For $5 \cdot 10^6 < N_L$ $\quad \sigma_P = \sigma_{P\,lim}$ See the equation (6.032)

6.8 Stress Concentration Factor (Y_β, Y_α)

6.8.1 Definition

In the paragraphs above we said that shape or section variations or shearing stresses create further stresses that add to the theoretical stresses calculated. At the beginning this consideration was just empirical as it was remarked that mechanical components break because of the presence of allowable stresses calculated where cutting effects occurred. Experimental studies which were carried out by photo-elasticity enabled to list the supplementary stresses. Analytical studies which have been recently carried out (for instance finished element calculation) allowed experts to improve their knowledges. Finished element calculations would surely provide better result but they require a lot of means (powerful computers) and they involve various problems (computer training taking a long time and difficulties in finding the limit conditions required by this calculation). Therefore it is preferred to use empirical formulas presenting an easier and quicker level of application. The stress concentration factor can be defined as the relation generated by shape distortion to the stress theoretically calculated in that same point. There is a static stress concentration factor which is represented by $Y_{\alpha k}$ and an allowable stress concentration factor (for a high number of load cycles which are periodically applied). Such factor is represented by $Y_{\beta k}$. A k index is adopted for each torsional stress, bending stress or normal stress. Factors relevant to these stresses show different values.

6.8.2 Notch Sensibility Factor (ρ^*)

Each material has a notch sensibility depending on its static breaking stress σ_R which is given by:

$$\rho^* = 0.054 - 6 \cdot 10^{-6} (\sigma_R - 300) \ . \tag{6.035}$$

6.8.3 Notch Gradient ($s_{\sigma k}$)

In proximity to the modification of the section of a solid, the stress varies with the distance x from where modification takes place. The cutting gradient is the relation between the tangent of the angle forming the tangent at the stress variation according to the distance of x and the maximum stress causing the perturbation. This gradient is represented by $s_{\sigma k}$, (k indicates the type of stress). Such factor is a typical characteristic of the perturbation of the section as the allowable stress concentration factor.

6.8.4 Notch Factor ($v_{\sigma k}$)

This factor is given by the ratio of the static stress concentration factor to the limit stress concentration factor. It will result that:

$$Y_{\alpha k} = v_{\sigma k} \ Y_{\beta k} \tag{6.036}$$

and

$$v_{\sigma k} = 1 + \sqrt{\rho^* \ s_{\sigma k}} \ . \tag{6.037}$$

6.8.5 Values of the Allowable Stress Concentration Coefficient and Notch Gradient

Cylindrical Shaft Without Notches. In this case we have $Y_\beta = 1.00$ and $S_{\sigma k} = 0,000$

Cylinder with Splines (or more than two Keyways for Keys).
- During bending

$$Y_{\beta b} = 3.9 \cdot 10^{-7} \sigma_R^2 + 4.9 \cdot 10^{-5} \sigma_R + 1.5 \quad . \quad (6.038)$$

- During torsion

$$Y_{\beta t} = 3.25 \cdot 10^{-7} \sigma_R^2 + 4.1 \cdot 10^{-5} \sigma_R + 1.3 \quad . \quad (6.039)$$

- During traction-compression

$$Y_{\beta n} = 1.00 \quad .$$

In all these cases:

$$S_{\sigma k} = \frac{2}{d} \quad (6.040)$$

where d is the external diameter of the shaft.

Cylinder with one or two Longitudinal Grooves. (b = bending, t = torsion, n = traction-compression)

For a keyway executed by means of thread milling machine:

$$Y_{\beta b} = 2.8 \cdot 10^{-7} \sigma_R^2 + 2.48 \cdot 10^{-4} \sigma_R + 1.7 \quad (6.041)$$

$$Y_{\beta t} = \frac{Y_{\beta b}}{1.2} \quad (6.042)$$

$$Y_{\beta n} = 1.00 \quad (6.043)$$

$$S_{\sigma k} = \frac{2}{d} \quad (6.044)$$

d is the external diameter of the shaft and $k = b$, t or n respectively.

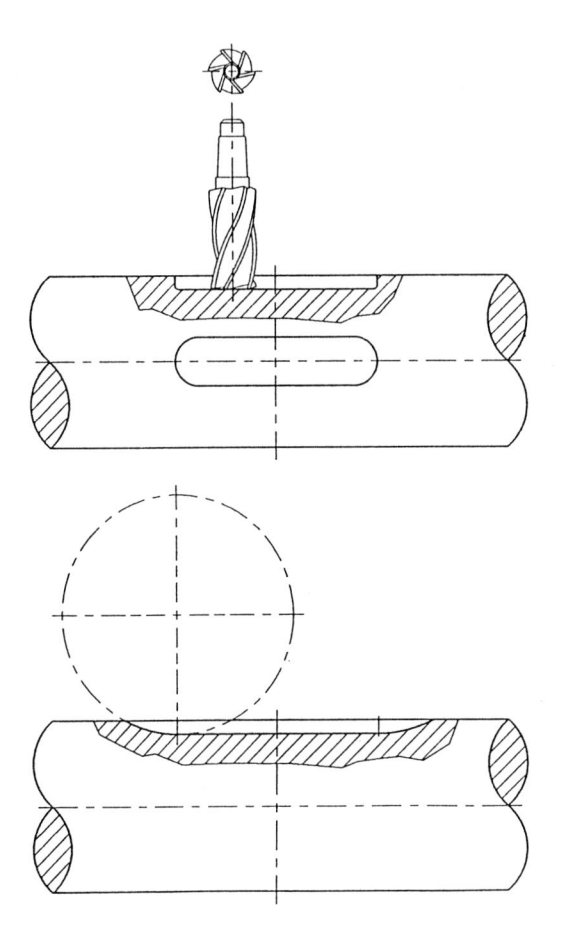

Fig. 6.5. Keyways for keys

For another kind of milling:

$$Y_{\beta b} = 2.80 \cdot 10^{-7} \sigma_R^2 + 2.48 \cdot 10^{-4} \sigma_R + 1.5 \qquad (6.045)$$

$$Y_{\beta t} = \frac{Y_{\beta b}}{1.2} \qquad (6.046)$$

$$Y_{\beta n} = 1.00 \qquad (6.047)$$

$$S_{\sigma k} = \frac{2}{d} \cdot \qquad (6.048)$$

Cylinder with Semicircular Grooves[1]

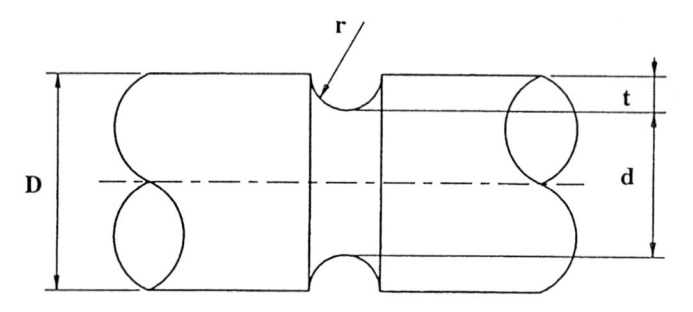

Fig. 6.6. Groove with semicircular profile

During bending $(k=b)$ $K_1 = 0.715$ $K_2 = 2$
During torsion $(k=t)$ $K_1 = 0.365$ $K_2 = 1$
During traction $(k=n)$ $K_1 = 1.197$ $K_2 = 1.871$

$$K_p = \sqrt{\frac{t}{r} \frac{d/D}{1-d/D} + 1} - 1 \qquad (6.049)$$

$$K_q = \sqrt{\frac{t}{r}} \qquad (6.050)$$

$$S_{\sigma b} = \frac{2}{r} + \frac{2}{d} \qquad (6.051)$$

$$S_{\sigma t} = \frac{1}{r} + \frac{2}{d} \qquad (6.052)$$

$$S_{\sigma n} = \frac{2}{r} \qquad (6.053)$$

$$\beta = \sqrt{\left(\frac{1}{K_1 K_p}\right)^2 + \left(\frac{1}{K_2 K_q}\right)^2} \qquad (6.054)$$

$$Y_{\beta k} = \frac{1+\beta}{\beta} \qquad (6.055)$$

---

[1] H. NEUBERT - Kerbspannungslehre (Springer - 1958)
R.E. PETTERSON (1974), HOTTENROTT (1952)

Cylinder with Single Shoulder[2]

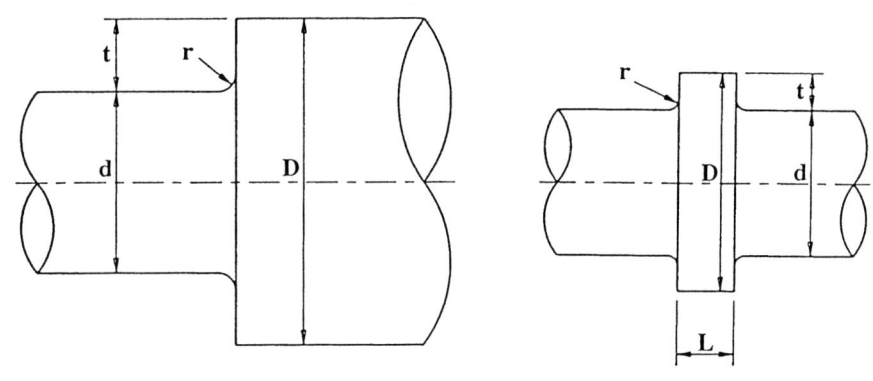

Fig. 6.7. Single shoulder and double shoulder

As per half-circular groove, but with:

$k=b$	$K_1=0.541$	$K_2=0.843$
$k=t$	$K_1=0.263$	$K_2=0.843$
$k=n$	$K_1=0.880$	$K_2=0.843$

$$S_{\beta b} = \frac{2}{r} + \frac{4}{d+D} \quad (6.056)$$

$$S_{\sigma t} = \frac{1}{r} + \frac{4}{d+D} \quad (6.057)$$

$$S_{\sigma n} = \frac{2}{r} \quad (6.058)$$

Double Shouldering Cylinder. If $L > 2d$, calculation must be done following the prescriptions indicated in the paragraph above; otherwise you must choose:

$$D = d + 0.3 \ L \ .$$

---

[2] Bending: M.M. FROCHT (1966), I.H. WILSON e D.J. WHITE (1973), I.M. ALLISON (1952).
 Torsion: E. HOTTENROTT (1952), M. FESSLER, C.C. ROGERS and J.P.STANLEY (1969).
 Traction: LIPSON and JULINVALL (1963), CHIH BING LING (1968), M. FESSLER, C.C. ROGERS and J.P. STANLEY.

6.8.6 Static Stress Concentration Factor (Y_α)

Using the values given by the paragraph above (allowable stress concentration factor and cutting gradient) it is possible to calculate the static stress concentration value as follows:

$$v_{\sigma k} = 1 + \sqrt{\rho^* \, s_{\sigma k}}$$

$$Y_{\alpha k} = Y_{\beta k} \, v_{\sigma k} \quad .$$

6.9 Form Factor (Y_F)

The allowable stress during bending or torsion is different from the allowable stress generated by static traction. Such empirical difference is considered by the form modification factor. Such factor depends on the form of the section (in shafts this section is circular), on the static stress concentration factor concerning the section and on the static tensile stress of the material. It will result:

$$Y_{Fk} = 1 + 0.75 \left(\frac{C}{Y_{\alpha k}} - 1 \right) \left(\frac{500}{\sigma_R} \right)^{0.25} \qquad (6.059)$$

where:

$C=1.7$ for a bent circular section
$C=1.3$ in case of torsion

in case of traction the value of the shape factor is equal to 1. If the calculation (see 6.059) provides a result lower than 1, we will adopt the value equal to 1.

6.10 Real Stress (σ_k)

The real stress depends on the theoretical stress and on the shape factor as well as on stress concentration. As the value of stress concentration varies with the number of cycles, the real stress depends on the number of applied cycles. The variation of the concentration stress factor varies from the static value to the allowable value σ_{Pk} following S-N curve.

For $N_L < 5 \cdot 10^3$

$$\sigma_{Pk} = \frac{\sigma_R \, Y_E \, Y_{Fk}}{Y_{\alpha k}} \qquad (6.060)$$

for $5 \cdot 10^3 < N_L < 5 \cdot 10^6$

$$\sigma_{Pk} = \frac{\sigma_R Y_E Y_{Fk}}{Y_{\alpha k}} \left(\frac{N_L}{5 \cdot 10^3}\right)^{\frac{1}{3} \log v_{\sigma k} y_R y_x y_c y_\phi} \quad (6.061)$$

for $5 \cdot 10^6 < N_L$:

$$\sigma_{Pk} = \frac{\sigma_R Y_E Y_{Fk} Y_R Y_x Y_c Y_\phi}{Y_{\beta k}} \quad (6.061\text{bis})$$

6.11 Safety Factors

6.11.1 Simple Stress

For each simple stress, the safety factor is given by the quotient of the allowable stress and the real stress:

$$S_k = \frac{\sigma_{Pk}}{\sigma_k} \quad (6.062)$$

We underline the fact that the index k is replaced by b in case of bending and by n in case of traction-compression.

In case of torsion:

$$S_k = \frac{\tau_P}{\tau} \quad (6.063)$$

6.11.2 Stress Reduced to Reference Stress

The allowable stress varies according to the type of solicitations applied to the shaft in a given section. In order to compare real stresses, we choose a unique arbitrary reference stress (σ_{ref}). The stress reduced to a reference stress is a theoretical stress providing the same degree of reliability with regards to reference stress as that provided by the real stress with regard to the allowable stress:

$$S_k = \frac{\sigma_{Pk}}{\sigma_k} = \frac{\sigma_{ref}}{\sigma_{red}} \quad (6.064)$$

It results that:

$$\sigma_{red} = \frac{\sigma_{ref}}{S_k} \quad . \quad (6.065)$$

In case of torsion:

$$S_t = \frac{\tau_P}{\tau} = \frac{\tau_{ref}}{\tau_{red}} \quad . \quad (6.066)$$

Thanks to the relation existing between the static tensile stress and the static shearing stress (torsion) we can affirm that:

$$\tau_{red} = \frac{\sigma_{ref}}{\sqrt{3}\, S_t} \quad .$$

$$(6.067)$$

6.11.3 Composition of Stresses Following the same Direction

Stresses that have the same direction can be algebraically added up. However, if we want to determine the safety factor for all the stresses, these stresses could not be compared with allowable stresses, as they varies according to the type of solicitation. To compare the sum with a common value, we must consider the reduced stresses. The resulting stress can be compared with the reference stress. It results that:

$$\sigma_{red\,x} = \sum_{i=1}^{n} \sigma_{red\,i} \leq \sigma_{ref} \quad . \quad (6.068)$$

According to the definition of reduced stress:

$$\frac{1}{S_x} = \sum_{i=1}^{n} \frac{1}{S_{xi}} \leq 1 \quad . \quad (6.069)$$

6.11.4 Stresses Having any Direction

A real solid is often characterized by a distributed stresses. The Figure 6.8 shows the stress distribution on a point of a section belonging to a shaft.

The stress σ_r represents all the stresses radial to the shaft on this point while the stress σ_a represents all the axial stresses in the same point. Following what we said in the previous paragraph we can affirm that these stresses are reduced stresses resulting from the sum of the different reduced stresses.

The stress τ represents the reduced stress resulting from torsional stresses.

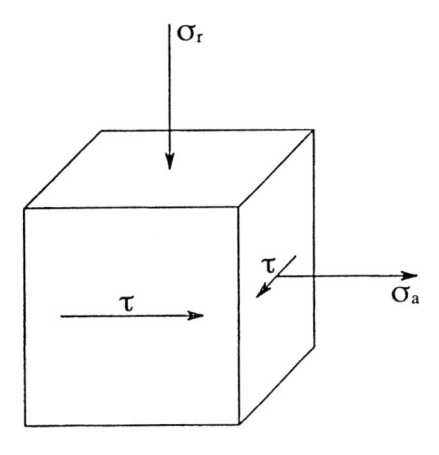

Fig. 6.8. Stresses on a Point

The comparison stress is given by HENCHY-VON MISES's principle:

$$\sigma_{red} = \sqrt{\left(\frac{\sigma_{ref}}{S_r}\right)^2 + \left(\frac{\sigma_{ref}}{S_a}\right)^2 - \left(\frac{\sigma_{ref}}{S_r}\right)\left(\frac{\sigma_{ref}}{S_a}\right) + 3\left(\frac{\sigma_{ref}}{\sqrt{3}\,S_t}\right)^2} \quad (6.070)$$

or

$$\frac{1}{S} = \sqrt{\left(\frac{1}{S_r}\right)^2 + \left(\frac{1}{S_a}\right)^2 - \left(\frac{1}{S_r}\right)\left(\frac{1}{S_a}\right) + \left(\frac{1}{S_t}\right)^2} \quad . \quad (6.071)$$

In shafts radial compressive stresses can be generated (interference mounting)

$$\frac{1}{S_r} = \frac{1}{S_{interference}} \quad .$$

This value will be null if there is no interference in correspondance to the point considered. The axial stress results from the stress produced by bending and from stress resulting from an axial stress load.

Given an index b for bending and an index n for the axial stress:

$$\frac{1}{S_a} = \frac{1}{S_b} + \frac{1}{S_n} \quad .$$

In the considered section there will be elastic stability of shaft if:

$$S \geq S_{lim} \quad . \tag{6.072}$$

6.11.5 Stresses Generated by a Spectrum of Loads

If a section must bear a stress caused by a spectrum of loads, it will be possible to calculate the value of the whole considering that this stress results from the application of an equivalent single load (Miner's rule).

The stresses caused by a spectrum of loads must be considered only if the stress is periodic. It is the case of a stress resulting from a periodically variable torque or of a stress due to bending under a non-periodical torque but on a shaft that turns in relation to the load. In this last case the bending stress passes from the positive tensile stress to a negative compressive stress. If the stress is not periodical, we should apply Miner's rule to the stress which is determined calculating the moment equivalent to the torques forming the spectrum (in case of torsion and bending) and the equivalent force that represents the forces of the spectrum (in case of traction-compression).

$$M_{equ} = \left[\sum_{1}^{j} \frac{\Delta N_i}{N_{equ}} M_i^{p'} \right]^{1/p'} \tag{6.073}$$

with

$$p' = \frac{3}{\log \dfrac{M_\alpha}{M_\beta}} \quad . \tag{6.074}$$

In this relation:

$$M_\alpha = 0.1 \, d^3 \, \sigma_{P\,stat} \tag{6.075}$$

and

$$M_\beta = 0.1 \, d^3 \, \sigma_{P\,lim} \quad . \tag{6.076}$$

If $M_i < M_\beta$ we do not consider this component of the spectrum. We only consider j components whose value is higher than that of the allowable torque.

Such argument is also valid for torsion and traction-compression if modications are performed on the basis of what we said previously.

The resulting equivalent stress is considered as a constant stress.

6.12 Stress Concentration Importance

If we calculate stress concentration on the basis of the empirical formulas mentioned above, we will obtain values that are similar to those obtained using finished elements. Therefore such values would be very near to the physical reality.

Let us take a 100 mm shaft made of steel 18 CrNiMo 6 and let us calculate shape and dimension distortions.

In case of keyways for keys worked by means of a thread milling machine we will have:
the stress concentration static factor = 2.37

$$\text{limit} = 2.30 \ .$$

For another type of milling:
the stress concentration static factor = 2.16

$$\text{limit} = 2.10 \ .$$

In case of grooves the values are 2.12 and 2.06 respectively.

If we consider a 110 mm shouldering (height of the shoulder = 5 mm) the table 6.4 will result as a function of the groove radii between the two capacities of different diameter.

Table 6.4 Stress concentration factor

Radius	Concentration factor		% Safety Factor	
	Static	Limited	Static	Limited
Smooth shaft	1	1	100	100
5	1.94	1.7	51.5	58.8
2	2.45	2.14	40	46.8
1	3.02	2.64	33	37.8
0.1	7.2	6.3	14	15.8

In these examples we notice the importance of these factors that must not be ignored; in addition we notice the importance of the groove radii in relation to the variation of diameter. A bad choise (or a bad execution) of such diameters can provoke a decrease in the safety degree causing undesired breaking.

6.13 Shaft Designing

Ideally the most resistant shaft is without hollows and modifications of diameter. Generally these requirements are not met. There are many imperative reasons that require engineers to modify the shape of the shaft. Let us list the most important ones.

First of all we must consider the working conditions. The different parts forming the shaft have special functions, therefore they require a different degree of precision during manufacturing. The parts built to house bearings or elements with a tendency to shrink require a higher degree of precision than those required by shafts produced to transmit torque. It would be very expensive if all the elements forming a shaft were to show the highest degree of precision. To limit the areas having different tolerances we must attribute to them different diameters having similar values.

Shoulders must be applied to the shaft to delimit the exact place of each element driven on the shaft, to ensure the transmission of axial stresses generated by driven units to the shaft and to avoid any displacement of the driven components as is the case of bearings and elements which have been shrunk.

Keyways and grooves are necessary to transmit the movement to the shaft or to enable it to receive the torques. Keys make disassembling easier while grooves make the mounting, the dismounting and sometimes the axial displacement of the driven unit easier. The designer must take the importance of stress concentration factors into account and reduce them as much as possible and not create the condition for sudden variations in diameter.

If the difference existing between two diameters is too high, variations of the diameters must be progressive. This is particularly suitable for connecting two cylinders having different diameters. In all cases particularly long groove radii must be taken into account. Some units require well defined shoulders which limit the opportunity of creating sufficient groove radii. In this case the best solution is to create a ring that will be placed between the shaft and the component considered. The radius of the shaft's dimensions is based on the resistance in order to form a shoulder having sufficient dimensions.

Narrow keys do not allow sufficiently wide grooves to be made. As in the case of keyways for circlips that are used to fix the shaft components axially.

Such hollows produce large stresses that may possibly be avoided. If they cannot be avoided, it will be necessary to try to concentrate them where the shaft is least stressed (for example near the housings in an area where torsion does not occur) and to place the circlips on the element by means of a free spacer ring. Single hollows are more negative than multiple hollows, which are placed near one another. Where a hollow is required, similar hollows may be used. In this way a high number of stress concentrations to the thread are avoided. It has been observed that stress tension concentrations can also occur near a unit connected to the shaft. To avoid it, even partially, it is advisable to have the hollows on the bore of the unit tightened on the shaft near the ends of the bore.

6.14 Distortion of the Shaft

The resistance allows a theory relative to the distortion of the shafts to be defined. The resulting calculations are often long and complex. An analytical method, derived from the general method, is that used for the original conditions. It will be possible to calculate the distortions occurring at a given point of the axis of the shaft directly, if the conditions of the torque, the shearing stress, the rotation of the section and the bending at the end considered as an origin, which is usually a support, are known. As to shafts, it is always possible to find a simple support at the end of the shaft, ignoring, if necessary, an overhanging element beyond the bearing, by which no stress causing the bending is tolerated. In this case the shearing stress, the bending torque and the bending at the end considered are null. Only the rotation of the section represented by θ_0 is still to be considered. On the side of the section where the origin is placed, the shaft is subjected to forces, the reaction to the bearing included, and to applied outside torque, given P_i et M_i respectively as the forces and the torques occurring on this side of the shaft and given a_i as the distance of the force P_i from the support and b_i as the distance from the torque. If distributed forces are applied, the distributed force will be represented by q_i. The abscissae of the end points of the interval are respectively c_i and d_i.

Subsequently the result will be:

$$A_1 = \Sigma M_i \frac{(x - a_i)}{1!} \tag{6.077}$$

$$A_2 = \Sigma P_i \frac{(x - b_i)^2}{2!} \tag{6.078}$$

$$A_3 = \Sigma q_i \frac{(x - c_i)^3}{3!} \tag{6.079}$$

$$A_4 = \Sigma q_i \frac{(x - d_i)^3}{3!} \tag{6.080}$$

$$B_1 = \Sigma M_i \frac{(x - a_i)^2}{2!} \tag{6.081}$$

$$B_2 = \Sigma P_i \frac{(x - b_i)^3}{3!} \qquad (6.082)$$

$$B_3 = \Sigma q_i \frac{(x - c_i)^4}{4!} \qquad (6.083)$$

$$B_4 = \Sigma q_i \frac{(x - d_i)^4}{4!} \qquad (6.084)$$

The rotation is therefore expressed in the following way:

$$\theta_x = \theta_0 + \frac{1}{EJ}(A_1 + A_2 + A_3 - A_4) \qquad (6.085)$$

and the deflexion is:

$$w_x = \theta_0 x + \frac{1}{EJ}(B_1 + B_2 + B_3 - B_4) \qquad (6.086)$$

with
E = elasticity module of the material.
J = moment of inertia related to the considered section.
$J = \pi d^4/64$.

For the shafts with a variable section, the same formula is applied by adding to each change of section the forces P_{si} and the moments M_{si} given by:

$$P_{si} = Q_{si}\left(\frac{J}{J_{i+1}} - \frac{J}{J_i}\right) \qquad (6.087)$$

$$M_{si} = M_{xi}\left(\frac{J}{J_{i+1}} - \frac{J}{J_i}\right) . \qquad (6.088)$$

Q_{si} being the shearing load and M_{xi} the bending moment of the section where the variation occurs, the moment of inertia J is the moment of inertia of an equivalent shaft with an arbitrary constant section and is used in the equation for bending and rotation, while J_{i+1} and J_i are the moments of inertia of the sections before and after the change. The resulting bending is the bending of the real shaft.

The calculation of the bending occurring in gear units can be interesting; in fact it has been observed that gears were influenced by distortions related to the real contact between the teeth. If the empirical formulas contained in the rules cannot be applied, it will be necessary to calculate such distortions. If longitudinal changes (crowning) are required, it will be also necessary to calculate the form of the distortion to which the torsional distortions are added. Such torsional distortions, for a cylinder with a constant section and a length equal to 1, are given by :

$$\psi_t = \frac{T_i \, l}{G \, I_0} \quad . \tag{6.089}$$

The angle is expressed in radiants, G is the module of transverse elasticity (80000 MPa for steels) and I_0 is the moment of polar inertia of the section ($I_0 = \pi \, d^4/32$).

7 Fits Between the Shaft and the Hub and Fits Between the Hub and the Shaft

7.1 ISO System Concerning Tolerances

The productions of shafts and hubs cannot strictly be made to a precise dimension without causing a rise in the production costs; therefore any variations in the indicated dimension, must be taken into account. The indicated dimension, which acts as a reference dimension, is the nominal dimension. The higher the required degree of accuracy, the smaller are the deviations relative to such dimension, even if a considerable price rise is involved. Another reason where the deviations from the nominal dimension are acceptable is that it is possible to produce connections made up of a shaft and a hub which can be mounted or dismounted or not. There is not a question of accepting tolerances on one of the components, but rather of producing tightened or clearance that can be mounted or dismounted according to working requirements. The tolerances have been standardized in system ISO based on the following principles:

(1) The qualities are classified on the basis of numbers ranging from 1 to 16; the lowest degree of quality corresponds to the highest number. The qualities worth values of 0 and 01 have been produced in case an extremely high degree of accuracy is required.

(2) Diameters are classified according to different intervals. The diameter calculated by the geometrical average of the interval is characterized by the calculation of the tolerances.

(3) For each quality and each interval of the diameters a tolerance interval is established. This interval is given by the difference between the maximum and the minimum dimensions tolerated for each interval of diameter and for each quality. The base of the calculation is represented by the tolerance unit expressed by the following formula:

7 Fits between the Shaft and the Hub and Fits Between the Hub and the Shaft

$$i = 0.45 \sqrt[3]{d_{ref}} + 0.001 \, d_{ref} \, . \tag{7.001}$$

For qualities, the interval of tolerance is a multiple of i (Table 7.1).

Table 7.1. Intervals of tolerance ISO

IT5	IT6	IT7	IT8	IT9	IT10
7i	10i	16i	25i	40i	64i

IT11	IT12	IT13	IT14	IT15	IT16
100i	160i	250i	400i	640i	1000i

The lower deviations are respectively ei and Ei for the shaft and the holes. The upper deviations are respectively es and Es for the shaft and the holes. Deviations are given by the difference between the real and the nominal dimensions.

(4) Lower and upper deviations are placed in relation to a line called line 0, which corresponds to the nominal diameter. The position of such deviations can vary for each quality from an extremely negative position to an extremely positive position. The position of the deviations is indicated by a letter of the alphabet ranging from a to z for shafts, from A to Z for the holes. For borings the deviations indicated by A are in the extreme positive position. The different positions to which precise letters correspond are usually expressed by formula. Nevertheless some peculiarities are observed. The deviation indicated by h for the shaft is positioned in order to make the upper deviation null. The lower deviation is therefore equal to -IT. For borings the position H corresponds to a null lower deviation and to an upper deviation equal to +IT. A particular deviation is that indicated by js. The upper deviation is equal to +0.5 IT and the lower deviation is equal to -0.5 IT.

The rule, however, provides detailed tables for all the deviations. The reference for a given dimension is of type 100 s6 for shafts and 100 H7 for holes. The result is: nominal diameter 100, position s for the shaft or position H for the hole and quality 6 for the shaft or quality 7 for the hole. These references are found in the tables of the standard for all the values of the upper and lower deviations which are (in the example provided):

For the shaft   Lower deviation (ei) = +71 μm
                 Upper deviation (es) = +93 μm
                 IT 6   = 22 μm

For the hole   Lower deviation (Ei) = 0 μm
                 Upper deviation (Es) = +35 μm
                 IT7   = 35 μm

If the two elements (shaft and hole) make up a connection, this latter will be indicated by: 100 H7s6. If tolerances are specified on a plane, it will be necessary that the final dimension of the shaft or of the hole ranges from the values $d + ei$ (or Ei) and $d + es$ (or Es). It is clear that it is more difficult to realize a working of quality 3 than a working of quality 8. The price of the mechanical unit rises as necessity dictates. When defining tolerances on a drawing it is advisable to consider what has been said above and apply precise tolerances only in case of particular working requirements.

For the connection of a shaft - hole the clearance given by the difference between the dimension of the hole and the dimension of the shaft is considered. If e is the real deviation of the shaft and E the real deviation of the hole, such difference will be the real clearance $j = (E - e)$. Therefore it is possible to determine a minimum clearance and a maximum clearance. The minimum clearance is given by the difference between the minimum dimension of the hole and the maximum dimension of the shaft. Consequently the result is:

$$j_{min} = Ei - es.$$

The maximum clearance is the difference between the maximum dimension of the hole and the minimum dimension of the shaft. Consequently the result is:

$$j_{max} = Es - ei.$$

For example, for the hole 100 H7s6 the result is:

$$j_{min} = 0 - 93 = -93 \ \mu m$$

$$j_{max} = 35 - 71 = -36 \ \mu m.$$

In this case its clearance is defined as a negative backlash. Consequently the result is:

$$S_{max} = -j_{min}$$

$$S_{min} = -j_{max}$$

Connections with positive backlash are connections whose elements can move in relation to one anothers. Consequently the rotating and the sliding, which are produced, are of course dismountable. Connections with negative backlash are made up of non-dismountable elements fixed in relation one to another. Connections with a negative minimum backlash and with a positive maximum backlash are uncertain; they are usually dismountable, when it is not possible to ensure the displacement of their elements.

7 Fits between the Shaft and the Hub and Fits Between the Hub and the Shaft   237

The principal intervals of the diameters are indicated in Table 7.2 below:

Table 7.2. Table of the diameters

Above	Up to
10	18
18	30
30	50
50	80
80	120
120	180
180	250
250	315
315	400
400	500

7.2 Measurement of Roughness

Non-perfectly smoothed surfaces can be obtained from mechanical workings.
If a line drawn on the obtained surface is considered, a series of protuberances and hollows will be noticed; they are characterized, however, by lower degrees of intensity than the geometrical deviations of shape. They are measured in microns (μm) and give rise to a particular aspect: roughness. To measure roughness different methods are used, each of which provides a different information.

Maximum roughness (R_t): on a given length (some mm) the highest and the lowest protuberances are determined. The difference between them, expressed in μm, is the maximum roughness R_t (Fig. 7.1 a).

Mean depth of roughness (R_z): the length divided into 5 congruent parts. Subsequently the maximum roughness of each length is measured (R_{ti}) and the average of the five measurements is calculated.

$$R_z = \frac{R_{t1} + R_{t2} + R_{t3} + R_{t4} + R_{t5}}{5} \quad . \tag{7.002}$$

See Fig. 7.1b.
Mean roughness (R_a): arithmetical mean of the ordinates in relation to the mean line of the determined profile.

$$R_a = \frac{1}{\lambda_m} \int_{x=0}^{x=\lambda_m} |y|\, dx \quad . \tag{7.003}$$

See Fig. 7.1c. The mean roughness is also defined as method CLA.

Between the measure R_z and R_a the ratio ranges from 3 to 7. In case of uncertainly a ratio of 6 can be selected.

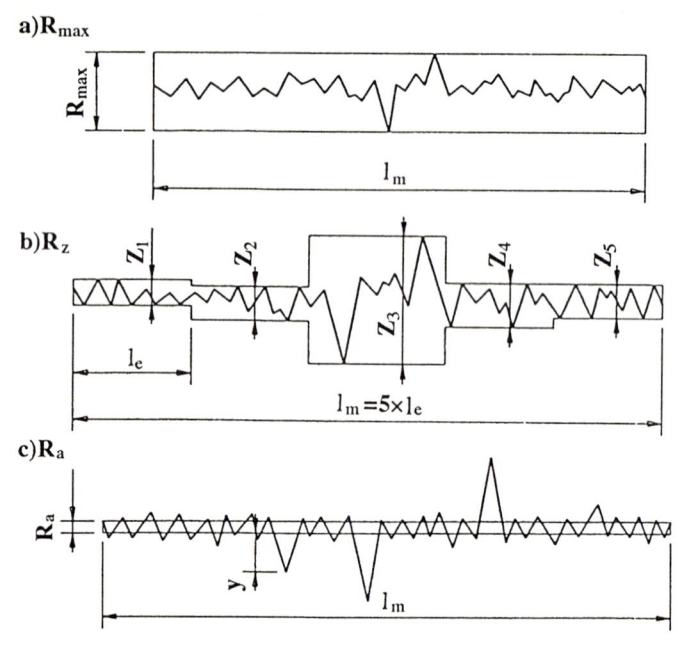

Fig. 7.1 a),b),c) Measure of the Roughness

7.3 Fixing by means of Prismatic Key (Figure 7.2)

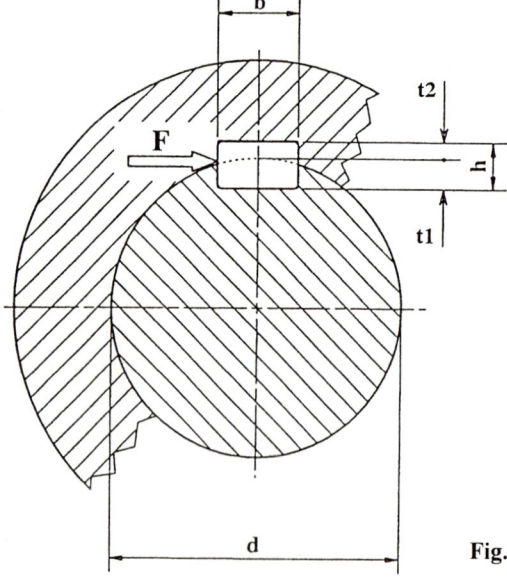

Fig. 7.2. Prismatic Keys

7 Fits between the Shaft and the Hub and Fits Between the Hub and the Shaft

The connection between shaft and hub is realized by means of an intermediate prismatic element (hollow) placed inside a spline of the shaft and inside a spline of the hub. The torque transmission is carried out by cutting the key. The dimensional principle is the following: the shearing stress inside the hollow is equal to the torsional stress at the periphery of the shaft. If d is the diameter of the shaft, L the length of the hollow, T the transmitted torque and b the width of the hollow, the result will be:

$$\frac{T}{0.1\, d^3} = \frac{2\, T}{d\, L\, b} \quad . \tag{7.004}$$

Consequently the result is:

$$b = 0.4 \frac{d^2}{L} = 0.4 \frac{d}{k} \tag{7.005}$$

if $L = k\, d$.

Table 7.3 provides standardized values based on the relation between the dimensions of the hollow and a series of values relative to the diameters.

Table 7.3. (figure 7.2) Dimensions and Tolerances

d >	=	b	h	t1	t2	Roun. max	Tollerance min	Shaft	Hub	k
30	38	10	8	5	3.3	0.4	0.25	-36	-18	1.3
38	44	12	8	5	3.3	0.4	0.25	-43	-21.5	1.3
44	50	14	9	5.5	3.8	0.4	0.25	-43	-21.5	1.35
50	58	16	10	6	4.3	0.4	0.25	-43	-21.5	1.35
58	65	18	11	7	4.4	0.4	0.25	-43	-21.5	1.35
65	75	20	12	7.5	4.9	0.6	0.4	-52	-26	1.4
75	85	22	14	9	5.4	0.6	0.4	-52	-26	1.45
85	95	25	14	9	5.4	0.6	0.4	-52	-26	1.45
95	110	28	16	10	6.4	0.6	0.4	-52	-26	1.45
110	130	32	18	11	7.4	0.6	0.4	-62	-31	1.5
130	150	36	20	12	8.4	1	0.7	-62	-31	1.55
150	170	40	22	13	9.4	1	0.7	-62	-31	1.6
170	200	45	25	15	10.4	1	0.7	-62	-31	1.65
200	230	50	28	17	11.4	1	0.7	-62	-31	1.7
230	260	56	32	20	12.4	1.6	1.2	-74	-37	1.75
260	290	63	32	20	12.4	1.6	1.2	-74	-37	1.75
290	330	70	36	22	14.4	1.6	1.2	-74	-37	1.75

The tolerances of depth of the splines are equal to 2 for $d =$ from 30 to 130 mm and 3 for the remaining.

Let us remember that the tolerances relative to the width of the splines are "reduced" and negative in order to ensure a good support on the sides and an easy mounting. The tolerances of depth of the splines are large and positive in order to ensure that on the upper surface of the key no contact occurs. With such dimensions the shearing of the key does not need to be calculated, but it is necessary to test the pressure along the sides. For this purpose a rather simple method is used. The surface of contact on the shaft is indicated with $L\,t_1$, while the surface of contact on the hub is expressed by $L\,t_2$. If σ_a is the permissible stress on the shaft and σ_m the permissible stress on the hub, the transmissible torque will be:

$$T_a \equiv \frac{\sigma_a\, L\, t_1\, d}{2000} \quad (7.006)$$

and

$$T_m = \frac{\sigma_m\, L\, t_2\, d}{2000}\,. \quad (7.007)$$

The lowest of such values is considered as an acceptable value (T_P) and if T is the torque that is to be transmitted, the number of the keys will be:

$$n > \frac{T}{T_P}\,. \quad (7.008)$$

The number of the keys is limited to 2. In this case if only 3/4 of the total section are considered, the pressure exerted along the sides will occur. It is necessary to verify that the resulting stress is lower than the permissible stress.

$$\frac{T}{1.5\, L\, d\, t_1} \geq \sigma_a \quad (7.009)$$

$$\frac{T}{1.5\, L\, d\, t_2} \geq \sigma_m\,. \quad (7.010)$$

If such conditions are not satisfied, 3 keys will be required, altough this may be in-advised. If more than 2 keys are required, it will be necessary to modify the dimensions of the shaft in order to select a larger key.

The permissible stresses are:

- for steel: mild-steel key $\sigma_{a,m} = 210$ MPa
    hard-steel key $\sigma_{a,m} = 280$ MPa
- for cast iron: mild-steel key $\sigma_m = 100$ MPa

If the torque is applied alternatively, 0.7 of such values will be selected.

7.4 Interference Mounting (Fig. 7.3)

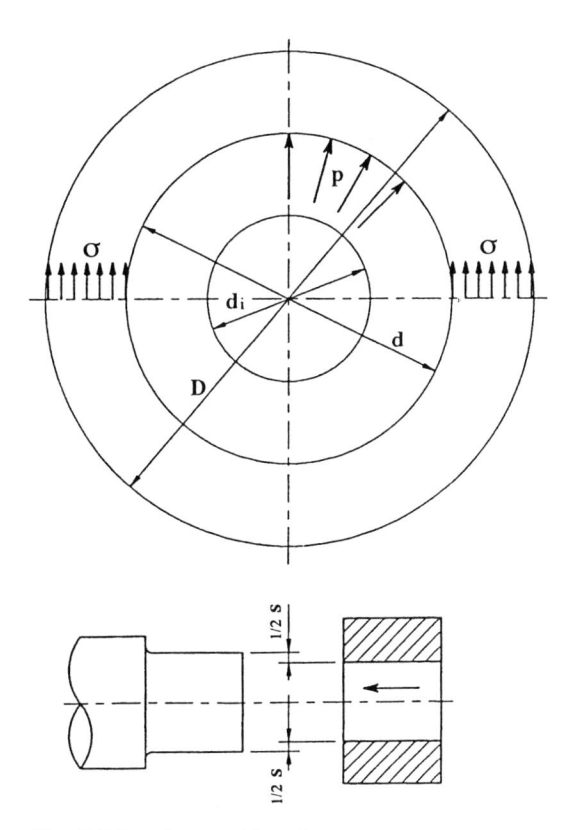

Fig. 7.3. Interference Mounting

The interference mounting is the positioning of a shaft inside a hole of small dimensions is made by heating the hub or by cooling the shaft. Once the temperatures return to their previous equilibrium, the hub tightens on the shaft with such precision as to allow transmission of the torque by means of a friction with a sufficient degree of safety.

If μ is the coefficient of friction, L the length of the coupling, d the diameter of the shaft, T the torque that must be transmitted in Nm and S_f the factor of safety of the coupling, the required pressure will be equal to:

$$p = \frac{2000 \, T \, S_f}{\pi \, \mu \, L \, d^2} \, . \tag{7.011}$$

For a hub with a nominal bore d and an outside diameter D, and for a hollow shaft with a nominal outside diameter d and an inside diameter d_i, there is a theoretical interference corresponding to the calculated pressure. Such interference is given by:

$$s = \frac{p\,d}{E_1}\left[\frac{d^2+d_i^2}{d^2-d_i^2}-v_1\right] + \frac{p\,d}{E_2}\left[\frac{D^2+d^2}{D^2-d^2}+v_2\right] \quad (7.012)$$

where:
- E_1 and v_1 are the module of elasticity and Poisson's coefficient of the shaft respectively,
- E_2 and v_2 are the module of elasticity and Poisson's coefficient of the hub respectively.

For a solid shaft and similar materials (steel) for the shaft and the hub the result is:

$$s = \frac{2\,p}{206000}\left(\frac{D^2}{D^2-d^2}\right) \quad (7.013)$$

The theoretical interference takes the roughness of the shaft and the roughness of the hub into account and its value is:

$$s_{th} = s + 0.5\,(R_{a1} + R_{a2}) \quad . \quad (7.014)$$

A coupling with a minimum interference higher than such value is selected. If the connection is made up of a bore H, the selection of a quality will allow the values of $Es = IT$ for the hole and the values of $Ei = 0$ to be determined.

For the shaft, tolerances are to be determined by calculating es and ei. The minimum interference is:

$$s_{min} = ei - Es$$

hence:

$$ei > s_{th} + IT_{bore} \quad .$$

Therefore it is possible to determine the tolerances of the shaft and to calculate the values of es and ei. By means of equations (7.013) or (7.014) it is possible to calculate a maximum pressure and a minimum pressure corresponding to the tolerances. With the minimum pressure it is possible to calculate the corresponding torque:

$$T_{min} = \pi\,\mu\,p\,d^2\,L/2000$$

and a real safety factor:

$$S_f = T_{min}/T\,.$$

7 Fits between the Shaft and the Hub and Fits Between the Hub and the Shaft

On the other hand the maximum pressure is applied to the shaft and should be lower than the elastic limit of the material of the shaft. Such pressure will take part of the shaft calculation; furthermore there is the risk to generate a maximum traction strenght in the hub given by:

$$\sigma_{max} = p \frac{D^2}{D^2 - d^2} \quad (7.015)$$

and to generate on the external surface f the hub a minimum stress:

$$\sigma_{min} = 2p \frac{d^2}{D^2 - d^2}. \quad (7.016)$$

The maximum stress will be lower than the elastic limit of the material of the hub. The connection will be made by heating the hub or cooling the shaft at such temperature that the hub distorsion or the shaft contraction are higher than the dimensions difference existing between shaft and hub hole. The temperature increase with respect to the environment is given by:

$$\Delta T = \frac{S_{max}}{\lambda\, d} \quad (7.017)$$

being λ the distortion coefficient of the hub or of the shaft, according to the method choosen. Such coefficient is equal to $12 \cdot 10^{-6}$ for steel. The friction coefficient for steel on steel is assumed to be equal to 0.12. For a cust-iron hub E will be equal to 180.000 MPa and a friction coefficient equal to 0.08 will be adopted. The safety factor will be between 3 and 6 since it is not possible to assume the state of the contact surface, which can be negatively affected by grease or lubrification powders, remarkably modifying the real friction coefficient.

Often, together by fixing through a prismatic key, it is made use of a slight interference whose function is not to guarantee the torque trasmission; the key calculation will be carried out taking into consideration that it transmits completely the torque. Its main function is to avoid corrosion, really harmful for shaft integrity, developing between two surfaces when micro displacements occur due to torsion or flexion. The existance of a sort of pressure between the two surfaces submitted to micro-displacements, gets the above risk lower.

7.5 Fretting Corrosion

According to Waterhouse, it is possible to calculate the fretting corrosion risk through the following calculation. Let us consider M_i as the bending moment in a shrinked section, Q_i as the shearing stress, L as the length of the shaft between the supports, D as the outside diameter of the hub, d as the diameter of the shaft, b as the width of the shrinking, n as the frequency of the deformations (in case of a shaft rotating in relation to its load it is the rotation frequency of the shaft, that is the speed in min^{-1} divided by 60) and s as the interference expressed in mm at such point.

Subsequently it is possible to calculate:

$$f = 0.6\ D + 1$$

$$K_1 = \frac{M_i\ f}{E\ D^3}$$

$$K_2 = \frac{Q_i\ f}{E\ D\ L}$$

$$\lambda = \frac{L}{d}$$

$$\delta = \frac{d}{D}$$

$$\varepsilon = s\,b\ \frac{1 - \delta}{1 - \delta^4 + 0.5\ \lambda}$$

$$\beta = \frac{1 - \delta^4 - 0.11\ \lambda}{1 - \delta^4 + 0.5\ \lambda}$$

$$K_{11} = \frac{K_1}{1 + 0.5\dfrac{\lambda}{1 - \delta^4}}$$

$$A = \varepsilon + K_2 + K_1\ \lambda^{1.5}\beta\ (1 + 45000\ K_1\lambda\ )$$

$$B = \varepsilon - K_1\lambda^{1.5}\beta\ (1 + 45000\ K_1\lambda\ )$$

$$K = K_{11}\left(\frac{1}{A} + \frac{1}{B}\right).$$

The higher is the factor K, the higher is the fretting corrosion risk.

7.6 Splines

Splines consist of keyways regularly distributed on a shaft where a rotating unit is to be set in rotation. They substitute keyways, but in comparison with the latter they have the advantage of being regularly distributed around the shaft and of making assembly and disassembly easier. Splines have also turned out to be particularly useful when the unit is to be set in motion during rotation and needs to slide freely on the shaft. The hub shows internal splines shaped exactly like the teeth that fit perfectly with the splines on the shaft. The shearing of the splines on the shaft can be carried out through different methods according to the shape of the splines. The internal splines of the hub can be accessed only through broaching.

There are three different types of toothing: parallel teeth (Fig. 7.4 a); trapezoidal teeth (Fig. 7.4 b); involute toothing (Fig. 7.4 c).

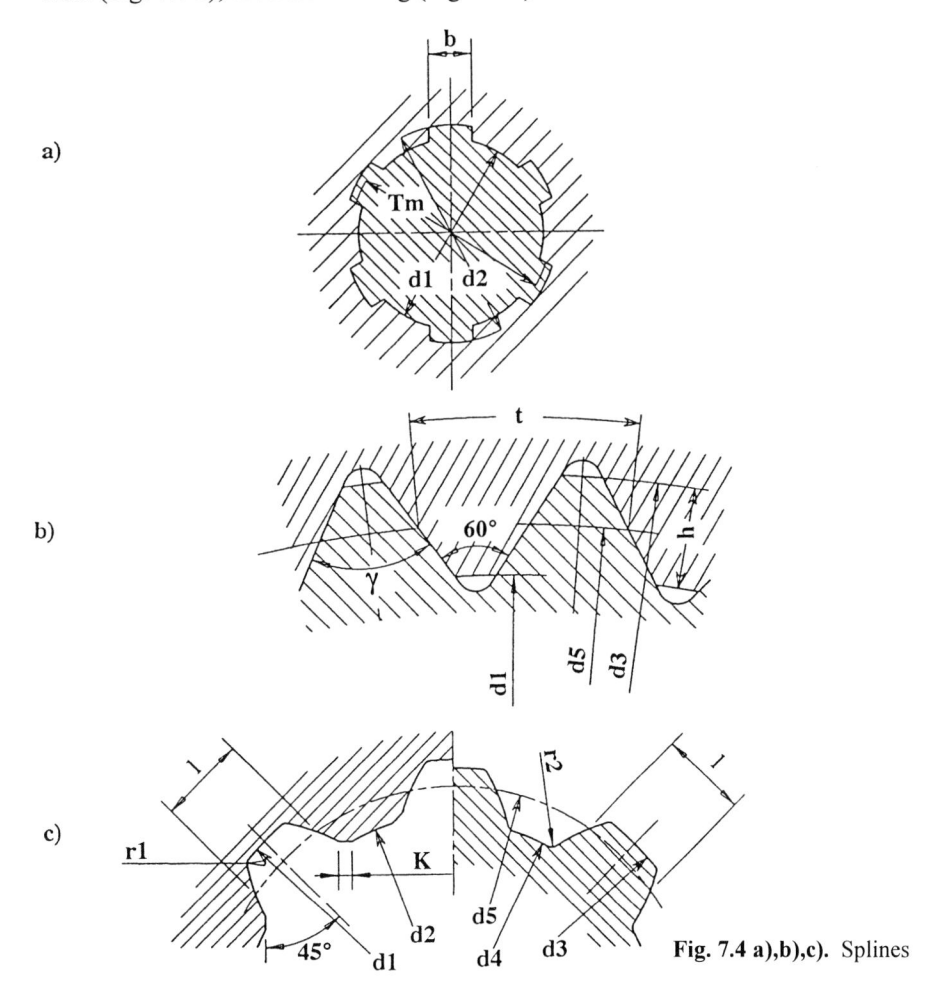

Fig. 7.4 a),b),c). Splines

For all the types of toothing the transmissible torque depends on the surface of contact on the flanks of the teeth. However, as the working shows a certain degree of imprecision, only 75% of the flanks are considered to be active. If h is the height of contact between the teeth, l the width of the hub, z the number of teeth, d_m the diameter at the middle height of the teeth and p the contact pressure, the result will be:

$$M_t = 0.75\, z\, h\, L\, p\, \frac{d_m}{2}. \qquad (7.018)$$

Steel, being the load constant, pressure p is limited to 100 MPa. In alternating loads 70% of such value has to be considered. For cast-iron hubs, 60% of the value adopted for steel hubs in the same conditions has to be considered.

Parallel splines. The number of splines ranges from 6 to 20 according to the diameter and the type of spline. The diameter d_l at the bottom of the spline is a whole number. The height of splines ranges from 1.5 mm (for small diameters and a light series) to 6.5 mm in a heavy series with 112 mm of diameter, (125 mm of shaft diameter). The spline width is given by the whole number lower than half the length that is obtained dividing the full diameter d_l by the number of splines. For example, a shaft with 88 mm of outside diameter has, in the light series, an 82 mm diameter at the bottom of the spline, and its splines are 3 mm high and 12 mm wide, 10 being the number of the splines. The maximum transmissible torque will be 96750 Nmm of width, if the hub is made of steel. This kind of spline turns out to be particularly suitable for the units that have to slide on the shaft (for example, in the gear-box of cars). There are particular regulations giving dimensions preferred.

Trapezoidal splines. For the series ranging from 7 to 60 mm the following dimensions are available (Table 7.4)

Table 7.4.

d1	d3	dm	z	h
7	8	7.5	28	0.5
8	10	9	28	1
10	12	11	30	1
12	14	13	31	1
15	17	16	32	1
17	20	18.5	33	1
21	24	22.5	34	1.5
26	30	28	35	2
30	34	32	36	2
36	40	38	37	2
40	44	42	38	2
45	50	47.5	39	2.5
50	55	52.5	40	2.5
55	60	57.5	42	2.5

The angle of the tooth in the hub is $\beta = 60°$ (50° for the higher diameter). The angle of the tooth of the shaft is equal to:

$$\gamma = \beta - \frac{360°}{z} \quad . \tag{7.019}$$

The pitch p of the toothing on the tooth diameter is:

$$p = \frac{\pi \, d_3}{z} \quad . \tag{7.020}$$

This kind of spline is the most resistant owing to its high number of teeth.

Involute splines. These show the typical teeth of the gears and have all their characteristics. The standard reference profile is a reduced profile in which the addendum and the dedendum are equal to 0.8 m. The backlash between the head of the tooth of the shaft and the dedendum of the tooth of the hub is equal to 0.1 mm. The standard modules are equal to 0.8 / 1.25 / 2 / 3 / 5 and 8. For a 300 mm diameter at the bottom of the hub and a module equal to 8, the number of teeth is lower than 300: 8 = 37.5, that is 36 teeth. The reference diameter will then be 288 mm. The root diameter of the shaft is equal to 300 - 0.2. 8 = 298.4. The head depth is therefore equal to 5.2 mm, that is 0.65 m. The toothing modification coefficient of the toothing is therefore:

$$0.65 - 0.8 = - 0.15 \, m \quad .$$

For such splines the depth h is equal to 2 m and the torque formula is the following:

$$M_t = 0.75 \, p \, L \, d_m^2 \quad . \tag{7.021}$$

Compared to the above mentioned type of spline, involute splines turn out to be less resistant due to the lower number of teeth but easier to be produced. They have standardized dimensions.

7.7 Couplings - Generalities

Couplings are devices that enable two shafts to be fixed at their ends. Although shafts should be aligned in relation to one another, this rigorous positioning is in many cases not possible. There is a radial misalignment (r) and an angular misalignment (β). During working axial position variations (a) and relevant rotations can occur. Couplings in which neither misalignments nor position variations occur are known as rigid couplings. On the contrary, couplings in which misalignments and deviations occur are known as flexible couplings.

7.8 Rigid Couplings

Rigid couplings include forged flange couplings placed at the end of the shaft or assembled flange couplings. The forged flanges at the end of the shaft are plates integral to the shaft and are kept together by bolts. Such bolts can be subjected to traction, bending or shearing. In the first case they cause a friction between the two tightened plates. In the second case they are fixed alternatively at the two plates, which are placed so that one bolt of a plate corresponds to one backlash hole of the other plate. The fixed bolts are started by a shaft and transmit the movement to the free bolts (Fig. 7.5a, 7.5b).

Couplings with sheared bolts (figure 7.5c) show hollows to which tolerances have been applied. Such tolerances characterize worked bolts and cause a small interference. The torque is transmitted by the pure shearing of the bolts. Plate fits (figure 7.5c) are identical to flange couplings, but have plates integral to the hubs which are fixed on the shaft by means of keys and have the advantage of allowing the dismantling of the flanges. The calculation of clutch couplings is carried out as follows: if D_m is the diameter of the circle where the axes of the bolts are placed, the applied torque M_t originates a force that is given by:

$$F = \frac{2 M_t}{D_m} \qquad (7.022)$$

Such force must be created by means of a friction between the two plates.
With a coefficient of friction μ the result is:

$$F_b = \frac{F}{n \mu} \qquad (7.023)$$

being F_b the traction force of the bolt. So it is possible to check the diameter of the bolts subjected to simple traction. For the calculation of sheared bolt fits, the shearing force per bolt is given by the force F divided by the number of bolts. Such force is a pure shear. For bent bolts, the force F, divided by the number of bolts, is multiplied by the distance c that separates the centre of the free end of the bolt from its seat. This torque is a bending torque. These fits do not allow any deviation between the shafts, because tensions would exeed the values allowed. Therefore such couplings are suited only to machines rigidly mounted on well-levelled planes and provided with perfectly aligned shafts. Although such couplings are very simple, their installation turns out to be extremely complex.

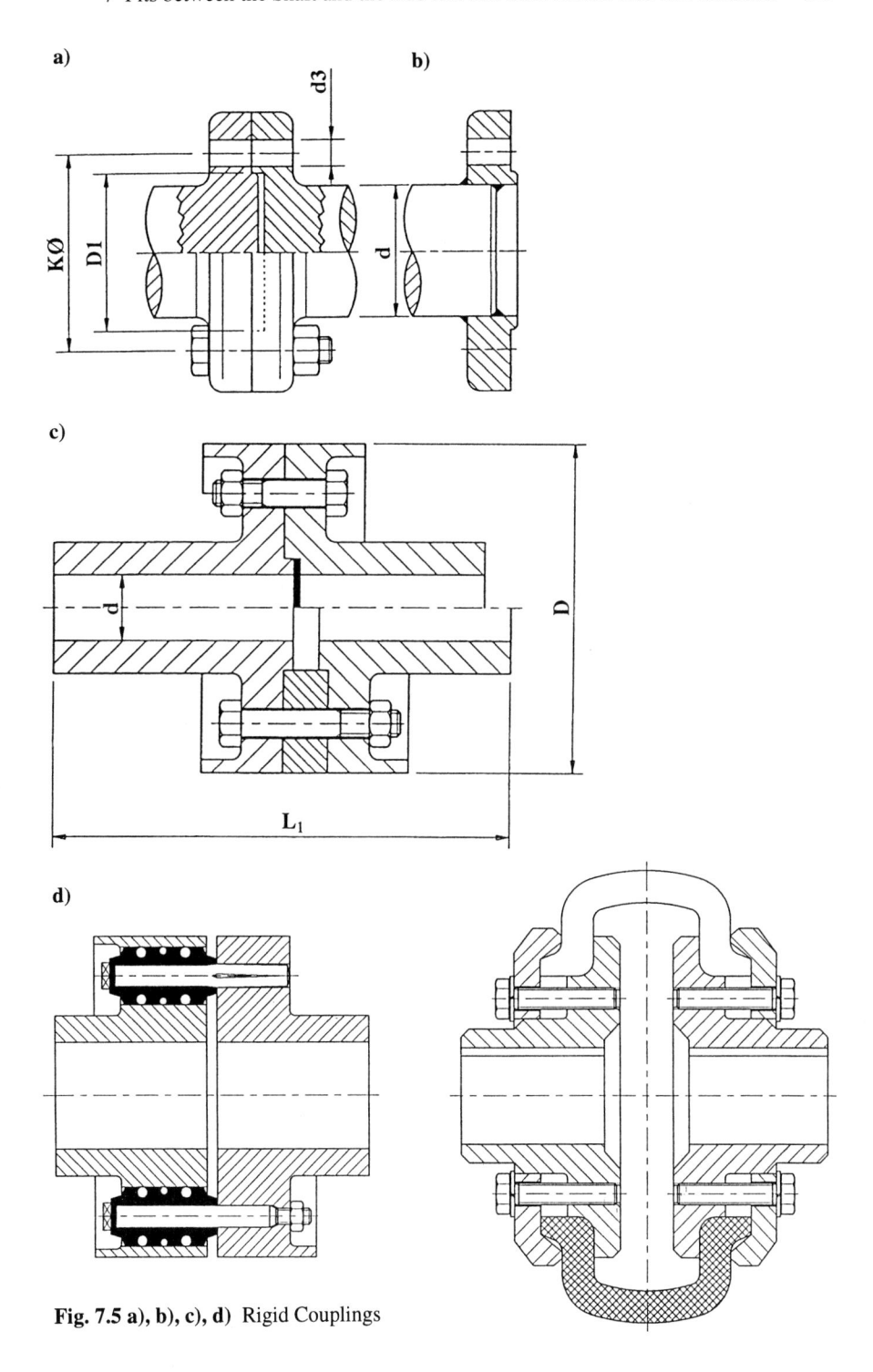

7 Fits between the Shaft and the Hub and Fits Between the Hub and the Shaft   249

Fig. 7.5 a), b), c), d) Rigid Couplings

7.9 Elastic Couplings

There is a wide range of different elastic couplings. Their selection depends on the quality of construction. In fact such couplings are usually made up of cast iron and require mass-production on a large scale, so that production costs can be amortized. For this reason such couplings are produced by specialized industries. The simplest ones derive from inflexed bolt rigid fits. The broach is surrounded by an elastomer sleeve put in the hollow of the other plate. Connection is determined by an elastomer, which allows a slight offset and an attenuation of the transmitted shocks. The other types of couplings are more complex; they are made up of two flanges put together by an elastomer fixed on one of these two. The torque is transmitted by the elastomer; as a consequence, a certain flexibility both in the alignment of the shafts and in the transmission of the shocks, which are widely attenuated by the hysteresis of the elastomer, is possible. Each type of coupling differs from the others in the fixing and in the shape of the elastomer. They show, however, a disadvantage, which lies in the relative fragility of the elastomers. For this reason elastic couplings have been devised, where the connection between the two shafts occurs by means of a steel spring. The fit known as Bibby, which is made up of two hubs characterized by hollows that flare one in front of the other, are shown in Fig. 7.6.

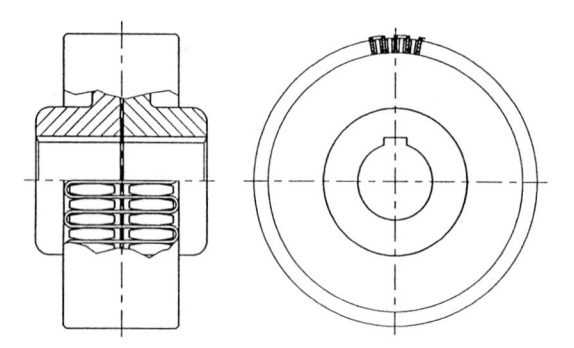

Fig. 7.6 Bibby coupling

In these hollows there is a steel spring turning all around the hub. The torque is transmitted by means of this spring, which can be twisted under the action of the elastic stress. The hollow shape allows approach of the bearing points when the distortion grows; consequently, a certain degree of rigidity, which grows with the applied torque, is produced. Such couplings are very resistant, therefore deformations do not easily occur.

7.10 Gear Toothing Couplings

Gear toothing couplings are made up of two hubs characterized by external involute toothing and by a bushing with internal toothing mated at each end. The external toothing of the driving shaft initiates the first internal toothing of the bushing, which starts the gear wheel which, in turn, starts the external toothing of the secondary shaft by means of the second internal toothing. External teeth are characterized by a very crowned profile. Such type of coupling allows a misalignment of the shafts. The latter are rigid during torsion, but allow axle and radial displacements during operation. Such couplings are extremely interesting; their calculation is particularly complex, as the friction between the teeth, which represent the main cause of a possible deterioration of the units, occurs in a very peculiar way. These fits are produced by experts in the field and are sold in catalogues with all the information concerning their characteristics and the transmissible torques relative to different applications.

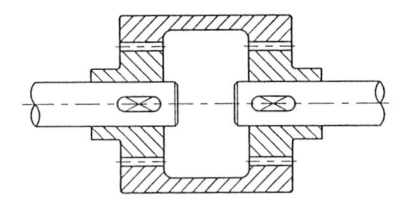

Fig. 7.7. Gear toothing couplings

7.11 Clutch Couplings (Fig. 7.8)

Clutch couplings are made up of alternate plates, the first of which are set in motion by a hub integral to a shaft, while the others are integral to a gear wheel started by the other shaft. Plates are pressed one against another and they are started by friction. The transmissible torque depends on the number of supporting faces, on the spring pressure, on dimensions and on the coefficient of friction between the surfaces of the plates. If F is the force exerted by the spring, r_1 the internal radius of the plate and r_2 the external radius, the pressure exerted on the plates will be equal to:

$$p = \frac{F}{\pi \left(r_2^2 - r_1^2\right)} \tag{7.035}$$

and the transmissible torque will be equal to:

$$T = \frac{2F}{3} \mu \frac{r_2^3 - r_1^3}{r_2^2 - r_1^2} (2n - 1) \tag{7.036}$$

being n the number of the pairs of plates.

Such type of coupling can function as a torque limiter, by regulating the force of the spring to a value equal to the maximum permissible torque. In overloading, the coupling transmits only the maximum torque. If the spring force is adjustable (by means of a mechanical or electromagnetic action exerted on the spring), the coupling will become a friction. Where the bearing bush is integral to a fixed structure, the resulting component could be a brake.

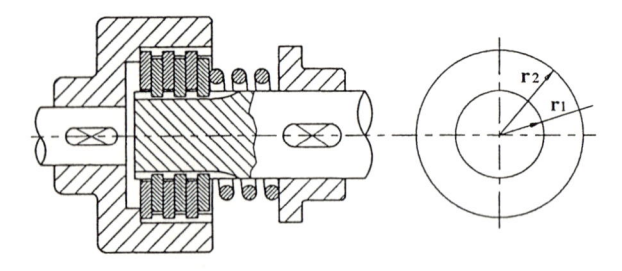

Fig. 7.8. Clutch couplings

7.12 Couplings by Means of Shrink Disks (Fig.7.9)

Such a coupling is used to fix the shaft of a machine in the hollow shaft of another machine. The solid shaft is worked with a tolerance equal to g6, while the hollow shaft is worked with a tolerance H6. The result is a coupling with a slight

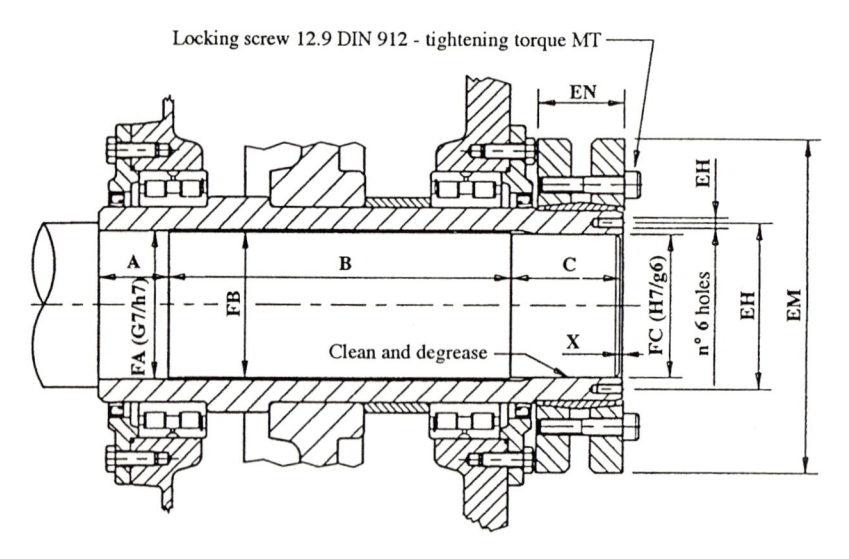

Fig. 7.9 Coupling by means of shrink disks (cfr. Bonfiglioli's Catalogue).

interference. A biconical ring nut is mounted on the hollow shaft; on this ring nut two flanges come closer to another by tightening the bolts. This approach causes a deformation of the hollow shaft, which clings to the solid shaft producing the fit necessary to transmit the torque. This latter depends on the tension in the bolt and therefore on the tightening torque of the bolt.

8 Bearings

8.1 Generalities

Shafts must be supported by bearings producing radial and axial bearing reactions, which must also be able to turn freely round the fixed elements of the bearing.

Bearings are units that can be present both in fixed and mobile elements; they are provided with devices that tend to reduce as much as possible any losses between such elements.

They are made up of three essential elements plus an accessory. The three essential elements are constituted by two rings, one fixed and the other mobile, characterized by sliding planes and separated by mobile elements known as rolling elements. The accessory element, although useful, is made up of a stand that keeps the rolling elements in the right position and at the same distance from one another.

8.2 Classification of Bearings on the Basis of the Rolling Elements

The rolling elements can be balls or rollers.

These can be cylindrical rollers, single row rollers, taper rollers or needle rollers, that is rollers with a very short diameter in comparison with their length. We will distinguish three important types of bearings: ball bearings, roller bearings and needle roller bearings.

8.3 Classification of Bearings on the Basis of their Positioning

There are also deep groove bearings, single row angular contact ball bearings, or self-aligning bearings. Deep groove bearings can be distinguished in single row deep groove ball bearings and double row deep groove ball bearings (non-dismountable) and in cylindrical roller bearings. Single row angular contact ball bearings are ball bearings in which the sliding planes provide contacts that are not on the perpendicular plane to the axis. We will describe single row angular contact ball bearings, double row angular contact ball bearings and taper roller bearings. Self-aligning roller bearings allow a deep bending of the shafts where bearings are placed. Such bearings can be double row deep groove ball bearings equipped with proper sliding planes, in order to enable the two rings to occupy the oblique planes, and single row self-aligning roller bearings. Each type of bearing will be thoroughly described in the following paragraphs.

8.4 Calculation of Bearings

Where there is constant load exerted on bearings, these latter undergo a stress as a result of their own working. Their resistance therefore follows a rule similar to Miner's law, known as Palmgren's law. According to such a rule, working life is inversely proportional to powers P of the loads. The exponent p for ball bearings is equal to 3, while the power P for roller bearings is equal to 10/3. In fact bearings are subjected to Hertz's pressure occurring between the elements in contact only at some points for ball bearings and in rectilinear contact for roller bearings. The catalogues on bearings give information about the dimensions of each bearing and also about two load values, one static and the other dynamic equal to 10^6 loading cycles. The static load is called C_0, while the dynamic load is indicated by C. Between the radial load, which stresses the bearing, and load C there is the following relation:

$$L_{10} = \left(\frac{C}{P}\right)^p \qquad (8.001)$$

where L_{10} is the operating life expressed in millions of revolutions. It corresponds to a reliability equal to 0.9 (that is 10% of the failure risk). With another degree of reliability it is possible to modify the value of L by writing:

$$L_r = a_1 \, L_{10} \qquad (8.002)$$

being the value of a_1 taken from table 8.1.

Table 8.1. value of a_1

Reliability	Symbol	$a1$
0.9	L10	1
0.95	L5	0.62
0.96	L4	0.53
0.97	L3	0.44
0.98	L2	0.33
0.99	L1	0.21

If special steel alloys are used or the lubrication and the running temperature in particular are considered, it will be possible to take other modification coefficients into account. There are specialized catalogues for consultation.

Load P is considered as a purely radial load. It is given by the loads that have been applied by the modification coefficients that follow the load classifications. These latter can produce dynamic stresses (shocks), which take into consideration an application factor. The literature by the producers of bearings gives application examples and advice on how to select such factors, from years of experience; if there are any doubts consult the producer. For to input and output belt drives on the shafts, it is necessary to consider the supplementary loads produced on the shafts by the belts. If the calculation is done taking the tangent load of the belt into account, such load will be multiplied by one of the factors indicated as follows:

Table 8.2. Factor for tangent load of the belt

Type of belt	Factor for tangential load
Notch gear belt	from 1.1 to 1.3
Trapezoidal belt	from 1.2 to 2.5
Flat belt	from 1.5 to 4.5

The loads on bearings are not always purely radial loads. It will now be shown of when and to what extent axial loads can be accepted in relation to each type of bearing and also explain how to transform the combination between the radial and axial loads in order to obtain an equivalent axial load.

8.5 Variable Loads

Bearings can be subjected to variable or continuous loads. With variable loads, according to the type of bearing, if P_i is one of these loads and if a number of cycles N_i corresponds to such value and if N is the sum of all the partial cycles, the result will be:

$$P_{equ} = \left(\sum \frac{P_i^p N_i}{N}\right)^{\frac{1}{p}} . \qquad (8.003)$$

If the load ranges in a constant way from a maximum value to a minimum value, the result will be:

$$P_{equ} = \frac{P_{min} + 2 P_{max}}{3} . \qquad (8.004)$$

8.6 Static Capacity

Given the maximum static load on the bearing, it is necessary to check that such load does not exceed a limiting value given by C_0/s_0. Such factor C_0 depends on the stress and the type of bearing; information on this will be given when we describe each bearing thoroughly.

8.7 Dimensions of Bearings

The dimensions of bearings are determined by their producers on the basis of specific standards (ISO 15 is for radial bearings and ISO 355 for taper roller bearings). Such standards determine classes of dimensions for outside diameters and for lengths (in case of radial bearings) or for the dimensions of diameters, widths and lead angles (in case of taper roller bearings).

For radial bearings, increasing outside diameters (series 7,8,9,0,1,2,3 and 4) and growing widths with the same series information correspond to each bore diameter. The smallest bearing with a given diameter d is expressed by 00 for a width series 0 and a diameter series 0. For the same bore diameter and the same outside diameter, the result is, for example, equal to 20, that is a diameter series 0 and a width series 3. The producers of bearings conform to these dimensions for such an extent that bearings turn out to be interchangeable.

For taper roller bearings the series of widths and diameters are replaced by letters to which the lead angle of the bearing is added. Standard ISO 582 gives the values recommended for roundings.

The catalogues on bearings also provide information about bearing tolerances and the tolerances required by the shafts and by bearing housings in fixed bearings. This subject will be dealt with when treating mounting.

8.8 Radial Deep Groove Ball Bearings

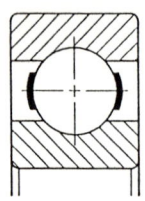

Fig. 8.1 Radial deep groove ball bearings

Radial deep groove ball bearings are made up of two rings both equipped with a toroidal sliding race. The rolling elements are balls. They are used for numerous applications. Their planning is very simple; they can work at high speeds and their maintenance is easy or not required at all. As they are mass-produced, their production costs are extremely low. As to loads, radial deep groove ball bearings can bear high radial loads and not so high axial loads.

Axial loads can function for both and can also be used at high speeds. Radial deep groove ball bearings can be single row or double row deep groove ball bearings. Single row deep groove ball bearings are characterized by silent working, while double row deep groove ball bearings turned out to be noisier.

Friction is slight, although more marked in double row deep groove ball bearings. The rotation of single row deep groove ball bearings is very precise, but less precise in double row deep groove ball bearings. During operation radial deep groove ball bearings do not tolerate any misalignment of the shaft and do not compensate the possible initial misalignments. Such phenomenon is more marked in double row deep groove ball bearings than in single row deep groove ball bearings. The permissible bending during operation, that is the rotation of the bent shafts in the bearings, is limited and cannot exceed a value ranging from 2 to 10 minutes of an angle (from 0.000582 to 0.002909 radiants).

This bending limit depends on numerous factors such as the circumferential backlash, the dimensions of the bearings, their internal structure, the stresses and the applied torques. Double row deep groove ball bearings are even more sensitive, as their bending is limited to two minutes. When an axial load is exerted, calculation must be done with a pure radial load equal to the combination between the applied radial load (F_r) and the axial load (F_a). The equivalent pure radial load (P) is calculated as a function of the relation between the axial load and the basic static load.

If the value of this relation is adopted and the values of e, x and y are taken from the table, the result will be:

$$\text{If: } \quad \frac{F_a}{F_r} \leq e \quad \Rightarrow \quad P = F_r \qquad (8.005)$$

If: $\dfrac{F_a}{F_r} > e \quad \Rightarrow \quad P = X F_r + Y F_a$. \hfill (8.006)

In the same way it is possible to determine an equivalent static load as follows:

$$P_0 = 0.6\ F_r + 0.5\ F_a \ . \hspace{2em} (8.007)$$

A purely axial load can be applied only if it does not exceed half the static basic load.

For small bearings and for the bearings with a light series the value is limited to a quarter of the static basic load value. For each bearing, precise tables provide geometrical information as well as dynamic and static basic loads.

Precise mounting tolerances are provided in the paragraph entitled "Mounting of bearings".

8.9 Double Row Self-Aligning Ball Bearings

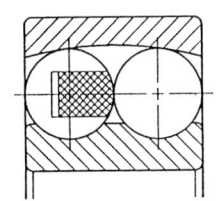

Fig. 8.2 Double row self-aligning ball bearings

In these bearings the external ring creates a spherical sliding plane common to two rows of balls, while the internal ring produces two toroidal sliding planes, one for each row of balls. This positioning allows for shifting of the plane of the external ring in relation to the race of the internal ring. Such bearings therefore able to tolerate a high-value bending, are particularly suitable for applications in which a misalignment of the shaft can occur and where wide bending angles are generated in the housings (not very rigid or long shafts). Bending can range from 1.5 to 3 angle degrees according to the dimensions of the bearing. Bending also depends on the mounting conditions and seal devices. This type of bearing tolerates a normal radial load intensity but a limited axial load. The permissible speed is lower than the speed tolerated by rigid radial bearings. The rotation accuracy is normal (not excessive). Such bearings do not show a high degree of rigidity. Their working is moderately silent; they produce a slight friction, which is, however, higher than that created by rigid bearings. The pure radial load is calculated by means of the following formulas.

If: $\dfrac{F_a}{F_r} \leq e \quad \Rightarrow \quad P = F_r + Y_1 F_a$ \hfill (8.008)

if: $\dfrac{F_a}{F_r} > e \quad \Rightarrow \quad P = 0{,}65 F_r + Y_2 F_a$. \hfill (8.009)

The values of e, Y_1, and Y_2 are shown in the tables according to each bearing, such values being different for each type of bearing. The equivalent static load is expressed by:

$$P_0 = F_r + Y_0 F_a \ .\qquad (8.010)$$

For each bearing, as in the previous cases, the tables provide geometric dimensions, mounting tolerances and the information relative to the calculation.

8.10 Cylindrical Roller Bearings

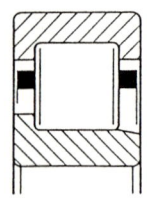

Fig. 8.3 Cylindrical roller bearing

Cylindrical roller bearings are made up of cylinder-shaped rolling elements. The sliding planes are cylinders. Such bearings are designed in different ways. In some the external ring creates shoulders at each end of the sliding plane, while the internal ring is smooth, without any shoulder. The unit made up of the external ring and the roller can slide on the internal ring in an axial way. For other bearings it is the opposite, the internal ring being integral to the rollers and able to slide on the external ring. And finally in some other types of bearings, the two rings are characterized by a shoulder and are altogether axially rigid. These bearings are rigid and their bending ranges from 3 to 4 minutes of an angle. Compared to ball bearings, they bear a purely radial load of high intensity. Axial loads as well as combined loads, must not be present, except in bearings with shoulders on the two rings. In such cases, however, the axial thrust must be limited, the working being considered not as a rotation but rather a rotation followed by frictions of high intensity occurring on the shoulders. So we can assume that axial thrusts are proscribed or at least accidentally tolerated. Such bearings tolerate high speeds; they show a considerable rigidity and their working is silent (less silent, however, than that of ball bearings). They are particularly suited to bearing very high radial loads. As for other bearings, the catalogues provided by the producers of bearings give all the required information. Where there is not enough space for a normal rotation, needle roller bearings are used. They are cylindrical roller bearings whose rolling elements show a very small diameter in comparison with the diameter of the sliding plane. They can be available with rings or not, as the assembled elements substitute the sliding planes. The load capacity evidently depends on the hardness of the sliding planes of the bearing.

8.11 Double Row Self-Aligning Roller Bearings (Fig. 8.4)

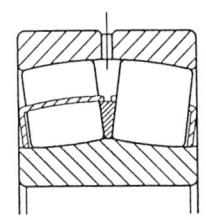

Fig. 8.4 Double row self-aligning roller bearings

Double row self-aligning roller bearings are provided with an external ring with a spherical sliding plane. The rolling elements are arc-of-a-circle generatrix cylindrical rollers that fit with the shape of the sliding plane. The internal ring creates two different sliding planes for the two rows of rollers. The unit made up of the internal ring and the rollers can oscillate freely in relation to the external ring. This kind of rotation allows a considerable bending, which ranges from 1 to 2.5 degrees of an angle. Such bearings tolerate high degrees of radial stresses and pure or combined axial stresses. It is possible to calculate the equivalent radial load by means of the following formulas.

$$\text{If: } \frac{F_a}{F_r} \leq e \quad \Rightarrow \quad P = F_r + Y_1 F_a \tag{8.011}$$

$$\text{If: } \frac{F_a}{F_r} > e \quad \Rightarrow \quad P = 0.67 F_r + Y_2 F_a \tag{8.012}$$

The values of e, Y_1 and Y_2 are shown in the tables. For calculation of the equivalent static load, given by Y_0 expressed in the tables, the result is:

$$P_0 = F_r + Y_0 F_a \tag{8.013}$$

Such bearings tolerate high axial and radial loads, allow a considerable bending degree and show a moderate rigidity. The tolerated speeds are, however, not excessively high and the friction degree is relatively high. Their working is less silent than that of ball bearings.

8.12 Angular Contact Ball Bearings (Fig. 8.5)

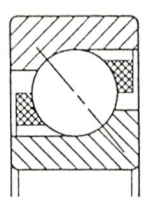

Fig. 8.5. Angular contact ball bearings

The sliding planes of the internal and external ring are staggered in order to generate an oblique contact in the right sense, in the case of axial loads. Under the action of a combined load, the contact reaction between the balls and the sliding races turns out to be angular in relation to the axis of the shaft. Consequently such bearings can tolerate axial loads only in one sense. They are always mounted in pairs and opposite one to the other on two different housings or on the same one. Such bearings are also dismountable. The internal ring with the balls is maintained by the stand parts from the external ring. There are single and double row angular contact ball bearings, and also special bearings of the same kind. They are bearings provided with four bearing points. Angular contact ball bearings allow a slight bending and tolerate therefore only slight misalignments or slight angular deformations. The reaction lines of bearings, being angular to the axis, do not meet such axis in the housing of the bearing. It is necessary to mount such bearings so as to enable reaction lines to take a shape similar to an X or an O; these are referred to respectively as X-mounting or O-mounting. The O-mounting allows to space more the support points, thus increasing the stability, but also reducing the stiffness. On the contrary, X-mountings increase rigidity and reduce stability. If such bearings are used with an axial load equal to K_a the action of this latter, combined with the radial loads, will produce in each bearing a resulting axial load, which can be calculated as follows.

Bearing A (with the corresponding indexes) is the one towards which the axial load is directed (O-mounting).

If: $F_{rA} \geq F_{rB}$

$$F_{aA} = 1.14\, F_{rA} \ ; \quad F_{aB} = F_{aA} + K_a \quad . \quad (8.014)$$

If: $F_{rA} < F_{rB}$

if: $K_a \geq 1.14\,(F_{rB} - F_{rA}) \Rightarrow F_{aA} = 1.14\, F_{rA} \ ; \quad F_{aB} = F_{aA} + K_a \quad (8.015)$

if: $K_a < 1.14\,(F_{rB} - F_{rB}) \Rightarrow F_{aB} = 1.14\, F_{rB} \ ; \quad F_{aA} = F_{aB} - K_a \quad (8.016)$

With the resulting axial loads it is possible to calculate the equivalent radial load as follows.

$$\text{If:} \quad \frac{F_a}{F_r} \leq 1.14 \quad \Rightarrow P = F_r \qquad (8.017)$$

$$\text{if:} \quad \frac{F_a}{F_r} \geq 1.14 \quad \Rightarrow P = 0.35\, F_r + 0.57\, F_a \quad . \qquad (8.018)$$

The equivalent static load is given by:

$$P_0 = 0.5\, F_r + 0.26\, F_a \quad . \qquad (8.019)$$

The other working features, different from those mentioned above, are the same as the features of double row self-aligning ball bearings.

8.13 Taper Roller Bearings (Fig. 8.6)

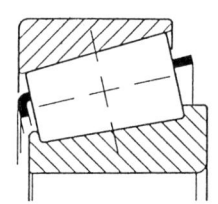

Fig. 8.6 Taper roller bearings

The two rings show conical sliding planes between which taper rollers are placed. The cones of the three elements have a vertex in common. The information given above about angular contact ball bearings is the same for conical roller bearings (K_a towards A bearing; O-mounting). The equations concerning the axial load K_a are expressed below.

$$\text{If:} \quad \frac{F_{rA}}{Y_A} \geq \frac{F_{rB}}{Y_B}$$

$$F_{aA} = \frac{0.5\, F_{rA}}{Y_A}$$

$$F_{aB} = F_{aA} + K_a \quad . \qquad (8.020)$$

If: $\dfrac{F_{rA}}{Y_A} < \dfrac{F_{rB}}{Y_B}$

if: $K_a \geq 0.5\left(\dfrac{F_{rB}}{Y_B} - \dfrac{F_{rA}}{Y_A}\right)$

$$F_{aA} = \dfrac{0.5\,F_{rA}}{Y_A}$$

$$F_{aB} = F_{aA} + K_a \qquad (8.021)$$

if: $K_a < 0.5\left(\dfrac{F_{rB}}{Y_B} - \dfrac{F_{rA}}{Y_A}\right)$

$$F_{aB} = \dfrac{0.5\,F_{rB}}{Y_B}$$

$$F_{aA} = F_{aB} - K_a \ . \qquad (8.022)$$

The equivalent radial load exerted on each bearing is calculated as follows.

$$\text{If: } \dfrac{F_a}{F_r} \geq e \Rightarrow P = F_r \qquad (8.023)$$

$$\text{if: } \dfrac{F_a}{F_r} < e \Rightarrow P = 0.4\,F_r + Y F_a \ . \qquad (8.024)$$

The equivalent static load is determined as follows:

$$P_0 = 0.5\,F_r + Y_0 F_a \ . \qquad (8.025)$$

The values of e, Y and Y_0 are given by the tables according to each type of bearing.

The other bearings, different from conical roller bearings, are made up of nickel-chrome steel and are mass-hardened.

Conical roller bearings are mass-treated as the other bearings, or are made of case-hardening steel.

8.14 Selection of Bearings

A crucial problem is the selection of the most adequate type of bearing. This is made according to technical and economic requirements. These latter are based on the cost of the bearing, although they can occur in a scrupolous production only if economic selection is made in order to support equivalent technical solutions. As selection is extremely difficult, it is only manufactured by experienced constructors of bearings. The selection can be made according to following method.

In Table 8.3 bearings are marked with numbers by designation. Such numbers correspond to their ability to perform in foreseen applications; they will be positive if the bearing is adequate, otherwise they will be negative. Such numbers increase according to the bearing fitness. If the precise qualities of the application are selected and the points for each bearing type are totalled, the resulting number will provide a classification in which the highest numbers correspond to the most adequate bearings. The columns are shown in table 8.3:

A possibility of a pure radial load
B possibility of a pure axial load
C possibility of a combined load
D compensation for errors of alignment during operation
E compensation for errors of alignment due to the construction
F slight noise
G slight friction (high efficiency)
H rigidity
J high speed
K accuracy of the rotation

Table 8.3. Selection of the most adequate type of bearing

Type	A	B	C	D	E	F	G	H	J	K
Single row deep groove ball bearing	1	1	1	-1	-1	3	3	1	3	3
Double row deep groove ball bearing	1	1	1	-2	-2	1	2	1	1	1
Double row self- aligning ball bearing	1	1	2	3	2	2	2	-1	2	2
Cylindrical roller bearing	2	-2	-2	-1	-1	2	2	2	3	2
Double row self-aligning roller bearing	3	1	3	3	2	1	1	2	1	1
Angular contact ball bearing	1	1	2	-1	-1	2	2	1	2	3
Taper roller bearing	2	2	3	-1	-1	1	1	2	1	2

For example, for the silent transmission of a pure radial stress with a high degree of stiffness, the result is:

Single row ball bearing	5
Double row ball bearing	3
Double row self-aligning bearing	2
Cylindrical roller bearing	6
Double row self-aligning roller bearing	6
Angular contact ball bearing	4
Taper roller bearing	5

As cylindrical roller bearings and double row self-aligning ball bearings have the same value, the selection of a specific type of bearing is determined by reasons of economy, mounting and availability. Single row ball bearings and taper roller bearings will be dealt with later.

In table 8.3 are shown, in decreasing order, angular contact ball bearings and double row ball bearings and double row self-aligning ball bearings. Such classification is valid only for the above required conditions.

8.15 Bearing Fixings

Bearings must be fixed on shafts and in fixed bearing housings. Such fixing must obey precise regulations in order to assure a good working and a life equal to that previously calculated.

For the axial fixing it is necessary to consider the bearing internal backlash, which must not, in any circumstances, become null, except for specific applications in which a pre-tightening is foreseen. As shafts and bearing housings show variations in dimensions due to the temperature, the bearing backlash being null or having a negative value. In order to avoid this, it is necessary to leave some space around the bearings so that they can move freely when the shaft expands. On the other hand, bearings must be well-positioned on the shaft and in the bearing housings. To meet such two conditions, bearings are fixed tightly to the mobile element and one bearing is fixed tightly to the fixed element, the other being able to move axially on the shaft.

Dismountable bearings (angular contact ball bearings and taper roller bearings) will have an internal backlash due to the positioning of the internal element in relation to the external ring.

For gear units, the mobile element is usually represented by the shaft while the fixed element is usually represented by the bearing housing. Therefore any bearings of a same shaft are mounted in each side of the shaft. Fixing is obtained by positioning the bearing against one shoulder of the shaft by means of one of its faces and then by fastening it at the other end by a nut or another fixing device. If the shoulder on the shaft turns out to be impossible, a tabular spacing bush will be placed from one bearing to the other, or when possible, from one bearing to another element on the shaft (for example a gear wheel).

Fixing by means of a nut can be replaced by a circlip carefully positioned so as to avoid being placed at a very stressed point of the shaft (stress concentrations). The height of the shoulder is shown in the catalogues for each type of bearing. The height is to be respected, as compressive stresses could occur where a height lower than that established in the catalogues; where a height is higher than that established in the catalogues undesired frictions could be generated on the fixed ring.

The fixed bearing is kept tightened at each side by means of a shoulder and a movable fixing device. The shoulder is mounted on the bearing housing or, if this is not possible, is replaced by a tubular spacing bush located on another point of the housing. External circlips can also be used. The fixing opposite to the shoulder is obtained by means of a circlip, an internal nut or a closing cover placed at the opening of the bearing housing. The other bearing must not be fixed in order to be able to move when the shaft expands.

Dismountable bearings (angular contact ball bearings and taper roller bearings) have an adjustable internal backlash. Therefore it is sufficient to shoulder each bearing just at one side, providing it with an adjustable backlash suitable for the mounting.

Radial fixing is obtained by coupling the bearing with the bearing housing. Initially the mobile ring in relation to the stress must remain tightly fixed on the mobile element (the shaft in general). This condition is only partially met, relative shifting of the ring and the shaft can occur, with possible wear and contact corrosion in particular. Therefore a tightening fit is used.

The fixed ring in relation to the force (usually the ring placed in the housing for gear units) must be free for one of the bearings; it forms a clearance fit together with the bore of the housing, since it must be able to move axially. Such clearance must be large enough to allow the free shifting of the bearing, but it must be narrow enough to ensure an accurate working during rotation.

Bearings are produced with specific tolerances. To obtain the desired tightening of the fixed ring or the required clearance of the free ring, the tolerances recommended by the producer are applied. The following Tables 8.4 and 8.5 provide the symbols recommended for the shaft and for the bore of the bearing.

Table 8.4. Selection of a fit on a steel solid shaft

Diameter of the shaft (mm)				
Ring that turns in relation to the load or in an indeterminate direction				
Conditions of use	Balls	Cylindrical or taper rolling bearings	Double row self-aligning roller bearings	Tolerances
Light or variable load	(18) to 100 (100) to 140	<40 40 to 100		j6 k6
Normal or heavy load	<18 (18) to 100 (100) to 140 (140) to 200 (200) to 280	<40 (40) to 100 (100) to 140 (140) to 200 (200) to 400	<40 (40) to 65 (65) to 100 (100) to 140 (140) to 280 (280) to 500 >500	j5 k5 m5 m6 n6 p6 r6 r7
Fixed ring in relation to the load				
Idle wheel				g6
Selection of a fit in a housing bore				
External ring rotating in relation to the load				
Hub of the wheel on the bearing				P7/N7
External ring fixed in relation to the load				
All the loads				H7
Normal or light loads in simple working conditions				H8

The bearing produces a rounding whose maximum value is shown. The bearing must lean on the shoulder for all its heigth. The rounding of the fillet of the shaft is therefore lower than the rounding of the bearing. Such procedure turns out to be detrimental owing to the tension concentrations on the shaft. The positioning shown in Fig. 8.7 is adopted which can be used with the dimensions shown in Table 8.5.

Table 8.5. Undercut

Rounding of the bearing	Undercut ba	Undercut ha	rc
1	2	0.2	1.3
1.1	2.4	0.3	1.5
1.5	3.2	0.4	2
2	4	0.5	2.5
2.1	4	0.5	2.5
3	4.7	0.5	3
4	5.9	0.5	4
5	7.4	0.6	5
6	8.6	0.6	6
7.05	10	0.6	7

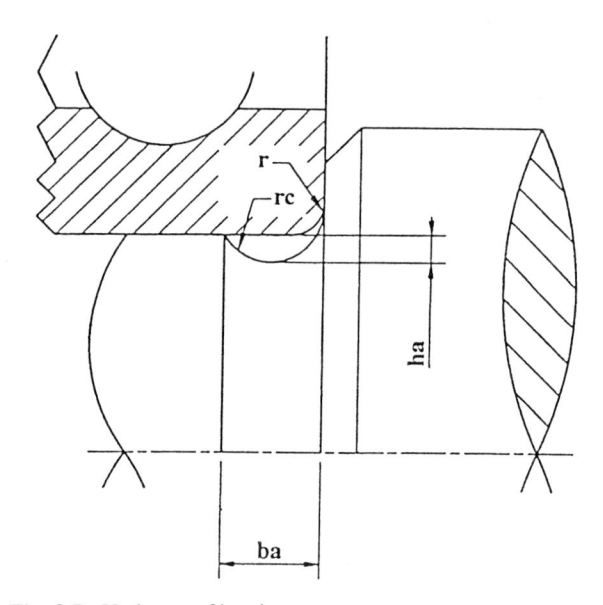

Fig. 8.7. Undercut of bearings

8.16 Dismounting of Bearings

It is necessary to predict the possibility of dismounting bearings in order to ensure their replacement. For this purpose it is necessary to provide the shaft with notches, in order to position the dismounting collet, or with screwed openings, in order to position the dismounting screws of the external clamping rings.

8.17 Maximum Speed and Number of Cycles

The application of bearings is limited by the maximum speed of rotation. This speed creates centrifugal forces that cause an expansion of the ring, so reducing the clamping of the shaft and producing additional stresses on the rolling elements. Maximum speed depends on the type of bearing and on its diameter. The larger are the mass of the rings for a given diameter and the base diameter, the lower is the maximum speed. The maximum speed is indicated in the catalogues according to each type of bearing. If the design of an element predicts the positioning of bearings, the maximum speed will be taken into account. The working speed of the designed and produced unit cannot be increased indefinitely without exposing bearings to serious problems if the speed limit is exceeded. In gear units, for example, we would be tempted to reduce the transmitted power in proportion with the increase in speed. The torque would remain constant and the working would be assured; nevertheless, if a certain limit is exceeded, this procedure will turn out to be dangerous, not only because the dynamic factor of the gearing risks assume too high a value, but also because the maximum speed risks might become exceeded. The number of cycles for which the bearing has been calculated is linked with the number of working hours and with the speed on the basis of the following relation:

$$L = 60 \ 10^{-6} \ h \ n$$

where L is the number in millions of cycles, h is the life expressed in hours and n the speed of rotation in min^{-1}.

8.18 Oil Seals

Bearings must be protected from the impurities present in the environment; we consider powders and other substances in particular, which can cause early wear to the rolling elements and the sliding planes. The lubricant must not seep from the gear unit. For this purpose oil seals, which are installed on the openings of the shafts, are used. For shafts inside the gear unit, seal covers, which insulate the inside from the outside ensuring a complete seal, are installed. There are friction or frictionless oil seals.

Friction Oil Seals. A soft gasket is fastened against the shaft. Owing to a slight pressure, this gasket produces a friction on the shaft and thereby the seal is made. The simplest gaskets are made up of a felt ring with a trapezoidal section; each ring is placed inside a spline that is concentric to the shaft placed on the fixed element. These gaskets are very simple, cheap and easy to be installed. However, as the gasket is only subjected to wear for a short time, it can only partially fulfil its function (Fig. 8.8 a).

Another method is to position an elastomer toroid ring inside a semicircular hollow, on the fixed element or on the shaft, which produces a friction on the element to which it is not fastened (O' rings)(Fig.8.8 b). These gaskets wear rather quickly, so they must be frequently replaced, although less frequently than felt gaskets. More complex gaskets with a better efficiency are also available. They

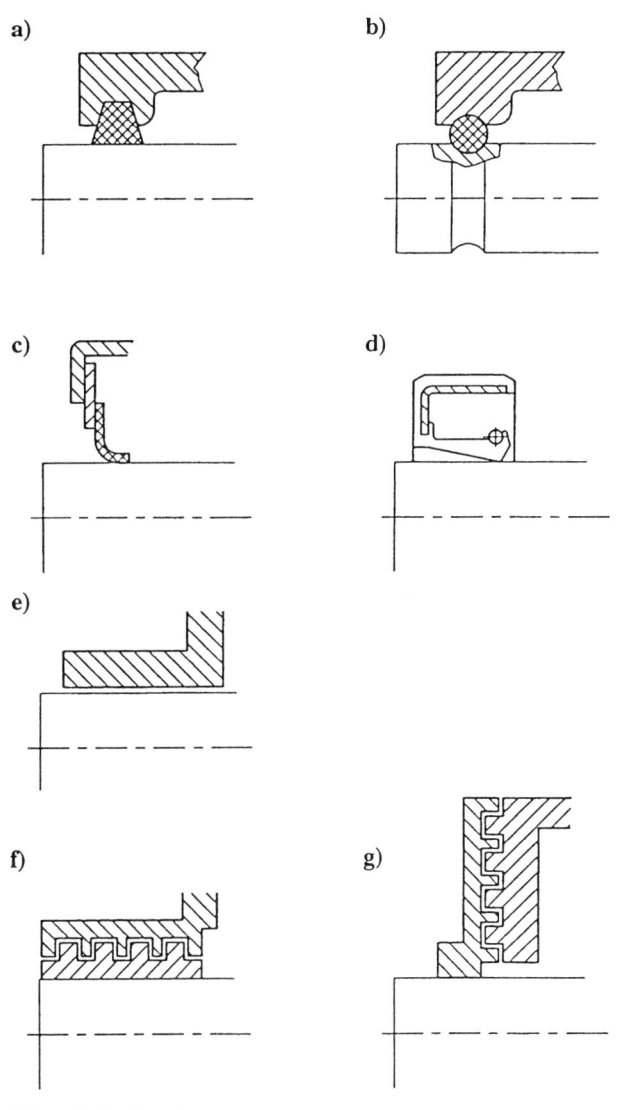

Fig. 8.8 Seal gaskets

are made up of an elastomer diaphragm whose central hole is smaller than the diameter of the shaft. These gaskets are installed on a light metal framework placed on the fixed element (bearing bore). Due to the pressure exerted on the gasket, this latter adheres to the shaft; wear is counterbalanced by deformation of the gasket. Such gaskets have a longer life than the other types of gaskets mentioned above (Fig. 8.8 c).

In order to achieve even better results it is possible to use elastomer gaskets with a more complex structure. The elastomer ring is placed on the shaft by an internal metal spring, which ensures a more durable pressure on the shaft independent of the degree of wear.

These gaskets are mounted on a light metal framework which can be fixed, without any difficulty, to the bore of the bearing; their replacement is easy (Fig. 8.8 d). The shape of the gaskets varies from one manufacturing industry to another, but they all follow the same principle. Being mass-produced by specialized industries which provide specialized catalogues, these gaskets are available at low costs.

Friction Gaskets. In order to be effective, friction gaskets produce relatively high degrees of friction with resulting reduction in the efficiency of the transmission. To avoid such frictions it is possible to use frictionless gaskets (with loss of and with the minimum overall dimensions). For these gaskets, in case some lubricant keeps out in a tight space, a loss of pressure occurs depending on the length and the section of the pipe.

If such a section is long enough and not excessively large and rigid, the loss of pressure will be sufficient to prevent lubricant from the seeping out and foreign bodies from coming in. This is facilitated by the presence of a lubricant in the pipe. The simplest method is to place a bore around the shaft, which has a reduced clearance provided the length is adequate (Fig. 8.8 e). The clearance must be as reduced as possible, but sufficiently large to prevent a contact between the fixed element and the shaft, when this latter warps during operation. The length required is a function of the section of the gasket. To obtain clearance cheaper, it is necessary for the length to be considerable since it affects the overall dimensions. To reduce the length, labyrinths are adopted. Instead of being continuous, the clearance is made up of two elements, sticking to one another (Fig. 8.8 f/g). Owing to the shape of the gasket, the length is increased a little and the change of direction at each indentation thereby increases the loss of pressure and, therefore, also the efficiency. It is possible to design gaskets concentric to the shaft (Fig. 8.8 f) or gaskets placed on a perpendicular plane to the shaft. Such gaskets are effective and help maintain the efficiency of transmission, but are detrimental to the overall dimensions, have a high cost and with additional difficulties during production. Therefore they are adopted only in particular cases, such as when the seal must be rigorous and the gasket that ensures the seal must have a long service life, without requiring to be replaced (either because the replacement is difficult due to access, or because the transmission must work without any supervision or lubricant losses).

9 Lubricants and Lubrication

9.1 Function of Lubrication

In the gear units lubricants have a multiple function. First of all they are absolutely necessary to reduce the coefficient of friction between the elements that slide in relation to another. They also facilitates heat exchanges between equipment by carrying the heat produced by the friction whereas the dissipation of such forms of energy is easier and more frequent. And finally they ensure that gaskets, which have an opening to the outside, are protected (between the rotating and fixed elements) against the entry of powders or other corrosives. Lubricants are essential to transmissions and can therefore be considered as a real mechanical unit. Knowledge of lubricants requires separate expertise, therefore it is not possible, in this book, to deal with every aspect concerning this subject. We will only treat what turns out to be more useful for a thorough understanding of the role played by lubricants in the working of mechanical units; we will also provide further information on the selection of lubricant.

9.2 Lubricants

There are two kinds of lubricant: mineral lubricants derived from oil and synthetic lubricants. Mineral lubricants can either be used pure or mixed with solid substances (greases) or other additives. Synthetic lubricants have a different composition and can be produced on the basis of precise working requirements.

9.3 Mineral Lubricants

The classification of lubricants is made on the basis of the following.
 (a) How they are obtained: mineral oils are obtained by means of oil distillation

and represent the heavier elements. After distillation they can be refined by a chemical treatment, for example with free sulphur, in order to eliminate possible impurities. Refined oils are the most used in transmissions to ensure good maintenance of the mechanical elements.

(b) Their destination: the properties of mineral oils can vary according to how these are produced. Such properties turn out to be particularly favourable in particular applications. Therefore oils are classified on the basis of their ultimate destination: there are oils for transmissions, oils for refrigerating machines, oils for turbines, etc..

(c) Their molecular structure (linked to carbon chemistry): there are open chain aliphatic compounds and closed chain aromatic compounds. Some oils are called paraffinic. Most oils are made up of different substances mixed together.

(d) Their additives: the properties of oils are modified by adding substances with specific functions. The main additives are used to:
- increase the load capacity: chemical organic compounds with chlorine, phosphorus and sulphur, e.g. paraffins or naphthene chloride, tributyl phosphates, zinc dithionic phosphates. The oils treated to withstand very high pressures are known as EP (extreme pressure);
  - improved oil properties (viscosity): metacrylic ester polymers, polyolefines, etc.

Such additives are:
- antioxidating: chemical compounds with phosphorus, zinc or nitrogen, e.g. phenothiazines, dithionic phosphates;
- anti-corrosive: organic esters of phosphoric acid;
- anti-foaming: silicones.

It is possible to improve both oil appearance (by adding dyestuffs) and oil aroma.

9.4 Greases

Greases are powdered solid elements containing mineral oils. The basic solid element can be made up of limestone, lithium or silicone. Greases are used for the lubrication of elements with a slight friction and a reduced load capacity. They can be also used for the lubrication of bearings, some of which are hermetic and lubricated with permanent greases. Greases are not used very often for the lubrication of speed reducing gears.

9.5 Synthetic Lubricants

Synthetic lubricants are different carbonate compounds.

9.6 Viscosity

9.6.1 Definition

Lubricants are characterized by their shearing resistance. Let us consider two sections distant from *dy* (Fig. 9.1).

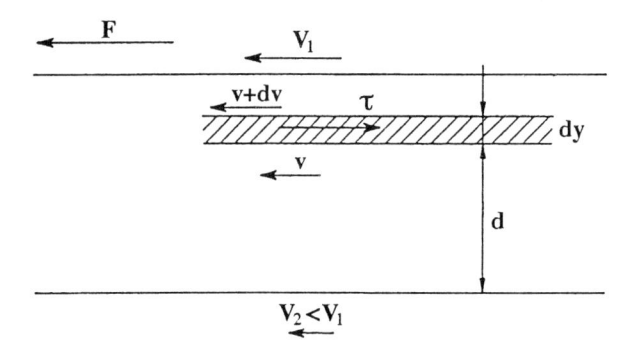

Fig. 9.1. Viscosity (Definition)

To move one of these sections in relation to the other with a relative speed, *dv* it is necessary to exert a shearing stress τ proportional to the speed variation in relation to the distance between the two sections. The factor of proportionality is given by the dynamic viscosity of the lubricant (Newton).

$$\tau = \mu \frac{dv}{dy} \tag{9.001}$$

where μ is the dynamic viscosity.

If the viscosity is constant in all directions and independent from speed, the fluid will be Newtonian. Lubricants are usually Newtonian. The dynamic viscosity is expressed in Ns/m^2 or in Pa s. There is also the kinematic viscosity ν given by the quotient of the dynamic viscosity μ divided by the lubricant density δ.

$$\nu = \frac{\mu}{\delta} \, . \tag{9.002}$$

The kinematic viscosity is expressed in m^2/s or more generally by a submultiple of such units, the mm^2/s or centistoke (cSt).

9.6.2 Variation of Viscosity by Temperature

Kinematic viscosity varies with temperature according to the nature of the lubricant. This variation is characterized by the Viscosity Index (VI). The viscosity of a lubricant at different temperatures can be expressed by a diagram (Fig. 9.2).

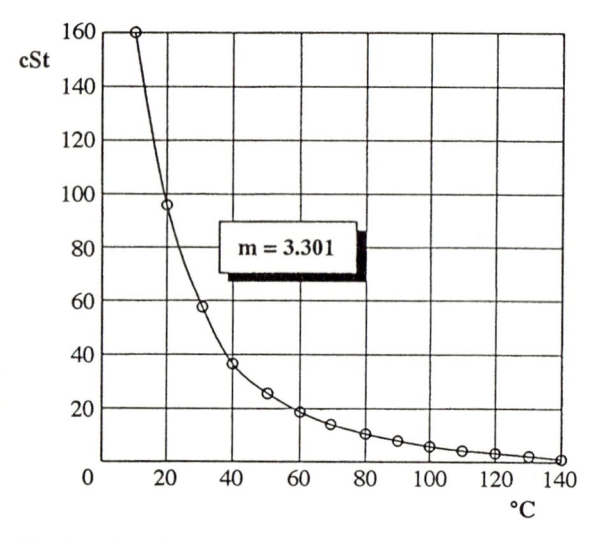

Fig. 9.2 Viscosity and temperature

According to Ubbelhode this curve can be analytically expressed by the function:

$$m = \frac{\log \log (v_1 + 0.8) - \log \log (v_2 + 0.8)}{\log T_2 - \log T_1} \quad (9.003)$$

Given the viscosity at a certain temperature (40°C for example) and the factor m characteristic of the lubricant, it is possible to calculate the viscosity at any temperature

$$v_t = -0.8 + (v_{40} + 0.8)^K \quad (9.004)$$

with

$$K = \left(\frac{313}{273 + t}\right)^m \quad (9.005)$$

The viscosity index VI is a function varying inversely to m. If $m = 3.6$ the viscosity index will be equal to 85. The oils usually used in mechanical transmissions have a viscosity index that is about 85 and an Ubbelhode constant m ranging from 3.2 to 3.6.

Synthetic oils have a higher viscosity index, that is they are less sensitive to temperature variations than to viscosity.

9.6.3 Viscosity Variation due to Pressure

The lubricants used in mechanical transmissions have a viscosity that varies little in relation to the pressure variations. At very high pressures the viscosity variation can, however, reach rather high values. As the viscosity variation in relation to pressure is very complex, it can only be expressed by empirical formulas.

According to Kuss it is possible to write:

$$\nu_p = \nu_1 \, e^{\alpha p} \qquad (9.006)$$

where ν_p is the viscosity at the pressure p, ν_1 is the viscosity at atmospheric pressure in MPa (N/mm²) and α is a constant characteristic of the lubricant. We assume that this constant has a value ranging from 1.1 to 1.6 . 10^{-5} for a mineral oil and that for a synthetic oil the value ranges from 1.5 5.0 . 10^{-5} So an oil with a viscosity index equal to 220 mm²/s at atmospheric pressure would have a viscosity index equal to 224 mm²/s at a pressure of 1500 MPa.

9.6.4 Measurement of Viscosity

Viscosity can be calculated directly by measuring the shearing stress for a given speed gradient. Some meters allow for the measurement of a conventional viscosity whose value depends on the characteristics of the meter used. So we calculate: Engler's degree (°E), Saybolt and Redwood's equivalent. All such measures are based on the egress of lubricant measured at a given temperature in a special measuring instrument; we consider the time for the oil's exit (Saybold) or we compare this time with that time for exit of water at the same temperature. The measurement most used in Europe is °E. The conversion of Engler's degree to kinematic viscosity in mm²/s is given by the following formula:

$$\nu \approx 7.85 \, E - \frac{8.32}{E} \, . \qquad (9.007)$$

A designation for viscosity is the abbreviation SAE , used especially in the car industry; it is not a viscosity measure but a simple standardized indication in which numerous properties of oils, such as oxidation resistance, anti-corrosive properties, viscosity at 0°F (-17.8°C) and at 210°F (98.9 °C) are considered.

9.6.5 Standardized Designation of Viscosity (ISO)

Standard ISO 3448 specifies how to symbolize viscosity and provides the preferential values relative to viscosity. Viscosity is calculated at 40°C and is expressed in mm²/s. A number preceded by VG specifies such viscosity. This number is a round figure and the designated viscosity index must not go beyond more than 10% of it. The viscosities preferred are VG 32, VG 68, VG 150, VG 220 and VG 320. The viscosity of an oil VG 220, for example, must be between 198 and 242 mm²/s.

9.7 Other Properties of Lubricants

Other properties of lubricants must be considered:
 - solidification point: this expresses the temperature at which the lubricant stops coming out because of gravity;
 - flash-point: this expresses the temperature at which an oil flares up under the action of an external flame;
 - spontaneous-combustion point: this expresses the temperature at which an oil flares up spontaneously.
The flash-point is 30-40°C lower than the spontaneous-combustion point;
 - specific heat: this property is important for heat exchanges. It can be evaluated in relation to density and temperature:

$$c = \frac{4.19}{\delta}(0.402 + 0.00081\,\theta) \qquad (9.008)$$

δ being the density in kg/dm³ and θ the temperature expressed in °C. The specific heat is measured in kJ/(kg. °K); for a normal mineral oil, the specific heat about is 1.85 kJ (kg °K).
 - oilness: this property is not measurable. It concerns adhering to the surfaces that must be lubricated and depends not only on the surface tension but also on the nature of the surfaces of contact;
 - different chemical properties: of course oils must not corrode the surfaces with which they come into cantact; they must remain stable as long as possible. However, with the passing of time they oxidize; oxidation is just one of the causes for which a frequent change of oil is required. Some oils (lubricating oils used in cutting tools) must be water-soluble.

9.8 Causes of Deterioration

There are many causes that give rise to oil deterioration and consequently frequent oil changes. The first cause is oxidation, followed by contamination from solid foreign substances (filings) or fluid foreign substances (condensation water). Such forms of deterioration have an important effect on oil properties which than deteriorate and risk becoming insufficient for lubrication.

9.9 Selection of Lubricants

Sometimes it is difficult to select the right lubricant because just one lubricant must carry out numerous functions. For example, the lubricant used to lubricate a gear is not necessarily the same as that used to lubricate a bearing, even if sometimes both a gear and a bearing can be lubricated with the same lubricant. The selection of a lubricant depends on the sliding speed of the elements that must be lubricated. Therefore several gears of a machine are often lubricated with the same lubricant. However, different lubricants should be used in relation to different speeds. A lubricant should also have certain properties when a slight friction is needed and the opposite properties when cooling is required. So the selection of a lubricant is often based on a compromise. Nevertheless such selection can influence the behaviour of some elements. For example, a lubricant suitable for a certain gear, could turn out to be less suited to bearings, as it affects their life negatively. For this reason sometimes it is necessary to consider the selection of a lubricant in the calculation of a gear. There are, however, some general rules: the selection of a lubricant for gears is carried out on the basis of its working temperature as well as on the basis of the sliding speed between the teeth (that is according to the reference speed).

The selection of lubricants can be made by following Tables 9.1 and 9.2.

Table 9.1. Use of mineral lubricants

Reference speed (m/s)	ISO degrees for ambient temperatures		
	from -40° to -5°	from 10° to 20°	from 10° to 50°
up to 10		150	320
from 10 to 20	see table 9.2	68	150
from 20 to 35	syntetic lubricants	32	68

Table 9.2. Synthetic lubricants for parallel gears and gear pair with intersecating axes

Properties	Information on ambient temperature			
	from -40° to -10°	from -30° to 10°	from -20° to 30°	from -10° to 50°
ISO degree	32	68	150	220
Minimum viscosity index	130	135	135	145

Synthetic lubricants are more advantageous than mineral oils, being usually stabler, having a longer life and working with a wider interval of temperature. Nevertheless synthetic lubricants are not always recommended. Each lubrication has its own properties and shows dangerous incompatibilities with other lubricating compounds such as an unexpected behaviour towards condensations due to humidity. Therefore before selecting a synthetic lubricant it is advisable to analyse its behaviour for the application considered and check that such lubricant does not dissolve some materials or coatings. Generally speaking, the user, lacking experience, will select a lubricant only after a consultation with the producer of the unit and the producer of the lubricant. If some elements of the elastomer unit are in contact with the lubricant, it will be necessary to test the actual dissolving properties of the lubricant and be careful.

9.10 Working at Low Temperatures

Gear units working at low temperatures must be filled with an oil that flows freely during the starting in order to avoid too high torques. Therefore it will be necessary to use an oil with a solidification point lower than the ambient temperatures and with a light viscosity in order to assure the right flow. In some cases a heating system for the lubricant is required in order to facilitate the setting in motion.

9.11 Working at High Temperatures

Working conditions and temperatures must not determine a heating above 95°C in the lubricant. The action of a lubricant at high temperatures on the elements or a gear unit, for example on elastomer gaskets, is to be considered.

9.12 Modalities of Lubrication of Gearing

Gears require dip lubrication or spray lubrication. Dip lubrication is obtained by dipping gear teeth into the lubricant, which is subsequently thrown on to the walls of the gear unit. The lubricant level must allow gear teeth to dip. Excessive lubrication could lead to additional stresses during transmission and the gears in a gear unit with more trains of gearing might not be able to remain at the same level. To avoid this, it will be necessary to mount devices that can keep lubricants at the right level for each gear. Generally speaking, dip lubrication also allows the lubrication of bearings. Spray lubrication is obtained by lubricating the seizing point by a nozzle and a pump. Lubrication can also be just for the front or the back part of the seizing point. Spray lubrication must allow for pipes in order to enable the bearings to be lubricated. Spray lubrication also allows a circuit to be installed for cooling the lubricant and consequently the gear unit. The lubrication of bearings must be assured by feed lines. Lubricants for spray lubrication have a viscosity lower than those used for dip lubrication. The selection between spray or dip lubrication depends on the centrifugal acceleration of the lubricant on the gear. If d is the diameter of the gear and N the speed of rotation expressed in $\min^{-1}$ the limit value of product $d.N^2$ will be approximately equal to $11 \cdot 10^7$ for rectified teeth and $9 \cdot 10^7$ for non-rectified teeth.

9.13 Elasto-Hydro-Dynamic State (EHD)

The lubrication state of gearing is called Elasto-hydro-dynamic state. Subjected to the speed and the pressure conditions of the tooth contact, the lubricant behaves as an elastic body. For minimum pressures, the behaviour is hydro-dynamic (as for example in sliding bearings). During hertzian contact between the two teeth, the pressure shows an increase from the entry to the exit and is characterized at the extremity by a peak followed by a rapid decrease (Fig. 9.3).

At this point, the level of the lubricant is minimum. According to H. Winter this level can be calculated by means of the following equation:

$$h_C \approx \frac{0.005 \left[\dfrac{u}{(u+1)^2}\right]^{0.3} (v_M \, v_t)^{0.7}}{\sigma_H^{0,26}} \qquad (9.009)$$

Viscosity v_M is the viscosity of the lubricant under temperature and pressure conditions in the point of contact. For a contact below 1500 MPa pressure of

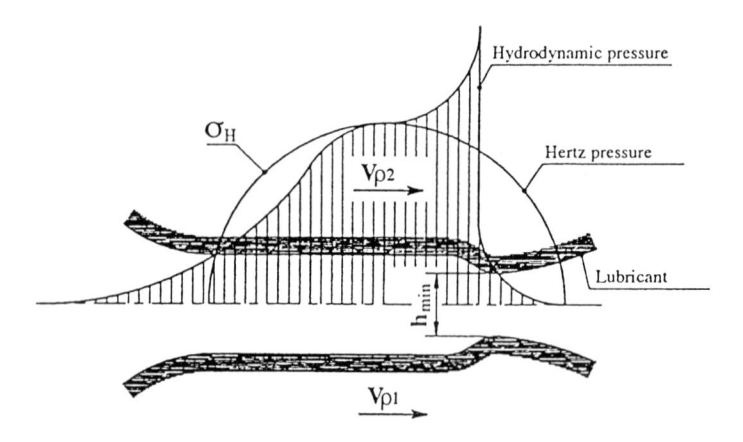

Fig. 9.3. Level of the fluid in EDH state

contact, a reference speed of 10 m/s and a local viscosity equal to 20 mm²/s, being the gear ratio equal to 4, the level of the lubricant is equal to 18 μm.

Notice that the level of the lubricant is in relation to roughness. The relation between this level and the roughness R_a is the thickness of the film of oil. Most gear units for general application have a specific thickness of the film of oil below 0.7. If the value of such thickness is higher than two, working will not produce any contact between the sliding elemertts: a positive condition as to wear but a negative condition as to lapping.

9.14 Scoring

If the elasto-hydro-dynamic state is established, a heat increase at the point of contact will occur. As temperature conditions vary according to geometry, the temperature at point of contact is given by Block's theory that establishes a flash temperature, which is expressed by the following equation:

$$\theta_{fla} = 0.62\, \mu_{inst}\, w_{bn}^{3/4} \left[\frac{E_{red}}{(1-v^2)}\left(\frac{1}{\rho_1}+\frac{1}{\rho_2}\right)\right]^{\frac{1}{4}} \frac{\left(\sqrt{v_{p1}}-\sqrt{v_{p2}}\right)}{B_M} \quad (9.010)$$

with:
    θ_{fla} = flash temperature
    μ_{inst} = coefficient of friction at the point considered
    w_{bn} = normal specific load on the teeth (N/mm)
    E_{red} = $2.E_1.E_2/(E_1+E_2)$ (N/mm²)

$\rho_{1,2}$ = toothing bending radii at the point of contact (mm).
v_{p1}, v_{p2} = rolling speed of the two teeth at the point of contact.
B_M = coefficient of heat resulting from contact = $13.6 N/(mm.s^{1/2}.°K)$.

It is very difficult to calculate the coefficient of local friction. The spontaneous flash temperature beyond a certain value, which is different for each lubricant, causes the immediate destruction of the film of oil and consequently the seizing even if the quantity of oil is enough.

This form of deterioration is extremely serious and there is no way either of predicting it or calculating it. A technical report by ISO tackles such problem and provides two different methods of calculation: the principle of the spontaneous flash temperature on the integral temperature. The first follows Block's rule exactly; however, hypothesis on the coefficient of local and instantaneous friction is required. The principle using integral temperature is based on the sum of the temperatures at the various points (average) and is determined by empirical factors resulting from tests or experience. It is not possible either to adopt definitely one of these methods or to reach a compromise between them, as there is no experience in its calculation. For this reason the ISO committee has decided to refer to the technical report and not to draw up a new standard, hoping for future progress for further applications of the formulae. In both the two cases the calculated temperature must have a value below that of a limit temperature (which is different for the two principles). These temperatures are calculated by friction and rolling tests carried out by specific testing instruments. However, as the thermic scuffing occurs only at speeds, there is practically no risk of scuffing at the speeds that prevail in general mechanics. But this risk must be considered in general turbines and navy engines.

10 Housings

10.1 Definitions and Functions

Housings are boxes with numerous functions. First of all they are used to support shafts by bearing housings. Second they contain lubricant and, third, they fix gear units in their working points. According to the first function they must be resistant as they must keep shafts rigidly in position. Therefore they are usually made of steel and their dimensions are calculated in order to bear the stresses transmitted with a minimum distortion. Secondly, they must be hermetic in order to avoid any leakage of lubricant. And, third, they must be equipped with carefully installed strong fasteners and support surfaces.

10.2 Materials

Housings are made up of iron steel and sometimes of light alloys (aluminium). The latter having small dimensions and being obtained by means of die-casting (extrusion). Steel housings are mechanically soldered. Therefore they are made of carbon steel available as a thick plate. Bearing housings are generally obtained with the working of forged carbon steels. The different elements are assembled by soldering. Cast-iron housings are made up of grey pig iron or by spheroidal iron. As they are cast, their shape must comply with the requirements of the elements casted. Light-alloy housings behave in the same way.

10.3 Primary Function: Support the Bearings

To support bearings and transmit the reactions to the "foundations", housings must bear such reactions without warping excessively. Distortions in housings create a misalignment of the shaft and consequently a reduction of the load capacity of gearing. As housings have a complex shape their calculation turns out to be rather difficult. Housings can be considered as plates subjected to traction or compressive stresses, the latter resolving in buckling. As the calculation of housings is approximate, it must be made with high values having a high safety margin. It is very difficult to predict the distortion in housings. Finally, there could be problems related to possible vibrations (see Chapter 12).

To determine the tensions and distortions occurring in housings, new calculation procedures are used. These are known as calculation with finished elements which produce results that are much more satisfying than the simple calculations based on the strength of materials.

10.4 Calculation with Finite Elements

It is not intended to develop this calculation procedure here but rather just report its basic principles. Sophisticated programmes, which enable experts in the field to make such calculation, have been developed.

The principle is to apply the theories of strength of materials not to solids as a whole but to small elements. Such elements, which make up the solid, are complete and are not infinitely small dimensions. Each element is characterized by a distribution of the stresses and the resulting distortions, as such stresses and distortions are influenced by the adjacent elements.

The solid is therefore hypothetically divided into elements whose co-ordinates of the extremes are subsequently determined. As each element functions as a link, such procedure is called FEM mesh.

The equations of the equilibrium of forces and stresses as well as these distortions are determined for each element progressively from one element to another. The resulting equilibrii provide numerous equations with many unknowns. Some constants are adopted in relation to fits and external forces. External forces are usually known in relation to their position on the solid. Sometimes it is difficult to determine binds as it is not possible to determine where tensions have specific values. In such cases it is necessary to go by experience and rely on previous calculations obtained by using the same procedures and subsequently advance some hypothesis. The equation system with multiple unknowns is solved electronically; this procedure allows tensions and distortions at any point to be determined.

The existing programmes facilitate calculations, automatically determining the FEM mesh by following the experience of the user. This latter must have an adequate knowledge of such calculations in order to be able to determine an appropriate FEM mesh in relation to the problems that must be solved. In particular the user must know the exact position of the most delicate points in

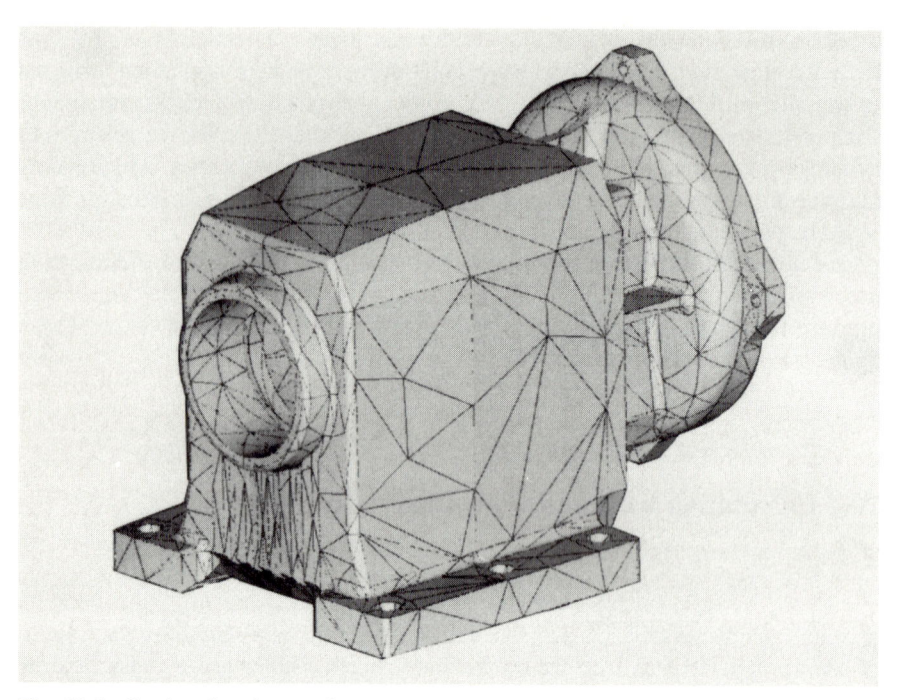

Fig. 10.1a Gearbox housing mesh

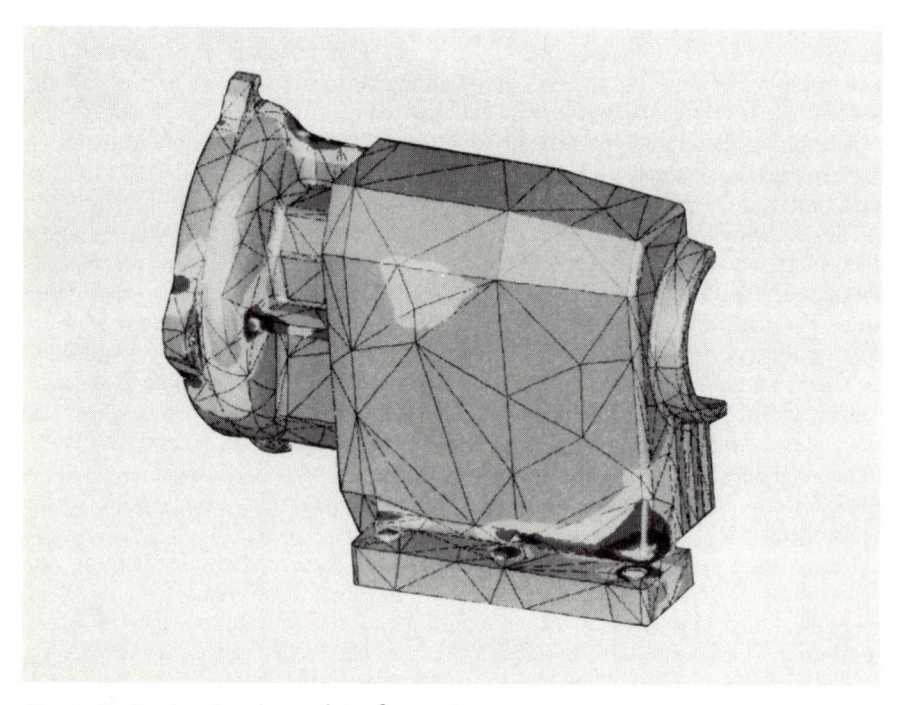

Fig. 10.1b Gearbox housing: points of max. stress

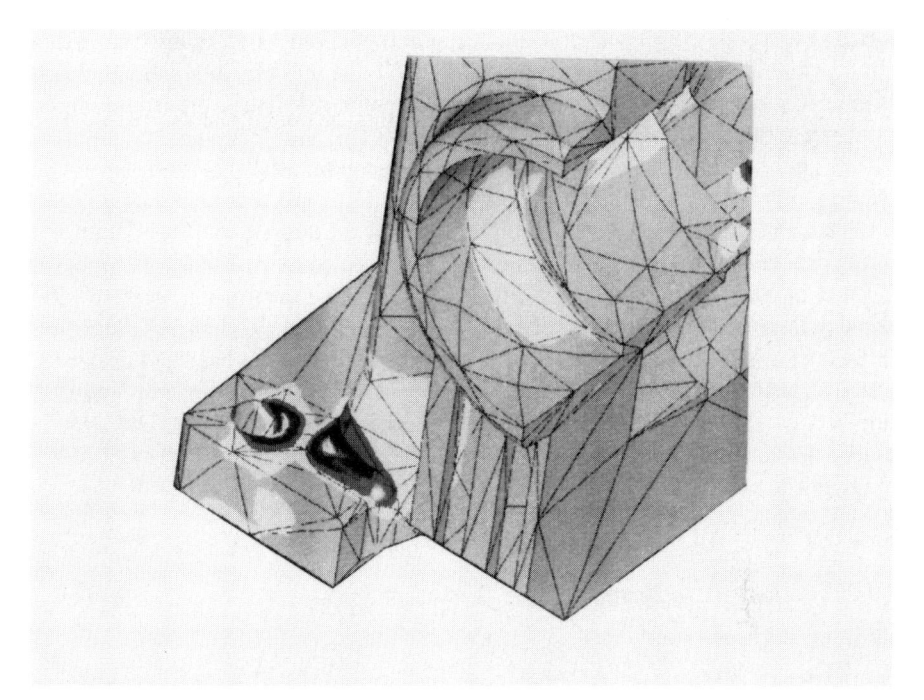

Fig. 10.1c Gearbox housing fixing area: max. stresses

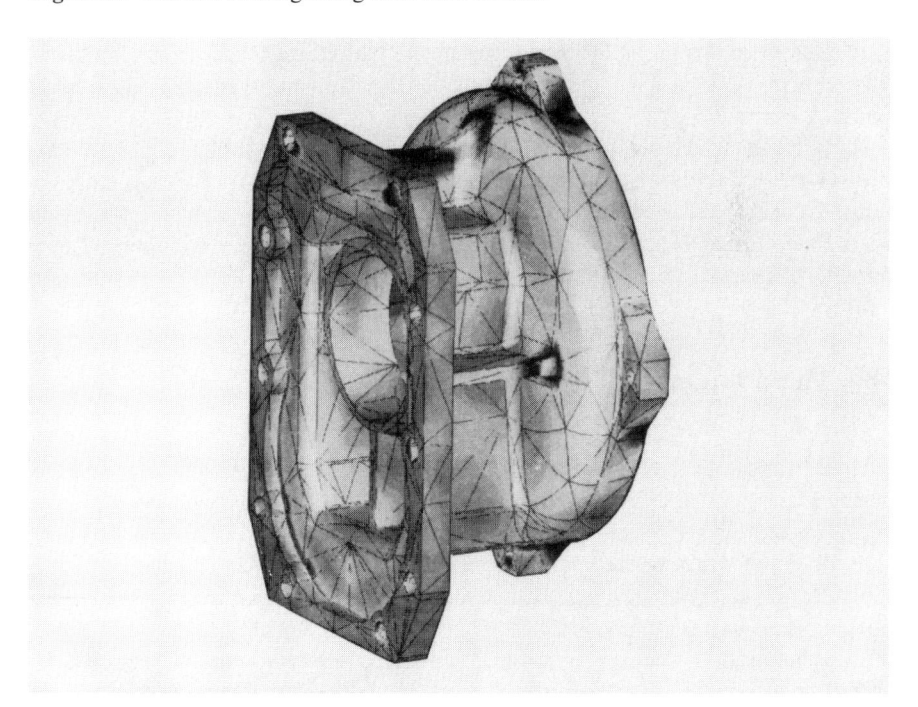

Fig. 10.1d Motor flange: max. stress

order to increase the FEM mesh at such points without being obliged to create another that is minute on the whole solid. If the FEM mesh is too minute where it is not necessary, saturation of the programme, length calculation procedures and a price rise will occur.

The operator must also determine the precise points where stresses will be exerted as well as the points where stresses are null in particular.

At the end of the calculation, the distribution of the stresses and their value and of the distortions are established; therefore it is possible to modify the FEM mesh, in order to make some points more precise, or to modify the working design of the housing in order to improve its behaviour. This results at the end of the calculation in a housing with the right shape for the purpose for which it has been designed, characterized by stresses and distortions with acceptable values. All the imperatives concerning manufacturing as well as those concerning welded or casted elements can be taken into account. Some examples of FEM mesh is shown in the Fig. 10.1a, 10.1b, 10.1c, 10.1d.

10.5 Second Function: Seal

Housings are made up of different elements in order to allow mounting of the internal elements.

Housings can have numerous shapes which vary according to design of the gear units; it is not possible to analyse all the manufacturing procedures. Apart from the housing shape, some elements, if not all, are subjected to stresses resulting from working.

The different elements are usually assembled together by bolts and nuts. The stresses exerted on bolts produce an elongation of the bolts. As the seal must be assured, the lips of the gasket, positioned between the different elements, must not open. Therefore bolts are subjected to pretensioning at the time of mounting. Later we will consider the calculation of bolts.

10.6 Third Function: External Fixing

Gear units must be fixed to units that support the gear unit, the conductor and the driven machine. Therefore it is necessary to predict housings with possible fits with "foundation", term used in its more general meaning. If the gear unit has been designed and produced for a precise purpose, its position in relation to the "foundation" is predetermined so that the bearing surface as well as the fixing openings can be accurately selected. Consequently the resulting gear unit has a precise function and is not used for other purposes. It will not be possible to predict precisely the fixing elements if gear units are designed independently from their intended application and are therefore used for general purposes. It is up to the expert to design housings evaluating all the possible positions that could occur in relation to the operational reliability of the gear unit (lubrication in particular must be considered). Even if the designing of gear units for general purposes differs from one producer to another, it allows, however, the fixing of the gear unit and consequently of its housing, on horizontal and vertical surfaces. Some gear units are designed as gear boxes. In this case the engine is an integral part of the unit and connecting flanges of the engine must be predicted. In other cases gear units must be able to rotate around the driving shaft, as there is a torque reaction arm whose function is to keep gear units fixed. In this case a fixing flange of the arm must be installed. The fixing of housings is obtained by means of bolts, which must bear the stresses that tend to separate the housing from the "foundations". Such stresses, which result from the working, are internal and external forces exerted on bearings.

10.7 Accessories

All necessary accessories must be incorporated into the design of housings. For example, it is necessary to determine charging and discharging devices for the lubricant. Such operations must however be easy and not require the dismounting of the housing. It is also necessary to incorporate a control system of the lubricant level during the working of the machine. Housings play a very important role in the maintenance of gear units. They are provided with devices strong enough to raise gear units. The positioning of gear units and their calibration in particular must be carried out accurately. For this purpose there are reference surfaces where control apparatuses and measuring instruments are installed. Such surfaces are used as reference surfaces during the mounting of housings on machine tools.

10.8 Working and Precision of Housings

Housings have rigorous imperatives due to their function, that is to support bearings and consequently axes. Bores must therefore be produced with the precision required by bearings (see section 8) in order to assure the parallelism of the axes (see section 4) and the precision of the tolerances required by the centre distances. Also the dimensions of the axes on the fixing plane must be respected in order to allow right mounting with the mating machines. The mounting surfaces of the different elements and of the covers must be accurately determined. They must have a right roughness degree, as roughness influences the efficiency of gaskets. The same applies to the mounting faces of housings on "foundations". In the mechanically soldered housings, weldings can create residual stresses, which while decreasing after the working process, can determine a series of distortions. Such distortions affect the resulting tolerances negatively and consequently also the correct general working. Therefore welded housings are produced in order to reduce the risk represented by the above-mentioned stresses; in many cases the welded housings are subjected to tempering in order to eliminate any stresses before working. Also casted housings can be characterized by residual stresses due to the casting process. Such stresses disappear for natural ageing. Housings are usually deposited (in a suitable place) for a period of time long enough to allow the elimination of stresses. Such ageing can also be carried out artificially by heating housings in special overns at low temperatures.

10.9 Mounting Screwed Elements (Bolts and Nuts)

Bolts are cylinders with one extremity screwed and with the other provided with a head that is usually hexagonal. In contrast, nuts have the same thread inside as bolts, but their external shape is like the bolt head. The hexagonal shape is not casual: it allows tightening by a special tool called a key. Bolts have been standardized by ISO both in terms of the properties of steels and relative designation and dimensions (ISO profile) and tolerances. The designation of steels is given by two figures. The first one characterizes the resistance class of steels (static breaking stress) and the second characterizes the tenths of the ratio between the elastic limit and the breaking stress. The usual classes are 4,5,6,8,10,12 and 14. The static breaking stress is obtained by multiplying the figure of the class by 100; so class 8 is characterized by a breaking static stress equal to 800 MPa. The second figures are 6.8 or 9 correspond to 0.6, 0.8 or 0.9 as the ratio between the elastic limit and the breaking stress. Bolts designated with the class 8.8 have a breaking stress equal to 800 MPa and an elastic limit equal to 640 MPa. The corresponding nuts are designated by the figure of the bolt class. According to this figure the highest degree of resistance of the nuts is the one that causes breaking of the bolts under such load. We will characterize by σ_R the breaking stress and with σ_E the elastic limit. The dimensions of bolts and nuts are provided starting from a standardized nominal diameter and a correspondent thread. The screw thread has a triangular profile with an angle at the vertex of 60°. The internal bore of the nut divides the point of the thread into sections. The thread sides of the bolt and the nut are fixed at the base by an arc of a circle. Based on the nominal diameter d (mm) and the thread P (mm) the principal dimensions are (Fig. 10.2):

Bolts:
Outside diameter (d)   = nominal diameter
Diameter of the core (d_3)   = $d - 1.22687\ P$
Medium diameter (d_2)   = $d - 0.64953\ P$

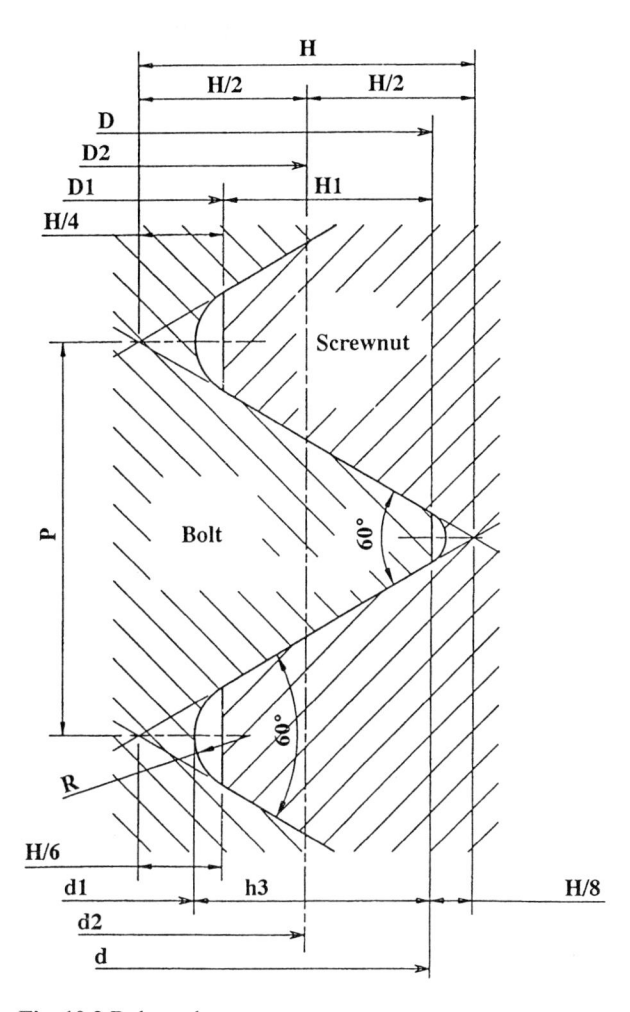

Fig. 10.2 Bolts and nuts

Nut:
Bore diameter (D_1) $= d - 1.08254\ P$

The diameter inside which the hexagonal prism constituting the head of the bolt and the nut is inscribed, is equal to $2d$. The height of the bolt has a value of about $0.8\ d$, while the value of the thickness of the nut is about d. The deviation between the two parallel faces of the nut or the two parallel faces of the head of the bolt is equal to $0.866\ d$.

The area of the minimum section (at the bottom of the thread) of the bolt is equal to:

$$A_{min} = 0.7854\ (d - 1.2267\ P)^2.$$

The area of the medium section of the bolt is equal to:

$$A_{med} = 0.7854 \, (d - 0.64953 \, P)^2.$$

Table 10.1 provides the standardized dimensions up to 30 mm of diameter.

Table 10.1 Standardized dimensions of bolts

Nominal Diameter	Pitch
3	0.5
4	0.7
4	0.8
6	1.0
8	1.25
10	1.5
12	1.75
16	2.0
20	2.5
24	3.5
30	3.5

In order to secure the seal of the gaskets and the stability of "foundations", bolts are subjected to pretensioning in order to allow for a force between the assembled elements, under the maximum force beared by the assembly. Let us consider the flanges with a thickness e mounted by means of a bolt. Under the tightening force, the bolt elongates to δ_1 and the flanges shorten to δ_2. It is possible to report the elongation and the shortening variations in relation to the force on a diagram. Such variations are expressed by a line. The common pretensioning force F_v is represented by Fig. 10.3, which shows on the left the elongation variation of the bolt in relation to the force, while on the right it is possible to observe the reduction of thickness of the flanges in relation to the same force.

With a force F_{max} lower than the pre-tightening force, the assembly remains unchanged. With a force F_{max} higher than the pre-tightening force, the bolt elongates and the force on the flange decreases. The force F_{max} is divided into two forces; F_{eff} that is the force exerted on the assembled gaskets and F_{flange} that is the residual force on the flange. As long as the force remains higher than 0, the flange remains tight. The tightening stress is expressed by:

$$\sigma_n = \frac{F_{max}}{0{,}785 \, (d - 0.64953 \, P)^2} \qquad (10.001)$$

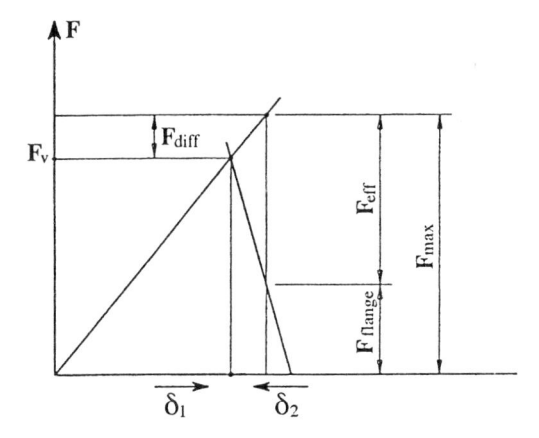

Fig. 10.3. Force present in the bolts

To apply the pretensioning F_v a torsional stress is exerted on the bolt. Such torsional stress is given by:

$$\tau = \frac{16 \mu \, d \, F_v}{\pi \, (d - 1.22867 \, P)^3} \quad . \tag{10.002}$$

The resulting stress is expressed by Hencky-Von Mises' principle.

$$\sigma = \sqrt{\sigma_n^2 + 3 \tau^2} \quad . \tag{10.003}$$

The value of the stress must remain below the elastic limit shown above. If care is not taken, the coefficient of friction μ could reach a value equal to 0.2. It is possible to decrease the coefficient of friction by using a lubrication grease (oil or wax applied on the base of the nut) or bolts provided with a special lining, such as for example galvanized bolts. A pretensioning force equal to 75% of the force corresponding to the elastic limit of the material is usually selected. Therefore:

$$F_v = (0.75 \quad 0.785) \, \sigma_E \, d_2^2 = 0.58875 \, \sigma_E \, (d - 0.64953 \, P)^2 \, . \tag{10.004}$$

Such pretensioning is sufficient to ensure the seal has a good fixing. Sometimes bolts work edge-on. In this case, in order to avoid an additional bending stress, it is necessary to make sure that bolts are placed in their housing with a minimum clearance. The stress allowed under the maximum stress is the elastic shearing stress τ_E, that is:

$$\tau_E = \frac{\sigma_E}{\sqrt{3}} \quad . \tag{10.005}$$

The best solution is to install the bolts in the housings so that they will be subjected to the minimum allowed stress during working. This is way well-designed housings have, when possible, a gasket perpendicular to the shafts. The other stresses, different from axial stresses, are absorbed by the housing at the point of obliquity in relation to the axis and not to the bolts. The covers that close the bearing housings are subjected to axial stresses of the latter. The bolts that close the bearing housings must be correctly dimensioned. Bolts are kept tight by means of special devices, which prevent them from collapsing under the action of the inevitable vibrations.

10.10 Economic Conditions Influencing the Selection of Housings

There are mechanically welded housings and cast-iron housings; the selection is based on purely economic principles. Mechanically welded housings are used for single elements or are produced in small series, since cast-iron housings show different characteristics from those of welded housings. However, the designing of cast-iron housings requires very expensive dies which must be amortized by mass-production, as for the mass-produced gear units sold in catalogue. Their production allows a minimum variety of shapes and dimensions to be made available; so in a foundry it is possible to produce many housings by means of the same die. As we will see further on when we will deal with the study of vibrations, cast-iron is very malleable and turns out to be a material of particular interest as it has more absorption properties than steel. Mechanically welded steel housings must also be worked starting from plates available on the market. For the mass-production of housings, it is therefore possible to use spheroidal cast-irons of good quality, which are usually better than rolled plates. Mechanically welded housings are used for very special applications for which it is not possible to find standard solutions.

11 Standard Gear Units

11.1 Definition and Fields of Application

To solve a problem of transmission it is possible to design a special gear unit as follows. Its study can be prolonged or rapid. In the first case the result is a good product characterized, however, by exorbitant production costs. In the second there could also be some problems during operation as the solution has been adopted in a too hurried way; the different possibilities may not have been considered; no improvements of any kind might have been studied or the improvements predicted have been applied in too hurried a way. The production costs of a gear unit, whose particular dimensions are linked to its future application, are extremely high; in fact in mass-production it is not possible either to consider the dampening factor of the material or to adopt better solutions, as it is necessary, for example, to use welded housings instead of cast-iron housings, which are usually considered preferable. So it is possible to produce a special gear unit for each specific application if such application cannot be solved by means of other methods owing to reasons of specificity or particular dimensions. If there are no problems of price, it will be possible to start the production, which is extremely difficult, of the particular gear unit. Gear units for the motor-car industry (gear boxes and transmission axles) are mass-produced and have a particular design. These transmissions always have a particular solution to specific the type of automobile, even if automobiles are mass-produced and consequently their costs are amortized. In the aircraft industry the high costs of the prototypes, which are subjected to prolonged laboratory tests and real tests due to safety requirements, are justified by the fact that passengers' life is involved. Under identical conditions the same rule of motor-car industry is valid for gears working at high speed (gearboxes for turbine or multipliers). However, they are often single elements and incur very high expense due to the installation costs. Nevertheless they are studied by experts who can use standardized production methods. In industrial gears (referring to various industrial applications at medium or low speeds) the two commonest problems are the reduction of speed from a shaft to another and the modification of the torques transmitted to the different shafts, there is no reason to adopt particular solutions. The speed ratios are in general acceptable with a high tolerance within precise limits. Dimension requirements, if considered, are not particularly rigorous. Standard gear units are mass-produced and are provided with elements that are mass-produced too. Such elements are interchangeable and being mass-produced they have low production costs. The producers of such gear units are experts as they have a great deal of knowledge in the field. Standard gear units can require lengthy study in order to be sure that all possible problems have been considered before starting their design and production. Production methods are more advanced, as buying prices and maintenance costs of the machines are always amortized by the mass-production or the production is limited to a range with well-determined dimensions.

11.2 Design Principle

To amortize the production costs of cast-iron or light alloy housings, a very small range of housings both as to shapes and dimensions, is used. Such housings have similar shapes but different dimensions. Similar housings are used for different products, within the limits ot the possible, where possible without affecting their functions in a negative way. Reduction ratios are usually in geometrical progression are often determined according to the Renard series. Being influenced by the whole number of teeth, their value is close to allowable theoretical value, so that best results as to the required progression can be obtained. The combination of more stages is in geometrical progression with the combination of progressions of the single stages. So it is possible to interchange stages: the last or the second stage of a low transmissible power gear unit (so with minimum overall dimensions) can become the first stage a high power gearbox (with large dimensions). Centre distances are determined by the desired single stage. The series of the centre distances is close to geometrical progression, as the dimensions of the housings are influenced by the centre distances. As the number of stages is reduced by the standardization of dimensions, the number of the considered elements also decreases and their study as well as their production turn out to be cheap. The same is valid for shafts, bearings and all the accessories. The studies carried out by skilled experts are amortized too by mass-production, consequently they can be very prolonged (and therefore complete) without influencing the buying price. Such studies enables high quality products to be obtained. The number of housings necessary to obtain the series is extremely small. Consequently it is possible to study housings carefully and to calculate the most adequate resistance. They can also be designed to be fixed on different bearings. With a few models it is possible to solve practically every problem.

11.3 Types of Gearboxes

Generally speaking different types of gear units are available on the market in ranges characterized by very large dimensions. The most used types of gear units now available on the market are:
- parallel shaft gear units with one, two or three stages in compliance with the required total speed reducing ratio (being the stage below 7.5);
- perpendicular shaft gear made up of a single tapered stage (right angle gearbox) or a tapered stage with one or two parallel stages. The tapered stage can be the input stage or the second stage in the case of two parallel stages;
- worm gearboxes. They are compact for a specific transmission and allow for speed reducing ratios ranging from 6 to 100 with only one worm gearing. Their mass-produced fit allows even more important speed reducing ratios with minimum dimensions. However, they show a worse efficiency and a lower heat-capacity. Another disadvantage is represented by their irreversibility in a high speed reduction ratio by means of worms with a small lead helix angle. There are torque limiters designed to avoid overloadings in case of abrupt stopping;
- planetary gear. These gears are considered interesting because of their minimum dimensions and their high mechanical capacity. It is possible to fix them easily to a mass-produced gear pair, so increasing their speed reducing capacities. They can be designed for housings with one or two gear stages. Despite their disadvantage of having a slightly reduced heat-capacity due to their compactness, they are characterized by a very large field of application as a consequence of the co-axial input and output shafts. Planetary gear stages are largely used in conveyor belts, in the individual drive of live axles, in the field of transports and agricolture and in capstans;
- gear units with a reduced backlash. Such gear units must be particularly accurate as their reduced backlash depends on the backlash of centre distances for which a high degree of precision during their production turns out to be fundamental. Consequently they are provided with gears characterized by an exceptional high degree of precision. In order to be used for many applications (radars, machine tools, etc.) an extremely accurate mean of positioning, the rotation of the output shaft must be perfectly proportional to the rotation of the drive shaft (engine) in the two senses of rotation. Therefore a high degree of rigidity of all the elements as well as a reduced backlash in the bearings are required.

There are also special gearboxes, such as those which are particularly flat and which can be mounted directly on the shaft of the driven machine, and two or three stage planetary in-line gearboxes. Also these are used in conveyor belts as well as in other applications.

For the applications of gear units see Chapter 14.

11.4 Selection of a Gear Unit

After selection of a gear unit to be adopted in relation to a particular application, the selection depending on specifications, the positioning of the conducting and the driven machines, and knowledge of special conditions, such as probable dimensions, it is necessary to determine the adequate dimensions for the gear unit.

Data required by the user are the transmission power and the desired input and output speeds (or input and output torques). In all these cases it is possible to determine an input power, an input speed and the speed reducing ratio required.

Beginning with output characteristics, the efficiency will need to be considered, if the input power is to be determined. The efficiency is calculated starting from the speed ratio, which influences the number of the probable trains. A 2% losses for each stage, a 0.96 efficiency for two stages and a 0.94 efficiency for three gear stages can be calculated approximately these figures being intended in terms of safety.

For worm gearing, producers usually provide efficiency curves in relation to the speed reduction.

The power influences the dimensions of the gear unit, which can be selected from a catalogue in relation to the power and to the speed reducing ratio nearest to the one required. It could be possible to follow the above-mentioned procedure, if use of the gear unit is made under the same conditions in which the calculation has been made by the producer, but as this happens very rarely the operation factor is introduced.

11.5 Operation Factor

In the calculation of gears, we have noticed that the operation factor was influenced by factors depending on the dimensions and the accuracy of gears as well as condition of use. These latter are represented by the application factor, which depends on the conducting and the driven machines or the applied load spectrum, the endurance factor, which depends on number of cycles to which the gearing is subjected, the reliability grade required and the safety factor in gearing for the application.

It is clear that at the time of the design and the production, the producer does not know the operation conditions of the client's gears. Consequently the producer will adopt precise conditions, which will enable him to select the suitable factors. The power indicated on the catalogue is in relation to such selection. In a gear, the result of the combination of these factors is a factor known as individual operation factor. Obviously there is only one individual operation factor for the pinion and the wheel of each train, for the contact stress and for the bending of each of these elements.

As there are also different operation factors for shafts and bearings, which can also be referred to housings, it is not possible to calculate a combined factor for each particular application by which the nominal power indicated in the catalogue can be multiplied or divided. The producer knows empirically the possible value of the operating factor for the different applications due to his experience in the field of particular cases, technical periodicals or tests on the site. The producer's experience can integrate with that of the user and consequently be enriched.

The factor described in this paragraph is called the operating factor (for a given application) and must not be confused with the application factor.

11.6 Equivalent Power

Once the real input power of the gear has been determined, it is possible to calculate an equivalent power by multiplying such power by the operation factor relative to the type of application considered. The equivalent power is the same as that obtained with a spectrum of loads. For such particular application the recommended dimensions of the gear unit are those of a gear unit capable of transmitting the nominal power (according to the catalogue) with a value immediately higher than that of the equivalent power.

11.7 Example of Calculation

Let us consider a machine, rotating at a speed of about 10 min^{-1}, which must be set in motion by an asynchronous motor rotating at 1400 min^{-1}

The output torque of the gear unit must be equal to 20000 Nm. The result is that the total speed reducing ratio must be equal to about 140 and consequently a three stage gear unit is required. As the efficiency considered is equal to 0.94, the value of the input torque is 20000 / (0.94 . 140) = 152 Nm.

The corresponding power, at 1400 min^{-1} is equal to 23 kW.

The application under consideration is thought to require an operation factor of 1.25 with a resulting equivalent power of 28.75 kW. For example, for an input speed of 1400 min^{-1} catalogues suggest a power of 26 kW and subsequently of 31 kW for a gear unit. This latter is the one selected. Dimensions are illustrated in the catalogue.

11.8 Factors that Influence the Operation Factor

The operation factor depends on conditions of use.
   Such conditions include:
   – the type of conducting machine: synchronous, asynchronous or direct current motor, multicylindrical or monocylindrical bursting motor, turbine, air motor or hydraulic motor, etc..
   – running time and running frequency: number of running hours per day, number of running days per year, expected service life expressed in days, etc..
   – severity during operation: machines using gear units work in different ways. It is not possible to compare a ball-mill to a conveyor belt of ground products with a constant loading; which is like comparing a ventilator to a machine for the production of paper, etc..
   Such conditions must be included in the operation factor.

11.9 Particular Precautions

The operation factor includes only the normal working conditions. It considers overloadings and the dynamic conditions that occur in normal conditions. Particular conditions produced by external conditions linked to the dynamic response of the unit comprise the conducting machine, the gear unit and the driven machine must be carefully examined. These external conditions depend up on other machines which are different from the gear unit. It is necessary to study the mounting conditions on the foundations as well. The torques at the starting are considered according to the limits of the type of engine used by the producer and up to a maximum of twice the nominal torque multiplied by the operation factor. These torques are limited, in terms of number, usually to 10000 for the whole service life of the gear unit. If such conditions are not satisfied, it will be necessary to select another gear unit or to mount a particular protection system. If radial loads are applied to the shafts fixed on the housings , it will be necessary to adopt special precautions and intervene with the equivalent power. These radial loads are generated by the transmissions of the overhanging pinion belt pulleys and chain gearing. The consequences of such radial loads are an increase in the stresses to which the shaft is subjected, additional distortions of the shaft resulting in a further misalignment of the teeth of the gear considered and a reduction of the load capacity.

11.10 Gear Motors

Manufacturers also supply gear motors, which are gear units mounted directly on a suitable motor. The advantage is that a motor is selected by the producer, on the basis of his experience and according to the conditions of calculation of the gear unit (torque at the starting, possible overloading, etc..). In addition, gear motors have the advantage of having a high degree of compactness and minimum dimensions in comparison with a gear unit with a separate motor. Another advantage offered by gear motors is their dynamic response to the vibrations due to torsion (see Chapter 12), which is well-defined in a unit made up of motor and gear unit; the only unknown factor being by the dynamic behaviour of the driven machine. The influence of this on general behaviour can be avoided by using an elastic bending fit.

12 Vibrations and Noises

12.1 Natural Vibration

If a unit made up of a spring and a mass is subjected to distorsion in the spring, when the unit is set free, it will than be subjected to a periodic sinusoidal movement in relation to its position of static balance (Fig. 12.1).

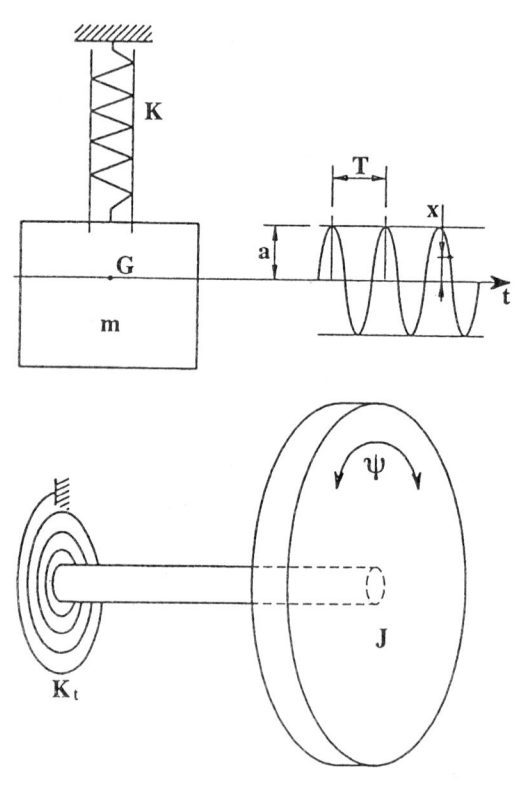

Fig. 12.1. Natural vibration

The frequency of the periodic movement is given by:

$$\omega_0 = \sqrt{\frac{k}{m}} \qquad (12.001)$$

m being the mass and k the elastic constant of the spring.

This frequency is the frequency found in an elastic system. This law is universal whatever system and type of considered distortion there might be. The movement obtained is termed the natural vibration of the system. For the vibration of a mass on a shaft between the housings, the elastic constant is the elastic constant of the bent shaft and the movement is called the lateral movement.

For a system subjected to vibrations due to torsion the elastic constant is that of a spring in torsion designated by k_t and the mass is substituted by the moment of mass-inertia J of the vibrating disk.

$$\omega_{0t} = \sqrt{\frac{k_t}{J}} \qquad (12.002)$$

12.2 Forced Vibration

If an elastic system is subjected to a sinusoidal excitation force, the system will be subjected to a periodic movement whose amplitude depends on the value of the ratio between the frequency of excitation and the frequency of the system. If such ratio is below 1, it will be said that the system is in the subcritical zone, while if ratio is above 1, it will be said that the system is in the supercritical zone. If the ratio is equal to 1 or near such value, it will be said that the system is in the critical zone. In this zone the amplitude of the movements increases rapidly with the passing of time and reaches values that produce breakage of the system. It is said that the system is in resonance. The amplitude of the vibrations in the critical zone decreases by a certain value due to the damping. The damping of metals is slight, while the damping of elastomers and of certain materials, such as felt, reaches very high levels. Metals have different damping levels. Grey pig iron, for example, has a degree of damping higher than that of steel.

12.3 Vibration Modalities

The vibration of an elastic system does not occur in such a simple way as in the previous paragraphs; there are many proper frequencies called harmonics of the fundamental frequency. If the frequncy of excitation is near an harmonic frequency, the resonance will occur. The bent elements can also take different shapes which can produce different proper frequencies. A shaft, for example, can take the shape of simple arc of a sinusoid, of a complete sinusoid, or of two sinusoids, etc..

Each distortion can occur at the time in which a proper natural or forced vibration takes place. Each of these distortions is called the modality of vibration.

Bent shafts are not the only ones characterized by vibration modalities. Flanges and housings also have different vibration modalities with proper frequencies for each modality.

12.4 Flexion Vibration

Let us consider a shaft on two bearings, without a proper mass provided with a disk noncentrated in relation to the axis of rotation. Given m as the mass of the disk and e as the non-centering; as a consequence of the rotation, the shaft is subjected to a bending under the action of the centrifugal force $m\omega^2(y+e)$, y being the distortion of the shaft. If k is the elastic constant of the shaft, the force $k\,y$ will counterbalance the centrifugal force and the result will be:

$$m\,\omega^2\,(y + e) = k\,y \qquad (12.003)$$

where the bending of the shaft under the disk is:

$$y = \frac{\omega^2 e}{\frac{k}{m} - \omega^2} = \frac{\omega^2 e}{\omega_0^2 - \omega^2} \qquad (12.004)$$

as k/m is the square of the unit made up of the shaft and the disk. It is observed that if the frequency of rotation is equal to the proper frequency of the system, the deflexion will become infinite independent of the value of the eccentricity e. Such frequency, which produces a resonance, is called critical frequency of the shaft. If y_{stat} is the static deflection of shaft near the disk, under the action of the weight of the disk $m\,g$ the result will be $k\,y_{stat} = m\,g$ and it can be assumed that the critical speed expressed in $\min^{-1}$ is equal to:

$$n_{crit} = \frac{30}{\pi}\sqrt{\frac{g}{y_{stat}}} = \frac{945.8}{\sqrt{y_{stat}}} \qquad (12.005)$$

If numerous disks are placed on the shaft, it will be possible to calculate the critical speed by following Rayleigh's hypothesis. Let us consider the inflexion shaft under the weight of the disks supported by the shaft. The weights for the overloading disks are considered in the opposite direction to their real direction, in order to obtain a uniform bending of the shaft after its distortion. Let us calculate the value of the bendings under the weights (see Chapter 6). Let us consider y_i as the bending of the shaft in these conditions under the disks with a mass of m_i It will be possible to apply equation (12.005), if the value indicated below is taken for y_{stat}.

$$y_{stat} = \frac{\sum_{1}^{n} y_i^2}{\sum_{1}^{n} y_i} . \qquad (12.006)$$

To obtain the proper mass of the shaft, this latter must be divided into more sections whose mass is calculated and which are considered as isolated disks in the central point of the section. For a shaft with a constant section (diameter d) and a length L between the non-overhanging bearings the result is:

$$n_{crit} = 3.8 \cdot 10^8 \frac{d}{L^2} . \qquad (12.007)$$

The critical bending speeds are usually above the working speeds of the gear units for general applications.

12.5 Torsion Vibration

The different shafts of a gear unit are kinematically jointed together to build a unit made up of inertial masses and torsion springs. These are the shafts themselves and the teeth of the gearing. The masses are all the rotating masses that take part in the movement. The mass outside the gear unit, which take part in the movement too, cannot be excluded. These masses are pertaining to the conducting and of the driven machines. It is because of the existence of all these masses and torsion springs that the unit has proper frequencies. The frequencies of excitation during torsion must be different from the above-mentioned frequencies. It is not possible to study the vibrations due to torsion for a gear unit that is not mated with its machines, that is the conducting and the driven machines. Therefore the gear unit will not have proper frequencies if it is not considered together with the unit whith which will form during operation. To utilize a gear unit the user must know all the masses and the elastic constants of the unit in order to include each unit the characteristics of the gear unit and to make a complete calculation.

The aim of the calculation is to take all conditions on a single shaft whose rotation speed is n (arbitrary). The masses and the constants from all the shafts of the unit are borne on the reference shaft by being multiplied by the square of the ratio between the speed of the real shaft (n_i) and the speed of the reference shaft (n). The reference shaft therefore bears a set of masses with a moment of inertia calculated by elastic calculaded constant. The equations of the movement constitute a set of equations with multiple unknowns that, if solved, will provide the proper frequencies of the unit.

It will be possible to consider the proper frequency of a gear unit taken separately from the conducting and the driven machines, if these latter are jointed to the gear unit by elastic fits with a sufficiently limited rigidity to torsion. In this connection gear motors turn out to be particularly interesting as, being a well-defined unit including fits included, they provide the dynamic response to the torsion of the unit.

12.6 Sinusoidal Vibration Function

The sinusoidal function that represents a simple vibration is given by the following equation:

$$x = A \sin(\omega t + \phi) \qquad (12.008)$$

x is called displacement
A is the amplitude,
ω is the pulsation,
ϕ is the phase displacement.
The pulsation and the frequency are linked together by the following equation:

$$f = \frac{\omega}{2\pi} \qquad (12.009)$$

The time for a complete cycle is T and is called a period. Between the period and the frequency there is the following relation:

$$fT = 1 \qquad (12.010)$$

The speed of a vibration is the derivative of the movement, while the acceleration is the derivative of the speed. Consequently the result is:

$$v = A\omega \cos(\omega t + \phi) \qquad (12.011)$$

$$a = -A\omega^2 \sin(\omega t + \phi) = -\omega^2 x \qquad (12.012)$$

12.7 Periodic Vibration Function (Fourier)

This vibration linked to movement is periodic. Fourier demonstrated that each periodic vibration could result in a sum of sinusoidal vibrations with growing pulsations. The vibration with the lowest degree of pulsation is the fundamental vibration; the other vibrations are harmonics of a growing degree. Each vibration has its own frequency, amplitude and phase displacement. The expression of a vibration is therefore:

$$x(t) = \sum [A_i \sin(\omega_i t + \phi_i)] \quad . \qquad (12.011)$$

However, Fourier transformation is a mathematical function that enables the amplitudes of the different harmonic functions of a given periodic vibration to be found. It is possible to program for this calculation computers. By adopting Direct Fourier Transformation, the calculation turns out to be very long; therefore Fast Fourier Transformation (FFT), a simplification of this method, is adopted in order to save time.

12.8 Modal Analysis

Each vibrating element can be subjected to different distortions. Each particular distortion taken by the warped element is called modality of vibration. A proper frequency corresponds to each modality of vibration. An electronic procedure derived from Fourier's analysis and based on the recording of the vibration on more points of the analyzed element allows the proper frequency of each modality to be determined and the warped structure by following this modality. Such method is called modal analysis.

12.9 Measurement of Vibrations

It is possible to measure the displacements of a sinusoidal vibration, whose maximum is represented by the amplitude, the speed or the acceleration and go from one characteristic to another by electronic derivation or electronic integration. Measuring instruments, usually accelerometers, are instruments sensitive to accelerations, used to take measurements. These instruments must be in contact with the element whose vibrations are to be measured. Their mass, being added to the mass of the element, can influence the registered and measured vibration. They must have a negligible mass in relation to measured mass. Some measuring instruments, which contact with the object analyzed is not relevant measure displacement, that is displacements whose amplitude can be stated. For measurements of real vibrations which are not simple sinusoidals, it is possible either to use filters which keep vibrations in a frequency band, or to record the vibration and transform it according to the FFT method.

For the measurement of vibrations of the gear units, it is possible to record the vibrations on the housing or on the shaft. The measurement taken on the housings will be considered sufficient, if shafts are mounted on bearings. In fact these bearings have a rather small end play so that the vibration of the shafts can be integrally transmitted to the housing. In this case it is advisable to take the measurement on the bearing housings. In reduced accessibility, the measurement must be taken according to the fixing points of the housing. Measurements are always taken following three ortogonal directions, two of which are on a perpendicular plane to the axis of the shaft considered. The measurements on the housing are taken by measuring instruments fixed rigidly to the housing. Also the measurements on the shafts are taken following three ortogonal directions, one of which is on the axis of the shaft. The measuring instruments do not usually have any point of contact. Measuring instruments capable of recording a range of frequencies ranging from 10 to 10000 Hz and more are used for the measurement on housings. Meters of frequencies ranging from 0 to 500 Hz are used for the measurement of the vibrations on the shafts. Measuring instruments have a good degree of precision; they can measure vibrations in frequency bands of 1/3 of octaves (an octave is the interval that divides the frequency from its double frequency; the third of an octave covers an interval ranging from one frequency to the frequency equal to the basic frequency multiplied by the cube root of 2, that is 1.26). Measurements are preferentially taken under conditions of use. If it is not possible to take any measurement on the unit of the transmission (conducting and driven machines included) no guarantees can be given on the influence of the dynamic response of the entire unit. For more details it is necessary to cunsult the standard ISO 8579-2.

12.10 Causes of Vibrations in Gearboxes. Frequency of Meshing

Vibrations in gear units are fundamentally linked to the movement of the elements of the gear unit. Other vibrations can be transmitted to the gear unit starting from the conducting and the driven machines. Such vibrations can be reduced by suitable elastic fits, even if these latter decrease the rigidity of the transmission where they are positioned; the final result is a decrease in the natural frequencies in torsion. The vibrations linked to movement, which is considered as a primary cause, can have several origins. First of all we can say that theoretical meshing is a source of vibrations. The vibration of the total overlap ratio causes a variable subdivision of the force on the teeth and consequently their distortion, which varies periodically. Such variations relative to the distortions of the toothing produce a non-continuous movement, which creates the vibration. On the other hand the deviations of the pitch and the profile as well as eccentricity produce periodical variations linked to movement; for each passage of teeth, variations, which are expressed by the measure of the accumulative tangential deviation, are produced. Such vibrations are linked to the rotation speed and to the number of teeth of the gear considered. The product of the rotation frequency of a gear, which is given by its speed divided by 60 and by the number of its teeth, is the frequency of a gear. It is noticed that this product is the same for the pinion and the wheel of a gear. The frequency of meshing is one of the main causes of vibrations. In a gear unit there are vibrations with meshing frequencies that characterize the different gears of the gear unit itself. By means of the centrifugal force, the imbalances produced by the non-coincidence of the geometrical axes and the axes of rotation of the parallel elements create vibrations whose frequency is the rotation frequency of the element considered. In gear units, the deviations of fabrication in the turning and milling phases produce vibrations whose frequency is given by the rotation frequencies of the different shafts.

Also bearings create vibrations by means of the passage of the rolling elements near the side of the load. These vibrations are linked to the rotation speed, the dimensions of the bearings, the type of bearings and the number of rolling elements. They are more complex to be defined as frequencies. All such vibrations are transmitted to the shafts, then to the bearings and finally to the housing. The bearing end play allows to the vibration to be absorbed and the transmission to be slightly reduced. Therefore it is possible to measure the vibrations on the shaft (usually on the overhanging ends of the shaft) or on the housings and preferably near the bearings.

12.11 Effects of Vibrations

Vibrations can produce negative effects on the machines, the environment and human beings. In minimum concentrations they can, for example, damage the machines. Damages will become even more serious if resonance phenemenon occurs, that is if proper frequency excitations of some elements take place. Also the environment can be damaged in the same way. In human beings vibrations can cause problems due to the resulting accelerations that produce forces in human organs, e.g. they can create serious lesions. Vibrations also cause negative and unwanted structural noises. If such vibrations have the proper frequency of some elements, such as housings and their covers, for instance, the effect of resonance will be to create unbearable noises (see the paragraph immediately below). Hence the important role played respectively by the measurement and the prevention of vibrations. In particular, the modal analysis on a prototype-housing at the time of mass-production allows changes in the structure and in the stiffening elements to be introduced useful for the reduction of vibrations.

12.12 Noise

Noise is the result of a physiological action of the air on the auditory organs. Noise is therefore subjective. It is caused by air vibration and pressure modulation. There are two different sources of noise: the noise produced by direct air vibration at the time of the starting of a machine; and the noise produced by air pressure resulting from the vibration of structures. The first source of noise is directly linked to vibrations. Decreasing vibrations reduce noise.

The study of structural noise is linked to the study of the vibrations. Let us consider in particular the noise produced in the air by the working of machines.

12.13 Measurement of the Noise

Noise intensity is the sound-power per unit of surface. Such intensity is a vector intensity in the sense that it has a dimension and a direction. The sound-power is a scalar magnitude in the sense that it has no direction. Some measurements are taken by double microphonic meters which allow the intensity to be determined by its mark. By surrounding the sound source with a series of microphones of the above mentioned type, it is possible to measure the starting and ending intensities as well as the intensity of the sound emitted by the source. This procedure is now becoming more and more widespread, as today very small microphones are available on the market; nevertheless it is still not possible to define it being current, since it is rather onerous in operation.

The most diffused procedure is the measurement of sound power by microphone. Generally speaking single power is not considered, but the ratio between the real power and the reference power. The value of this power of comparison is conventional 10^{-12} Watt. The logarithm of this ratio and not its value is now considered. The sound intensity is expressed in decibels and is given by the following relation is:

$$L_W = 10 \log \frac{W}{10^{-12}} . \qquad (12.014)$$

It is noticed that if the sound power is doubled, the measure will be increased by 3 decibels (10 log 2 = 3). It is possible to express the sound intensity in relation to pressure, with the following result:

$$L_W = 10 \log \frac{p^2}{p_0^2} = 20 \log \frac{p}{p_0} \ . \qquad (12.015)$$

If an arbitrary reference pressure which is not linked to the reference power (pressure usually determined by the threshold of human audibility, that is 20 µN/m²), is selected, the value of the sound pressure is the following:

$$L_p = 20 \log \frac{p}{p_0} \ . \qquad (12.016)$$

Between the two values given by equations (12.014) and (12.016) there is the following relation:

$$L_W = L_p + 10 \log \frac{S}{S_0} \qquad (12.017)$$

where $S_0 = 1$ m².

12.14 Well-Considered Scales (dB$_A$)

The human ear is not equally sensitive to all sound frequencies. It is less sensitive to low and high frequencies than to those ranging from 1000 to 5000 Hz. The measurement of the pressure by a meter equally sensitive to all frequencies cannot

Table 12.1 Corrective values for dB$_A$

Frequency	Corrective value
125	-16.1
250	-8.6
500	-3.2
1000	0
2000	1.2
4000	1
8000	-1.1

provide the intensity of the noise as it is perceived by humans The instruments for the measurement of noise intensity are usually equipped with filters which consider the measure in relation to the frequency. The system that is nearest to the human ear is system A, expressed by an attenuation curve whose value is equal to -50 dB per 20 Hz, reaches 0 dB per 1000 Hz, arrives to a maximum of 1.3 dB per 2500 Hz and falls again to -11 dB per 20000 Hz. Table 12.1 provides the correct values for system A for the medial points of the octaves usually used for the measurement of the noise produced by gear motors.

The sound level is usually expressed in dB_A.
To obtain an idea of the noise-measurement unit let us consider some examples:

rustle of leaves	30 dB
library	35 dB
office	from 40 to 45 dB
medium road traffic intensity	70 dB
motorcycle (at 7 m)	85 dB
noisy factory	from 95 to 115 dB
aeroplane taking-off	130 dB

If humans are subjected to noise sources ranging from 30 to 65 dB, they will not suffer any ill-effects. Prolonged exposures to noise sources ranging from 65 to 90 dB can cause some problems in the nervous system. Exposures to noise sources ranging from 90 to 120 dB can damage hearing. Exposures to noise sources above 120 dB can seriously damage humans. The noise produced by gearing ranges from 65 to 90 dB

12.15 Measurement of the Noise Produced by Gear Units

The measurement of the noise produced by gear units in a working place cannot be considered as a reliable measurement since many other noises, which are called parasitic and which can neither be eliminated nor evaluated, interfere with the noise produced by the gear unit. To measure the noise emitted only by the gear unit, the measurement must be taken in an anechoic chamber, that is a chamber whose walls, its floor excluded, do not reflect noise. The gear unit is isolated inside such chamber (the machines are left outside the chamber, except for the conducting machine where they are an itegral part of the gear unit - gear motor).

To compare the noise of a gear unit with the noise produced by other similar gear units, the measurement must be taken in identical conditions. As the noise decreases with distance, a standardized parallelepiped-shaped surface may be defined around the gear unit, whose total area, base excluded, is calculated.

Therefore it is possible to measure the value of L_S:

$$L_S = 10 \log \frac{S}{S_0} \qquad (12.018)$$

where S_0 is equal to 1 m². The mean value of the level of the sound pressure due to the gear unit is determined by measurements which must be taken on the reference surface with a sound level meter regulated on an octave band or on a band of thirds of octave (L_{pA}). Consequently the level of sound power is given by:

$$L_{WA} = L_{pA} + L_S \ . \qquad (12.019)$$

The same procedure is used for all the octave intervals or the intervals of thirds of octave in a total band ranging from 125 Hz to 8 kHz. For each octave or third of octave it is therefore possible to calculate the reference value of the noise produced by the gearing. The measures as a whole represent the noise spectrum. The values obtained can be corrected in order to include the reflectability of the walls. To gauge the measures it is possible to install a source adjustable in sound level frequency in the place that is taken by the centre of the gear unit at the time of the measurement. The frequency and the sound level of the source can be regulated by measurements like those taken on the gear unit, it is possible to evaluate the degree of reliability of the measurement itself. For more details consult the standard ISO 8579-1.

13 Thermal Power of the Gearing

13.1 Definition

Gear units are calculated in order to bear the stresses to which the elements are subjected. Nevertheless the power loss that follows the working of a gear unit is due to different frictions that have not been taken into account. Also the fact that the power lost turns into heat has not been considered. The heat is dispersed in the environment by different means of heat transmission. At a certain temperature, however, a balance intervenes depending on the working conditions, the structure of the gear unit and the surfaces of exchange. If the temperature exceeds the limiting value of the safety temperature, special measures will be required to eliminate overheat or limiting the transmitted power. The maximum power that can be transmitted by a gear unit without any risk to its elements (lubricant included) is the minimum power of the two calculated values, that is the thermal power and the mechanical power of the gear unit.

13.2 Losses in Gearboxes

The power losses caused by the frictions present in gearboxes occur at different levels. First of all there are the losses in gears. It has previously been considered that gears work by means of frictions and that their efficiency is lower than the unit. There are losses in bearings that do not work by pure rolling but are subjected to actual non-negligible frictions. There are then the relatively important frictions on the seals. Also the mixing of the lubricant represents a source of mechanical losses with consequent heat exchange.

13.3 Losses During the Meshing

Losses in gears have been defined in Chapter 3 by equations (3.114) and (3.115), which if combined will give the following result:

$$P_f = P \mu_m \pi \frac{u+1}{z_1 u} \frac{\varepsilon_1^2 + \varepsilon_2^2}{\varepsilon_a} \qquad (13.001)$$

The coefficient of friction is given by the Elastio-hydro-dynamic theory and is expressed by the following equation:

$$\mu_m = 0.045 \left(\frac{F/b}{V_{\Sigma C}\, \rho_C} \right)^{0.2} \eta_{oil}^{-0.05} X_R\, X_L \qquad (13.002)$$

The factor X_R is the roughness factor and its value is:

$$X_R = 3.8 \left(4\sqrt{\frac{R_a}{d_1}} \right) \qquad (13.003)$$

Roughness R_a is the average of the roughness of the pinion and of the wheel.

The factor X_L is the lubricating factor and a constant that depends on the type of lubricant. Its value is equal to 1.0 for mineral oils. The term $V_{\Sigma C}$ is the sum of the rolling speeds of the two teeth around the respective points tangent to the base circle. This factor depends on the reference speed v_t which will be limited to 50 m/sec. if this value is exceeded. The factor ρ_C is the bending radius reduced in the reference point ($1/\rho_C = 1/\rho_{C1} + 1/\rho_{C2}$).

The specific load F/b will be limited to 150 MPa if its value is lower than this number, the load F being the force tangential to the base circles on the transverse plane ($F = F_{bt}$).

Such equations are valid for cylindrical gear pairs and for bevel gear pairs. As to these latter, the equations are applied to the equivalent cylindrical gears. The same equations are applied to the wheels for worms, except for the coefficient of friction, which is expressed as follows:

$$\mu_z = \mu_{z0}\, Y_W \sqrt{\frac{v_{gm}}{V_{\Sigma C}}}\, 4\sqrt{\frac{R_z}{R_{z0}}} \qquad (13.004)$$

Coefficients with an index 0 are obtained by testing the friction and the rolling between the two disks. The value v_{gm} is equal to about 2.5 times $V_{\Sigma C}$. The value of Y_W is near the unit. The efficiency is indicated in paragraph 5.6.6, the angle of friction being the angle whose coefficient of friction is the tangent. In a gear unit the total loss due to the seizing is given by the sum of the losses of each gear train and each seizing per train. The losses for worm gear pairs are higher than those occuring in parallel gears and in gear pairs with intersecting axes.

13.4 Losses due to Dipping

The dipping creates a friction between the wheel and the lubricant. This friction produces an internal circulation of the lubricant and causes a loss independent from the load. Such loss can be expressed (according to Mauz) by the following equation, where T_H is the mechanical torque corresponding to the loss (in Nm) and v_t is the reference speed.

$$T_H = C_{Sp}\, C_1\, e^{C_2 \cdot v_t} \quad . \qquad (13.005)$$

The factor C_{Sp} takes the projecting direction of the lubricant into account; the lubricant follows the shaft rotation of the wheels. If the lubricant is taken directly at the point of contact, such coefficient will have a value of 1.0. If the lubricant reaches the point of contact after its projection on the walls of the housing, this coefficient will have a lower value depending on the depth of immersion of the wheel, the position at the point of contact and the length of the distance covered by the lubricant. The value ranges from 0.75 to 1.0. The coefficients C_1 and C_2 are given by:

$$C_1 = 0.0063\, e_1 + 0.0128\, 10^{-3}\, b^3 \qquad (13.006)$$

$$C_2 = \frac{e_1}{8000} + 0.02 \qquad (13.007)$$

e_1 being the depth of immersion of the wheel in the lubricant and being b the width of the wheel.

n being the rotation speed of the shaft of the wheel, the lost power is:

$$P_{vz0} = \frac{T_H\, n}{9550} \quad . \qquad (13.008)$$

In a gear unit, the total loss is the sum of the losses relative to all the immersed wheels. Such losses do not have very high values and will increase exponentially together with the depth of immersion especially if this latter gives $e_1/8000$ a significant value in relation to 0.02. Therefore it is advisable not to fill gear units with lubricant excessively and to respect the level suggested by the producer. Overlubrication is useless and damages the efficiency and the heat power of gear units.

13.5 Losses in Bearings

Bearings cause two kinds of losses: those independent of the load and losses proportional to the load.

The losses independent from the load depend on the product of the kinematic viscosity of the lubricant under conditions of working temperature and rotation speed. If this is equal to 2000 or higher than such value, the result will be:

$$T_0 = 10^{-7} f_0 (\nu n)^{\frac{2}{3}} d_m^3 \qquad (13.009)$$

while if the product is below 2000, the result will be:

$$T_0 = 160 \cdot 10^{-7} f_0 d_m^3 \ . \qquad (13.010)$$

Where T_0 is the dissipated torque (in Nm), d_m is the mean diameter of the bearing $((d+D)/2)$ and f_0 is the factor provided by Table 13.1 below. As we have said before, n is the rotation speed of the shaft in $\min^{-1}$ and ν is the kinematic viscosity of the lubricant at the working temperature.

$L_1 = (P_0/C_0)^{0.55}$;

$L_2 = (P_0/C_0)^{0.4}$

$L_3 = (P_0/C_0)^{0.33}$

$A = F_r$
$B = F_a$

(1) If $F_p/F_a < Y_2$ $1.35\ Y_2\ A$

If $F_r/F_a \geq Y_2\ B\ [1 + 0.35(Y_2\ B/A)^3]$

The dissipated torque proportional to the load is given by:

$$T_1 = f_1\ P_1^a\ d_m^b \qquad (13.011)$$

Table 13.1. Factor f_0

	Type of lubrification			
Type of bearing	Grease	Air-oil	bath of oil	jet of oil or vertical shaft
single row deep groove ball bearing	0.75...0.2	1	2	4
double row deep groove ball bearing	3	2	4	8
double row self-aligning ball bearig	1.5...2.0	0.7...1.0	1.5...2.0	3.0...4.0
single row angular contact ball bearing	2	1.7	3.3	6.6
double row angular contact ball bearing	4	3.3	6.6	13
cylindrical roller bearing				
series 2, 3, 4	0.6	1.5	2.2	4.4
series 22	0.8	2.1	3	6
series 23	1	2.8	4	8
double row self-aligning roller bearing				
series 213	3.5	1.75	3.5	7
series 222	4	2	4	8
series 223	4.5	2.25	4.5	9
taper roller bearing	6	3	6	8...10

Table 13.2 Factors f_1 and P_1

	f1	P1
Deep groove ball bearing	0.0009 L1	3 A - 0.1 B
Double row self-aligning ball bearing	0.0003 L2	1.4 Y2 A - 0.1 B
Single row angular contact ball bearing	0.001 L3	A - 0.1 B
Double row angular contact ball bearing	0.001 L3	1.4 A - 0.1 B
Cylindrical roller bearings		
series 2	0.0003	A
series 3	0.00035	A
series 4	0.0004	A
Double row self-aligning roller bearngs		
series 213	0.00022	
series 222	0.00015	-1
series 223	0.001	
Taper roller bearings	0.0004	2 Y B

The factors f_1 and P_1 are indicated in Table 13.2; the exponents a and b are equal to 1, except for double row self-aligning roller bearings. In this case they assume the values indicated in Table 13.3

Table 13.3 Values of *a* and *b*

Double row self aligning roller bearing	a	b
series 213	1.35	0.2
series 222	1.35	0.3
series 223	1.35	0.1

The total loss is the sum of two different losses. The total torque for a bearing is therefore:

$$T = T_1 + T_0.$$

The power lost in the bearing is therefore:

$$P_B = \frac{Tn}{9550}. \qquad (13.012)$$

In some particular conditions (C/P = about 10, good lubrication and normal working conditions), it is possible to calculate *T* directly as follows:

$$T = 0.5 \, \mu \, F \, d \qquad (13.013)$$

where μ is the coefficient of friction indicated in Table 13.4, *F* is the load on the bearing and *d* is the diameter of the bore of the internal ring.

Table 13.4 Values of μ

	m
Deep groove ball bearings	0,0015
Double row self-aligning ball bearings	0,001
Single row angular contact ball bearings	0,002
Double row angular contact ball bearings	0,0024
Cylindrical roller bearings	0,0011
Double row self-aligning roller bearings	0,0018
Taper roller bearings	0,0018

13.6 Losses in Sealings

Frictionless sealings are characterized by almost insignificant energy losses. Sliding seal show energy losses that depend on the nature of the ring and on the diameter of the shaft.

The loss of an oil seal, expressed in kW, is given by:

$$P_S = \frac{3 \cdot 10^{-3} \, d_{sh} \, n}{9540} \qquad (13.014)$$

d_{sh} is the diameter of the shaft, while n is its rotation speed.

13.7 Total Losses

To obtain the value of the total losses it is necessary to calculate and sum all the losses previously analysed. The total loss is represented by P_{tot}.

13.8 Dissipation of Heat

The heat produced in gear units circulates by means of the lubricant and is finally transmitted to the housing, which disperses it into the atmosphere. There are different forms of dissipation of heat: conduction by means of solid materials, convection by gases and fluids and radiation in the vacuum. The value of the dissipations by radiation is insignificant and the dissipation by conduction can be included in the dissipation by convection, which depends on the environmental conditions and the temperature of the housing. The term "environmental conditions" includes the ambient temperature, the speed of the air near the gear unit and the condition of the gear unit near an intense source of radiation. The temperature of the housing can be compared to that of the lubricant, whose maximum value is usually established to be 95°. Sometimes such temperature is a function of heat dissipations and heat exchanges. For equilibrium of the temperatures, the lost heat Q is equal to that produced by the dissipations. As such an equation is a function of the same variables in the two terms, it can be solved through interation by selecting a transmitted power, by calculating the losses and by comparing it with the dispersion power. If the losses are lower than the power that can be dispersed by the housing, owing to the maximum temperature of the lubricant, it will be possible to increase one power and calculate equilibrium again. If the heat produced exceeds the heat that can be actually dispersed, it will be necessary to reduce the power and repeat the calculation.

13 Heat Power of the Gearing

The equilibrium heat power is the thermal power of the gear unit.
The power that can be dispersed is calculated as follows:

$$Q = A_{effective} \; k \; \Delta T \tag{13.015}$$

$A_{effective}$ being the surface of the housing that can transmit the heat by convection. This surface is the outside area of the housing without the surface in contact with the foundations and without the excesses caused by lumps and stiffening elements. ΔT is the difference in temperature between the housings and the environment; k is the coefficient of heat-transmission. Factor k has a value ranging from 0.010 to 0.014 (kw/m²K°) for the speed of air in close proximity to the housing, below 1.4 m/s (corresponding to a limited space without ventilation).

If the speed exceeds this value up to 3.7 m/s, the factor will be multiplied by 1.40; but if such speed is above 3.7 m/s, the factor will be multiplied by 1.90. The higher value of k can be accepted only if justified by proof. The heat power resulting from the equilibrium of the exchanges is usually calculated or determined using standard conditions of air speed and ambient temperatute. Those conditions are often a low air speed (< 1.40 m/s) and an ambient temperature of 40°C. For other ambient temperatures the thermal power is multiplied by one of the following factors:

Ambient temperature of	
10°C	1.39
20°C	1.25
30°C	1.13
40°C	1.00
50°C	0.81

If the gear unit works intermittently with abrupt stopping allowing a significant cooling, it will be also possible to increase the thermal power of the gear unit. If the thermal power is higher than the mechanical power used as a basis for the calculation of the elements of the gear unit, it will be insignificant; but if this power is lower than the mechanical power of the gear unit, the transmitted power will be limited to such value or measure which tend to increase it, that is air-cooling, oil-cooling or cooling by refrigerator immersed in the oil bath. For external cooling by ventilation, the calculation is carried out by factor k, which is indicated in Table 13.5.

In the case of oil-cooling, being the difference of temperature between the entry and exit of the oil Δt and ΔQ being the difference between the heat power produced and the power that can be really dispersed, the required quantity of oil m_{oil} is given by:

$$m_{oil} = \frac{\Delta Q}{c \; \Delta t} \tag{13.016}$$

Table 13.5

Air speed m/s	k
2.05	0.015
5	0.024
10	0.042
15	0.058

k in kw/m²K°.

c being the specific heat of the oil at the running temperature as defined in paragraph 9.7.

13.9 Efficiency of Gearboxes

The efficiency in gear units is the ratio between the horse-power used and the power that has been spent, that is the power on the input shaft of the gear unit, produced by the engine, but decreased by all the losses that have been calculated and by the additional energy consumptions required by the working. Among these consumptions included are lubricant circulation pumps, in pressure feed and the power absorbed by the lubricant refrigerators, if used.

13.10 Measurement of the Efficiency of Gearboxes

Instead of calculating the losses by an analytic method, if there is a gear unit it will be possible to use an experimental method of calculation. There are two methods: open circuit measurement and closed circuit measurement.

13.10.1 Open Circuit Measurement

Such measurement, which is characterized by an open circuit in relation to the power, is based on the starting of the gear unit by an engine and on the mating of this egine with a mechanical or electric brake. The mechanical brake turns the output power into heat that is then dispersed by the action of a refrigerant. The electric brake is made up of a generator that discharges the current in the resistances or in an electric circuit. The input and output powers are measured.

The difference represents the mechanical loss, and this efficiency is obtained by dividing the output power by the input power. Instead of measuring the power, it is possible to measure the torque and the speed. Such methods allow the heat power of an existing gear unit to be determined, experimentally by determining which power will provide the maximum tolerated by the running temperature.

This method is apparently simple, but in reality has serious difficulties. First of all the measurement for high power reducing gears turns out to be particularly difficult owing to the dimensions of the gear units and of the brakes; in fact it is the brakes, or the resistance, that are necessary to disperse all the transmitted power turned into heat. As the motor starter must have the nominal power of the gear unit, high costs are required.

On the other hand the efficiency of gear units ranges from 0.98 to 0.93 and the losses reach values with an increase ranging from 2% to 7% of the value of the transmitted power. Therefore it is necessary to measure high powers and subtract them as the accuracy of these measures in order of magnitude of the difference contribute towards the losses. Consequently extremely accurate meters are required to take such measurements. For instance, if there is a difference of 5%, the measure of the input and output powers will require an accuracy of 0.25% in order to have a sufficient accuracy relative to such measure (for example, an accuracy in the order of magnitude of 5%). If the measurement is taken by gauging the speed and the torque, the accuracy of the measurements relative to each of these two magnitudes will require an even higher level, since the accuracy of the power is given by the sum of the accuracies of the measurements relative to the torque and the speed. This last objection is not valid as to the determination of the heat power, since in this case it would not be necessary to consider the efficiency. Therefore this method is suitable for the evaluation of the power in low and medium power gear units, but it is absolutely unsuitable for the measurement of the efficiency.

13.10.2 Back to Back Measurement

To make this measurement, two identical gear units are installed, one opposite the other, that is to say that the low speed shafts as well as the high speed shafts are mated (see Fig. 13.1).

A static torque is applied to the shaft of the gear unit, given T as the torque applied to the high speed shaft. The system is started by an engine mated with the high speed shaft. The power that stresses the elements of the gear unit is the power resulting from the applied static torque and the rotation speed, even if the engine transmits only the power lost in the two coupled gear units. The advantage is that the value of the required power is much lower and consequently it is not necessary to create a loss of energy by braking. The measurement of possible losses can only be taken by measuring a torque and the output speed of the engine directly. It is possible to measure the torques and the speed of the engine as well as the

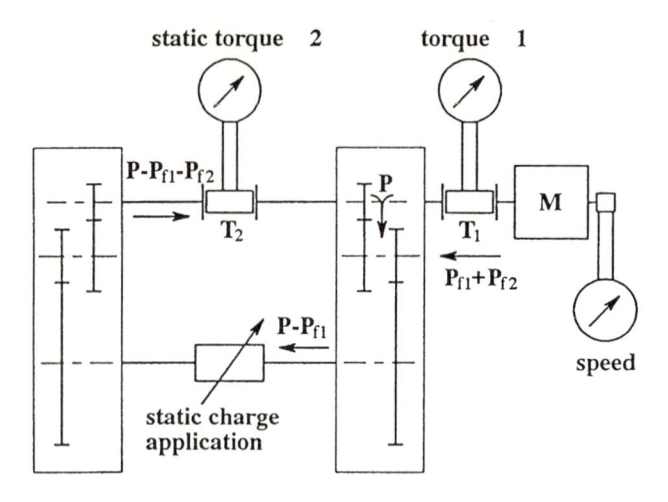

Fig. 13.1. Back to Back Measurement

introduced static torque. In this way the lost power P_f can be directly determined. The transmitted power P corresponds to the applied torque. In order to modify this power and to carry out a test with a spectrum of charges, special devices can be used during the testing.

Given P_{f1} and P_{f2} the powers lost in gear unit 1 and gear unit 2. Power P transfers gear unit 1 and power $(P - P_{f1})$ is transferred. Such power is transferred to the gear unit 2 and power $(P - P_{f1}) - P_{f2}$ is transferred.

The lost power is $P_f = P_{f1} + P_{f2}$.

If η_1 is the efficiency of the first gear unit and η_2 is the efficiency of the second gear unit, the result will be

$$P_{f1} = (1 - \eta_1)P \quad \text{and} \quad P_{f2} = (1 - \eta_2)\eta_1 P$$

Therefore the result will be:

$$P_f = P_{f1} + P_{f2} = (1 - \eta_1)P + (1 - \eta_2)\eta_1 P = P(1 - \eta_1\eta_2)$$

If the two gear units have the same efficiency, the result will be:

$$\eta = \sqrt{1 - \frac{P_f}{P}} \,. \tag{13.017}$$

Since the value of the efficiency is near the unit, a sufficiently precise value will be obtained in the following way:

$$\eta = 1 - \frac{1}{2}\frac{P_f}{P} \qquad (13.018)$$

13.11 Comparison Between the Different Types of Gear Units

The different types of gear units show a particular behaviour as to the heat power. Gear units with the best heat power are parallel shaft gear units and intersecting shaft gear units, that is gear units provided with the largest housing for a known mechanical power and a known speed reducing ratio. Such gear units have maximum dimensions in terms of volume. Parallel and intersecting gears have a smaller volume and a smaller exchange surface for the same working conditions. Despite their lower heat capacity, they offer the advantage of having small dimensions. Planetary gears have a volume and an exchange surface that are extremely small compared to the other parallel gears; consequently they show a lower heat capacity. Worm gearings are characterized by a considerable loss of heat due to their low efficiency and to their small exchange surface. Therefore they have a low heat capacity. The more the heat capacity risks being low the more it is necessary to consider such capacity. It is necessary to be careful in the selection of the solutions as possible advantages could turn out to be disadvantages. The importance of heat capacity is not a quality criterion, but just one of the characteristics that must be taken into account in the selection of the different solutions.

14 The Manufacturing of Gearboxes

14.1 Introduction

It is clear that manufacturing conditions contribute towards the cost and quality of products. However, much the study and the calculation has been undertaken, if manufacturing is badly made and organized, it can lead to differences in the relative dimensions and positions of the various components. In this way the functioning conditions may be different from those expected, thus causing lower efficiency and bringing about stresses that can lead to deteriorations or breakages.

It is also clear that the quality of manufactured products is conditioned by the attention made during manufacturing. But the equipment, i.e. the stock of machines and the tools utilized, are equally essential for final quality. Anyone studying the product has to take into account the processing opportunities available that depend on the equipment of the manufacturing shop. The improvements the manufacturer wanted to introduce in the product during the studying phase can be put into practice only if the material is available or if the manufacturing of the product studied can be supported by an investment.

The manufacturing of a unique piece or of a small-quantity series involves considerable investment, even if this is necessary to increase the quality of the product. At the same time, it does not allow for the investment of equipment that could increase the quality of the product.

In this way we personally ascertain the great advantage of mass-production, because adequate tools and machines may be chosen according to quality and economy.

This chapter is not an enumeration of the various machine tools or a technological study of the machines now available on the market. Our aim is to show, through an example, the various processing necessary to build a gearbox, the difficulties that may be encountered, for different operations and machines.

14.2 The Raw Material

The raw material will be used in the parts of cases, shafts and the blanks of wheels.

- Cases are made both of welded mechanical parts and of foundry pieces. Cases of welded mechanical parts are used for unique pieces or for very small quantity pieces which therefore cannot redeem the costs of the models or the foundry equipment. The vibrations and the noise emitted are higher than with casts because they are made of welded plates. Their cost is quite high because of the labour needed. Casts are made up both of cast iron and aluminium alloy; the latter are really only profitable for small-size pieces since aluminium pours very easily into relatively thin walls.

There are two types of pouring: pouring into sand moulds and pouring under pressure into mechanical moulds.

Sand moulds are made on the basis of models that take the outer shape of the pieces to be manufactured. These models are made of wood or steel. Models of wood are subject to wear and thus are convenient for only limited series. Models of steel are very durable and are used for large series production where they are often joined to a plate that will be used for their positioning during automatic casting. They are called model-plates. Sand is pressed around the models inside several frames. After having pressed the sand and having taken the model away, the imprint of the outer part of the piece to be cast is obtained. The inner part (the empty space) is made by a core of sand that has been made out of the preliminary.

The model should also have some elements able to make several pins in which the cores can be fitted. The melting metal is then poured into the empty space left between the imprint of the model and the core. After cooling and hardening, the melt piece is taken out of the mould. This latter is then destroyed and should be remade for every piece. There are several automatic systems for very large scale production which allows the creation of the mould with the core, as in the casting operation in the stripping on a transfer chain. This type of processing could be used for the moulds of metal that can be recovered after every operation, but, of course, the mould casting temperature should be markedly higher than that of the melt material. This kind of processing cannot be used for steel or cast iron pieces in moulds of steel or cast iron; aluminium is more suitable for this procedure. In this case, owing to the great fluidity of cast aluminium, the procedure used is moulding under pressure. The liquid metal is injected under pressure into a mould of steel with the core also made of steel. Of course, pieces have to be designed so that cores can be taken out without damaging them.

- Shafts and gear blanks

Shafts and gear blanks are made of steel (except for worm gears that are generally made of bronze which we will analyze later). They can be constructed both from bars of steel laminated by flaking and from pressed raw pieces. These latter have one advantage: their shape is very similar to the final shape apart from the necessary processing overthicknesses. We can easily understand the difficulties to be found in some of these pieces because the cylinder-shaped raw bars have a shape quite different from the final shape: pressing cannot then be used as a method as far as further shavings removal and total processing duration are concerned. But pressing has a drawback: it needs much equipment consisting of forging dies that can be amortized only through the mass-production of the same piece. In the case of large gears, the rims of alloy steel can be forged and welded to the cores of more ordinary steel or of grey iron with several intermediate sheet metals. The bronze rims of worm gears are often melted onto steel cores that have been previously processed to form fixing notches. Bronze is melted to an excess in order to eliminate the impurities and the blow-holes caused by shrinkage. Afterwards, the excess material should be eliminated. In some cases the bronze rims are produced by centrifugation and are hooped onto the core of steel or grey iron.

14.3 The Processing of Carters

Carters are generally made up of various pieces in order to allow the assembly of the parts into the gearbox. Every piece should be manufactured separately, but some processing procedures should be carried out using the assembled gearbox. These operations include the:

(a) levelling of the assembly benches on the frame or on the flanges of the motors;
(b) boring of journals and gaskets;
(c) drilling, on cases or flanges, of holes for the assembling or fixing of bolts;
(d) the screw threads of fixing holes for the same utilizations.

So we need several machines to make levelling operations (milling machines or planers), boring operations (boring machines), drilling (drills) and threading operations (threaders). These operations should be made separately for unique pieces or for small-quantity series. Nowadays classical machines are not widely used and are often replaced by numerically controlled machines that can be programmed. However, for very small-quantity series, every piece should be programmed which demands a long time. These numerically controlled machines can be programmed for a very high number of pieces (in the case of a large series) which does not affect the cost of the piece. These machines are often multi-axeled, which is an advantage because it allows different operations to be carried out on various plans and in appropriate directions by mounting the piece only once. In this way the positioning adjustments and the controls at every assembly can be avoided since the piece is assembled only once on the machine and all the operations are carried out during this single assembly. Modern machines are also able to control the position of the piece and modify the tools' position accordingly.

The processing made on a single position is of interest, for example, because of the necessity of having a perpendicularity with thin tolerances between the assembly side on a flange and the journal's boring axes.

For the processing of large-quantity series, other types of machines can be used. Here more specialized machines are used that carry out particular operations but are assembled on a production isle. The piece is transferred from one machine to another by specially-designed pliers that automatically position the piece on each machine. So, the following positionings are extremely accurate and the different parts of the piece are processed under the best conditions. These machines are specialized for one or a few limited operations, their advantage being that are then inflexible and perfectly set for this operation. The accuracy obtained is independent of the assembly. It is clear that pliers should be designed and produced for every kind of piece and that these pliers are very expensive; their cost can be amortized only for large-quantity series. The final cost is lower than the cost of the assembly and setting operations of single pieces or when processing a unique piece.

The processing of journals (boring) affects the positioning of axes of gears and, consequently, toothing alignment. This has great importance for the load capacity of gears (see Chapter 5). In a single operation these journals need to be bored on a unique piece. If in the carter the rolling bearing seats of the same shaft are situated on different parts, then will have to be assembled before being bored so that the two seats can be concentric. It is useful to put the rolling seats of the same shaft on the same piece in order to avoid one assembly during processing. But this operation is not always possible because of further assembly or for foundry reasons.

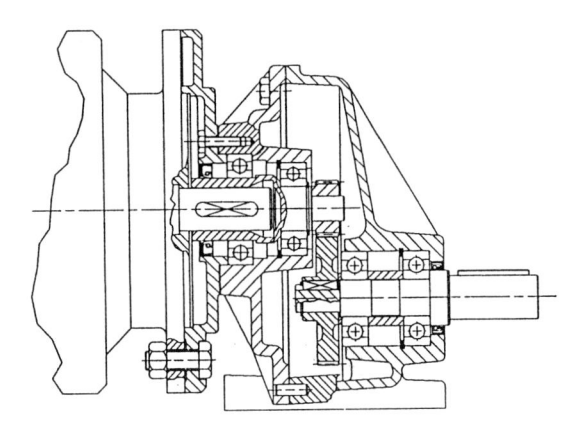

Fig. 14.1. General scheme

The scheme in Fig.14.1 shows the importance of a processing that ensures the perpendicularity of the axes of the rolling seats with the surfaces that will form the joint plane of the two parts of the carter. If perpendicularity is not guaranteed, the axes of the two wheels of the gear will not be parallel and the toothing seat will be worse than predicted by calculations. The gearbox loading capacity will consequently be reduced; this is also valid for the output axis for the joint plane of the motor flange. If this perpendicularity is not achieved, the motor shaft will be inclined to the boring of the output shaft and some additional bending stresses will affect the shafts. We also notice the importance of the concentricity circumference where the assembly belts are situated. If this circumference is not centred the toothing seat and the seat of the additional bending stresses on the shaft could be modified.

14.4 The Processing of Shafts

The processing of shafts is essentially a turning, but also includes the grinding for the rolling bearings' seats and the threading for the keyways. In the case of hollow shafts, there is also boring for the inner seat and broaching for the keyways. The example of Fig. 14.2 which shows the input shaft of the gearbox in Fig. 14.1, illustrates the complexity of the turning operation caused by the different seats.

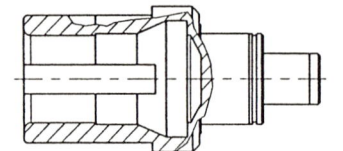

Fig. 14.2. Input Shaft

These seats are justified by the different degrees of accuracy required for their functioning. On a simple shaft, turning is made in a single operation with only one assembly of the machine. The processing of this kind of shaft using classical methods would be very expensive and the necessary change of position on different machines could affect tolerances. For mass-production, we could use either a processing on a multiple-broach lathe, where every broach ensures its particular function for only one assembly of the piece, or several machines on an isle with automatic feeding and automatic positioning of the piece by suitable pliers. The grinding of rolling bearing seats should be made on a cylindric grinder separate from the lathe. However, boring and broaching should be made on specialized machines. Grinding, in particular, should be made on machines set for one piece only and equipped with size-control devices before and after grinding. All these devices can be used only for mass-production.

In order to guarantee the parallelism of the gear's axes, the parallelism of the boring axes, the rolling seats and the pinion seat should be guaranteed, too. This parallelism for prescribed tolerances can be achieved by processing on different machines with a manual assembly for every operation. This requires an accurate and expensive setting that can be avoided by a mass-processing operation on machines designed and set for only one piece with suitable assembly and transfer pliers. The keyways should also be arranged on shafts. These grooves are produced by milling. One can use ball end two fluted-mill (with an axis perpendicular to the shaft's axis and concurrent with it) with a diameter equal to the width of the groove or disk milling machines with an axis orthogonal to the shaft's axis but not-concurrent with it. The first system creates grooves ending in two half-circumferences whose diameter is the groove's width. The second system creates grooves with cylindrical ends; the diameter of these grooves is that of the miller. The groove's length depends on the miller's diameter.

14.5 The Working of Gear Blanks

As gears are an integral part of shafts (shafted pinions), their processing is similar to that of the shaft because the gear is a particular seat of the processed shaft.

The processing of gears that will be put on the shaft (or of the inner gears that will be fixed to the carter) is independent of that of the shaft (Fig.14.3).

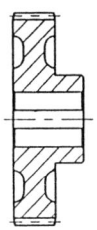

Fig. 14.3. Gear's blanks

The operations required will be the turning of the head diameters and of the diameter of the reference cylindric surface, the boring of the inner seats for fixing the shafts and the facing of the two side faces of the gears and also the finishing of the reference surface perpendicular to the axis. These operations can be made on different machines with an assembly and a disassembly every time with all that it involves: mistakes in assembly and, above all, a very high percentage of spoilt products. These operations can also be made on different machines where every machine carries out one or more operations but with a system of assembly pliers in order to guarantee an easy and quick fixing on the following machines. This second method eliminates times when the machines are not used and guarantees a higher processing accuracy but the cost of the processing pliers (design and production) really requires mass-production for economic reasons. A third method consists in carrying out all the operations on a single machine, often on a multispindle lathe or a processing unit. With this method the dead time is reduced and assembly mistakes are avoided. This method can be made on numerically-controlled machines for average or mass production. As for side faces processing, it is clear that the pressed raw pieces should be used as they reduce the volume of material rejected since the shape of the raw piece has in practice the shape of the finished piece.

14.6 Cutting Cylindrical Spur and Helical Gears

14.6.1 Particulars

There are two main proccessing methods: milling of the shape and by generation.

14.6.2 Processing by Milling with Form Tools

The aim of this kind of processing is to produce some grooves with the shape of the space between teeth on the gear blank by milling. So for this kind of operation one needs a milling machine able to accomplish the indexing, i.e. the shifting of the gear according to the number of teeth in order to produce every following groove. This adjustment is made by an indexing equipment that can be driven by the machine's motor through a specific adjustment for every gear with a definite number of teeth. Today numerical control can ease this adjustment since it is made by the number of teeth. Previously the problem was that the groove's shape was different for every tooth number and module, so a really high number of cutting millers was needed. To simplify this, at first a method was developed for a reduced number of millers effective in a defined space of teeth number. This did not meet the accuracy required by the actual gears. The use of sticked plate millers can solve this problem more easily than single cutting plates that have to correspond to the required number of teeth because the miller's disk can be used for a range of one or more modules. Moreover, the plates of sintered carbide allow production before sintering on a soft material with a dimensional check on an optical comparator. There are some devices equipped with diamond-shaped tools that carry out the same processing on ended plates; these tools are expensive, so must be used on quite large-quantity series. This type of milling is not profitable for gears that do not have to undergo any other finishing operation.

14.6.3 Processing by Generation. Principle

The principle regulating the processing by generation uses a tool whose cutting edges represent a gear a kind of involute engaging with the gear to be cut. The cutting machine, as well as the movements necessary for cutting, give to the piece and the tool a movement of generation; a relative movement causing an engaging among the various parts. The machine then gives a movement equal to that obtained by the engaging of the tool with the finished piece. Since cutting is made during engaging, the cut flanks will be the envelope of the cutting edges. These flanks will be flanks in involutes.

There are three main processes by generation:
1 Cutting by rack-tool;
2 Cutting by pinion-tool;
3 Cutting by lead-miller hob.

14.6.4 Cutting by Rack-Tool

The tool is made up of a sharpened surface whose profile is the reference plan of the gear processed. In order to create a slotting tool called a rack-tool, some cutting and release angles are situated on the normal fasteners on the sharpening face. The nominal size refers to a module that will be that of the gear to be cut. There is only one tool shape for all the gears of the same module, independent of the number of teeth.

The profiles of the cutting fasteners can differ from the reference plan as far as some details such as the cutting protuberances are concerned (see grinding). Because of the sharpening and the cutting and shift rakes, this profile decreases slightly after every sharpening even if the module of the tool is not exactly the desired module. Since the check in manufacturing is made on the teeth thickness and not on the relative position of the tool and of the wheel, this does not represent a drawback.

The cutting machine is driven by different movements. The tool is supported on a tool-trolley driven by an alternate rectilinear movement whose truck is slightly longer than the toothing width following teeth inclination.

Cutting is made in one direction when the tool's fastener appears. In the phase of return, when the tool cannot cut any more, it moves away from the processed piece.

During this time, the processed gear is driven by the movement of minimum rotation and by translation in parallel with the tool in the apparent section of the gear. The movement of translation has the same extent as that of the movement of rotation following the initial circumference of the gear to be cut. The conjugation of movements is then a movement similar to engaging (the initial circumference of the gear rolls without sliding on the initial line of the reference plan composed of the tool's fasteners).

When the tool starts to descend, it comes into contact again with the gear and removes some material. By repeating this operation, the cut of a flank made up of facets whose envelope is a flank with a profile in involute is achieved. The adjustment of the machine according to the number of teeth and the module

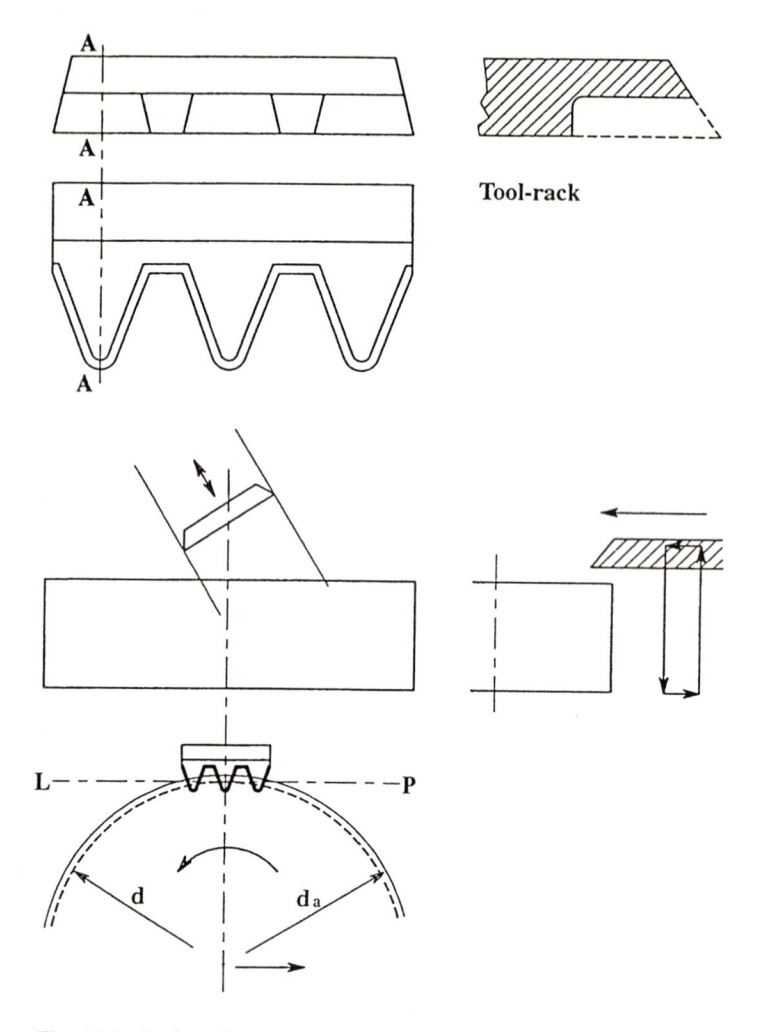

Fig. 14.4. Rack-tool

allows conjugation of two movements of rotation and translation relative to the number of teeth.

The translating shift parallel to the tool, the tool having a finished length, enables the gear to be cut. A supplementary operation then is required. After the cutting of one or several teeth (according to the machine, the tool and the working conditions) the tool is detached from the gear and this latter is brought to its original position but with the rotation corresponding to the same number of teeth of the translation. The operation can then start again. This set of operations is repeated until the gear has been completely processed.

The setting of the different movements depends on the module and on the number of teeth of the gear. The various shifts can be adjusted with a set of gears that has to be assembled on the machine according to the gear that has to be made. The movements are then automatically made by a single motor. Modern machines are set numerically.

The required parameters are set up and the speeds and the different engines that control the individual shifts are electronically adjusted. In this way the machines' setting is made easier.

Processing is not continuous on these machines. The processing time can be relatively long because of all the necessary movements that interrupt cutting. But the simplicity of the tool made up only of plans allows a very high accuracy to be reached. The surface state of the gears flanks is quite good even if it is made up of facets and is not of an even surface. But this is not an important factor if processing is to be followed by finishing operations.

14.6.5 Cutting by Pinion-Tool

Generation is made by a tool whose sharpening face is a pinion with a reference module and plane found in the gear to be cut. Thus, the movement of generation is similar to that of the movement of engaging of the gear to be cut and the pinion-tool cutter.

Processing involves an alternative movement of the tool's broach, rotation of the broach and rotation of the gear to be cut through the gears' engaging and a rotation of the tool during its rectilinear track to produce the toothing helix. As for cutting by rack, the position of the tool and gear is different according to the ascent or the descent of the tool.

During descent, the tool is in contact with the gear to be cut and, by its slotting movement, removes the top. During this time all rotations of the gear and tool are helicoidal. When the tool reaches the processing end of its stroke, it moves slightly radially away from the gear.

During its ascending phase, the pinion-tool and the gear make a slight movement of generation because the two shifts are bound by the number of teeth of the two bodies. When the tool reaches the upper end of its stroke, it is brought once more against the gear so that it will take several tips away from the gear to be cut. The movement can be continuous until the gear has been completely cut. The relative shifts of the tool and the gear to carry out engaging are made either by a set of adjusting gears defined by the number of teeth of the gear and the tool, or by a numerical check.

This processing is quicker than the previous one, so is more suitable for mass-pieces. It is specially convenient for the processing of inner gears and is often the only one possible. The surface obtained is a sequence of small facets. Here accuracy is lower due to the tool's complexity and because of this complexity the tool is more expensive than the pinion tool (Fig. 14.5).

PINION Fellows

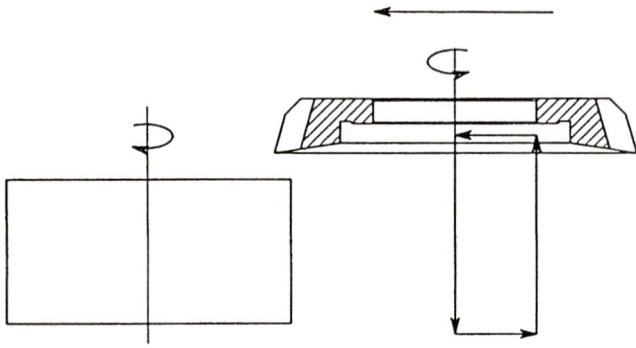

Fig. 14.5 Pinion-tool

14.6.6 Processing by Lead-Miller (Hob)

In explaining the theory of gears and worms, it was noticed that the gear in its medial section was a geometric gear (with involute profiles) conjugated with the reference rack defining the worm. Let us consider a cutting tool made from a worm by placing it perpendicular to the worm thread of the devices defining cutting fasteners and modifying worm threads in order to create the cutting and shifting angles. Let us put this tool into contact with the gear to be cut (Fig. 14.6).

If the gear and the worm turn at engaging speeds, i.e. at relative speeds conditioned by the number of the worm threads and the number of the gear's teeth, on the plane of this gear (a plane corresponding to the axial plane of the screw-tool) a gear by generation will be created. If the worm shifts following the gear's width, this geometric gear will be reproduced on all the planes and so the result will be a gear with involute flanks. The movements are complex. The lead-miller is inclined on the tool-carrier in order to create the helix angle of the gear to be cut. This inclination makes the flank of the worm in the direction of the gear's tooth. The worm is driven by a movement of rotation around its axis. At the same time, the gear is driven by a movement of rotation around its axis at a speed equal to that of engaging with the worm. The tool is driven by a very slow movement of translation with respect to the other two in the direction of the toothing width. This movement allows the tool's descent to create the following plate wheels inside the apparent planes of the gear to be cut. The worm being inclined to the gear's axis, a relative compensating movement between the gear and the tool has to be created. It is a relative movement of rotation of the gear according to the helix angle. All these movements are created by kinematic chains which are adjustable relating to the tools and the gears to be cut from a single engine for conventional machines. As for numerically-controlled machines, the various movements are made by different engines whose speeds are electronically

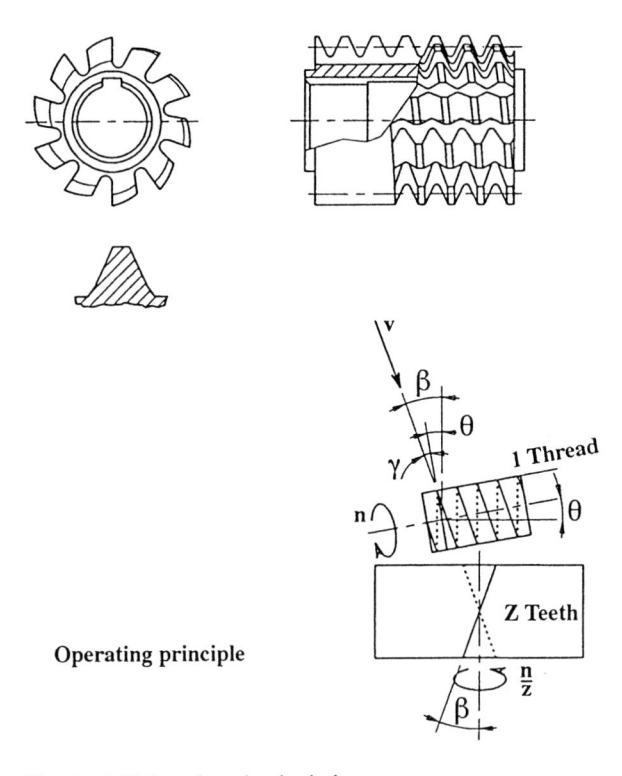

Fig. 14.6 Hob and cutting by hob

adjusted by the setting programme. These machines have a high production efficiency because they work continuously. The tool is very complex and, consequently, difficult to create and very expensive. It is the machine for excellence for the mass-working of outer toothing gears, but is not suitable for cutting of inner toothing gears. The quality of processing is good even if the flank does not have the ideal shape but is only a sequence of facets whose envelope has the ideal shape. These facets are created by the unevenness of cutting caused by the sequence of the miller's equipment.

14.6.7 Processing by a Tool of Carbide or Fired Clay

The above mentioned processings are traditionally made by tools of worked steel or carbide fixed plates. These processings are made under lubrication in order to avoid extreme overheating of the tool and processed piece. This kind of processing is made at relatively low cutting speeds for the tool's resistance. The most modern trend is to use cutting tools of metal carbides or of fired clays because these tools are very hard and thus they allow dry processing at a very high cutting speed. In reality, they are tools with plates fixed on a body of steel. The advantages of this cutting method are on one hand the saving of processing time and, on the other, the fact that working without cutting oil is cleaner, sounder and less expensive. So, during processing, nothing is changed but the tool's material.

14.7 The Processing of Bevel Gears

Independent of the kind of gear cut, the processing procedure is always based on the plate fixed gear. Flat disk-shaped tools have linked teeth (usually 3 by 3) able to create the desired profile. The flat gear thus created and the gear to be cut are assembled on the machine in their relative position depending on the angle of the reference cone of the gear to be cut. A combined movement of the two gears creates the tooth profile on the transverse plane and on the cone. Of course, the movements depend upon the tooth outline according to the flat gear definition and so from the type of toothing.

14.8 The Processing of Worm and Worm Wheels

Worm processing is made either by lathe threading or by milling with the disk-miller or with the ball end two-fluted mill.
Processing by threading is made with the lathe because the tool has the shape of a rack tooth with the desired pressure angle. The tool's cutting fasteners are situated on the plane and match the shape of the reference plane. If this plane is confused with a radial plane on the worm, the cut profile is the profile ZA. If this plane is confused with a plane tangential with the base cylinder of the worm, the cut profile is the profile ZI. If the tool is inclined so that its cutting plane is perpendicular to the worm thread helix, the profile is ZN. The helix angle is made by adjusting the lathe for a threading while the tool shifts parallel to the worm's axis with an axial pace of the helix at every round of the worm. Multi-thread worms can be cut by a simple tool (a tooth of the reference plane) repositioned for every worm thread or by a multiple tool (the number of teeth is equal to the number of worm threads) processing the worm just once. This last method guarantees a higher cutting accuracy avoiding the intermediate adjustments and the resulting mistakes. Fig. 14.7 shows this cutting procedure.
The disk-miller is made by a sequence of radial planes where the cutting fasteners are situated which form a tooth of the reference plane. They are monobloc or fixed plates millers. The miller is assembled on an axis around which it turns and shifts parallel to the worm's axis. The miller's axis is parallel to the axis of the worm for profiles ZA and ZI and is parallel to the real section (perpendicular to helix) for the profile ZN. For profile ZA the cutting fasteners are arranged so that

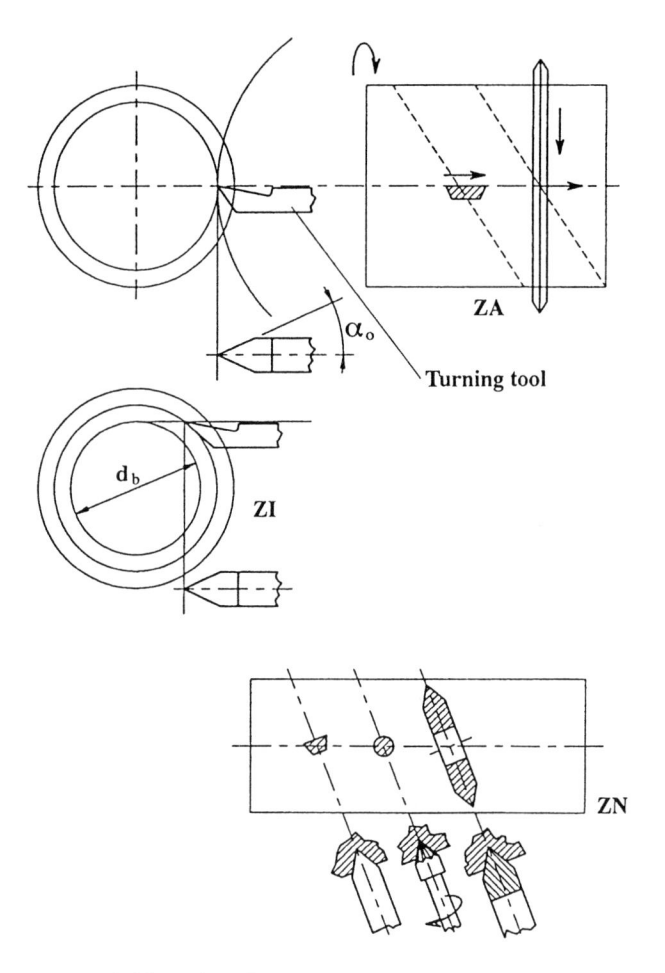

Fig. 14.7. Worms' cutting

they are mixed with a radial plane of the worm. For the profile ZI, fasteners are mixed with the plane tangent to the basic cylinder of the worm. Fig.14.8 shows the miller and Fig.14.7 shows the processing arrangements.

Profile ZN can be created by a cone-shaped miller with an angle equal to double the worm's pressure angle. Its axis is situated on a plane perpendicular to the real plane of the worm (Fig. 14.7).

Furthermore, profile ZI can be created by a cutting procedure of the helicoidal gears illustrated in paragraph 14.6. Actually, these worms are helical worms whose helix angle is the worm's complementary angle of the helices inclination.

The worm-gears processing is made by the same machine used for the helical gears by the lead-miller because the movement following the toothing width cannot be carried out. A lead-miller with an adequate helical angle comes into contact with the gear radially while the gear and the tool are driven by a

Disk cutter

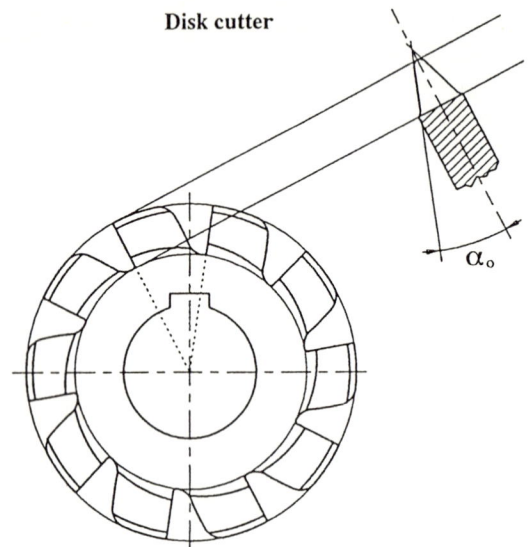

Fig. 14.8 Disk-miller

movement creating the exact engaging between the worm and the gear. The worm can also be replaced by a simple tool whose movement is that of the cutting fastener of the lead miller.

14.9 The Finishing of Cylindrical Gears

According to the material, the desired surface and the accuracy required, the helicoidal gears can be finished in different ways if one of the above mentioned procedures does not finally create a satisfactory gear cut. We should notice that thermal treatments are often the cause of deformations that need a supplementary finishing procedure. In this case, our aim is more the accuracy of the shape than the surface state.

The means used for this are grinding, skiving and honing, respectively.

14.9.1 Grinding

Grinding of cylindrical gears is made by generation or with shape grinding profiles.

These are now more frequent due to the opportunities offered through electronic devices that can give the desired profile to the disk-grinding wheels. Moulds are able to be put into the space between teeth thus creating the desired profile on the two flanks. The gear to be rectified shifts by following a very careful division so that the space between teeth will be placed in front of the grinding wheel in the right position. The flank's plan and the division of the rectified gear are made electronically on specially-designed numerically checked machines.

Grinding by generation can be made by following the principle of the reference rack or by following a procedure identical to the cutting lead-miller.

The grinding following the reference rack technique can be made with one or two disk-grinding wheels working on their flat face. A single grinding wheel gives the gear and the tool a relative movement of generation simulating the involute and every flank is cut separately. One operation cuts the right side, another cuts the left side. The grinding-wheel is flattened after a certain number of operations by a diamond processing. The grinding-wheel position for the tooth is adjusted by automatic devices according to the wear of the grinding wheel.

If we use two grinding wheels, they can be parallel to each other or inclined. If the grinding wheels are parallel, the machine places them so that the theorical contact points with the flanks to be rectified are on a line tangent to the basic circle of the gear. If the engaging movement is achieved, the two grinding wheels remain tangent to the two anti-homologous wheels and teeth are grounded to the desired shape. If the two grinding wheels form an angle, they constitute a reference rack shifting, without gliding, on a circumference of the gear to be rectified. The diameter of this circumference depends on the gear considered and the angle between the two grinding wheels. These machines have to be set according to the number of teeth, the module and the geometrical conditions of the toothing to be rectified. Traditional machines are adjusted by the appropriate division of gears (time difficulties and duration). Numerically-checked machines can be set electronically by sight using a set values. During grinding the plate-grinding wheels could come into contact, with the junction profile of the flanks. In this case, they would create several cuts into this profile situated in the dangerous section of the tooth for the bending stresses which would create concentrations of harmful stresses. This can be avoided by cutting the gears with a bulging tool before grinding. This tool corresponds to that used when processing over thickness on the flank of the cut teeth. This over thickness will be completely or partially taken away by grinding without any scratches of the junction profile. Rectification by screw-shaped grinding wheels is another method to rectify cylindrical gears. These grinding wheels are worms with one or more threads, made of an abrasive product. They have a large diameter (in comparison with the module) so that the profile can be compared to a line for case when straightening the flanks. The grinding wheel works as a lead-miller except for the fact that the movement of generation is continuous and is not interrupted during processing. So the rectified flank is a continual involute and not as in cutting using the lead-miller a sequence of small facets provoked by the processing discontinuity. This method is interesting because the operation is continuous and there is no division between operations as in the other procedures where the operation is interrupted for the need to shift the gear versus the grinding wheels after the grinding of a tooth.

14.9.2 Shaving

Shaving is a finishing operation that takes the place of grinding. The tool is a toothed gear with grooved flanks (Fig. 14.9).

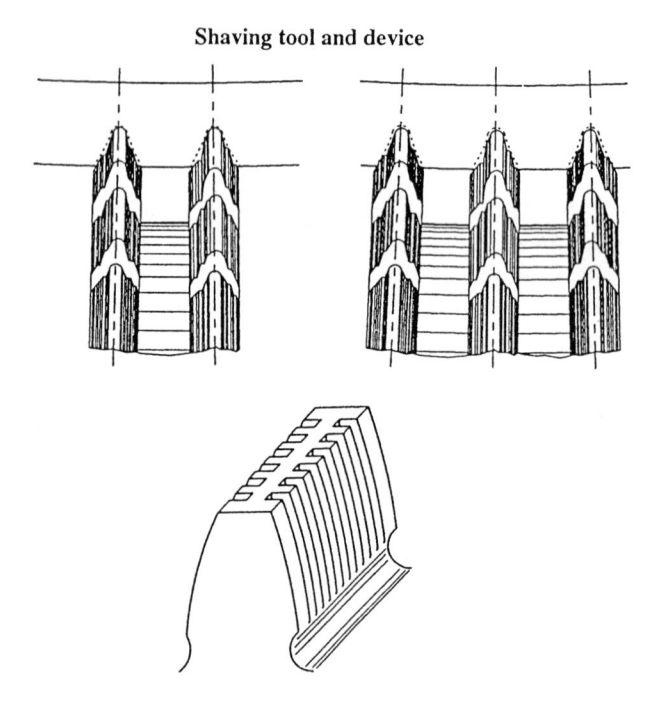

Shaving tool and device

Fig. 14.9 Shaving

The tool and gear are assembled with left axes and are driven by the same movement of that resulting from the working of the two gears on the left gearing. The grooves' fasteners of the gear-tool cause slight eradication of material as well as a local compression on the flank of the processed gear that can generally be sufficient for processing quality. This procedure can be interrupted and is quicker than grinding it is often used in mass-productions. The operation can be made after case-carburizing or surface-hardening or before thermal treatment. In this case, the surface-imperfections caused by processing are eliminated, but the deformations resulting from the thermal treatment persist.

14.9.3 Honing

Toothing honing is a newly developed finishing procedure. It is a type of superfinishing used in place of grinding after cutting (often made dry by a fired clay tool used at a high speed) and case-carburizing. The result is very high accuracy and a considerable reduction of the noise of the gears. It has been integrated with many technological solutions, but the principle is the same. An inner gear represents the tool, and the gear to be processed is assembled as an helicoidal gear with a crossed axis with the tool-gear. The inner toothing gear is, in practice, a carborundum grinding wheel. The abrasive action of the gear-grinding wheel acts as a consequence of the sliding caused by the engaging and the left position of the two axes. The toothing profile is made by the engaging. Honing is also used after a grinding in order to reduce noise.

14.9.4 Skiving

This operation, recently introduced on the market, is similar to shaving. The tool is not a grooved tool, but similar to the traditional cutting tool. The machine also works by generation as a traditional cutting machine. A very small quantity is removed, at a high speed, by a tool made of carbide or fired clay. The finished product is thus very good. The operation is generally carried out before thermal treatment.

14.10 Finishing of Bevel Gears

Until some time ago, bevel gears could not be ground. The two conjugated bodies were only geared by means of a grinding blend on special machines. This procedure was not satisfactory. Now grinding machines for bevel gears have been on the market for a few years. The principle employs the use of generation by a flat gear, the flat gear being a grinding wheel representing the complement of the gear to be processed, i.e. whose teeth are the hollows of the gear to be processed. It functions following the same principle of a cutting machine. These machines are checked numerically so that the engaging can be accurately made.

Skiving is still being developed and will be applicable to bevel gears.

14.11 Grinding of Worms

The grinding of worms is based on the same principle as disk-miller cutting. The disk-miller is replaced by a plate grinding wheel whose marginal section is a reference profile.

14.12 Assembly

Assembly is a sequence of operations of great importance. The different pieces have to be assembled in order to create a functional body. Assembly is completely manual for unique pieces or small-quantity series. Sets are generally made separately (for example, the shaft will be equipped to the connected wheels, bearings and the possible ferrules). These elements will be assembled by following the scheme as accurately as possible. The duration and so the cost of assembly can be reduced by having regard for the processing tolerances. The different by-bodies will then be assembled on the carter.

For mass-production, assembly can be made automatically. These plants have the advantage in being able to avoid any adulteration and to prevent any mistake or oversight. They should be very carefully organized, not as for assembly but also for any kind of manufacturing. In particular, the feeding of pieces into the assembly should be taken into account when processing so that these pieces can be placed on feeding pallets which fit the devices taking them perfectly. These automatic assemblies are a warranty for good organization and for methods, their perfect automatic checking avoiding the inevitable mistakes and oversights of manual assembly.

15 Applications

15.1 The Application of Gears

Gears are present in all industrial applications. Whenever a speed of rotation or a mechanical torque has to be transformed, or the operational security and flexibility increased, a gear is needed. They are used in machines of light and heavy industry as well as in the very complex mechanisms of robotics. Several examples can be given as in the maintenance or lifting apparatuses, the multiple instruments of chemical, iron and steel, and paper and cement industries, the mining industry and by the structures such as dams or movable bridges etc.

Nautical industries make use of gearboxes to reduce the high speeds of turbines to the very slow speed of screw propellers; these gearboxes are so large that they could fill the rooms of a small apartment.

High-speed equipments have been developed in the field of gas turbines and centrifuges.

Gears of course are widely used in the motor-car industry: transmissions reducing the motor speed to the hub speed, bevel gears ensuring the transmission of the universal shaft to the transverse hub, the differential gear compensating the different speeds of the two torques when the vehicle turns or gearboxes are just a few. The great reliability and automaticity of these mechanisms should not be forgotten.

The aircraft and space industries follow the motor-car industry as far as several gear transmissions are concerned. All the flaps on the wings are equipped with a control device with its own motor and gearbox of course these gears have to be very light and during take-off, large and important gearboxes are used. The gear chain, called the kinematic chain, of helicopters is a very complex and fascinating mechanism that not only has to make rotors operational, but also the blades turn around their own axis. Gear transmission is equally widely used in railway traction. Some smaller fields create particular problems as in the overdrive of the wind wheels. Wind moves these blades so slowly that they seem to stand still, so gears have to increase this quite low speed until it reaches the speed needed by the electric generator.

This list is not complete, but is enough to show the different applications and their varying conditions.

The field of application of mass-produced prefabricated gearboxes is generally limited to industrial drives. This chapter will deal with this and show the possible but not the only applications of these gearboxes. Together with speeds and powers, the gearboxes can form all the possible combinations in an infinite number of fields.

15.2 Hoisting Instruments

15.2.1 Bridge Cranes

Bridge cranes are the typical hoisting instruments present in every industry. These machines have three main movements: the hoisting of loads, the translation of the truck and, that of the bridge. Hoisting of a drum is made by a motor using alternating or direct current, where the lifting cable is rolled up and a gearbox whose role is to transforme the motor speed of rotation in the drum's speed. The latter depends on the lifting speed, i.e. on the rolling speed of the cable on the drum. The translation of the truck and of the bridge is simple. An engine transmits its movement of rotation directly to the sliding rollers through a gearbox (necessary to achieve a speed in conformity with rollers) and a rigid shaft linking the two opposite rollers. All these are gearboxes with parallel axes. To carry out the lifting, the engine and the drum are assembled on the same side of the gearbox in order to occupy the smallest volume of space since the mobile mechanisms should be as small as possible so that the bridge can cover the largest useful space in the depository where it is assembled. In the case of the lifting winch, the engine and the drum are the two mechanisms regulating the volume of space occupied. The gearbox chosen will have to allow for the positioning of these two essential mechanisms and will be chosen according to the size resulting from the volume of the engine and the drum. The gearbox will have to guarantee the resistance necessary to lifting, a resistance that depends on the load, speed and cranking torque of the engine. The size of the gearboxes for translation will have to meet the requirements of the total load (including the weight of the whole transported): the rolling resistance and the friction on the rollers' axis as well as inertia. For translation, according to the position of the engine chosen for the smallest volume occupied, the gearboxes will have parallel or perpendicular axes. All gearboxes have two or three stages, a drum and rollers. The development of special gearboxes such as planetary or co-axial gearboxes is not necessary since it is not the gearbox that affects the total volume occupied.

15.2.2 Cranes

Cranes have multiple shapes. However, the movements needed are always the same: the lifting movement itself, the lifting of the jib and the rotation of the jib. Mobile cranes have the additional movement of translation that can be made on rails or tracks. The lifting winch is equipped with a parallel axes gearbox. The rotation of the jib is made around a vertical axis and is generally achieved by a special train composed of a pinion and a big wheel with inner or outer toothing. The pinion's movement of rotation can be achieved by an engine through a coaxial gearbox assembled vertically or by a parallel gearbox with vertical axes. The movement of translation on rails can be achieved with the intervention of a central gearbox or by one or more gearboxes, each operating on a sliding roller. The

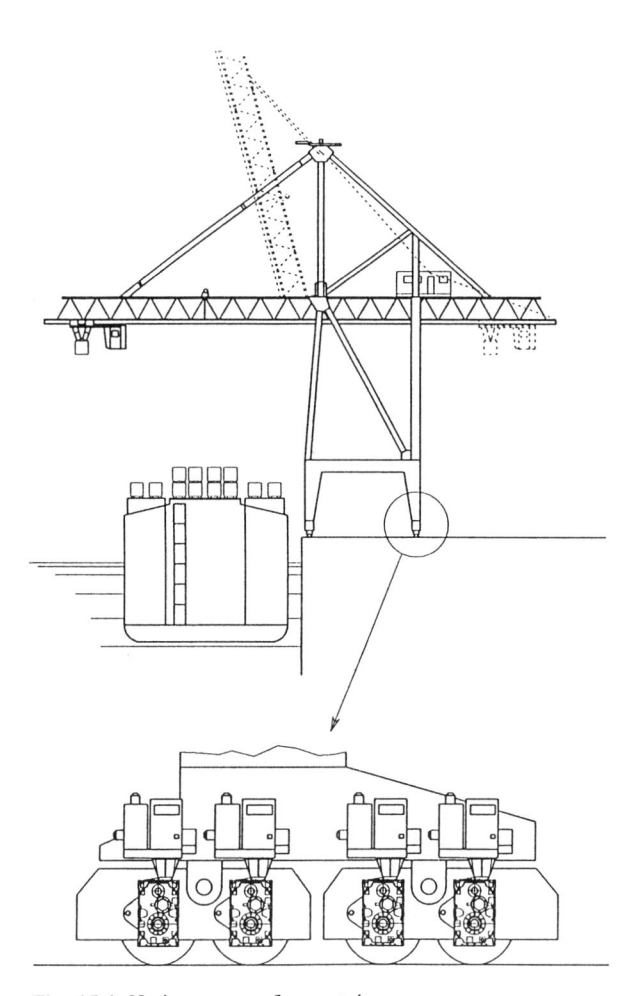

Fig. 15.1 Harbour-crane for containers

translation concerning tracked gearboxes is based on the same principle and will be considered in the paragraph about special vehicles.

Fig. 15.1 shows the transmission for the translation of a portal for a harbour-crane specially designed for the maintenance of containers for loading on boats. This large crane is assembled on two-axelled bogies each bogie being made up of a roller driven by an independent engine and a gearbox. In total, there are 16 rollers on the right and 16 rollers on the left, using gearboxes with perpendicular axes. They are assembled perpendicularly, which means that the outer axis of the gearbox is assembled directly onto the axle. During rotation they are kept still by an arm of reaction linking them to the frame. Every driving engine uses direct current with a power of 25 kW. The reduction of each gearbox is equal to 36. The rollers have a diameter of 800 mm, giving a greater flexibility. A single engine would have required a very complex transmission especially because the rollers are assembled on independent bogies. For application at sea, special precautions have to be taken to prevent corrosion. Gearboxes need to be covered with a thick coat of special paint to avoid corrosion. Movements of this harbour-crane will be operated by gearboxes with perpendicular axes.

15.3 Conveyors

15.3.1 Belt Conveyors

These conveyors are made up of a belt, generally of rubber-coated cloth, moved by a motorized drum standing on equidistant supports that keep it trough-shaped by three cylindrical rollers, the first is horizontal and the other two have a 20° inclined axis on the horizontal (Fig. 15.2).

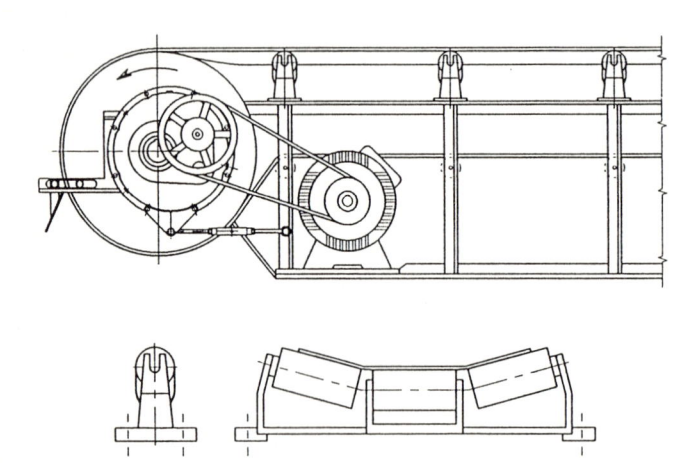

Fig. 15.2 Belt conveyor

The belt is continual and goes back to the motorized drum supported by plane rollers. These belts can transport merchandise in bulk both horizontally or on an inclined plane. The maximum inclination of these planes depends on the merchandise, but it varies between 15° and 20°. They are often assembled on frames and the volume of the order as well as its weight should be reduced. Often the solution is the use of small pendular gearboxes contained in a cylindrical carter. The order is made by a driving belt starting from a pulley assembled on the engine and another assembled on the input shaft of the gearbox. An arm of reaction compensates the torque on the carter during operation. If the engine were to stop, the load on the belt during an inclined transport would tend to make the belt sag with all the risks involved; so, in this case, the gearbox will be equipped with a no-recovery system. This system, which is assembled on the cap, is made up of a device assembled on the driving shaft and enables the movement to proceed in the direction required, but can stop it through locking of the shaft in the wrong direction.

15.3.2 Chain Conveyors

These conveyors have multiple uses. Traction is made by caterpillar wheels moving two chains where joined plates may be assembled for merchandise in bulk

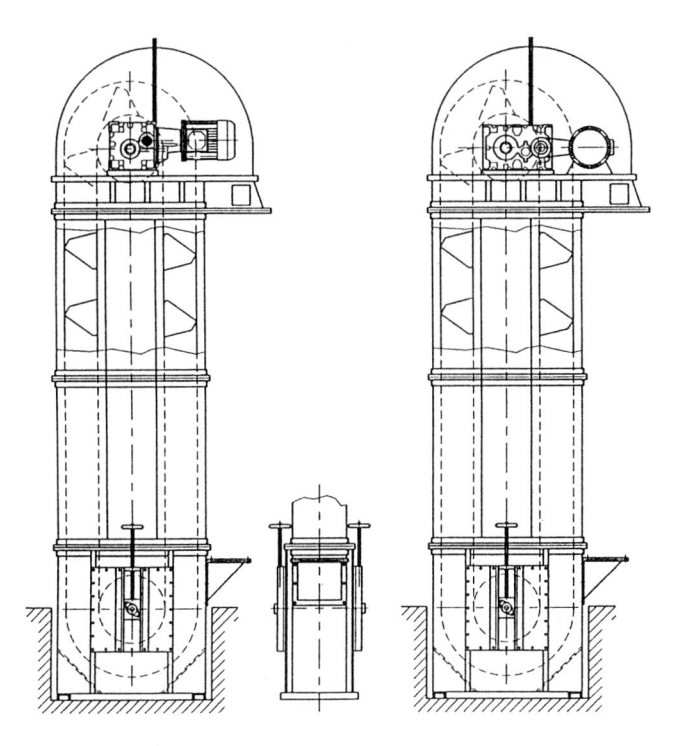

Fig. 15.3 Bucket conveyor

or for pieces of a defined size. Hanging hooks can also be useful for the transportation of products; they are used for the industrial transportation of raw materials, for the transportation of pieces in automatic assembly plants or in packing and sorting plants, etc. Their advantage on belt conveyors is that the distance covered can have any kind of inclination up to the vertical. A typical example is the bucket conveyor as in Fig.15.3.

Buckets, assembled at a regular distance on the chain, are filled to the lower level of the product to be conveyed. They are brought to the upper level and, at the moment of rotation on the upper pulleys, they are again thrown on the conveying belt by centrifugal force. This method allows the transport of granulated solid products or powders. The gearboxes have orthogonal axes directly driven by the motor (or motorized gears with orthogonal axes) or gearboxes with parallel axes driven by the motor through a driving belt and an hydraulic coupling. In all cases, the gearbox should be equipped with a no-recovery system so that if the motor should stop, the load in the riding buckets does not drive the system in the opposite direction.

15.3.3 Roller Conveyors

For long products, such as semi-finished products in the siderurgical industry, roller conveyors are used, also utilized in the packages movement. They consist of parallel rollers, near one to the other, some of which are half-moving. Products are driven by them and drive the other for effect of friction. The half-moving rollers are equipped with in-line gearboxes directly assembled on the roller shaft and on the motor. Such gearboxes are external to the rollers or are placed inside the same. They can be parallel shaft or planetary gearboxes. The latest ones are often used because of their dimensions.

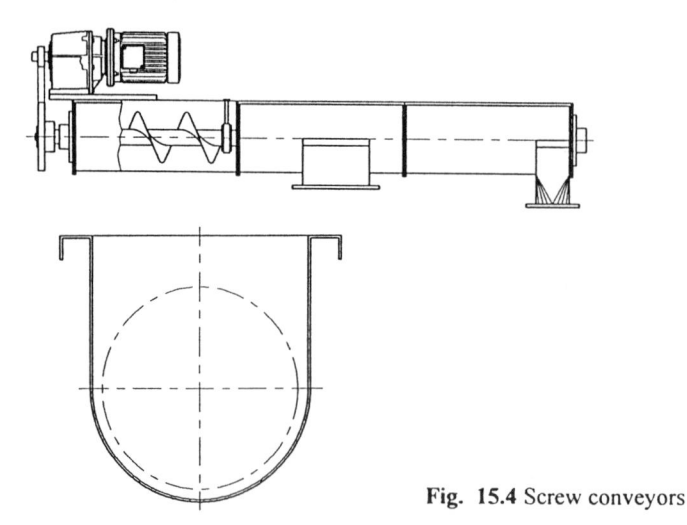

Fig. 15.4 Screw conveyors

15.3.4 Screw Conveyors

Screw conveyors are used for the horizontal transport of products which are dusty.

Figure 15.4 shows one of these conveyors. The principle is that of an Archimedes' screw. The screw is situated into a hopper with a circular bottom and covered by a plane cap. The gearbox, that in reality is a co-axial motorized gearbox, is assembled on the cap and is linked with the screw's axis through a driving belt and several pulleys. This type of conveyor is widely used for the transport of non-abrasive and non-binding powder products such as maize or flours or barley meals; of corns such as wheat, soy-bean or cocoa and also for the transport of powder abrasive products such as foundry sand.

15.4 Waste and Treatment of Used Water

15.4.1 Waste Treatment

Bulky waste should be reduced in volume, by generally being split up and made compact in order to reduce its size. The compact products are then eliminated or used as fuels or briquettes for building. An example is given on Fig. 15.5

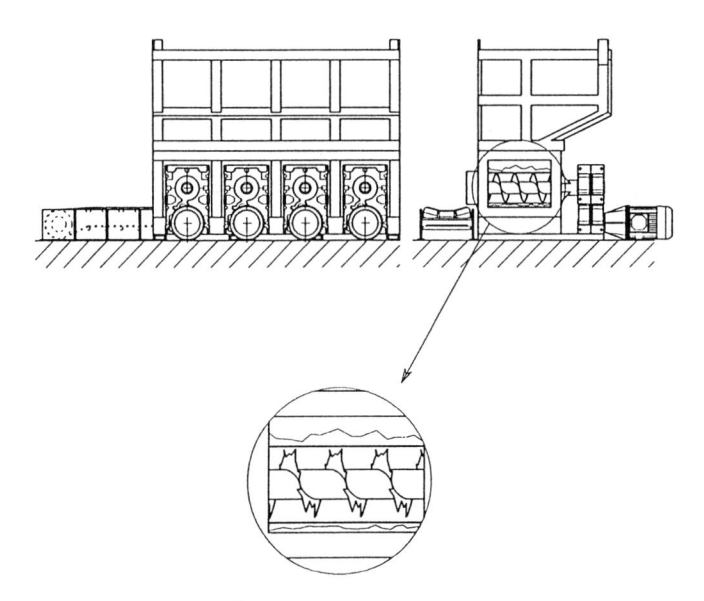

Fig. 15.5. Waste crusher

This plant is suitable for slashing pallets, wood or wood-paper waste. When the waste emerges, a conveying band brings the slashed waste towards a compacting machine that produces briquettes for the heating of public buildings. The crushing system is a screw system (with 2 to 6 screws) where every screw is driven by a fixed motorized gearbox with parallel axes whose ratio is equal to about 100. The outlet is below the inlet (plan of vertical axes). The system is compact and, according to the number of screws, can process from 10 to 150 m^3/h of wood. Due to its compactness, it can be assembled on an independent frame for transport.

15.4.2 The Treatment of Used Waters

This treatment requires several specialized machines for the excitation of water, the retention of floating products, for clarification and pumping.

The retention of floating products is made through grids that can be cleared only if they have a certain movement. The grids used can be either vertical or rotating. The first are assembled so that they can move like a belt driven by a pulley. The

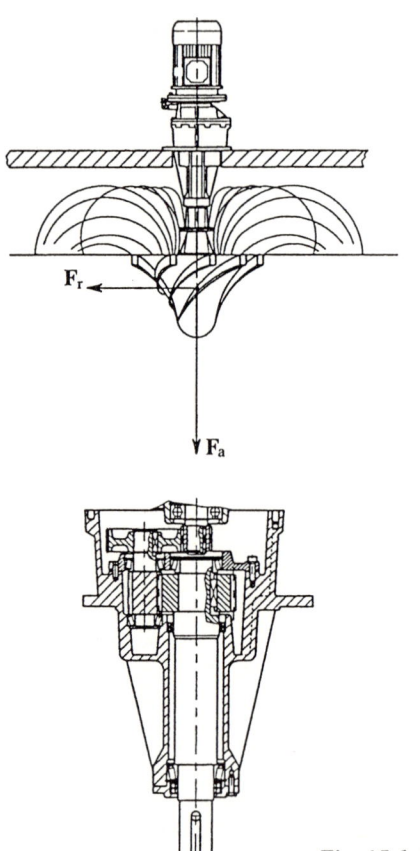

Fig. 15.6. Surface fan

movement is achieved by a worm screw gearbox assembled with a friction coupling that protects the motor in case of overloading or stoppage of the grid. The second grids are assembled on a rotating drum driven by a co-axial gearbox. They are often equipped with an automatic grid-removing device driven by the same movement of the grid.

Mud thickeners and decanters are represented by swing bridges. They are driven either by screw gearboxes whose gear has a vertical axis and is situated on the axis of rotation, or by co-axial gearboxes situated at the bridge's end which drive a sliding roller on a circular distance at the bridge's edge.

Fig. 15.6 shows the operation of a surface exciter (fan).

A rotor supporting pallets (similar to a centrifugal pump) is driven by a movement of rotation on the surface of the water to be purified. The system is assembled floating on the basin so that the rotor is perfectly immersed. The axes are vertical. The gearbox is a co-axial gearbox giving a reduction equal to 30 (rotor's speed of rotation of about 50 min^{-1}) equipped with an extended shaft. The gearbox is a classical co-axial of standardized manufacture on which is situated a cap consisting of two bearings supporting the end of the extended shaft towards the rotor. This cover is also of standardized manufacture. The bearings are shaped so that they can support the vertical and horizontal reactions coming from the rotor. The system is completely noiseless and can be used in the purification plants near towns.

Water pumping for low heights is made by Archimedes' screws driven by co-axial motorized gearboxes assembled on the screws' shaft above the screw.

15.5 Special Vehicles and Farm Machines

Tracked vehicles are operated by the difference in speed between the two driving tracks. This difference can be achieved in a merely mechanical way but there are many complications. An easier way is to drive every track by an hydraulic motor with an incorporated gearbox and to use the speed of these motors to carry out the necessary operations. Thus it is a thermic motor driving a pump that gives the necessary pressure to the power in the liquid. This liquid is contained in a tank and is canalized and brought from the container to the hydraulic motors which are combined with gearboxes on the driving drum of the tracks. Because of their co-axiality and small size planetary gearboxes can be extremely useful and also because, by using a different number of planets at different levels, it is then possible to create a gearbox with two or three levels in a spur carter. For hydraulic motors, consult the relevant appendix.

Another application of this kind can be found for trucks carrying cement. Cement is contained in a rotating drum. The truck is driven by a thermic motor. The power needed for the drum's rotation is achieved through this motor. The vehicle's speed is achieved by the variation of the speed of rotation of the thermic motor. The speed of rotation of the mixing drum cannot be affected by the speed of rotation of the thermic motor which drives a pump feeding a tank and the drum is driven by a hydraulic motor coupled with a planetary gearbox. This hydraulic motor has a constant pre-set speed independent of the truck's speed. The drum's speed can be set according to the operation (transport, emptying, filling or cleaning) by adjusting the speed of the hydraulic motor.

Farm machines are driven by a thermic motor and often require hydraulic power to carry out jack operations. Other operations are made by a movement of rotation. The speeds of the force required should be independent of the speed driving the thermic motor. Some farm machines (such as tractors for the work on vineyards) are assembled on a mobile frame with the shape of a shed; this particular shape complicates the transmission from the motor to the wheels. So, hydraulic motor linked with a planetary gearbox is placed on the wheel's hub and these motors are driven by a pump situated on the thermic motor. Planetary gearboxes, that are quite small, may be placed on the spindle of the wheels.

15.6 Equipments Used for Intermittent Operation

Some equipment operate intermittently. For example, equipment used for positioning that does not require great accuracy or transferance on the assembly line, or parking or highway barriers. These barriers are widely used to block the entrance to private parks, garages and toll highways. They are remote-controlled by an operator or automatically with the introduction of money or magnetic cards. Worm gearboxes are perfectly suitable for these applications. In fact, they can be very small for such high reductions and being able to operate intermittently and the powers used being small they cool naturally; their thermic power will then be equal to their mechanical power in spite of their low efficiency.

Fig. 15.7 shows the usual arrangement of these barriers.

Worm gearboxes are equipped with a torque limiter to avoid the motor overloading when the barrier is stopped.

15.7 Precision Positioners

Precision positioners are used when the positioning precision has to be exact and when it is checked by the rotation of the driving motor. They are used in servo-mechanisms where the object set has to be in a precise position defined by the angle of rotation of the motor. An important application is the control of radar

① Rod

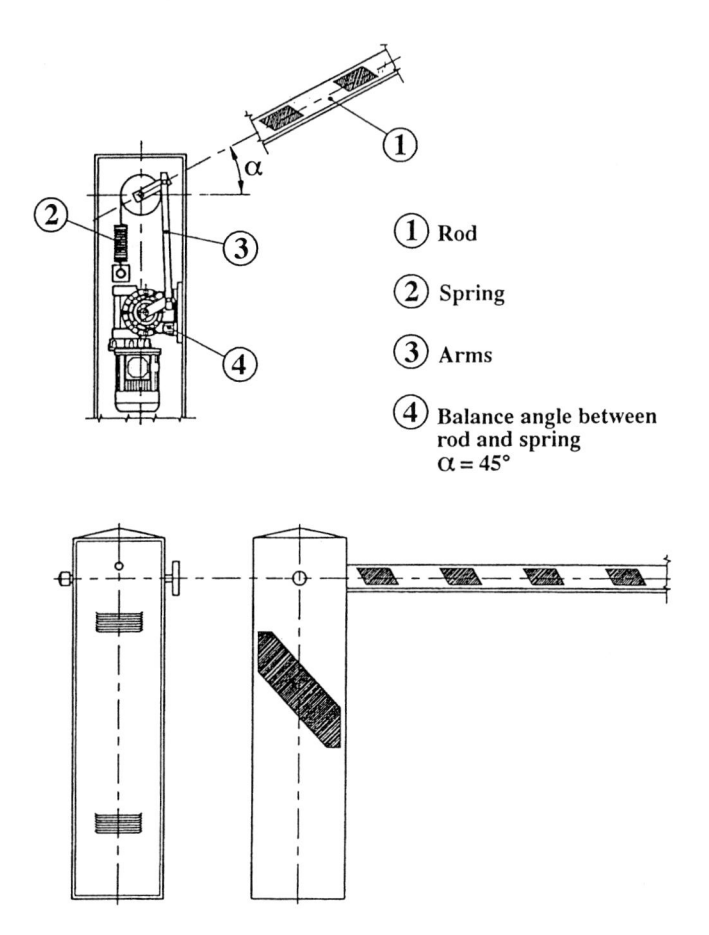

② Spring

③ Arms

④ Balance angle between rod and spring
$\alpha = 45°$

Fig. 15.7 Highway barriers

aerials; the position of an object is taken up by the radar according to the angular position of the aerial, but also by the angle drawn by the rotating aerial and the time between the wave's emission and its picking up by the aerial. This period of time is reliable only if the aerial's rotation is perfectly regular. So the transmission cannot in any case have any clearances or deformations beyond the minimum depending on the required position. Another important application is found in the mechanical controls of robots. These machines have an essential function of positioning independent of the particular utilization of the robot.

The characteristics of these gearboxes are the quality of their gears and the shafts' solidity. Gears should be very precise because they have to work with a very small clearance. As in the chapter on gears, operation without clearance is made by contact on the two flanks and, in this case, the processing deviations turn into variations of the centre distance if it is free, or into supplementary forces on the teeth (in this case they will be very high). These deviations can be avoided by very careful processing of toothings. On the other hand the response of the movement of the gearbox output shaft versus the input shaft is good only if the transmitted torques do not give very strong deformations. This imposes a higher size upon shafts for the same transmitted torque as well as for a good transmission. Transmissions for robots have either parallel or orthogonal axes. For these latter, the bevel gears have to be designed with all the above-mentioned considerations about quality and solidity. Planetary gears are particularly effective for parallel axes. In effect, their compactness, co-axiality and solidity, due to the division of the torque, make them particularly compact and little less able to be deformed; however, case has to be taken when toothed wheels are involved. For mass-production, the different components are processed with a certain precision and, consequently, with relatively high tolerances. However, within the range of production, some gearboxes will have a minimum clearance, and others maximum clearance with all the intermediate range. A selection of the gearboxes with a minimum clearance will give clearances suitable for most positioning equipments. So the problem lies in the selection of minimum global clearance products within the range produced, bearing in mind the profit on precision-made goods and the cost price of mass-production.

Appendix. Hydraulic Motors and Variators

A.1 Hydraulic Motors

There are various types of hydraulic motors. The best known are the radial piston, orbital, gear, blade and axial piston motors with an inclined plate or with inclined bodies.

- Radial pistons motors (Fig. A.1).

These motors have a high efficiency, a regular functioning speed, also at a low speed, a high starting torque, but they are also large in weight and size. They generally run at slow speed (80 to 600 min^{-1}).

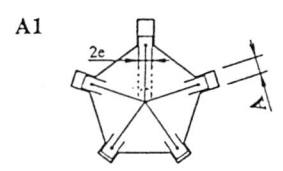

A1

Fig. A.1. Radial piston motors

- Orbital motors (Fig.A.2).

A toothed rotor can turn in an hollow stator if the number of hollows is equal to the number of teeth plus one. This rotor is linked to a shaft by a cardan coupling. The introduction of oil under pressure between the rotor and the stator creates a torque and the rotor turns simultaneously inside the stator and around its own axis. These motors are slow (200 to 800 min^{-1}). They are easily produced and they have good mechanical and volumetric efficiencies. The starting torque is weak. The sizes are reduced and the cost not too high.

- Gear motors (Fig. A.3).

The principle is simple: two toothing wheels constitute a gear, the edge of whose teeth have a very weak play with the carter. The fluid under pressure is

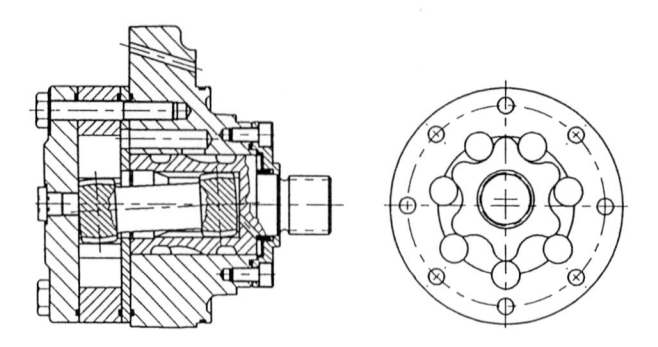

Fig. A.2. Orbital motors

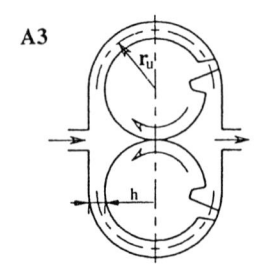

A3

Fig. A.3. Gear motor

tangentially brought to the two wheels on one side and an outlet is tangentially arranged on the other side. The pressure is exerted on the teeth between the spaces left free by engaging, so there is the creation of a torque that makes the motor turn, as shown in Fig. A.3. These motors have a high speed (1500 to 3000 min^{-1}). They are easily produced, their efficiency is good of a small size but the noise is high at high speeds. They are very sensitive to impurities in the hydraulic liquid.
- Blade motors (Fig. A.4)

A rotor out of centre with reference to a stator is equipped with radial plates supported by the stator's edge. The fluid pressure is different on the faces of the plates and creates the torque. These motors are quite noiseless and they function quite regularly. They are very sensitive to the peaks of pressure and to the impurities in the hydraulic liquid. Their speed ranges from 500 to 3000 min^{-1}.

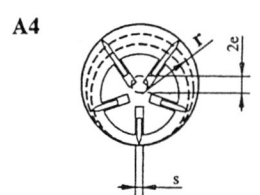

A4

Fig. A.4. Blade motors

- Axial piston motors.

Several pistons are placed axially in a turning keg. They are set on a plate inclined with regard to the keg's axis. There are two possible versions. The first is equipped with a fixed bearing plate, the second with a mobile rotating plate. The second is called "with an inclined piston's body".

Both versions have a high speed (1000 to 1500 min^{-1}). These motors have an optimal efficiency, a small size, but they are noisy at high speeds. The starting torque is higher for the version with an inclined piston body. Speed is regulated by the modification of the inclination of the plate. This latter can be modified by the relative inclination of the axis of the pistons and of the bearing plate. These motors have a quite high ratio power/weight.

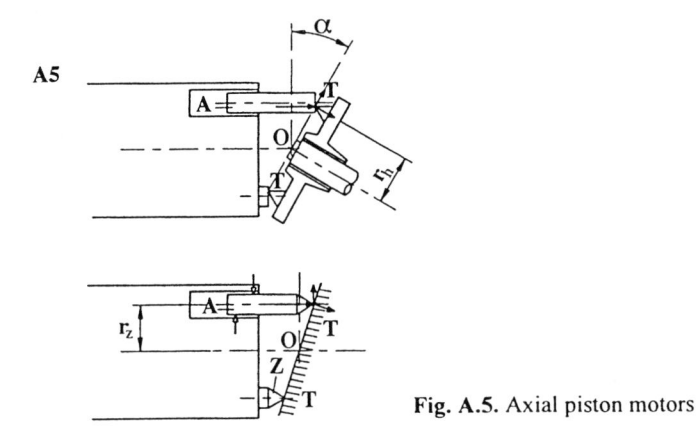

Fig. A.5. Axial piston motors

A.2 Speed Variators

Speed variators are based on driving by friction. Their function is to ensure a progressive variation of speed between the input and the output. Gearboxes allow only one reduction. To variate speed in a particular field and progressively, the gearboxes should be supported by a variator.

There are two types of speed variators: one is based on the friction of a belt on several pulleys with varying diameters and the other is based on the friction between the wheels forming a planetary train whose diameters can vary.

A.2.1 Belt Variators

A belt with a trapeizoidal section is driven by a pulley in two parts and it drives another pulley in two parts. The two parts of each pulley are cones with their apex on the axis of rotation. If the two cones are moved away from the driving pulley, the belt moves into a smaller diameter. The pressure on both the cones of the driven pulley loosens and, pressed by an axial spring, they approach until the pressure can balance the force of the spring again. The rolling diameter of the belt on the driven pulley is increased, so the reduction is also increased. If the driving cones are approached, the belt unrols on a bigger diameter, it stretches and divides the two driven cones. The diameter on the driving pulley increases and the diameter of the driven pulley decreases. Therefore the transmission ratio is also decreased (Fig. A.6).

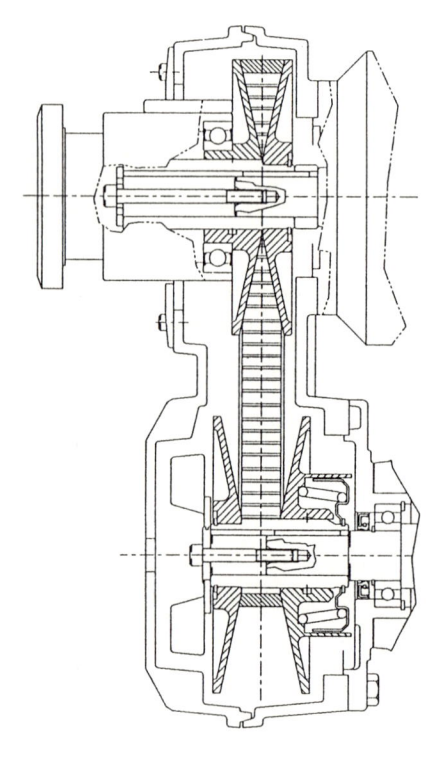

Fig. A.6. Belt variator

In some cases the trapeizoidal belt is replaced by a steel trapeizoidal trim. This latter configuration slightly increases the tranmissible power to the detriment of the volume.

A.2.2 Planetary Variators

The planetary variator is a friction variator following the principle of the planetary trains. Fig. A.7 shows the principle of this type of variator.

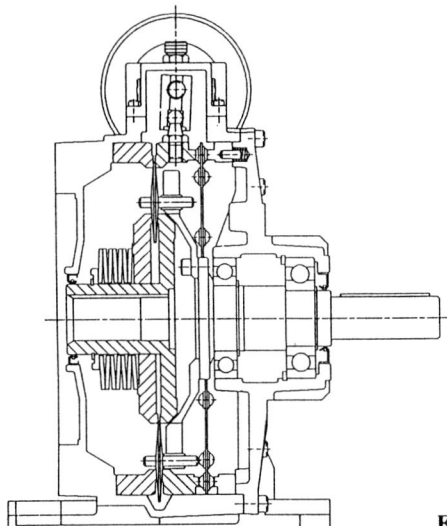

Fig. A.7. Planetary variator

Two co-axial wheels are assembled on the drive axle: one is fixed, while the other can move on the axle. These two wheels are two cones with a large angle at the top. Several biconical wheel, that are planets of the double gear the latter having the function of a solar, are clasped by the two driving wheels. The axles of the two wheels are assembled on a gear co-axial to the two driving wheels including the output shaft. The planetary disks are, in their turn, wedged between the two conic fixed disks; these are integral with the scaffolding for rotation and one can move on the axle. These two disks have the function of a ringwheel. The two solar cones drive satellites and oblige them to move by driving the output shaft. If the fixed disks undergo an axial movement, planets will have to move towards the centre by rejecting the two solar cones through a contrary action on their spring. Let us call d_3 and d_1 the average diameters of contact of the planets with the fixed ringwheel and with the solar in a precise position, respectively. If the position is modified, the two diameters will become d_3+x and d_1+x. We know that the ratio of the input and output speeds is the ratio of the diameters of the gearwheel and of the increased solar of a unit. When x is positive, the ratio $(d_3 + x) / (d_1 + x)$ is smaller than the ratio d_3/d_1. The ratio of speeds as then decreased. When x is negative, the ratio of the speeds has increased. By a progressive action, with the help of a rotating cam, the output speed is varied on the relative position of the rotating fixed disks. This kind of variator is more compact than others having the same efficiency. It is very interesting because it can be integrated in a prefabricated gearbox.

The ISO Standards Relating to Gears (TC 60)

On 1 January 1995 according to the Technical Committee of the ISO, the standardization for gears will be the following:
   Published standards:

ISO 677	Straight bevel gears for general engineering and heavy engineering - Basic rack.
ISO 678	Straight bevel gears for general engineering and heavy engineering - Modules and Diametral.
ISO 1328-1	Cylindrical gears - ISO system of accuracy - Part 1 Definitions and allowable values of relevant to corresponding flanks of teeth gears.
ISO 1340	Cylindrical gears - Information to be given to the manufacturer by the purchaser in order to obtain the gear required.
ISO 1341	Straight bevel gears - Information to be given to the manufacturer by the purchaser in order to obtain the gear required.
ISO 4468	Gear hobs - Single start - Accuracy requirements.
ISO 8579-1	Acceptance code for gears: Part 1 - Determination of the airborne sound power level emitted by gear units.
ISO 8579-2	Acceptance code for gears: Part 2 - Determination of mechanical vibrations of gear units during acceptance testing.

   Existing standards under revision:

ISO 53	Cylindrical gears for general and heavy engineering - Basic rack.
ISO 54	Cylindrical gears for general and heavy engineering - Modules.
ISO 701	International gear rotation - Symbols for geometrical data.
ISO 1122-1	Glossary of gear terms - Part 1:Definitions related to geometry.
ISO 1122-2	Glossary of gear terms - Part 2: Metric dimensions of speed reducing worm gears
ISO 2490	Single start solid (monobloc) gear hobs with tenon drive or axial keyway, 1 to 40 module - Nominal dimension.
TR 4767	Addendum modification (rack shift) of the teeth of external cylindrical gears for speed-reducing and speed-increasing gear pairs.

New standards in the course of publication:

ISO 6336-1	Calculation of the load capacity of spur and helical gears - Part 1: Basic principles, introduction and general factors of influence.
ISO 6336-2	Calculation of the load capacity of spur and helical gears - Part 2: Calculation of surface durability (pitting).
ISO 6336-3	Calculation of the load capacity of spur and helical gears - Part 3: Caculation of tooth bending strength at the tooth root.
TR 10495	Cylindrical gears - Calculation of service life under variable loads - Conditions for cylindrical gears according to ISO 6336.
ISO 10825	Gears -Wears and damage to teeth - Terminology.

Standards awaiting the vote of ISO to be accepted as International Standards:

DIS 6336-5	Calculation of the load capacity of spur and helical gears - Part 5: Quality of materials and endurance limits.
DIS 14104	Surface temper edge after grinding.

Standards awaiting the vote of TC 60 to be accepted as DIS:

CD 1328-2	Cylindrical gears - ISO system of accuracy - Part 2: Definitions and allowable values of deviations relevant to radial composite deviation and runout.
CD 9084	Gears - Simplified calculation of cylindrical gears - Application standard for high speed and similar requirements.
CD 9085	Gears - Simplified calculation of cylindrical gears - Application standard for industrial gears.
CD(TR) 10064-2	Cylindrical gears - Code of inspection practice - Part 2: Inspection of radial composite deviations, runout and tooth thickness allowance.
CD 10300-	Calculation of load capacity for bevel gears - Part 1: Introduction and factors of general influence.
CD 10300-2	Calculation of load capacity of bevel gears - Part 2: Calculation of surface durability (Pitting).
CD 10300-3	Calculation of load capacity of bevel gears - Part 3: Calculation of tooth bending strength at the tooth root.
CD 10347	Geometry of worm gears - Information on the manufacture of various worm flat form.

CD(TR) 13989-1   Gears -Calculation of scuffing load capacity of cylindrical, bevel and hyphoid gears - Part 1: Flash temperature method.
CD(TR) 13989-2   Gears - Calculation of scuffing load capacity of cylindrical bevel and hyphoid gears-Part 2:Integral temperature method.

Standards under study in the work groups:

9082	Gears - Simplified calculation of cylindrical gears - Application standard for vehicle gears.
9083	Gears - Simplified calculation of cylindrical gears - Application standard for marine gears.
(TR) 10063	Accuracy of cylindrical gears - Function groups, test groups tolerance families.
(TR) 10064-3	Cylindrical gears - Part 3: Recommendation relevant to blanks, centre distance, parallelism of axes.
(TR) 10064-4	Cylindrical gears-Part 4:ISO system of accuracy and recommandations relative to surface texture and tooth contact pattern checking.
(TR) 10826	Rack shift of the teeth of internal spur and helical pairs.
10827	Calculation of industrial bevel gears.
10828	Geometry of profile of worm gears.
13593	Gears - Enclosed gear drives for industrial gears.
13691	Gears - Specifications for high speed enclosed gear units.
14179	Gears -Thermal ratings of gears.

Standards recently registered in the work programme:

?	Calculation of worm gears.
?	Testing of gear lubricants.

Abbreviations: ISO : International Standard Organization;
DIS : Draft International Standard;
CD : Committee Document
TR : Technical Report;
(TR) : Future Technical Report;
  ?   No number has yet been assigned.

Bibliography

- Gears:

Dudley D. and Towsend D.P.
    Gear Handbook,
    2nd edition, McGraw-Hill, New York , (1992)

Henriot, G.
    Traité pratique et théorique des engrenages,
    Vol. 1 (1985)
    Vol. 2 (1981)
    Bordas (Paris)

Niemann, G. and Winter, H.
    Mashinenelemente,
    Vols1. 2 and 3,
    Springer-Verlag, Berlin, (1983)

- The calculation of shafts:

Brand, A.
    Calcul des pièces à la fatigue. Méthode du Gradient.
    CETIM, Senlis (1980)

Brand A. , Flavenot J.F. , Gregoire R. and Tornier C.
    Recueil des Données technologiques sur la fatigue.
    CETIM , Senlis (1977)

Farie J. P., Monnier P., Niku-Lari,
    Guide du Dessinateur: Les concentrations de contraintes.
    CETIM, Senlis (1977)

Niemann, G. (Hirt, M.)
    Maschinenelemente,
    Vol. 1
    Springer-Verlag, Berlin (1975)

- For the calculation of bearings:

General catalogue SKF, FAG and NSK.
Terms, Symbols and units.

Part IV

Electric Motors and Drives

Professor Dr.-Ing. Dr-Ing. h.c. D. Schröder
Lehrstuhl für Elektrische Antriebstechnik
Technische Universität München

1 Electric Machines-Standards and Definitions

1.1 Introduction and Standards

Electric machines produce in the operating point torque in a speed range determined on the one hand by the type of motor, and on the other hand by the load characteristic. Generally speaking, we must distinguish between direct-current - and alternating-current - or rotating field-machines; this type of distinction concerns the electrical power supply of the machine. Another type of distinction is based on the speed - torque characteristic. For example, here we distinguish between the series excitation characteristic, thus the speed increases clearly as the torque diminishes; the shunt excitation characteristic, thus the speed decrease slowly as the torque increases; or the synchronous characteristic, thus the speed is constant (not the angular position) with variable torque. Further distinctions are made based on the type of construction, the fields of application (classes of protection) or the control. Requirements and standards have been developed in order to unify the various additional conditions for electric motors, such as electrical connection, operating conditions, and types of constructions.

There are several normative references like:

IEC 27	Letter symbols to be used in electrical technology
IEC 27-1	1992, Letter symbols to be used in electrical technology Part 1: General
IEC 27-4	1985, Part 4: Symbols for quantities to be used for rotating electrical machines
IEC 34	Rotating electrical machines
IEC 34-2	1972, Rotating electrical machines - Part 2: Methods for determining losses and efficiency of rotating electrical machinery from tests (excluding machines for traction vehicles)

IEC 34-3	Rotating electrical machines - Part 3: Specific requirements for turbine-type synchronous machines
IEC 34-5	Rotating electrical machines - Part 5: Classification of degrees of protection provided by enclosures of rotating electrical machines (IP code)
IEC 34-6	1991, Rotating electrical machines - Part 6: Methods of cooling (IC code)
IEC 34-12	1980, Rotating electrical machines - Part 12: Starting performance of single-speed three-phase cage induction motors for voltage up to and including 660 V
IEC 34-18	Rotating electrical machines - Part 18: Functional evaluation of insulting systems
IEC 38	1983, IEC standard voltages
IEC 50	International Electrotechnical Vocabulary (IEV)
IEC 50 (411)	1973 International Electrotechnical Vocabulary (IEV) Chapter 411: Rotating machines
IEC 72	Dimensions and output series for rotating electrical machines
IEC 85	1984, Thermal evaluation and classification of electrical insulation
IEC 279	1969, Measurement of the winding resistance of an AC.machine during operating at alternating voltage
IEC 364	Electrical installations of buildings
IEC 364-4	Electrical installations of buildings Part 4: Protection of safety
IEC 364-4-41	1991, Electrical installations of buildings Part 4: protection for safety - Chapter 41: Protection against electric shock
IEC 445	1988 Identification of equipment terminals and of termination of certain designated conductors including general rules of an alphanumeric system
IEC 449	1973, Voltage bands for electrical installations of buildings
IEC 971	1989, Semiconductor converters - Identification code for converter connections
IEC 497	1973, Guide to the choice of series of preferred numbers and of series containing more rounded values of preferred numbers

Indeed there are also national standards, which have to be considered.

1.2 Duty and Rating

The CEI, IEC 34- 1 is an internal standard for rotating machines and should be discussed here, especially the duties and rating are very important.

Duty types Duty. The duty may be described by one of the duty types defined later or by the specification of another duty by the purchaser.

Declaration of duty. It is the responsibility of the purchaser to declare the duty as accurately as possible.

In certain cases where the load does not vary or where it varies in a known manner, the duty may be declared numerically or with the aid of a time sequence graph of the variable quantities.

If the time sequence is indeterminated a fictitious time sequence (duty types S2 to S8) not less onerous than the true one, shall be selected, or the duty type S9 shall be applied.

In the duty is not stated, duty type S1 (continuous running duty) applies.

Rating. The rating is assigned by the manufacturer by selection of one of the classes of rating defined. The class of rating selected shall normally be maximum continuous rating based on duty type S1 (continuous running duty), rating with discrete constant loads rating based on duty type S10 (duty with discrete constant loads) or short-time rating based on duty type S2 (short-time duty).

If this is not possible, a periodic duty type rating based on one of the duty types S3 to S8 (periodic duty) or the non-periodic duty type rating based on the duty type S9 (non-periodic duty) shall be selected.

Selection of a class of rating. Where a machine is manufactured for general purposes, it shall have maximum continuous rating and be capable of performing duty type S1.

If the duty has not been specified by the purchaser, duty type S1 applies and the class of rating assigned shall be maximum continuous rating.

When a machine is intended to have a short-time rating, the rating shall be based on duty type S2 as defined and as designated .

When a machine is indended to supply varying loads or loads including a period of no-load or periods where the machine will be in a state of rest and de-energized, the class of rating shall be a periodic duty type rating based on a duty type selected from duty types S3 to S8.

When a machine is indented non-periodically to supply variable loads at variable speeds, including overloads, the non-periodic duty type rating shall be based on duty type S9.

When a machine is indented to supply discrete constant loads including periods of overload or periods of no-load, or periods where the machine will be in a state of rest and de-energized, the class of rating shall be a rating with discrete constant loads based on duty type S10.

In some cases, a test at the actual or estimated duty may be arranged by agreement between manufacturer and purchaser but such a procedure is not generally practical.

In the determination of the rating:
- for duty types S1 to S8, the specified value(s) of the constant load(s) is taken to be the rated output and is expressed in Watts for motors and Volt-Amperes for generators.

Duty types. Generally speaking, the following definitions are valid:
Definitions: operation at constant load: operating time t_b;
N(IEC34-1); time constant: T_b
Starting duration: t_a;   D (IEC 34-1))
pause, at rest and de-energized: t_p;
R (IEC 34-1), time constant T_p
operation on no load: V (IEC 34-1)
electric braking duration: t_{Br}; F (IEC 34-1)

(a) Continuous running duty - Duty type S1. Operation at constant load of sufficient duration for thermal equilibrium to be reached (see Fig. 1.1)

$$\frac{t_b}{T_b} > 3 \ ; \quad \frac{t_p}{T_p} > 3.$$

The factor 3 results from thermal dynamic response with e^{-1/T_9}.
the final stationary value (95%) is reached with $t \approx 3T_9$

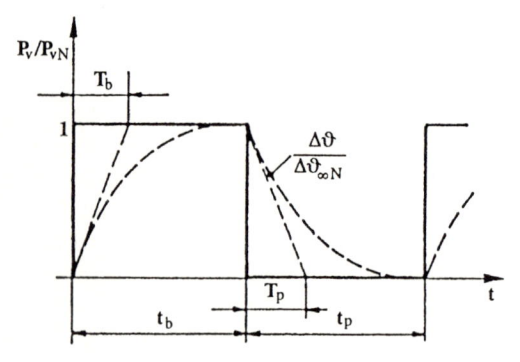

Fig. 1.1. Continuous running duty (S1)

Thermal load allowed: $\dfrac{\Delta\vartheta_\infty}{\Delta\vartheta_{\infty N}} \leq 1; \quad \dfrac{P_V}{P_{VN}} \Rightarrow 1;$

(b) Short-time duty - Duty type S2. Operation at constant load during a given time, less than that required to reach thermal equilibrium, followed by a rest and de-energized period of sufficient duration to reestablish machine temperatures within 2 K of the coolant (see Fig. 1.2).

$$\frac{t_b}{T_b} < 3; \quad \frac{t_p}{T_p} > 3.$$

Characteristic: the final stationary operation temperature is not reached; but during rest period the coolant temperature is always reached ($\Delta\vartheta = 0$).

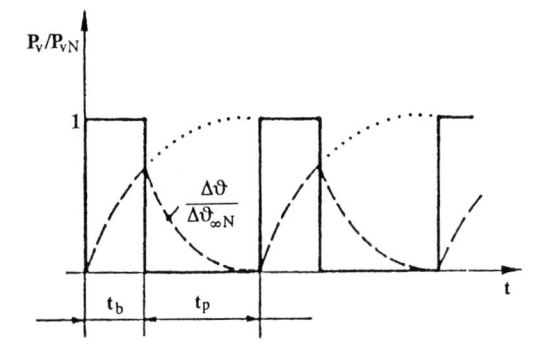

Fig. 1.2. Short time duty (S2)

$$\frac{\Delta\vartheta_\infty}{\Delta\vartheta_{\infty N}} = \frac{1}{1-e^{-t_b/T_b}}; \text{ Overloading of the machine in this particular type of operation}$$
$$\Downarrow$$

$$\frac{\Delta\vartheta_\infty}{\Delta\vartheta_{\infty N}} = \frac{P_V}{P_{VN}} \geq 1 \rightarrow i_{adm} = \sqrt{\frac{1+v}{1-e^{-t_b/T_b}} - v}.$$

(c) Intermittent periodic duty - Duty type S3. A sequence of identical duty cycles, each including a period of operation at constant load and a rest and de-energized period (see Fig. 1.3). In this duty, the cycle is such that the starting current does not significantly affect the temperature rise. Note - Periodic duty implies that thermal equilibrium is not reached during the time on load.

$$\frac{t_b}{T_b} < 3 \quad \frac{t_p}{T_p} < 3.$$

Characteristic: the temperature "oscillates" between $\Delta\vartheta_1$ and $\Delta\vartheta_2$ during operation and rest, there is a transient response additionally.

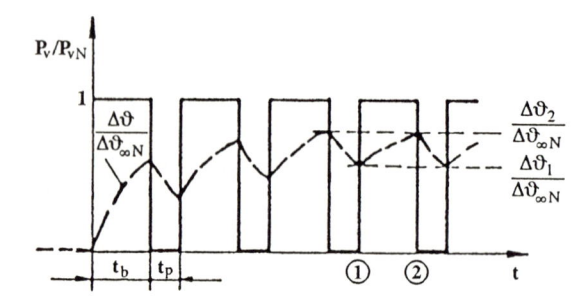

Fig. 1.3. Intermittent periodic duty (S3)

Cycle period $t_s = t_b + t_p$.

Normalization $\tau_b = \dfrac{t_b}{T_b}; \ \tau_p = \dfrac{t_p}{T_p}, \ \varepsilon = \tau_b / (\tau_b + \tau_p)$.

Resulting in:

$$\frac{\Delta\vartheta_2}{\Delta\vartheta_\infty} = \frac{1-e^{-\tau_b}}{1-e^{-(\tau_b+\tau_p)}} \leq 1.$$

For periodic operation:
$$\begin{cases} \dfrac{\Delta\vartheta_2}{\Delta\vartheta_\infty} = \dfrac{1-e^{-t_b/T_b}}{1-e^{-(t_b/T_b+t_p/T_p)}} \leq 1 \\[2mm] \Delta\vartheta_1 = \Delta\vartheta_2 \, e^{-t_p/T_p} \end{cases}$$

thermal load allowed (being $i = I/I_N$):

$$\frac{\Delta\vartheta_\infty}{\Delta\vartheta_{\infty N}} = \frac{1-e^{-(\tau_b+\tau_p)}}{1-e^{-\tau_b}} = \frac{P_V}{P_{VN}} = \frac{i^2+v}{1+v} \geq 1$$

⇒current load allowed:

$$i_{adm} = \sqrt{(1+v)\frac{1-e^{-(\tau_b+\tau_p)}}{1-e^{-\tau_b}}} - v \geq 1$$

with $\tau \ll 1$ and $T_b = T_p$:

$$i_{adm} = \sqrt{\frac{1+v}{\varepsilon}} - v$$

$$\frac{\Delta\vartheta_\infty}{\Delta\vartheta_{\infty N}} = \frac{P_V}{P_{VN}} \geq 1; \quad i_{adm} = \sqrt{(1+v)\frac{\Delta\vartheta_\infty}{\Delta\vartheta_2}} - v \geq 1 .$$

(d) Intermittent periodic duty with starting - Duty type S4. A sequence of identical duty cycles, each cycle including a significant period of starting, a period of operation at constant load and a rest and de-energized period (see Fig. 1.4)

Note - Periodic duty implies that thermal equilibrium is not reached during the time on load.

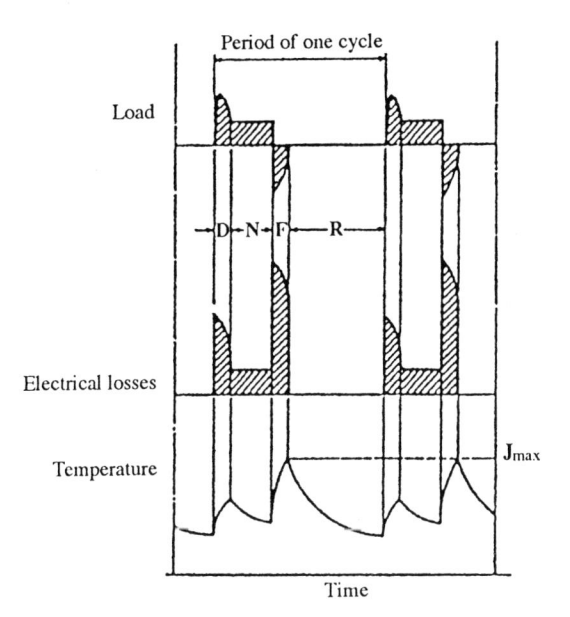

Fig. 1.4. Intermittent periodic duty with starting (S4)

(e) Intermittent periodic duty with electric braking- Duty type S5. A sequence of identical duty cycles, each cycle consisting of a period of starting, a period of operation at constant load, a period of rapid electric braking and a rest and de-energized period (see Fig. 1.5)

Note - Periodic duty implies that thermal equilibrium is not reached during the time on load.

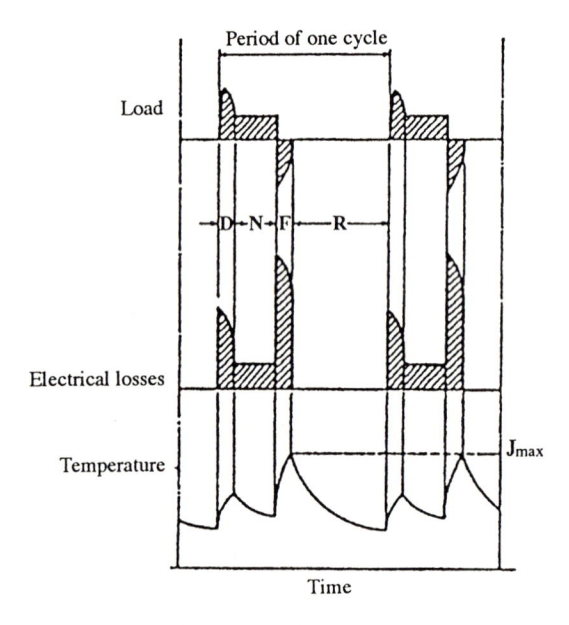

Fig. 1.5. Intermittent periodic duty with electric braking- Duty type (S5)

(f) Continuous-operation periodic duty- Duty type S6. A sequence of identical duty cycles, each cycle consisting of a period of operation at constant load and a period of operation at no load. There is no rest and de-energized period (see Fig. 1.6)

Note - Periodic duty implies that thermal equilibrium is not reached during the time on load.

(g) Continuous-operation periodic duty with electric braking - Duty type S7. A sequence of identical duty cycles, each cycle consisting of a period of starting, a period of operation at constant load and a period of electric braking. There is no rest and de-energized period (see Fig. 1.7)

Note - Periodic duty implies that thermal equilibrium is not reached during the time on load.

(h) Continuous-operation periodic duty with related load/speed variations - Duty type S8. A sequence of identical duty cycles, each cycle consisting of a

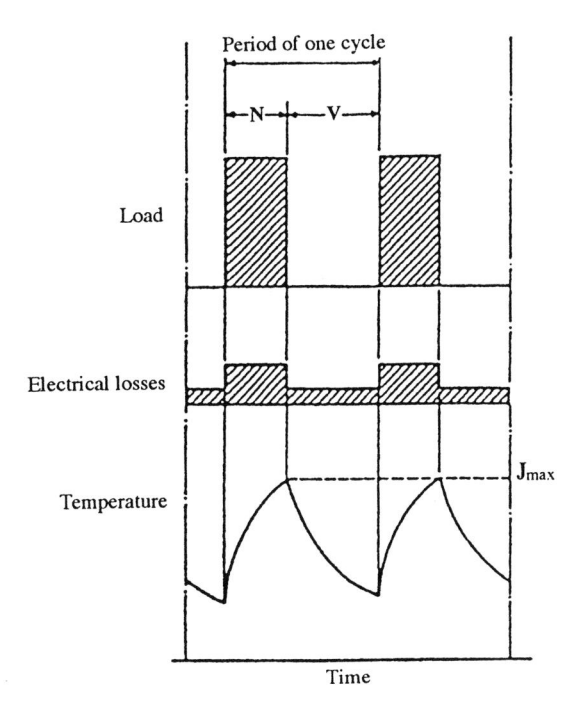

1 Electric Machines-Standards and Definitions 377

Fig. 1.6. Continuous operation periodic duty (S6)

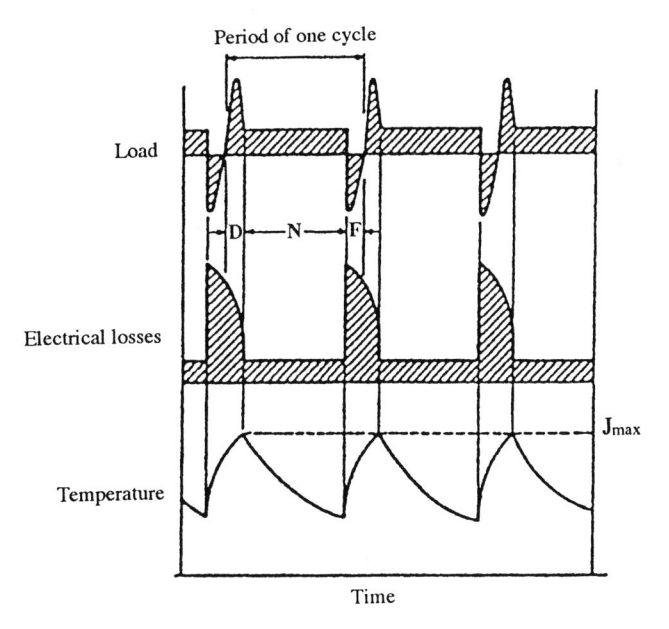

Fig. 1.7 Continuous-operation periodic duty with electric braking (S7)

period of operation at constant load corresponding to a predetermined speed of rotation, followed by one or more periods of operation at other constant loads corresponding to different speeds of rotation (carried out for example by means of a change of the number of poles in the case of induction motors). There is no rest and de-energized period (see Fig. 1.8).

Cyclic duration factors:

$$\frac{(D+N_1)}{(D+N_1+F_1+N_2+F_2+N_3)}\ 100\%$$

$$\frac{(F_1+N_2)}{(D+N_1+F_1+N_2+F_2+N_3)}\ 100\%$$

$$\frac{(F_2+N_3)}{(D+N_1+F_1+N_2+F_2+N_3)}\ 100\%\ .$$

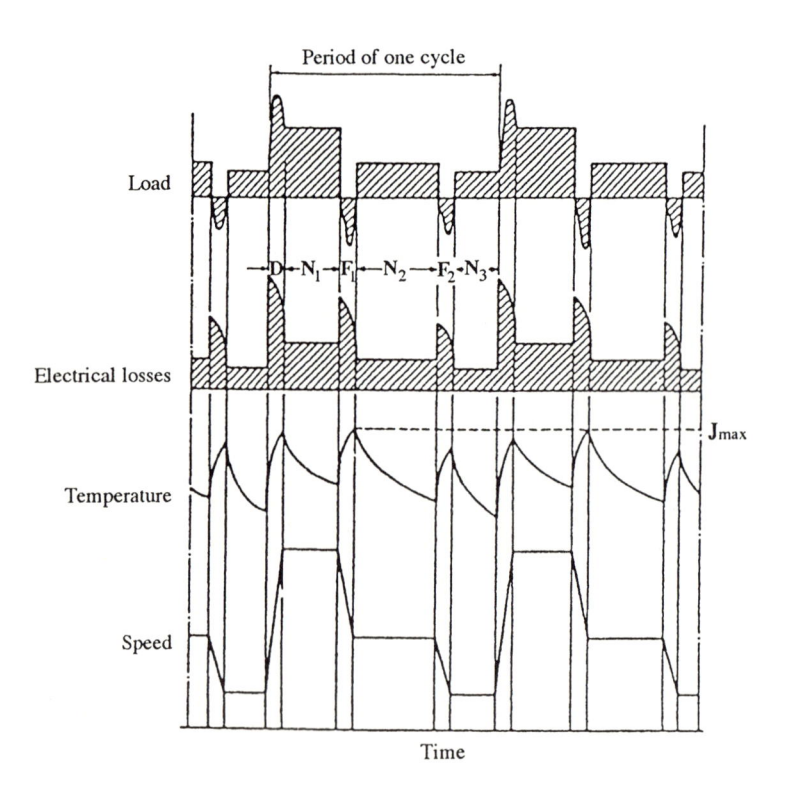

Fig. 1.8. Continuous operation periodic duty with related load speed variations(S8)

1 Electric Machines-Standards and Definitions   379

Note: Periodic duty implies that thermal equilibrium is not reached during the time on load.

(i) Duty with non-periodic load and speed variations - Duty type S9. A duty in which generally load and speed are varying non-periodically within the permissible operating range. This duty includes frequently applied overloads that may greatly exceed the full loads (see Fig. 1.9)

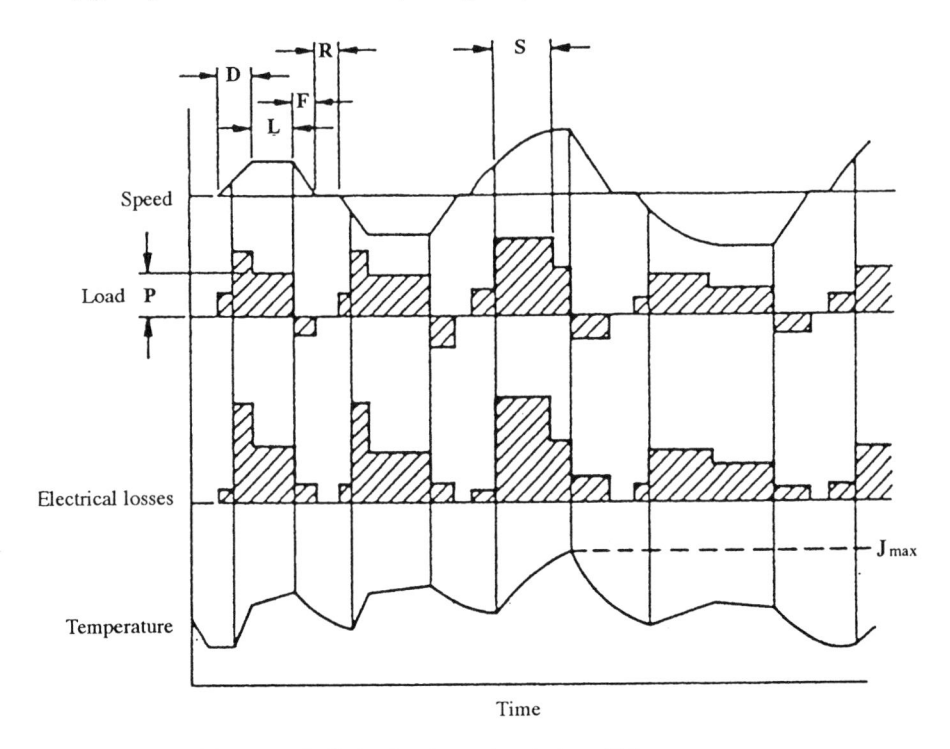

Fig. 1.9. Duty with non periodic load and speed variations (S9)

(j) Duty with discrete constant loads - Duty type S10. A duty consisting of not more than four discrete values of load (or equivalent loading), each value being maintained for sufficient time to allow the machine to reach thermal equilibrium. The minimum load within a duty cycle may have the value zero (no-load or rest and de-energized).

Notes
(1) The discrete values of load will usually be equivalent loading based on integration over a period of time. It is not necessary that each load cycle be exactly the same, only that each load within a cycle be maintained for sufficient

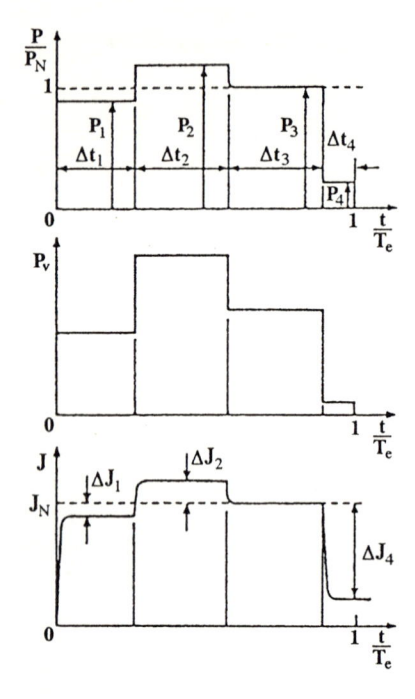

Fig. 1.10. Duty with discrete constant loads - Duty type (S10)

time for thermal equilibration to be reached, and that each load cycle be capable of being integrated to give the same thermal life expectancy.

(2) For this duty type, a constant load appropriately selected and based on duty type S1 should be taken as the reference value for the discrete loads (equivalent loading).

1.3 Machines with Different Duty Types

(a) Machines with different rated speeds. In machines with different rated speeds, the corresponding duty cycle and rating must be set for each operation.

(b) Machines with changing parameters. If a parameter (power, voltage, speed) can have several values, or constantly changes continuously between two limit values, the operation is set for these values or limits. This does not hold for voltage fluctuations of ±5%, nor for the star connection during star-delta starting.

1.4 Operating Conditions: Altitude, Ambient Temperatures and Coolant Temperatures

Altitude. The height above sea level does not exceed 1000 m.

Ambient temperature. The temperature of the air at the operating site does not exceed 40° C.

Temperature of the water at inlet to water-cooled heat exchangers. The temperature of the water at inlet to the heat exchanger does not exceed 25° C.

Minimum ambient and coolant temperature. The minimum temperature of the air at the operating site is -15° C.

The above applies to all machines except the following: a) A.C. machines with rated outputs exceeding 3 300 kW (or kVA) per 1 000 r/min, machines with rated output less than 600 W (or VA) and all machines having a commutator or sleeve bearings. For these machines, the minimum ambient temperature is mutator or sleeve bearings. For these machines, the minimum ambient temperature is +5° C.

(b) Machines having water as a primary or secondary coolant. The minimum temperature of the water and the ambient air is +5°C. If an ambient temperature lower than that given above is to be expected, the purchaser shall specify the

assumed maximum room temperature

altitude mm	temperature of the cooling air in grad C° class of isolation (temperature)					
	A	E	B	F	H
1000	40	40	40	40	40
2000	34	33	32	30	28
3000	28	26	24	19	15
4000	22	19	16	9	3

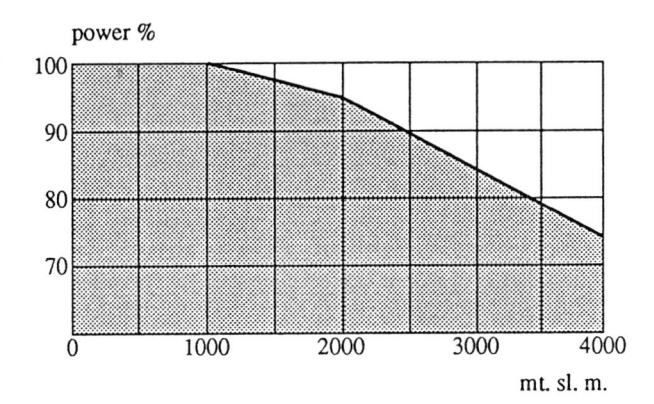

power %

mt. sl. m.

minimum ambient temperature and whether it applies only during transport and storage or after installation.

If the altitude exceeds 1000 m above level, thus the assumed maximum ambient temperatures must be reduced depending on the thermal classification of the machine. Another solution is to reduce the rated power depending on the altitude above sea level (table 1.1 and diagram).

1.5 Electrical Conditions

Electrical supply. A.C. machines within the scope of this standard shall be suitable for three-phase, 50 Hz or 60 Hz, with voltages derived from the nominal voltages given in IEC 38. In deriving rated voltages for machines, it is necessary to take into consideration the differences between distribution and utilization system voltages.

Form and symmetry of voltages and currents. Machines shall be so designed as to be capable of operating under conditions detailed as follows:

A.C. motors shall be suitable for operation on a supply voltage having a harmonic content as limited by the requirements of a) below, and, for a polyphase motor, from a supply system where the voltage unbalance is defined by the requirements of b) below.

Should the limits in (a) and (b) occur simultaneously in service at the rated load, this shall not lead to any deleterious temperature in the motor and it is recommended that the excess resulting temperature rise or temperature related to the limits specified in tables 1, 2 and 3 (specified in IEC 34-1) should be not more than approximately 10 K.

(a) Three-phase A.C. motor (including synchronous motors, but not design N motors) and A.C. single-phase motors shall be suitable for operation on a supply voltage having a harmonic voltage factor (HVF) not exceeding 0.02 unless the manufacturer declares otherwise.

Design N motors (see IEC 34-12) shall be suitable for operation on a supply voltage having an *HVF* not exceeding 0.03.

The *HVF* shall be computed by using the following formula:

$$HVF = \sqrt{\sum \frac{u_n^2}{n}}$$

where:

u_n is the per unit value of the harmonic voltage (referred to rated voltage *UN*),

n is the order of harmonic (not divisible by three in the case of three-phase A.C. motors).

Usually it is sufficient to consider harmonic orders $n \leq 13$.

In temperature-rise testing as specified in IEC 34-1, the *HVF* shall not exceed 0.015.

(b) A polyphase voltage system is deemed to form a virtually balanced system of voltages if its negative-sequence component does not exceed 1% of its positive-sequence component over a long period, or 1.5% for a short period not exceeding a few minutes, and if the voltage of its zero-sequence component does not exceeds 1% of its positive sequence component.

In temperature-rise testing as specified in IEC 34-1, the negative-sequence component of the system of voltages shall be less than 0.5% of the positive sequence component, the influence of the zero-sequence system being eliminated. By agreement between manufacturer and purchaser, the negative sequence component of the system of currents may be measured instead of the negative-sequence component of the voltages, and shall not exceed 2.5% of the positive-sequence component of the system of currents. Note - In the vicinity of large single-phase loads (e.g. induction furnaces), and rural area particularly and mixed industrial and domestic systems, supplies may be distorted beyond the limits set out above. Special arrangements will then be necessary between manufacturer and purchaser.

In the case of a D.C. motor supplied from a static power converter, the pulsating voltage and current affect the performance of the machine. Losses and temperature rise will increase and the commutation is more difficult compared with a D.C. motor supplied from a pure D.C. power source.

It is necessary, therefore, for motors with a rated output exceeding 5 kW, intended for supply from a static power converter, to be designed for operation from a specified supply, and, if considered necessary by the motor manufacturer, for an external inductance to be provided for reducing the modulation. The static power converter supply shall be characterized by means of an identification code, as follows

$$CC\ C - U_{aN} - f - L$$

where:

CCC is the identification code for converter connection according to IEC 971;
U_{aN} consists of three or four digits indicating the rated alternating voltage at the input terminals of the converter, in volts;
f consists of two digits indicating the rated input frequency, in Hertz;
L consists of one, two or three digits indicating the series inductance to be added externally to the motor armature circuit, in millihenrys. If this is zero, it is omitted.

Motors with rated output not exceeding 5 kW, instead of being tied to a specific type of static power converter, may be designed for use with any static power converter, with or without external inductance, provided that the rated from factor for which the motor is designed will not be surpassed and that the insulation level of the motor armature circuit is appropriate for the rated alternating voltage at the input terminals of the static power converter.

Note - By stating the identification code or, alternatively, in the case of motors with rated output not exceeding 5 kW, the rated form factor and the rated alternating voltage at the input terminals of the static power converter, the ability of the D.C. motor armature to carry subsequent undulating currents and eventually to be designed for a higher than usual dielectric test voltage is characterized.

Voltage and frequency variations during operation. For AC-machines, combinations of voltage variations and frequency variations are classified as being either zone A or zone B, in accordance with Fig. 1.11 for generators and Fig.1.12 for motors.

For DC-machines, when directly connected to a normally constant DC bus, zone A and B apply only to the voltages.

A machine shall be capable of performing its primary function continuously within zone A but need not comply fully with its performance at rated voltage and frequency (see rating point in Fig.1.11 and 1.12), and may exhibit some deviations. Temperature rises may be higher than at rated voltage and frequency.

A machine shall be capable of performing its primary function within zone B, but may exhibit greater deviations from its performance at rated voltage and frequency than in zone A. Temperature rises may be higher than at rated voltage and frequency and most likely will be higher than those in zone A. Extended operation at the perimeter of zone B is not recommended.

Thermal classifications of machines. A thermal classification in accordance with IEC 85 shall be assigned to the insulation systems used in machines. The classification of the insulation systems shall be by means of letters and not by temperature values.

It is the responsibility of the manufacturer of the machine to interpret the results obtained by thermal endurance testing as appropriate to his machine type and application.

Thermal classification of machines. The following classes are agreed upon:

        65 K insulations of windings class A

        80 K insulations of windings class E

        90 K insulations of windings class B

        115 K insulations of windings class F

        140 K insulations of windings class H

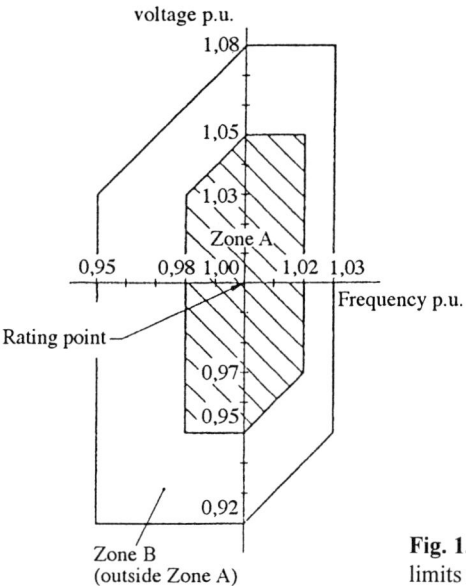

1 Electric Machines-Standards and Definitions 385

Fig. 1.11. Voltage and frequency limits for generators

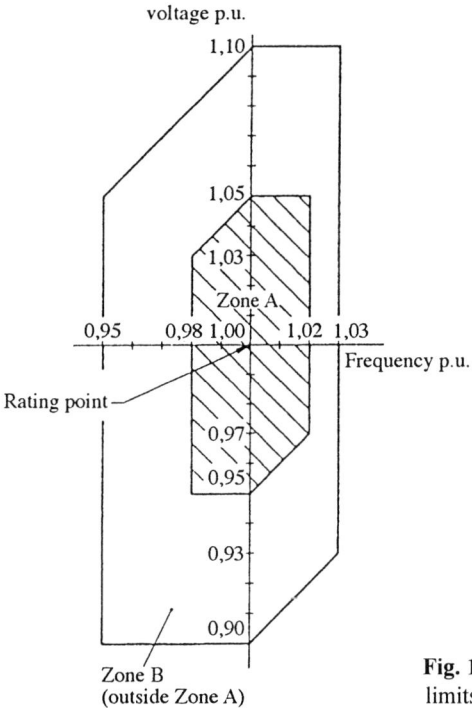

Fig. 1.12. Voltage and frequency limits for motors

1.6 System: Production Machine-Driving Machine

In the following sections normalized variables for the torque, the speed or the active power are assumed. (Normalized variables are written in small letters). This assumption is used to be independent of the real speed, torque or power rating of the component considered. Normalization is explained in section 3.1 "Normalization" for the DC-machine for example.

1.6.1 Stationary Behaviour of the Production Machine

Only in exceptional cases the production machine needs a constant driving power from the driving machine (motor). The driving load power or torque depends generally on the resulting duty of operation from the technology, which changes according, for example, to the speed, the position, the time or other parameters. In addition the dynamic behaviour is important, but this - should not be considered now.

The stationary behaviour of production machines is generally represented by families of characteristics $t_W = F(n,v,\varphi,x,t)$.

t_W = **const.** In all production machines and technologies where a lifting, a friction - or a cutting operation-torque is the load torque t_W mainly, this will be constant and independent from the speed. Examples: lifting devices, elevators and winches, as well as lathes or planing machines. Considering friction or shaping, at the time of a speed reversal the load-torque will also change direction: $t_W = \text{const} \cdot sgn(n)$, for example in valves, gate valves, throttle valves, in bagger and crane chassis or tool machine with cutting operation.

$$t_W = \text{const.} \qquad\qquad t_W = |t_W|_0 \, sgn(n)$$
$$p_w \sim n \qquad\qquad p_w \sim |n|$$

$t_W = f(n,v)$. (a) Only relatively few production machines produce a linear increasing load torque $t_W \sim n$ with respect to the speed n: calander drives for paper-, textile-, plastic- and rubber-films; electrodynamic brakes arid generators working on constant load resistance.

(b) If it is necessary to overcome the resistance of the air and liquids, the load torque must increase at the square of the speed $t_W \sim n^2$. Examples: fans, centrifugal pumps, condensers, centrifuges and boat propellers.

$t_W = f(\varphi)$. In addition to loads depending on the speed, in some production machines the load behaviour depends on the angle of rotation. In displacement compressors, for example, the load torque changes as the stroke changes.

Mold work, crank presses, shears and textile looms also require load torques depending on the angle of relation.

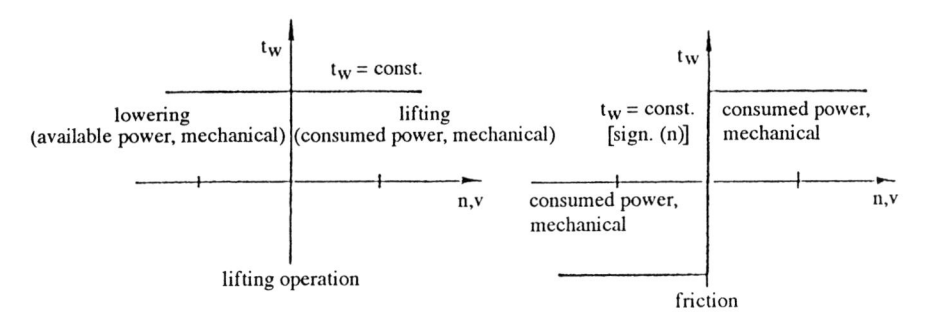

1 Electric Machines-Standards and Definitions   387

t_W | $t_W = $ const.
lowering | lifting
(available power, mechanical) | (consumed power, mechanical)
→ n,v
lifting operation

t_W | $t_W = $ const. [sign. (n)] | consumed power, mechanical
consumed power, mechanical | → n,v
friction

Fig. 1.13

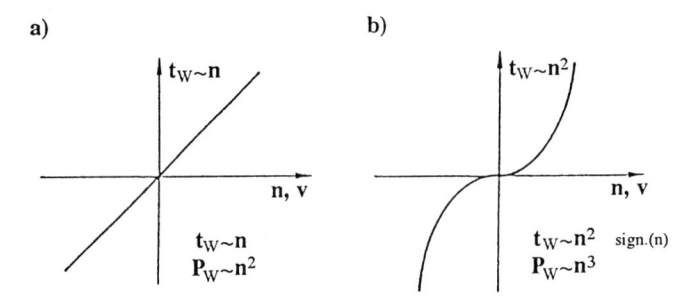

a) $t_W \sim n$
    → n, v
    $t_W \sim n$
    $P_W \sim n^2$

b) $t_W \sim n^2$
    → n, v
    $t_W \sim n^2$ sign.(n)
    $P_W \sim n^3$

Fig.1.14.

t_W (torque pulsations)
⊓⊔⊓⊔⊓⊔ compressor
→ φ

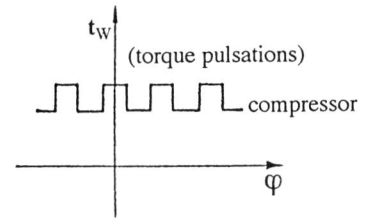

Fig.1.15

$t_W = f(r)$. In center drive winders for paper, metal or other materials, in the material the tension should remain constant for example even in the case of a declining or rising radius r, and with the constant peripheral speed v, the same is true in turning machines with facing cut. This lead to a load torque $t_W \sim r \sim \frac{1}{n}$.

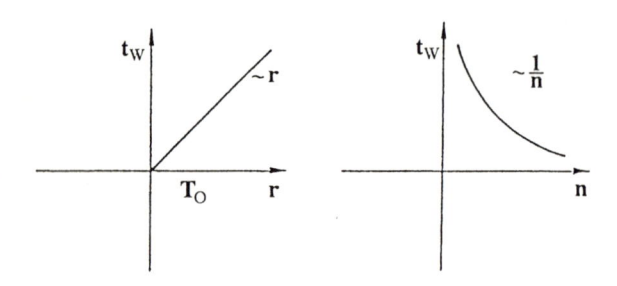

Fig.1.16.

$t_W = f(t)$ (also see the duty types S1-S8). In many production machines there is a time pattern of the required load torque $t_W(t)$. For example, a driving program is required often for electric railway- or conveyor- drives.

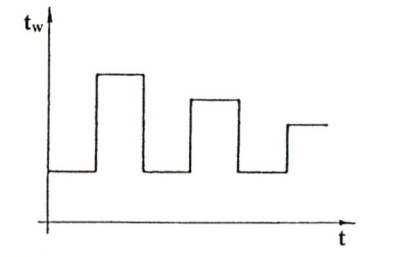

Fig.1.17.

1.6.2 Stationary Behaviour of the Driving Machines: $t_M = f(n, \varphi)$

As for production machines, it is also possible to give characteristics for driving machines (again neglecting dynamic behaviour), that describe the basic relationship between speed and torque.

All electric driving machines may be divided according to the following three cases:

Shunt characteristic (asynchronous behaviour).

$$t_M = f\left(\frac{1}{n^2}\right)$$

series characteristic "R",

1 Electric Machines-Standards and Definitions

$t_M = f(n_0 - n)$

for example, shunt-excited DC-machine " N "
" stiff" / "soft" separately-excited DC-machines;
(section 3) induction machines (IM)(section 5.2ff)

(n_0: no-load speed).

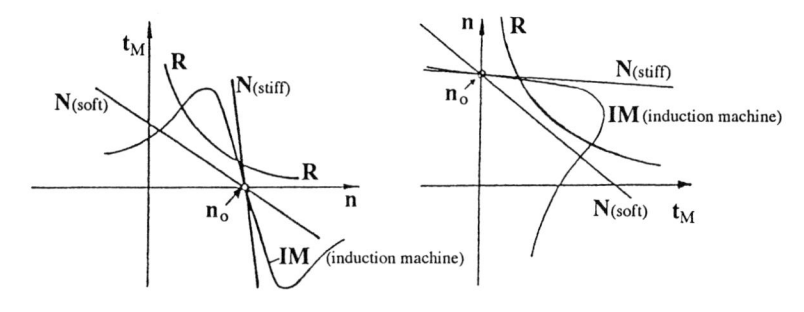

Fig.1.18.

Shunt characteristic is characterized by the fact that the speed decreases as the motor torque increases. If the speed variation is small, we speak of a "stiff" characteristic; if this variation is considerable, we then have a "soft" characteristic.

Constant - Torque behaviour.
$t = \text{const.} \cdot \text{sgn}(n_0 - n)$
hysteresis motor "H"
DC-machine with $R_A = 0$
(superconducting armature winding).

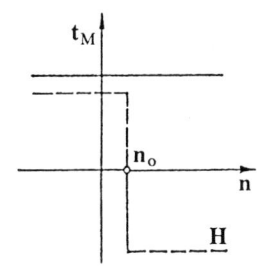

Fig.1.19

Synchronous behaviour.

$n = f(f_{mains}, Z_p)$

$t_M \neq f(n)$

Synchronous machine:

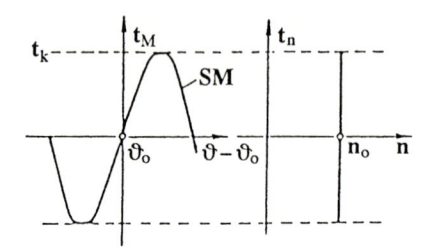

Fig.1.20.

$t_M = f(\vartheta)$ ϑ: rotor displacement angle

speed n: constant
rotor displacement $\vartheta = f(t_M)$ (t_K: pull out torque)

thus the position φ is elastic.

Synchronous behaviour is characterized by the fact that the speed, regardless of the torque, remains constant up to a maximum value (pull out torque). The torque variation is therefore related to the rotor displacement angle.

For stable behaviour, the rotor displacement angle $\vartheta < 90°$ must always be true, otherwise the system production machine- driving machine gets unstable: thus the motor torque t_M reduces, causing the speed to leave the operating point n_0.

1.6.3 Static Stability at the Operating Point

The balanced load condition of equilibrium for the operating point is according to the arrow system definitions:

$$t_M - t_W = 0$$

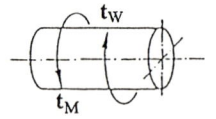

Fig.1.21.

By examination the static stability, we can see whether the point of equilibrium is in each case stable, unstable or indifferent (open loop control only, no closed loop control).

Graphic method. (Examination with the family of characteristics t/n, open loop control only).

1 Electric Machines-Standards and Definitions   391

Example: given: normalized characteristic of the driving machine $t_M = f(n)$; normalized characteristic of the production machine $t_{W1,2,3} = f(n)$. The examples shown examines the stability of three given operation points 1,2, 3.

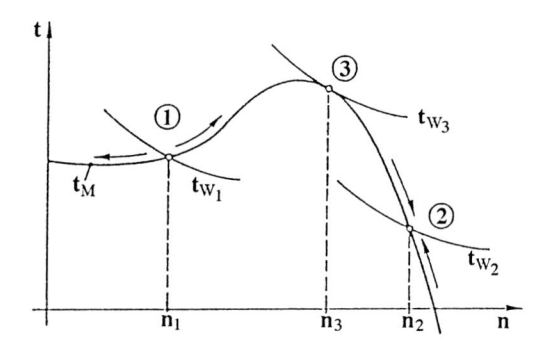

Fig. 1.22.

Point 1:   If $n > n_1 \rightarrow t_M > t_{W1}$,
       $\rightarrow$ further increase in speed
       If $n < n_1 \rightarrow t_M < t_{W1}$
       $\rightarrow$ further decrease in speed
       further in- or decrease of speed starting from the point 1 of operation:
       $\rightarrow$ point 1 is therefore an unstable operating point

Point 2:   If $n > n_2 \rightarrow t_M < t_{W2}$ $\rightarrow$ decrease of speed
       If $n < n_2 \rightarrow t_M > t_{W2}$ $\rightarrow$ increase of speed
       speed-de-and increase act towards point 2 of operation:
       $\rightarrow$ point 2 is therefore a stable operating point.

Point 3:   Limiting case between the point of the stable and unstable operating:
       If $n > n_3 \rightarrow t_M < t_{W3}$ $\rightarrow$ stable behaviour
       If $n < n_3 \rightarrow t_M < t_{W3}$ $\rightarrow$ unstable behaviour
       can not be used

During the examination of the static stability of the speed with the family of characteristics, we assumed that $t_M = f(n)$ and $t_W = f(n)$, thus depend solely on the speed and do not depend on angular acceleration.

Numerical Proof of Stability. The numerical proof of stability may be carried out in several ways. The first solution is solving the differential equation.

$$T_{\Theta N} \frac{dn}{dt} = t_M - t_W = t_B$$

with $t_M = f(n)$, $t_W = f(n)$, at the operating point.

Another method is the proof of stability in the Laplace domain. If the poles of the denominator polynomial of the transfer function are in the left half-plane of the complex s-plane (Laplace-domain), the system is stable.

1.6.4 Rating of the Electrical Drive System

Four essential aspects are essential factors for the rating of the drive system:

- power required
- torque characteristics
- speed characteristics
- type of construction

In examining these, we shall consider both static and dynamic behaviour.

Production machines.
1. *Family of characteristics* (steady state operation)
First of all, we consider the range of operation of the production machine and represent it as a family of characteristics T_W - N.

Example: $T_W = f(N)$.

We thus determine the speed-torque range in steady state operation. Furthermore we consider, if torque - or speed - reversal are necessary; this may influence the power supply of the driving machine. (1)

2. *Additional rating for dynamic operation*
Additionally to the steady state torque-speed range we have to distinguish:
a) acceleration and braking of the drive system

$$T_B = J \; 2\pi \; \frac{dN}{dt} \gtrless 0 \qquad (2)$$

(b) as discussed before, there are production machines, which produce torque pulsations. (There are driving machines having torque pulsations too).

Therefore an additional range of torque may be necessary. (3)

Examples: (Fig.1.23.)

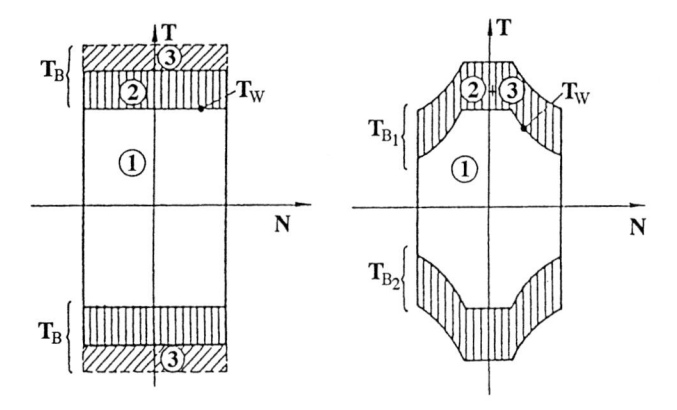

Driving machines. In choosing a driving machine, one must first consider the operating points of the production machine. The necessary torque-speed characteristic can be derived from

$$T_M = \underbrace{T_W}_{(1)} + \underbrace{T_B}_{(2)+(3)}$$

The requirement of a certain speed characteristic of the driving machine if the torque varies

$$\Delta N = f(\Delta T_W)$$

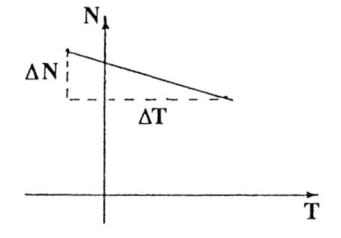

Fig.1.24

may also affect the selection of the characteristic of the driving machine or type of the driving machine.

In any case, the family of characteristics of the driving machine $T_M = f(n)$ must be set so that the family of characteristics $T_W(n)$ is within the boundaries of the family of characteristics of the driving machine (including safety factors). At this point it may be convenient to adapt the characteristic of the driving machine to the characteristic of the production machine (by open- or closed-loop control for example). This also sets the limits N_{max}, and T_{Mmax} and the *N-T*-torque-speed range.

Example:(Fig.1.25)

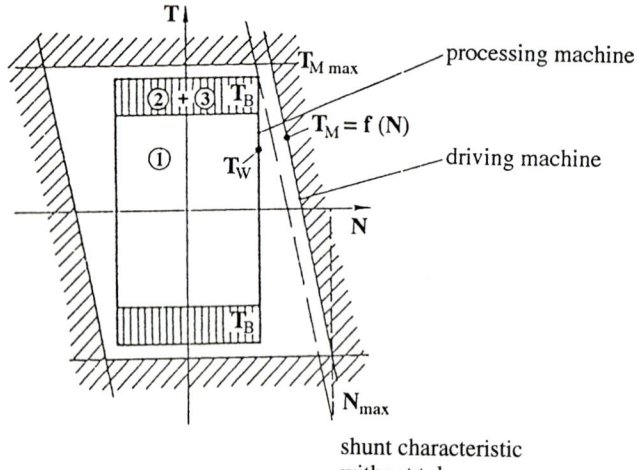

shunt characteristic
without tolerance

For the thermal rating of the driving machine, one must consider the duty type (see section 1.2).

Example:(Fig.1.26)

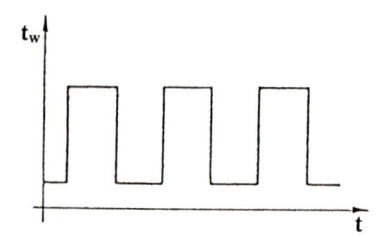

The rating of the driving machine must be chosen so that the drive is not overloaded thermally during operation. However, it is allowable to briefly exceed the duty-type rating. In addition to the criteria, it is again useful to check approximately the stability of the drive system (production- and driving machine) as described in Chapter 1.6.3. Of particular interest are starting or torque overload for the driving machine.

2 Electric Machines

2.1 Direct-Current Machine

The direct-current machine is an electric machine which is easily adjustable in speed and torque thanks to power-electronic converters available today. The mechanical construction of a direct-current machine is depicted in Fig. 2.1 The stator is made up of the yoke and the main and auxiliary poles, bearing concentric windings. Most machines have several pairs of poles, so that the succession of main and auxiliary poles is repeated several times. In any case, the principle of operation of direct-current machines having one or more pairs of poles is, generally speaking, identical. The rotor (armature) have windings inserted in slots. The ends of the armature windings are connected to the collector; brushes are provided between the main poles in the neutral zone (NZ) for the armature terminals.

The principle of operation of a direct-current machine may be explained by referring to its elementary structure. The stator of the direct-current machine is the magnetic return path for the magnetic flux. Fig. 2.1 shows that the main pole is

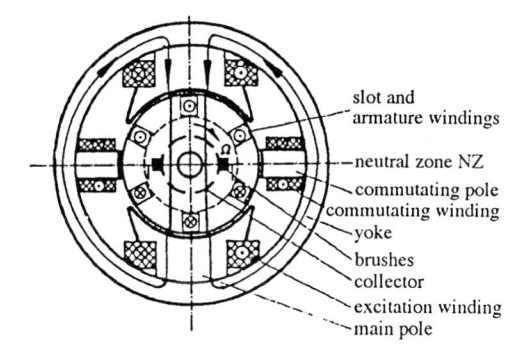

Fig. 2.1. Basic mechanical construction of a direct-current machine

the excitation pole with the excitation windings (field windings). If the field winding is supplied by a direct voltage U_E, then there will be the excitation current I_E -after a transient of approximately $3T_E = \frac{3L_E}{R_E}$ (L_E: inductance of the excitation coil, R_E: ohmic resistance of the excitation coil) - and the excitation flux. If we now apply an armature voltage U_A on the armature coil, an armature current I_A will result after approximately $3T_A = \frac{3L_A}{R_A}$ (L_A: inductance of the armature coil, R_A: ohmic resistance of the armature coil). Based on the direct-current machine design, the directions of the armature current and magnetic flux are vertical to one another. In this way, according to Fleming's rule, a force is obtained on each armature winding in the sector of the pole shoes and thus the resulting torque. If the direction of the armature- or the excitation-current is changed, the direction of the force - and thus that of the torque - will also change. However, these are the most fundamental considerations for producing torque.

As can again be seen from Fig. 2.1, there is a collector K and the brushes. These components are necessary in order to ensure equal direction of force and torque for the upper and lower sectors of the armature. The collector therefore has the task of reversing the current direction in the neutral zone (no field, current reversal). From Fig. 2.1 it can be derived that the collector and the brushes cause short circuits of the armature coils under the brushes. Through the impressed relationship dI_A/dt for the armature current during reversal, voltages are induced that delays the current reversal. This causes a high density of current on the end of the brush, which may cause sparks on the brushes. These sparks damage the collector and the brushes and must therefore be avoided. To prevent the sparks on the brushes, one must either keep dI_A/dt low - which limits the speed range - or include auxiliary pole windings. These auxiliary pole windings are supplied by the armature current and are orthogonal to the direction of the armature windings (concentrated lumped inductances assumed). This produces a magnetic field that in turn produces a voltage in the short-circuited armature coils to counteract the original induced voltage; thus the current reversal is no longer delayed and no sparks occur on the brushes.

The basic diagram also shows that the excitation winding produces a magnetic field shown in the air gap. The armature current also produces a corresponding magnetic field in the air gap of the pole shoes. The two fields superimpose and lead to an increase of the flux on one corner of the pole shoe and to saturation based on the characteristic of the iron; therefore, the resulting excitation field is weak end. This field distortion and thus the field weakening may be avoided by means of a compensation winding inserted in the main pole shoe, which is also supplied by the armature current I_A.

Today direct-current machines are generally supplied by converters. Since the voltages - and thus the currents - of these converters have harmonics, the armature and excitation circuits must use laminated iron.

The operation of the direct-current machine is fully described in chapter 3 "Direct-current machine". In this Chapter the four basic equations and also the signal-flow graph is explained. This signal-flow graph may be used, in general, to deduce the operating behaviour depending on the armature voltage U_A, the flux ψ, or the load torque T_w. In addition to the signal-flow graph, the following Chapter includes the transfer-functions. The transfer-functions are system equations that describe both the static and dynamic behaviour. In order to fully understand the transfer-functions, one must recall that a transfer-function considers an input signal and provides information about how the chosen output signal behaves if the input signal is varied. If one requires only stationary behaviour, then all of the terms with s in these transfer-functions can be set to $s = 0$, providing the transfer-function in the stationary state.

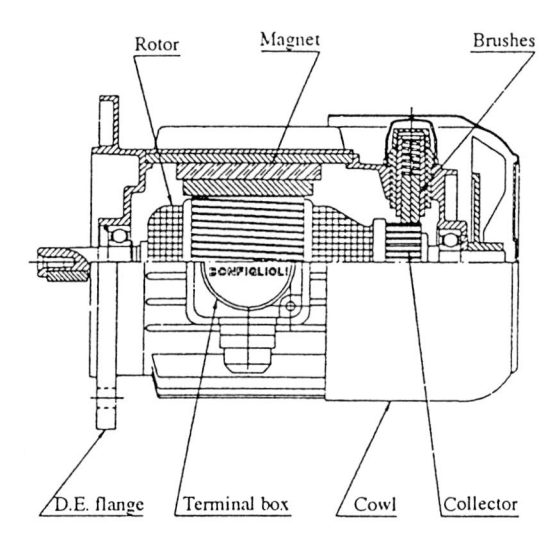

Fig. 2.2. Type of construction of a DC-machine

In addition this Chapter gives basic definitions describing the transient behaviour in the dynamic state.

2.2 Rotating-Field Machines

Rotating-field machines may be induction machines or synchronous machines. The principle of operation of the induction machines will be described in detail in Chapter 5, and that of the synchronous machines in Chapter 6.

This Chapter merely illustrates the basic structures and operation of the two types of machines. In both machines there is a symmetrical three-phase winding system in the stator, which produces - supplied from a symmetrical three-phase mains system - a spatially rotating magnetic field.

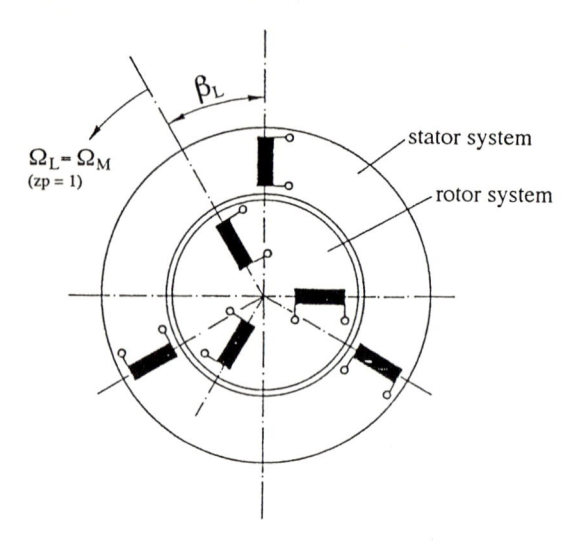

Fig. 2.3. Basic construction of the general rotating-field machine

The induction machine may have a symmetrical three-phase or poly-phase (cage rotor) winding system in the rotor. In the first solution, the three-phase rotor winding system has a star-shaped configuration, and the free ends of the three windings are accessible by slip-rings. The slip-rings may either be short-circuited, or connected to resistors or to a power electronic converter.

In the second solution, the poly-phase systems are actual conductors in the rotor slots which are short-circuited at the two ends by means of a ring conductor (cage). Generally, both the rotor- and the stator-irons are laminated. Instead of the three-phase winding systems in the stator, in converter drives with squirrel cage rotor induction machines one partially uses five-phase winding systems or two-three-phase winding systems, displaced electrically by 30°. This reduces the torque harmonics in certain converter drives.

There are two main versions of the rotor in the synchronous machines. The first is the salient pole rotor version. In this case, the rotor has salient poles, and there is a winding system on the rotor to which a direct excitation current is applied (I_E) Another version is depicted in Fig. 2.5; here, in addition to the excitation winding, the Fig. also shows a three-phase damper winding system (L_D, L_Q; carthesian d-q winding system).

This damper winding system is important for some converter drives with synchronous machines in order to improve load commutation of the stator currents from one stator winding to another (shorter commutation interval).

Another basic version of the rotor is in the synchronous machine with non-salient pole rotor. In this case, the poles are not salient as in the salient pole machine, but the rotor is cylindrical. However, as in the salient pole machine, either a single excitation winding or an excitation winding and damper winding system are present.

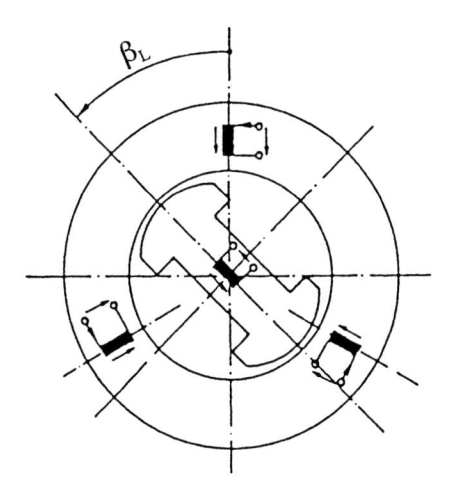

Fig. 2.4. Synchronous machine (salient pole rotor machine) without damper winding system

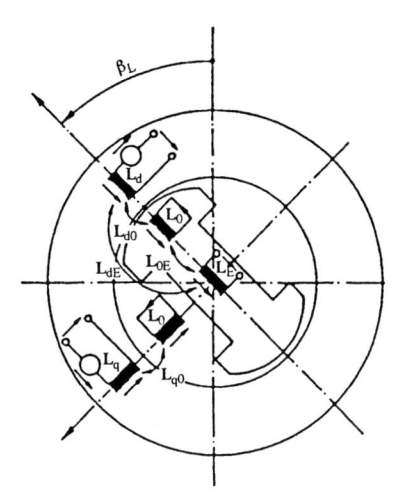

Fig. 2.5. Synchronous machine (salient pole machine) with damper winding system (d-q system)

A modern version of the rotating-field machine are permanent magnet machines. In these machines, instead of an excitation by a winding system and a direct current, permanent magnets are mounted on the rotor, so that a magnetic field is produced. Depending on the mode of operation, we can distinguish between the permanent magnet synchronous machine or the brushless direct-current machine. A more detailed explanation is given in Chapter 9.

2.2.1 Fundamental Principle of the Rotating-Field Machines

If a symmetrical three-phase voltage system supplies a symmetrical three-phase stator winding system, a symmetrical three-phase current system will result in the stator system after the transient response. For understanding the principle of the rotating field machines it is essential to remember, that the currents in the stator windings have an electrical displacement of 120° electrically to each other on the one hand and - very important - again 120° electrically a spatial displacement, due to the geometric design of the windings on the other hand (lumped windings assumed generally). Due to the electrical and spatial displacement a spatial current vector (phasor) will result from the stator currents that rotates in the air gap with the angular stator frequency Ω_1 (stator frequency F_1) (for the detailed description, see section 5.1). This current phasor will produce a magnetic flux phasor.

In the induction machine with a squirrel cage rotor, the rotating magnetic field will induce voltages in the short-circuited windings of the rotor, and these voltages will lead to a rotor voltage phasor and thus to the corresponding rotor current phasor of the rotor currents. Due to the combined action of the magnetic phasor and the rotor current phasor, a torque will be produced.

In these considerations, it is essential that voltages can be induced in the rotor windings only, if a relative velocity exists between the rotor windings and the magnetic field phasor of the stator, rotating with the frequency F_1. This means that in motor operation, the rotor speed must always be lower than the synchronous speed $n_S = 60F_1/Z_p$ (Z_p: number of pole pairs);

Z_p	1	2	3	4	5	6	
synchronous speed n_S							
50Hz	3000	1500	1000	750	600	50	min.$^{-1}$
60Hz	3600	1800	1200	900	720	600	min.$^{-1}$

indeed, at n_s there is no relative velocity. Relative velocity is indicated by the slip s:

$$s = \frac{n_s - n}{n_s} \ 100 \ \text{in \%}.$$

Therefore, generally when $s = 0$, there is no torque available from the induction machine. As the slip s increases, thus the relative velocity, the induced rotor voltage phasor and thus the rotor current phasor will also increase in amplitude and frequency, and thus in general the torque as well. In the result "increasing torque with increasing relative velocity", one must observe, that at an increasing relative velocity the frequency F_2 of the induced voltages and currents in the rotor also increases. Thus in addition to the rotor resistor R_2, one must also consider the reactive impedance X_2, which is a function of the frequency F_2. Based on the complex impedance the torque may not increase with increasing frequency F_2 and slip continuously, but there will be a maximum torque (breakdown torque) and then the torque will diminish when the slip, thus the frequency F_2, will continue to rise.

From the equations above, we can deduce an additional possibility for speed variation. If Z_p of the motor is changed, the synchronous speed will therefore be adjusted according to the table above. Generally speaking, it is therefore possible to use machines that have pole changing and thus have different synchronous speeds n_S.

Up to now we discussed operating speeds lower than the synchronous speed. But when the induction machine is operated in the generator mode, then the speed will be greater than the synchronous speed n_S and there again will be a "relative velocity" between the flux phasor Ψ_1 and the speed of the rotor again. Therefore the induction machine can also be operated at speeds greater than the synchronous speed. The torque-speed characteristic of the induction machine must be symmetrical to the operating point of n_S. At speeds lower than n_S the induction machine is in motor operation and at greater speed in the generator operation, supplying energy to the mains.

The principle torque-speed characteristic is shown in Fig. 2.6. The characteristic depends on the construction of the machine and the pairs of poles Z_p and very

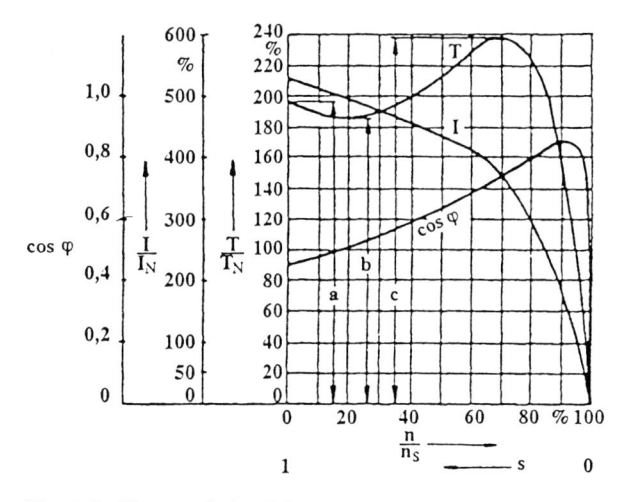

Fig. 2.6. Characteristic of the current, torque and power factor of a squirrel cage induction motor

important on the construction of the slots and the windings of the rotor. As explained earlier, the torque increases as the slip s increases, up to the breakdown torque T_K and the breakdown slip s_K. At $s > s_K$ the torque diminishes when the slip continues to increase, and may rise again - after the pull-up torque has been reached - to $s \approx 1$. Whether and to what degree the torque rises or falls to $s \simeq 1$ depends on the rotor slot and windings construction (Fig. 2.7).

A behaviour with a rotor, having a deep bar cage, is always desirable, if the induction machine with squirrel cage rotor is supplied by an inverter (without speed sensor, if possible), because with increasing torque there is only a small reduction in speed. The behaviour with a high resistance squirrel cage rotor, or better yet with double squirrel cage rotor, is advantageous, if the machine is directly connected to the three-phase power voltage system and frequent starting cycles are necessary. During starting cycles the starting current is a multiple of the rated current. In this case star-delta starting is recommended, thus the stator windings first have a star and then a delta configuration. This offers the advantage of reducing the starting current; however, due to the reduced current of the factor $\sqrt{3}$ in the windings, the available torque, which diminishes to the voltage square, is thus reduced to one third of the starting torque.

T_A = locked rotor torque
T_S = pull up torque
T_K = breakdown torque

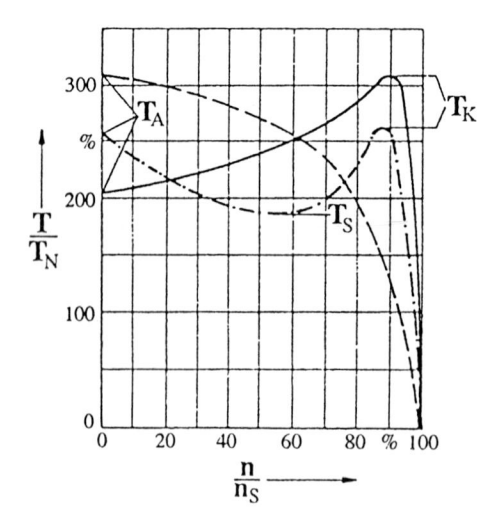

Fig. 2.7. Family of characteristics for various squirrel cage asynchronous motors

a = locked rotor torque
b = pull-up torque
c = breakdown torque

One must check, to see whether this procedure is still allowed in the case of a load torque.

The same situation results for the torque during reductions of the mains voltage, and thus should be considered if there are voltage reductions of more than 5%.

A controlled voltage reduction may also be obtained through the three-phase AC-power controller, thus with three anti-parallel pairs of thyristors in the three lines of the stator windings. By changing the control angle α of the thyristors from $\alpha = 0$ to $\alpha = 180°$, the voltage on the stator winding can be reduced from the maximum value to zero, and thus change the torque speed characteristic as mentioned above. By controlling the angle α, however, only voltage sectors of the three-phase voltage system are supplied to the windings of the induction machine, and thus the harmonic content increases considerably. This is a disadvantage due to the increased losses, to the point, that the induction machine cannot work at the rated power in the full range of operation.

Another possibility for starting is to use pole-changing motors.

A further possibility lies in the use of slip-ring motors, in which resistors are connected to the slip-rings in order to shift the breakdown torque to the point $s = 1$ (see section 5.3).

The most widely-used and advantageous solution from the technical system standpoint lies in the use of inverters. In this case, the characteristic of the induction machine remains, thus starting from $s = 0$ through $s = s_{nom}$ or $s = s_K$ at any speed, the characteristic may always be shifted below the synchronous speed.

However, the requirement in this case is that the magnetic flux remain constant in the machine. This can be achieved only if, as the speed increases, the inverter produces a three-phase voltage system with increasing amplitude and frequency.

Additional details on the operation of the induction machines per se or inverter operation are given in Chapters 5 and 7.

In conclusion, we point out that three-phase induction motors may also be supplied by a single-phase system (AC-system). In this case, the third winding must be connected to the alternating voltage by means of a condenser. Through this condenser, a three-phase voltage system may be produced; however, this is true only for one operating point.

2.2.2 Synchronous Machines

The synchronous machine, like the induction machine, has a system of stator windings, generally three-phase and symmetrical, supplied by a symmetrical three-phase voltage system. Thus the same conclusions hold as those for the stator flux phasor.

Also, as mentioned earlier, the rotor has an excitation winding supplied by direct current. The rotor system "metal structure of the rotor and excitation winding supplied by direct current" thus acts as a magnet. Since the stator flux phasor rotates with F_1/Z_p in steady state, the magnet will be affected by the

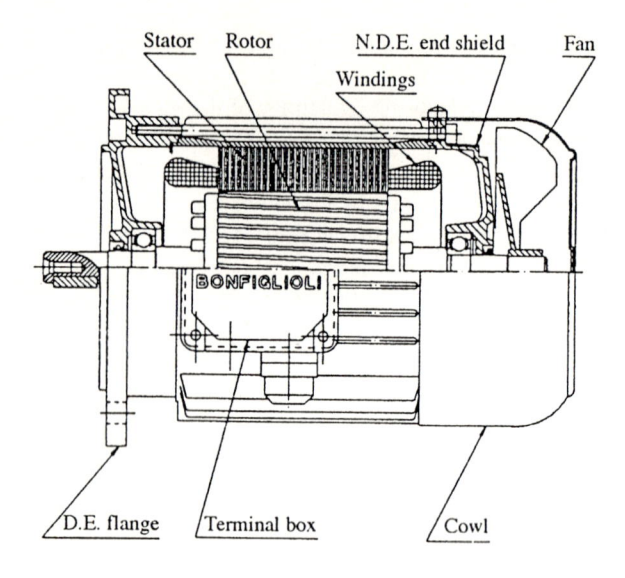

Fig. 2.8. Type of construction of an induction machine

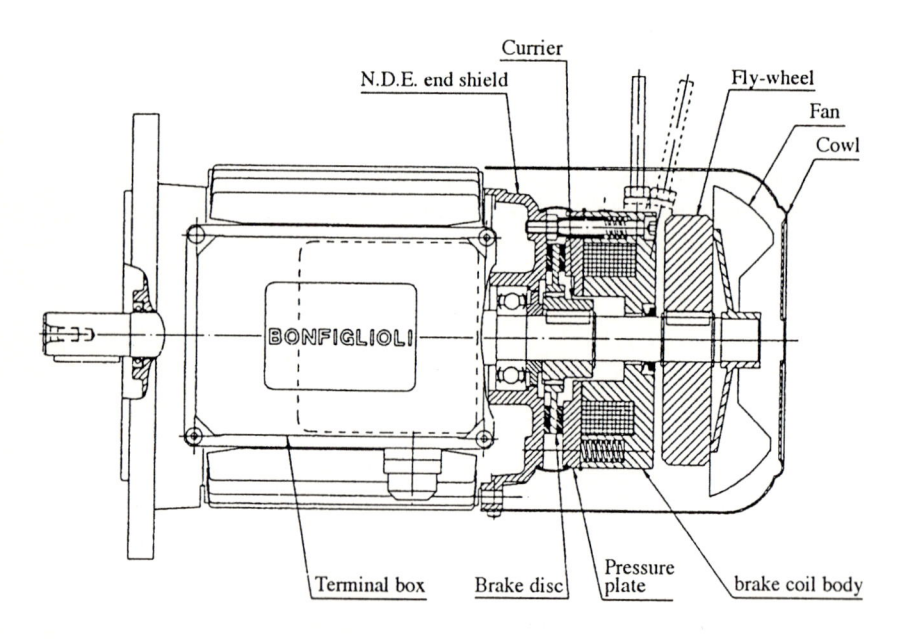

Fig. 2.9. Type of construction of an induction machine with mechanical brake

rotating stator flux. The speed is therefore:

$$N = \frac{F_1}{Z_P}$$

One can also understand that with a load torque on the shaft of the synchronous machine, the torque will produce a phase displacement with respect to the flux phasor; thus a constant speed is indeed given, but not a constant angle position with respect to the stator flux phasor. The angle between the rotor position and the stator flux phasor may be at a maximum of 90° electrically, since the breakdown torque is produced at 90° electrically.

A more specific mathematical representation follows in Chapter 6.

3 Direct-Current Machine

As mentioned in the previous chapters, the direct-current (DC-) machine is an electric machine which is easy to control in torque and speed. In principle control is realizable by correcting variables in the armature circuit or in the field circuit. In the case of control of variables in the armature circuit, the field circuit is generally at the rated point (rated excitation current, rated excitation flux) in the stationary state, thereby controlling the armature voltage U_A or the armature current I_A. The situation is reversed in the case of control in the field circuit, where only the excitation current I_E and therefore the flux ψ is controlled, while the armature voltage is kept constant. However, since simulation is widely available, the signal-flow graph of the separately excited DC-machine (separate excitation) will be obtained first.

3.1 Signal-flow Graph of Separately Excited DC-Machine, Armature Circuit

Set-up: ideal separately excited DC-machine,
i.e. no voltage of brushes (collector),
no armature reaction,
no loss through friction or ventilation,
no inductance saturation.

no saturation in the excitation circuit
U_A : armature voltage
I_A : armature current
I_E : excitation current, Ψ interlinking flux
E_A : e.m.f
N : speed

3 Direct-current Machine 407

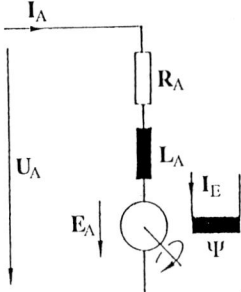

Fig. 3.1. General circuit diagram of separately excited DC-machine

The following basic equations are applied (armature circuit):

$$U_A = E_A + I_A R_A + L_A \frac{dI_A}{dt} \quad (1)$$

$$E_A = C_E N \Psi \quad (2)$$

$$T_{Mi} = C_M I_A \Psi \quad (3)$$

$$C_M = \frac{C_E}{2\pi}$$

$$T_{Mi} - T_W = J \frac{d\Omega}{dt}; \quad J = \text{const.} \,.\, (4) \,.$$

In order to exemplify the procedure for obtaining the signal-flow graph, the functional blocks of the transfer-functions in the Laplace domain (s) must be obtained from the equations in the time domain. Using these functional blocks, their transfer-functions are then used to calculate and discuss the overall transfer-functions which are necessary for control. The three examples are the following functional blocks which are described in equations (1), (3) and (4).

1. Equation:

$$U_A - E_A = I_A R_A + L_A \frac{dI_A}{dt}$$

$$L[U_A - E_A] = L\left[I_A R_A + L_A \frac{dI_A}{dt}\right]$$

$$U_A(s) - E_A(s) = I_A(s)[R_A + sL_A] \qquad I_A(+0) = 0 \,.$$

The relevant transfer-function of the functional block, and therefore the signal-flow graph, are obtained through the conversion of the equation to the Laplace domain.

Transfer-function:

$$G_A(s) = \frac{I_A(s)}{U_A(s) - E_A(s)} = \frac{1}{R_A + L_A s} = \frac{1}{R_A} \cdot \frac{1}{1 + \frac{L_A}{R_A} s}$$

$$G_A(s) = \frac{1}{R_A} \cdot \frac{1}{1 + T_A s} \qquad T_A = \frac{L_A}{R_A} : \text{time constant, armature circuit.}$$

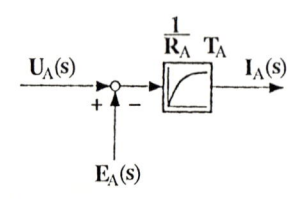

Fig. 3.2. Armature current signal-flow graph

2. Equation:

$$T_{Mi} = C_M I_A \Psi$$

$$T_{Mi}(s) = C_M \Psi(s) * I_A(s)$$

complex convolution $\Psi(t)\ I_A(t)\ \circ\!\!-\!\!\bullet\ \dfrac{1}{2\pi j} \displaystyle\int_{x-j\infty}^{x+j\infty} f_1(\tau) f_2(s-\tau) d\tau$.

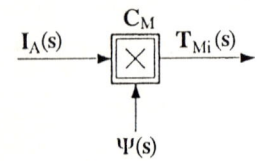

Fig. 3.3. Signal-flow graph of the torque-generation

3. Equation:

$$T_{Mi} - T_W = J \frac{d\Omega}{dt}$$

$$T_{Mi}(s) - T_W(s) = J s \Omega(s) \qquad \Omega(+0) = 0$$

$$G_M(s) = \frac{\Omega(s)}{T_{Mi}(s) - T_W(s)} = \frac{1}{J s}.$$

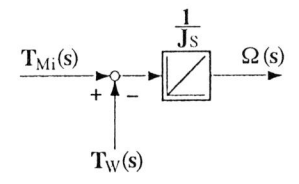

Fig. 3.4. Signal-flow graph of the mechanical part

The second equation is transformed in exactly the same way as the third equation.

This produces the signal-flow graph for the armature circuit of the separately excited DC-machine.

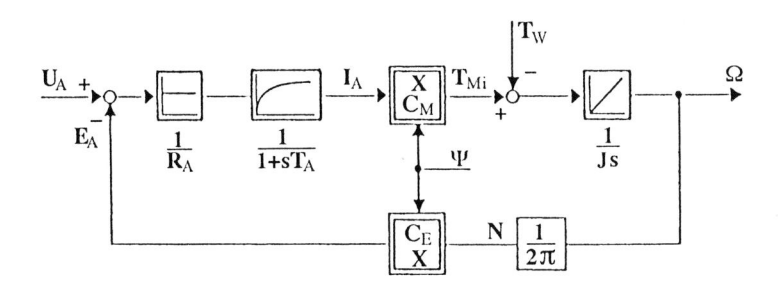

Fig. 3.5. Signal-flow-graph: separately excited DC-machine

This signal-flow graph represents the static and dynamic interdependencies in the armature circuit. If for example, based on a state of equilibrium with Ω_0 (operating point), the armature voltage U_A increases, then the armature current I_A, the torque T_{Mi} and the acceleration torque T_B will increase. This increase in T_B leads to an acceleration of the armature rotor (moment of inertia J) and therefore to an increase in the angular velocity Ω and in the e.m.f. E_A. As a result the $U_A - E_A$ voltage difference will decrease and at the same time I_A decreases. In the final stationary state for instance (with $T_W = 0$), the e.m.f. E_A will be equal to the armature voltage U_A, i.e. $U_A - E_A = 0$, and it follows that $I_A = T_{Mi} = T_B = 0$ and $\Omega = U_A / C_M \psi$. Similarly the same considerations can be derived for T_W and Ψ.

3.1.1. Normalization

In the simulation of a system the equations must be normalized since only numerical values are available on the computer.

The reference values chosen are the rated values of the separately excited DC-machine:

U_{AN} rated armature voltage
I_{AN} rated armature current
Ψ_N rated interlinked flux
T_{iN} rated air-gap torque
N_{0N} ideal idling- rated speed
        (with $U_A = U_{AN}, \Psi = \Psi_N$ and $T_{Mi}=0$)
$\Omega_{0N} = 2\pi N_{0N}$ ideal idling - rated angular velocity
$P_{0N} = U_{AN} I_{AN}$ rated electrical power
$R_{AN} = \dfrac{U_{AN}}{I_{AN}}$ reference resistance

Several rated values are not reference values:

$T_N = \eta_{mech} T_{iN}$ rated torque ($\eta mech$ mechanical efficiency)
$N_N = \eta_{el} N_{0N}$ rated speed (ηel electrical efficiency)
$P_N = 2\pi N_N T_N = \eta P_{0N}$ rated power ($\eta = \eta mech \cdot \eta el$: armature efficiency)

The relation between the normalized variables and the machine variables is obtained by inserting the rated values in the non-normalized equations of the separately excited DC-machine. This results in non-dimensional equations:

$$E_{AN} = C_E \Psi_N N_{0N} = U_{AN} \bigg|_{I_A = 0}$$

$$T_{iN} = C_M \Psi_N I_{AN} = \dfrac{T_N}{\eta_{mech}}$$

$$\Omega_{0N} = 2\pi N_{0N} = 2\pi \dfrac{U_{AN}}{C_E \Psi_N}\bigg|_{I_A = 0} .$$

Normalization 1. Equation

$$U_A - E_A = I_A R_A + L_A \dfrac{dI_A}{dt}$$

3 Direct-Current Machine

$$\frac{\dfrac{U_A}{U_{AN}} - \dfrac{E_A}{U_{AN}}}{\left(\dfrac{I_{AN}}{U_{AN}}\right) R_A} = \frac{I_A}{I_{AN}} + \frac{L_A}{R_A} \frac{d\dfrac{I_A}{I_{AN}}}{dt}$$

$$\frac{U_A}{U_{AN}} = u_A; \quad \frac{E_A}{U_{AN}} = \frac{E_A}{E_{AN}} = e_A; \quad \frac{I_A}{I_{AN}} = i_A$$

$$\frac{R_A}{\dfrac{U_{AN}}{I_{AN}}} = r_A; \quad \frac{L_A}{R_A} = T_A \qquad T_A: \text{armature time constant.}$$

Therefore the following holds true:

$$\boxed{\frac{u_A - e_A}{r_A} = i_A + T_A \frac{di_A}{dt}} \qquad \rightarrow G_A(s) = \frac{1}{r_A} \frac{1}{1 + sT_A}$$

differential equation first order (non-dimensional)
transfer-function

time domain solution in the: $i_A(t) = \dfrac{|(u_A - e_A)|_0}{\tau r_A} \left(1 - e^{-t/T_A}\right)$

with $|(u_A - e_A)|_0 \, \sigma(t)$ input signal.
The other equations must be normalized in the same way. We obtain:

2. Equation

$$\boxed{\frac{N}{N_{0N}} = \frac{\Omega}{\Omega_{0N}} = n = \omega} \qquad \frac{\Psi}{\Psi_N} = \psi; \quad \frac{E_A}{E_{AN}} = e_A$$

therefore:

$$\boxed{e_A = n\,\psi = \omega\,\psi} \qquad \text{non-linearity, } NL'$$

3. Equation

$$\frac{\Psi}{\Psi_N} = \psi; \quad \frac{I_A}{I_{AN}} = i_A; \quad \frac{T_{Mi}}{T_{iN}} = t_{Mi}$$

$$\boxed{t_{Mi} = \psi\, i_A}$$

non-linearity, *NL'*

4. Equation

$$\frac{J\Omega_{0N}}{T_{iN}} = T_{\Theta N}$$

therefore:

$$t_{Mi} - t_W = t_M - t_W = T_{\Theta N}\frac{dn}{dt} = T_{\Theta N}\frac{d\omega}{dt}.$$

Transfer-function:
$$\frac{n(s)}{t_{Mi}(s) - t_W(s)} = \frac{1}{sT_{\Theta N}} = G_M(s).$$

Solution in the time domain:

$$n(t) = |(t_M - t_W)|_0 \cdot \frac{t}{T_{\Theta N}} \qquad \text{with } |(t_M - t_W)|_0 \cdot \sigma(t) \text{ input signal}.$$

This gives the normalized signal-flow graph.

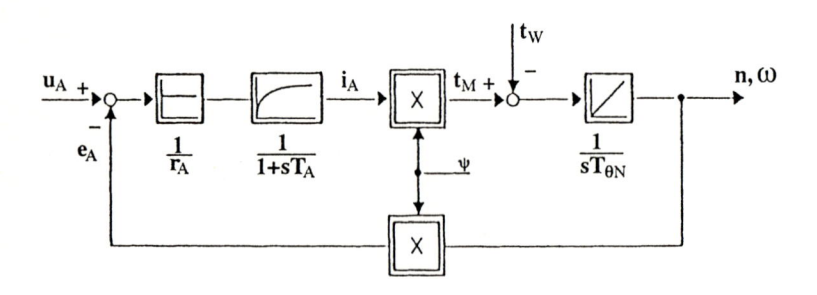

Fig. 3.6. Normalized signal-flow graph

Note: if $\Psi = \Psi_N$, then $i_A = t_M$ and $n = e_A = \omega$.

3.1.2 Field circuit - Excitation Circuit

In the field circuit there must be a non-linear relation between the excitation current I_E or the magnetic field strength H and the flux density B or the flux Ψ. Therefore a distinction must be made between direct-current values and alternating current values.

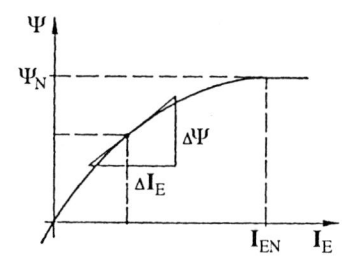

Fig. 3.7. Static magnetizing curve of flux without hysteresis

$\Psi = L_E I_E$ non-linear function

$L_{EN} = \dfrac{\Psi_N}{I_{EN}}$ rated inductance (direct-current value)

$L_{Ed} = \dfrac{\Delta\Psi}{\Delta I_E} = f(\Psi)$ differential inductance (alternating current value)

(linearization at working point)

In order to consider this non-linear relation, the following excitation circuit equation has been used:

$$U_E = I_E R_E + \frac{d\Psi}{dt} = I_E R_E + \frac{d}{dt}(I_E(t) L_E(t))$$

After normalization with the rated values U_{EN}, I_{EN}, Ψ_N we obtain:

$$\boxed{U_{EN} = I_{EN} R_{EN}} \quad ; \quad \frac{R_E}{R_{EN}} = r_E \qquad \frac{U_E}{U_{EN}} = u_E$$

$$\frac{\Psi_N}{U_{EN}} = \frac{\Psi_N}{I_{EN} R_{EN}} = \frac{I_{EN} L_{EN}}{I_{EN} R_{EN}} = T_{EN} = const.! \; [s]$$

therefore:

$$\boxed{u_E - T_{EN}\frac{d\psi}{dt} = i_E\, r_E}\quad ;\quad \text{non-dimensional excitation equation}$$

$$u_E(s) - T_{EN}\, s\, \psi(s) = i_E(s)\, r_E$$

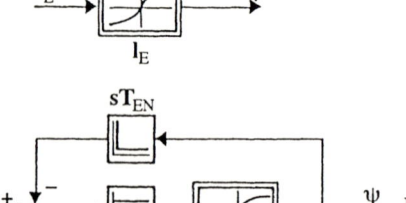

stationary relation of non-linear functions (without eddy-currents).

Fig. 3.8. Excitation circuit signal-flow-graph *without influence of eddy-current i.e. Laminated iron*

The whole normalized signal-flow graph of the separately excited DC-machine can be drawn as shown in Fig. 3.9:

3.2 Transfer-functions - Transient Responses

This chapter describes the transfer-functions and the resulting transient responses for the various input variables. A precondition of this is a basic understanding of the control theory. However, in the first example of the variable command transfer-function the process is represented by means of an example. The reason for this approach lies in the fact that there are equations for the transfer-functions of a closed-loop control that again require a unit feedback $G_r(s) = 1$. However, this is not the general situation and therefore using such a rule results in erroneous transfer functions.

3.2.1 Variable Command Behaviour and Variable Command Transfer Function

$n(s) = G_1(s)\, u_A(s);$ with $t_W(t) = 0$ (no disturbance variable):

$\psi = \text{const.} = \psi_0$
$0 < \psi_0 \le 1.$

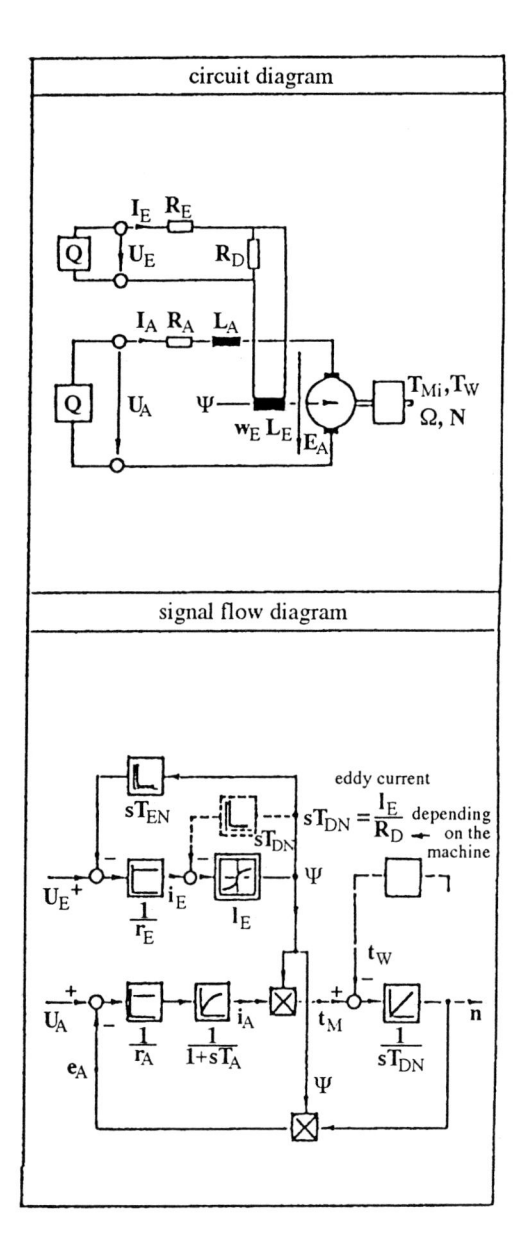

Fig. 3.9. Signal-flow graph of separately excited DC-machine

The following holds true:

$$G_1(s) = \frac{1}{\dfrac{1}{G_V(s)} - G_r(s)} = \frac{G_V(s)}{1 - G_V(s)\, G_r(s)} \quad \text{with} \quad G_r = -\psi_0 \,!$$

$$\text{and} \quad G_V(s) = \frac{1}{r_A} \, \frac{1}{1 + sT_A} \, \psi_0 \, \frac{1}{sT_{\Theta N}}$$

$$G_1(s) = \frac{\dfrac{1}{r_A} \dfrac{1}{1+sT_A} \psi_0 \dfrac{1}{sT_{\Theta N}}}{1 + \dfrac{1}{r_A} \dfrac{1}{1+sT_A} \psi_0 \dfrac{1}{sT_{\Theta N}} \psi_0} = \frac{1}{\psi_0 \left[1 + (1+sT_A)\, sT_{\Theta N}\, \dfrac{r_A}{\psi_0^2} \right]}$$

$$\text{with} \quad T_{\Theta st} = T_{\Theta N}\, \frac{r_A}{\psi_0^2} \quad ; \quad T_{\Theta N} = \frac{J\Omega_{0N}}{T_{iN}}$$

therefore:

$$\boxed{\,G_1(s) = \frac{1}{\psi_0\left(1+(1+sT_A)sT_{\Theta st}\right)} = \frac{1}{\psi_0\left(1+sT_{\Theta st}+s^2 T_A T_{\Theta st}\right)}\,}$$

and the signal-flow graph (Fig. 3.10).

$$\underbrace{\left(\dfrac{1}{\psi_0}\right)}_{} \quad \underbrace{\left(\dfrac{1}{1+sT_{\Theta st}+s^2 T_A T_{\Theta st}}\right)}_{}$$

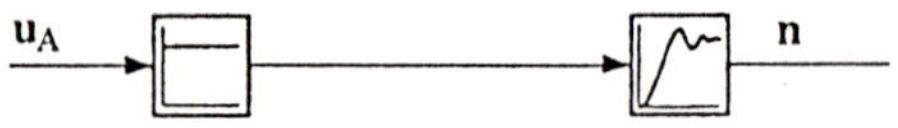

u_A → □ → ⌐ n

Fig. 3.10. Signal-flow graph

The second order transfer-function generally has a pair of conjugate complex poles, i.e. resonant behaviour if we only observe the DC-machine, and two real poles, i.e. aperiodic behaviour, if the DC-machine is connected to the load machine. The following transient responses result:

Transients

Input signal

Step response

(1) $T_A = 0$

(2) $T_A \leq \left(\dfrac{T_{\Theta st}}{4}\right)$

(3) $T_A > \left(\dfrac{T_{\Theta st}}{4}\right)$.

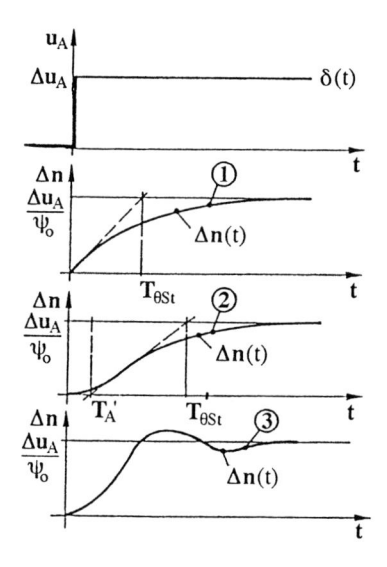

Fig. 3.11. Step response for different armature time constants

$$T'_{A_{1,2}} = +\dfrac{T_{\Theta st}}{2} \pm \sqrt{\dfrac{T^2_{\Theta st}}{4} - T_A T_{\Theta st}}$$

$$T'_{\Theta st} = \dfrac{T_A T_{\Theta st}}{T'_A}$$

for $T_A \ll \left(\dfrac{T_{Qst}}{4}\right)$ the following applies: $T'_A \cong T_A$ and $T'_{\Theta st} \cong T_{\Theta st}$.

3.2.2 Load Behaviour and Disturbance Transfer-Function

$n(s) = G_2(s)\, t_W(s);\quad u_A(t) = \text{const. and } \psi = \text{const.} = \psi_0$
The following applies:

$$G_2(s) = -\dfrac{\dfrac{1}{sT_{\Theta N}}}{1+\left(\dfrac{1}{1+sT_A}\dfrac{1}{sT_{\Theta N}}\dfrac{\psi_0^2}{r_A}\right)} = -\dfrac{r_A(1+sT_A)}{\psi_0^2\bigl(1+(1+sT_A)sT_{\Theta st}\bigr)}.$$

From which we obtain in this case ($T_A \ll T_{\Theta st}$):

$$\boxed{\,G_2(s) \cong -\dfrac{r_A}{\psi_0^2}\dfrac{1+sT_A}{(1+sT_A)(1+sT_{\Theta st})} = -\dfrac{r_A}{\psi_0^2}\dfrac{1}{1+sT_{\Theta st}}\,}$$

$$G_2(s) \cong -\dfrac{1}{\dfrac{\psi_0^2}{r_A}+sT_{\Theta N}} \qquad \text{Initial tangent of transient response,}$$

independent of r_A, ψ_0, T_A.

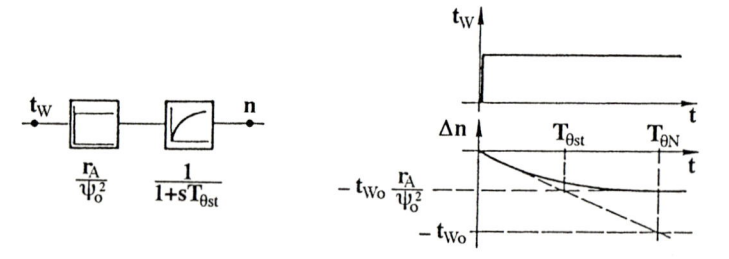

Fig. 3.12. Simplified signal-flow graph and transient response (load, step function)

3.2.3 Influence of ψ on n (Field Weakening)

The two poles of the characteristic polynominal are the two zeros of characteristic polynominal, the poles are $T'_{A1,2}$ and $T'_{\Theta st}$, these values are different to T_A and $T_{\Theta st}$.

The following problem is presented: $\quad G_3 = \dfrac{n(s)}{\psi(s)}$

$$u_A(t) = \text{const.}$$
$$t_W(t) = 0.$$

3 Direct-Current Machine

From Fig. 3.7 it can be seen that in the case of weakening of the flux Ψ both the multipliers are effective and, as in the previous two examples, a reduction to a proportional transfer-function is no longer possible. In order to calculate the transfer-function (linear) in this case, linearization is necessary at the operating point:

(a) Linearization: $t_M = i_A \psi \quad \rightarrow \Delta t_M \cong i_{A0} \Delta \psi + \psi_0 \Delta i_A$

(b) Linearization: $e_A = n \psi \quad \rightarrow \Delta e_A \cong n_0 \Delta \psi + \psi_0 \Delta n$

χ_0 = value of working point

with: $i_{A0} = \dfrac{t_{M0}}{\psi_0}$ (value of working point)

and $T_{\Theta st} = \dfrac{T_{\Theta N} \cdot r_A}{\psi_0^2}$.

After the elementary manipulation we obtain:

$$\Delta n(s) \cong \Delta \psi(s) \frac{1}{\psi_0} \left[\frac{t_{M0} \, r_A}{\psi_0^2}(1+sT_A) - n_0 \right] \frac{1}{1+(1+sT_A)sT_{\Theta st}}$$

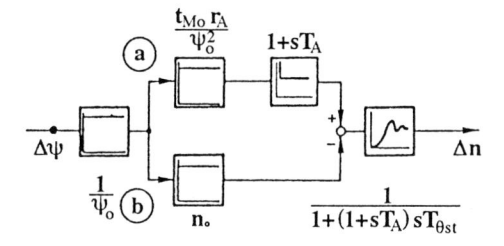

$\qquad\qquad\qquad\qquad\qquad$ (a) $\qquad\qquad\qquad$ (b)

Fig. 3.13. Linearized signal-flow graph

$$G_3(s) = \frac{\Delta n(s)}{\Delta \psi(s)} = \frac{(1+sT_A)\dfrac{r_A}{\psi_0^2} i_{A0} - \dfrac{n_0}{\psi_0}}{1+(1+sT_A)sT_{\Theta st}}.$$

Simplification, approximations:

with $\dfrac{i_{A0} \, r_A}{\psi_0} \ll n_0$: $\quad G_3(s) \approx -\dfrac{n_0}{\psi_0} \dfrac{1}{1+(1+sT_A)sT_{\Theta st}}$

and also $T_A \ll \left(\dfrac{T_{\Theta st}}{4}\right)$: $G_3(s) \approx -\dfrac{n_0}{\psi_0} \dfrac{1}{(1+sT_A)(1+sT_{\Theta st})}$.

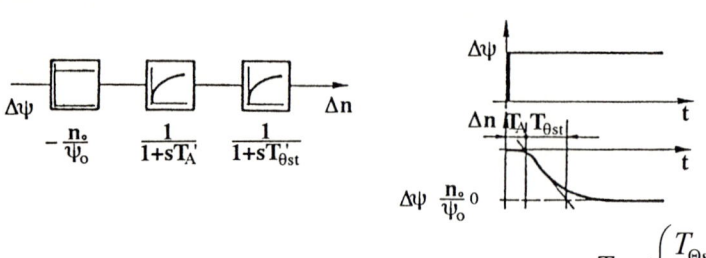

Fig. 3.14. Simplified signal flow-graph and time response for $T_A \leq \left(\dfrac{T_{\Theta st}}{4}\right)$

3.3 Open-loop Speed Control

3.3.1 Open-loop Speed Control by Means of Armature Voltage

The following holds true:

$$G_1(s) = \dfrac{n(s)}{u_A(s)} = \dfrac{1}{\psi_0} \dfrac{1}{1+sT_{\Theta st}+s^2 T_A T_{\Theta st}} \quad \Big|\, \begin{array}{l} \psi_0 = \text{const.} \\ t_W = 0 \end{array}$$

$$G_2(s) = \dfrac{n(s)}{T_W(s)} = \dfrac{r_A}{\psi_0^2} \dfrac{1+sT_A}{1+sT_{\Theta st}+s^2 T_A T_{\Theta st}} \quad \Big|\, \begin{array}{l} u_A = \text{const.} \\ \psi_0 = \text{const.} \end{array}$$

With approximation $\left(T_A \ll \dfrac{T_{\Theta st}}{4}\right)$: $G_2(s) \approx -\dfrac{r_A}{\psi_0^2} \dfrac{1}{1+sT_{\Theta st}}$.

The open-loop signal-flow graph (by means of superposition) is as follows:
$n(s) = G_1(s)\, u_A(s) + G_2(s)\, t_W(s)$ (exact).

With $T_A \ll \dfrac{T_{\Theta st}}{4}$:

$$n(s) \approx \dfrac{1}{\psi_0} \dfrac{u_A(s)}{(1+sT_A)(1+sT_{\Theta st})} - \dfrac{r_A}{\psi_0^2} \dfrac{t_W(s)}{1+sT_{\Theta st}}.$$

Open-loop signal-flow graph (linearized, superimposed, simplified).

3 Direct-Current Machine

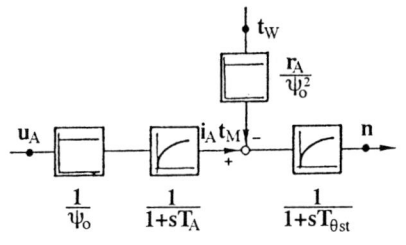

Fig. 3.15. Simplified signal-flow graph for $\left(T_A \ll \dfrac{T_{\Theta st}}{4}\right)$

Stationary state: $\dfrac{d}{dt} = 0; \quad s \to 0;$

$$n(t) = \dfrac{1}{\psi_0} u_A(t) - \dfrac{r_A}{\psi_0^2} t_W(t) \quad \text{quasi-stationary operation}$$

In the armature control range u_A is variable and $\psi_0 = 1 = \text{const.}$

Power source:

$e_Q = u_A + i\, r_Q = 1 + r_Q \quad$ (at rated point)

$i_Q = i_A = 1 \quad$ (at rated point)

e_Q: voltage of the source

r_Q: internal resistance of source

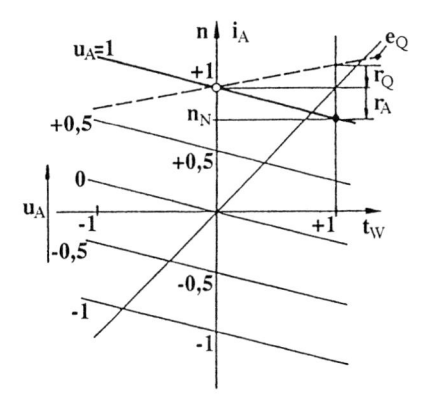

Fig. 3.16. Armature control range, family of characteristic curves (parameter u_A)

3.3.2 Open-loop Control Through Field Weakening

By modifying the operating point ψ_0 for the flux with $\psi_{min} \leq \psi_0 \leq 1$ and $u_A =$ const., $\Delta\psi_0 = 0$ holds true for quasi-static behaviour:

$$n(t) = \frac{1}{\psi_0} u_A(t) - \frac{r_A}{\psi_0^2} t_M(t).$$

In the field weakening range, $u_A(t) = u_N =$ const., and therefore:

speed: $\quad n = \underbrace{\frac{u_N}{\varphi_0}}_{\text{no load speed}} - \underbrace{\frac{t_M\, r_A}{\psi_0^2}}_{\text{reduction of the speed in the case of load}}$

current: $\quad i_A = \dfrac{t_M}{\psi_0}$

Illustration: $n = f(t)$ with $u_A = u_N = 1$ (field weakening).

$$\Rightarrow n = \frac{1}{\psi_0} - \frac{t_M\, r_A}{\psi_0^2}$$

Note: the decrease in n by r_A/ψ_0^2 increases by the order of 2 as ψ_0 decreases; the current requirements $i_A = t_M/\psi_0$ increase linearly.

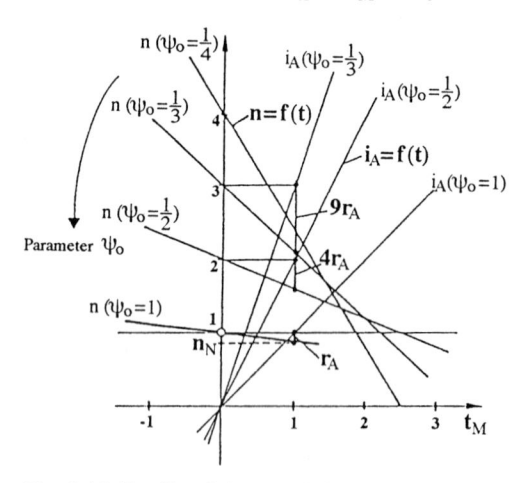

Fig. 3.17. Family of characteristics in case of field weakening

3.3.3 Control Through Armature and Field-Voltage

Stationary Behaviour, Characteristics.

$$n = \frac{u_A}{\psi_0} - \frac{r_A}{\psi_0^2} t_M$$

Armature control range: $0 \leq U_A \leq U_{AN}; \psi_0 = 1 \quad p_0 = u_A i_A$
$\rightarrow$ linear gradient

Field weakening range: $0 < \psi_0 < 1; U_A = U_{AN} \quad p_0 = p_{0\max} = u_A i_A$
$\rightarrow$ constant

Advantage of field weakening:

Increase in the speed range without overrating the machine and power supply.

Disadvantage:

Declining torque a controller and power supply are necessary for the flux or equivalent for i_E.

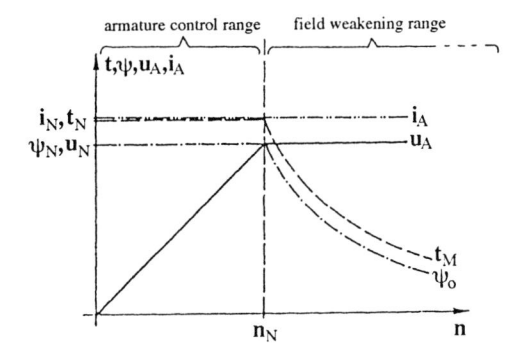

Fig. 3.18. Armature-current, voltage, flux and torque in dependency from speed

3.3.4 Control with a Series-Resistor in the Armature Circuit

The behaviour, speed and torque, of the separately excited DC-machine can be altered by a series-resistors R_V in the armature circuit. This solution was frequently used in the past when power electronics was still in its early stages. This method of modifying behaviour leads to substantial losses and as a result the efficiency of the system seriously declines.

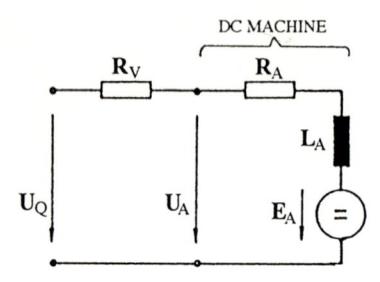

Fig. 3.19. Separately excited DC-machine with series resistor in the armature circuit

Normalization: $R_A^* = R_A + R_V = R_A \left(1 + \dfrac{R_V}{R_A}\right) \Rightarrow r_A^* = r_A \left(1 + r_V\right)$.

Speed: $\quad n = \dfrac{u_Q}{\psi_0} - t_M \dfrac{r_A}{\psi_0^2} \left(1 + r_V\right)$.

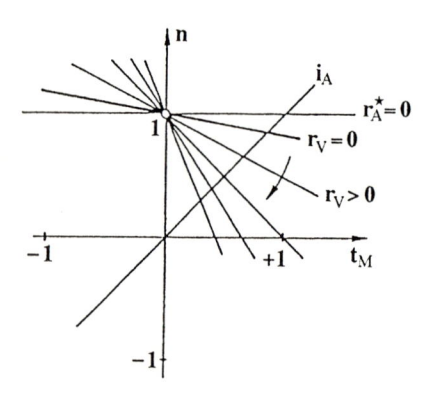

Current: $\quad i_A = \dfrac{t_M}{\psi_0}$.

Fig. 3.20. Family of characteristics for control by series resistor with $u_Q = 1$ and $\psi = 1$

Control behaviour: unilateral;
depends on load, i.e. on torque demand;
with losses.

Time response: $T_A^* = \dfrac{L_A}{R_A + R_V} \Rightarrow T_A \dfrac{1}{(1+r_V)}$ decreases

$T_{\Theta st}^* = T_{\Theta N} \dfrac{r_A^*}{\psi_0^2} \Rightarrow T_{\Theta N} \dfrac{r_A(1+r_V)}{\psi_0^2}$ increases.

For example, if various series-resistors with fixed armature voltage U_A are used, the machine can be started from zero speed to the operating point between a maximum and a minimum torque (armature current).

4 Converters and DC-Machine Control

It was found from the transfer-functions of the separately excited DC-machine that modification of the torque needs control of the armature current and modification of the speed requires control of either the armature voltage or the excitation voltage. Changing the speed-torque characteristic using series resistors is possible in principle but leads to losses; the torque and speed cannot be adjusted independently from each other and the control velocity is relatively low, especially when using mechanical switches. Power electronic converters offer considerable advantages. There are two basic solutions of the power supply of DC-machines. The first solution is the transformation of a constant direct voltage into the direct voltage required by the DC-machine: these are the DC-DC-converters. The second solution is the conversion of an alternating or three-phase-voltage into the direct voltage required, in which case the circuitry are line commutated converters or naturally commutated converters. This chapter deals with the principles of operating these two types of converters and their open- and closed-loop control.

4.1 DC-DC-Converter

4.1.1 DC-DC-Converter: Principle

The DC-DC-converter (buck converter) basically is an electronic switch which periodically connects a user to a voltage source or short-circuits the user. Figure 4.1. shows the principle of this DC-DC-converter. This converter applies voltage to the load and therefore is a voltage source converter.

$$\text{Assumption: } Z = R - L \ Load; \qquad \frac{L}{R} >> T \ .$$

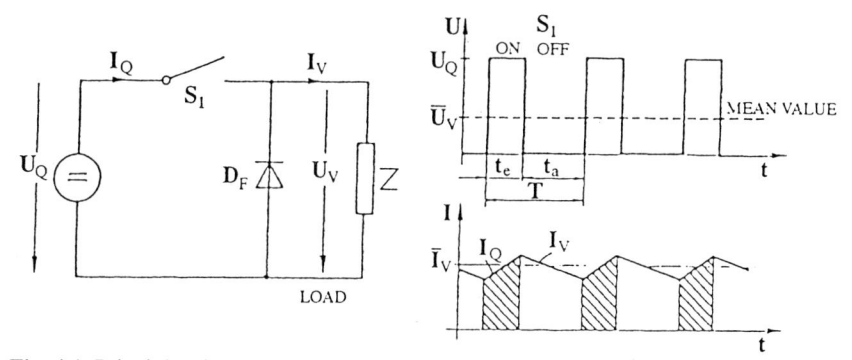

Fig. 4.1. Principle of a DC-DC-converter (buck converter)
a) circuit with mechanical switch S_1
b) voltage and current waveforms at an ohmic-inductive load

With the switch S_1 closed, a current I_Q flows from the voltage source U_Q to the load Z which should contain a large inductivity, resulting in $\frac{L}{R} \gg T$. If the switch S_1 is opened, the current flows through the diode D_F The voltage drop should be assumed to be zero. The current I_V is abated in the load circuit during this state. At a very high time constant in the load circuit the current is virtually constant. The mean value of the voltage U_V is obtained by the on-period t_e, the off-period t_a and the period $T = t_e + t_a$:

$$\overline{U}_v = U_Q \frac{t_e}{T} = U_Q a$$

where $a = \frac{t_e}{T}$ is the switching ratio. At constant load current I_V, the mean value of the current I_Q is:

$$\underbrace{\overline{I}_Q}_{\text{source}} = \underbrace{I_V}_{\text{load,user}} \frac{t_e}{T} = I_V a \qquad \begin{array}{c}\text{Mean value for}\\ \frac{L}{R} \to \infty\end{array} \qquad \frac{L}{R} \gg T \quad .$$

The power delivered from the source is:

$$P_Q = U_Q \overline{I}_Q = U_Q I_V a$$

and the power at the load is:

$$P_V = U_V I_V = U_Q a I_V \quad .$$

In DC-DC-converters without losses the following holds:

$$P_Q = P_V \quad .$$

The ideal DC-DC-converter thus converts power from the source to an user without causing losses. In the real DC-DC-converter the switch S_1 is replaced by a controllable semiconductor switch. In case of low output power, power transistors, for example of the MOSFET type, are used at medium power bipolar power transistors were used once and now IGBTs are commonly used, while at high power semiconductors of the GTO variety are used. Since these power semiconductors and the diode D_F undergo conduction losses, and the controllable power semiconductors also sustain switching losses, efficiency rates of approximately $\eta = 92\%$ to 97% are achieved.

DC-DC-converters therefore have a special advantage when the power source is a battery, as in the case of battery powered vehicles or when a DC-rail or DC-catenary or DC-generator is used. The circuit shown above represents a buck converter, in which the output voltage is lower than the one from the supply source.

A modification of the buck converter is the boost-converter:

$$\overline{U}_V \geq U_Q$$

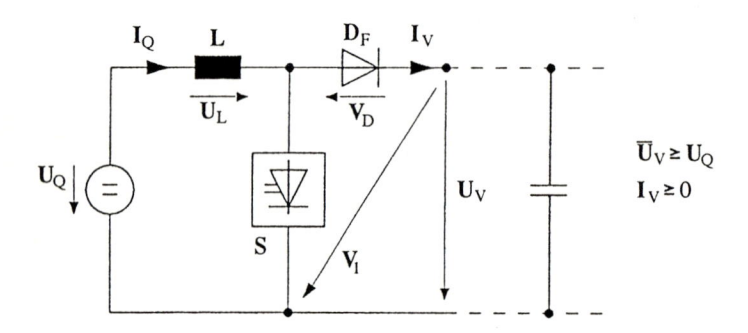

$$I_V \geq 0 \quad .$$

Fig. 4.2. Boost- Converter

The purpose of the boost converter is to achieve a voltage $\overline{U}_V$ which is greater than U_Q. The operation is described as follows:

when the switch S is switched on, the diode D_F is in the off-state. Therefore the reactor with the inductivity L is connected to the voltage U_Q approximately and the following applies:

$$U_Q = L\frac{dI_s}{dt} = L\frac{dI_Q}{dt} = U_L (S \text{ on}) \ .$$

Therefore the current will rise linearly. At a maximum current the switch S is switched off by control. Consequently a reactor (coil) voltage will be generated. Then we obtain:

$$U_Q - U_L (S \text{ off}) = U_V$$

$$U_Q - L\frac{dI_V}{dt} = U_V$$

and an electrical circuit is formed by the diode D_F and the load.
Taking again the voltages as in the buck converter, the following holds true:

$$m = \frac{\overline{U}_V}{U_Q} = \frac{\overline{V}_1}{V_1 - \overline{V}_D} = \frac{1}{1 - \dfrac{\overline{V}_D}{V_1}} = \frac{1}{1-a} \ .$$

A precondition in these equations is that the reactor (coil) L is ideal and therefore no DC voltage drop occurs. With $a = 0$ (i.e. switch is constantly open) we therefore have $\overline{U}_V = U_Q$, with $0 < a < 1$ which is the voltage $\overline{U}_V > U_Q$.

This type of circuit is used in the braking operation of a DC-machine. Other modifications are the buck-boost converters and the four quadrant converter.

4.1.2 Control Strategies for DC-DC-Converters

As shown the mean value of the output voltage depends on the switching ratio a.

$$a = \frac{t_e}{t_e + t_a} = \frac{t_e}{T}$$

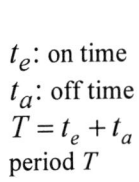

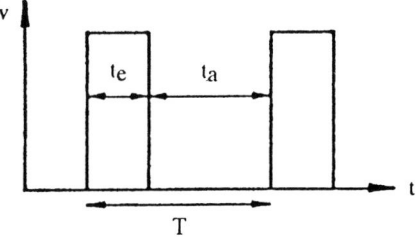

t_e: on time
t_a: off time
$T = t_e + t_a$
period T

Fig. 4.3 Output voltage U_V

Pulse-Width Control (T = const.). The pulse-width control is mainly applied in systems in which changing frequencies may affect the signal circuits; this is avoided by pulse width control. The method is based on the fact that the duration of the pulse or the pause can be controlled, while the duration of the period T of the cycle and therefore the frequency remains constant. The following relation applies here:

$$\overline{U}_V = \frac{U_Q t_e}{T}; \qquad t_e = T - t_a \qquad \text{variable.}$$

Neither t_e nor t_a can be set zero, since this is not possible due to the dynamic properties of the power semiconductors and their snubber circuits.

One circuit diagram of the pulse-width control is shown in Figure 4.4.

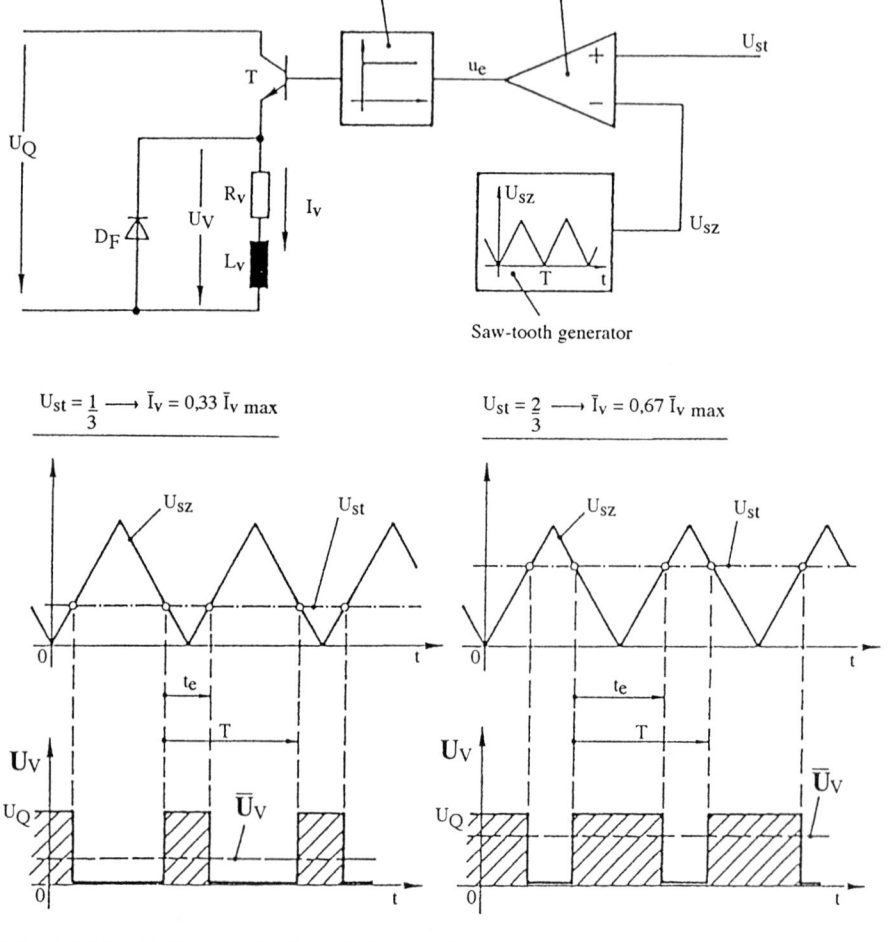

Fig. 4.4. Circuit diagram of pulse-width control

Pulse-frequency control. With this type of control the pulse-width t_e is constant, the pause duration is variable and therefore the duration of the periods of the cycle or frequency are variable.
From $T = t_e + t_a$ we obtain:

$$F = \frac{1}{t_e + t_a} = \frac{1}{T} \qquad (t_e = \text{const., period } T \text{ variable})$$

$$t_e \leq T < \infty \ .$$

What makes this frequency control (also known as pulse-repetition control) interesting is its low complexity in the signal processing.

$$F_{min} = F_{max} \frac{\overline{U}_{V\,min}}{U_Q}$$

With the equations $F_{max} = \dfrac{1}{t_e}$ and $F_{min} = \dfrac{\overline{U}_{V\,min}}{U_Q\, t_e}$

the frequency control range is obtained: $\dfrac{\overline{U}_{V\,min}}{U_Q\, t_e} < f < \dfrac{1}{t_e} = F_{max}$.

Whereas maximum output voltage is given with F_{max} and $t_e = T$, F_{min} is obtained by calculation from the minimum desired output voltage $\overline{U}_{V\,min}$ and the chosen pulse-duration t_e with assumed input voltage U_Q. Generally it must be observed that at low operating frequencies a high rating of the smoothing circuit is necessary, namely expensive inductors - in this case discontinuity of (armature) current can be avoided.

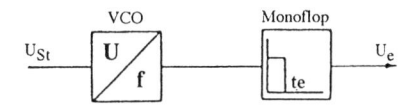

Fig. 4.5. Circuit diagram of pulse-repetition control

Closed-loop Control of a DC-DC-Converter. Two-step Controller (converter with impressed current). This type of control is used for load current control and both pulse-duration and pulse-frequency are variable. The corresponding on- and off-gate-signals are produced by a hysteresis controller as soon as the real current leaves the permitted hysteresis band.

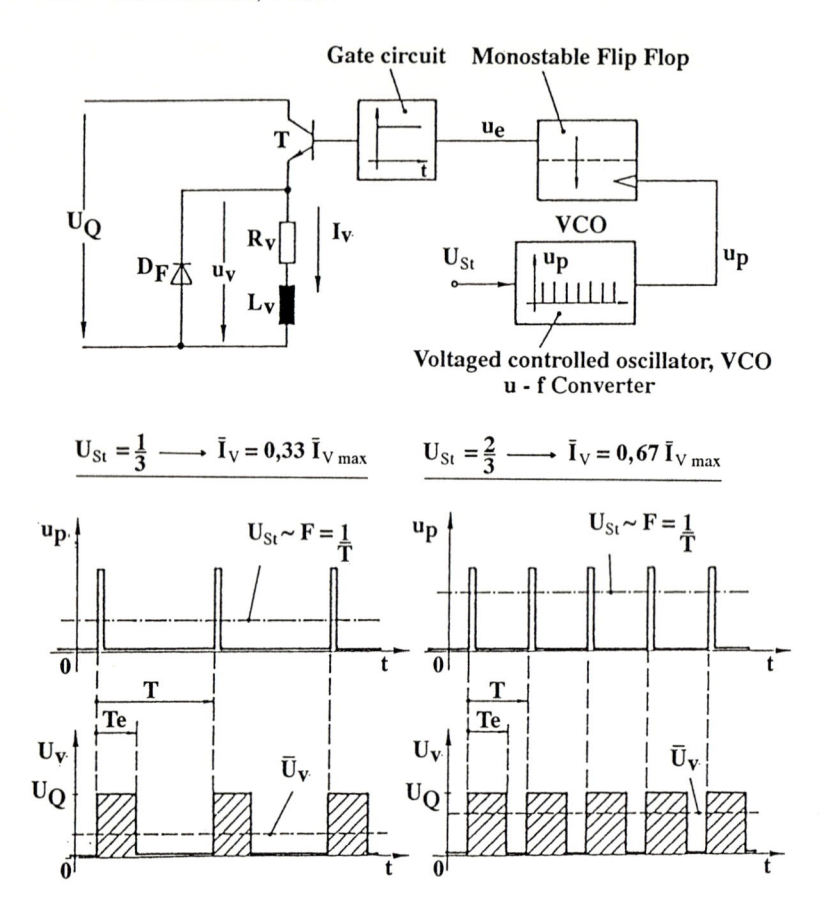

Fig. 4.6. Operation of pulse-repetition control

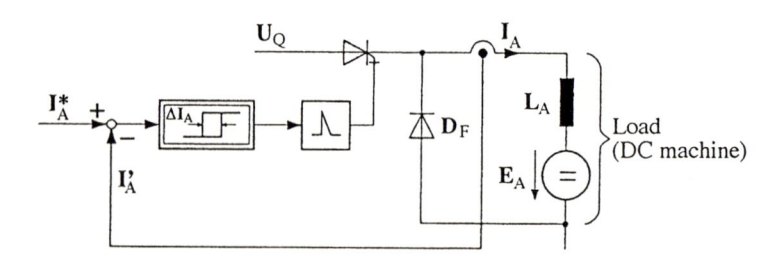

Fig. 4.7. Principle circuit diagram of two-step control

E.g.: separately excited DC-machine $E_A = a\, U_Q$ with $R_A = 0$
if the switch is on:

$$U_Q = L_A \frac{dI_A}{dt} + a\, U_Q \quad \rightarrow \quad \frac{U_Q}{L_A}(1-a) = \frac{dI_A}{dt}.$$

We obtain:

$$\Delta I_A = \frac{U_Q(1-a)}{L_A}\, a\, T \qquad aT = t_e;\ \ \bar{I}_A \text{ and } I_{d\min} \text{ from}$$

exact calculation including R_A
$\rightarrow$ differential equation

When the switch is off: $U_Q = 0$ therefore $E_A = a\, U_Q = -L_A \dfrac{dI_A}{dt}$; duration $(1-a)$.

$$\left[\frac{t_e}{T} = \frac{1}{3}\right] \qquad\qquad \left[\frac{t_e}{T} = \frac{2}{3}\right]$$

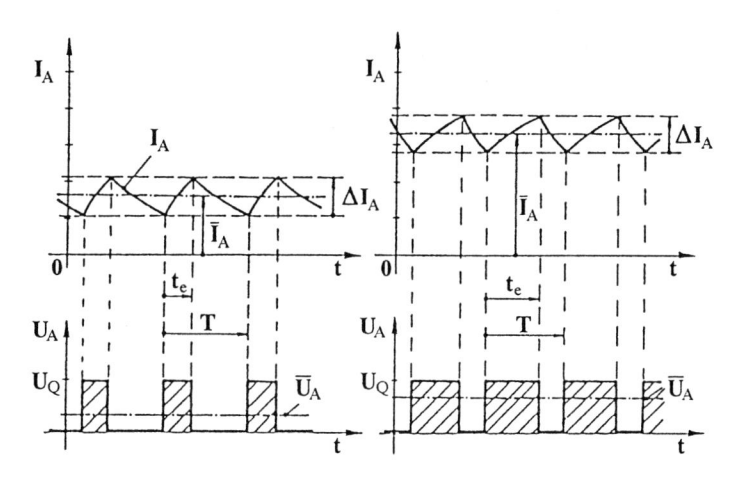

Fig. 4.8. Output signals of two step control for $R_A \neq 0$

4.1.3 DC-DC-Converters for one or Multiple Quadrant Operation of DC-Motors

Motor Operation of a Separately Excited DC-Machine. The buck converter can be directly used as power supply for a separately excited DC-machine. The load is R_A, L_A and the e.m.f. E_A is used as the electromotoric force.

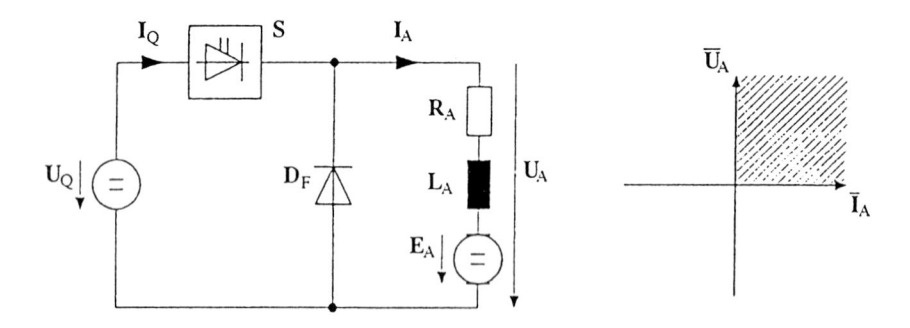

Fig. 4.9. Motor operation: circuit diagram and operating range

Under the assumption of continuous armature current, the mean values of U_A and I_A in the pulse-width control mode are:

$$\overline{U}_A = a\, U_Q \quad \text{with } a = \frac{t_e}{T}: \text{ on period} \quad (0 \le a \le 1)$$

$$\overline{I}_A = \frac{1}{R_A}(\overline{U}_A - E_A) = \frac{1}{R_A}\left(a\, U_Q - E_A\right).$$

Due to the freewheeling diode D_F we have to accept $U_A \ge 0$. S and D_F prevent a I_A polarity into the negative direction. The direction of the power flow is only possible from the source to the motor.

If the switch S is off and the armature current I_A is zero (after freewheeling) the armature voltage U_A over this period of time acquires the value of E_A and not zero. Therefore in case of discontinuous current the following also applies for the mean armature voltage $\overline{U}_A \ge a\, U_Q$. As soon as S is on, I_A increases and the output voltage $U_A = U_Q$.

Thus the motor cannot operate in the generator mode and thus cannot brake electrically since the armature voltage direction is fixed by the voltage E_A and therefore a reversal of current is necessary during braking.

Where a negative load torque is possible, a mechanical braking device is required.

Special case: in the case of positive load torque and short on periods t_e (i.e. $\overline{U}_A \to 0$), the speed may become negative ($n = e_A < 0$). Both the motor and the DC-DC-converter supply power to the armature resistance which converts it into heat (reverse current electrical braking).

Braking Operation of a Separately Excited DC-Machine. With the boost converter a power transfer from the load to the voltage source (energy recovery) is possible. The input and output terminals are changed in relation to the buck converter (Figure 4.10). To achieve energy recovery during stationary operation (with the DC-machine in generator-mode) the load must be a series connection with an inductivity (L_A) and a voltage source (e.m.f. E_A). If an inductivity is the load only, the energy in the inductivity can be recovered.

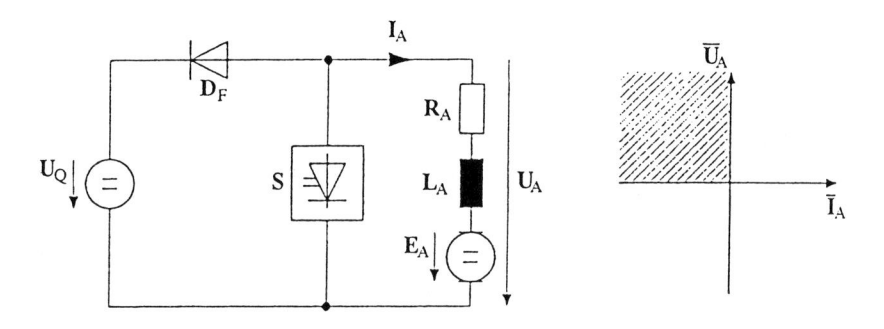

Fig. 4.10. Braking operation: Circuit diagram and quadrants of operation

Principle of operation. By switching S on (i.e. $U_A = 0$) an armature current $I_A < 0$ is generated. Due to the inductance L_A, once S is switched off, the armature current is not interrupted but switched by D_F to the "source". The voltage U_{LA} of the inductance adds to E_A and produces a voltage of approximately U_Q. For continuous current the following holds true:

$$\overline{U}_A = (1-a) U_Q > 0$$

$$\overline{I}_A = \frac{1}{R_A} (\overline{U}_A - E_A) = \frac{1}{R_A} ((1-a) U_Q - E_A) < 0$$

Two-Quadrant Operation. By combining the circuits for first-quadrant motor operation and generator (braking) operation, operation in two quadrants is possible. In this case either the armature current or the armature voltage can be reversed.

Two-Quadrant Operation with Reversal of the Armature Current. The solution with contactor (Fig. 4.11) is economically convenient if the dynamic requirements in the transition from the first to the second quadrant are low.

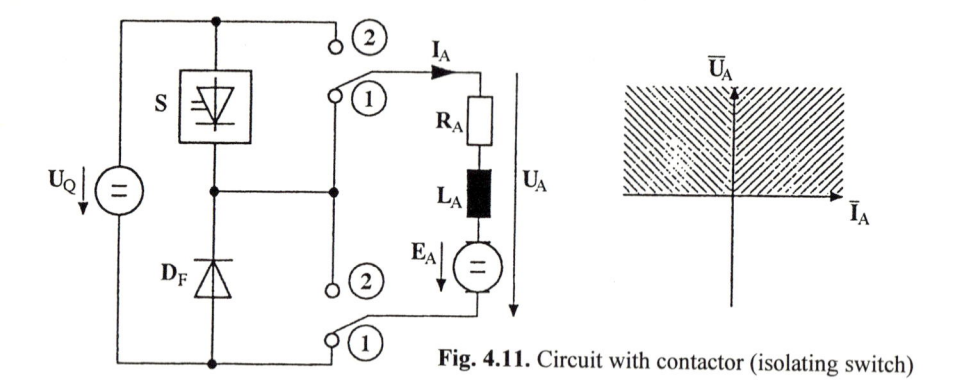

Fig. 4.11. Circuit with contactor (isolating switch)

Switch position 1: the circuits behave as a buck converter. The diode D_F is in parallel and the switch S is in series with the DC-machine. Position 1 is therefore provided for motor-operation.

Switch position 2: the switch S is in parallel and the diode D_F is connected in series with the DC-machine. This is the solution for the braking operation.

The alternation between the two positions of the mechanical switch should occur without current by means of control, prior to switchover of switch S.

A solution without mechanical switches is possible by using two diodes and two controllable power semiconductors which can be switched on and off (for example transistors, IGBTs, GTOs) (Fig. 4.12).

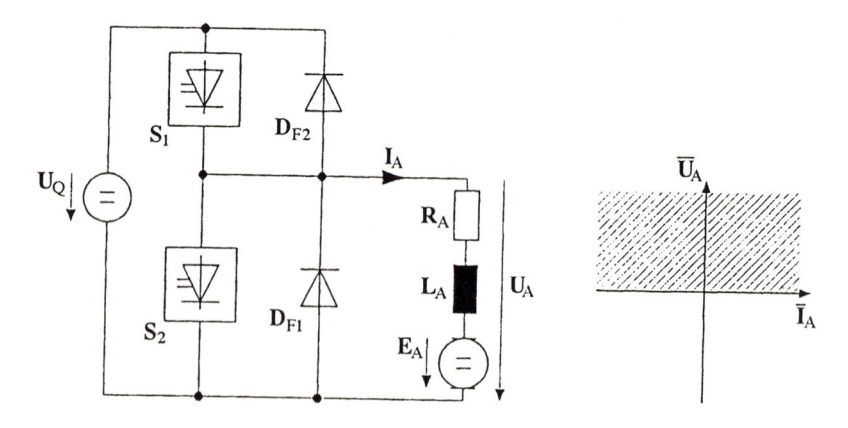

Fig. 4.12. Circuit for armature current reversal without contactor

Example of application: vehicles with one direction of movement (motoring and braking).

Two-Quadrant Operation with Armature Voltage Reversal.

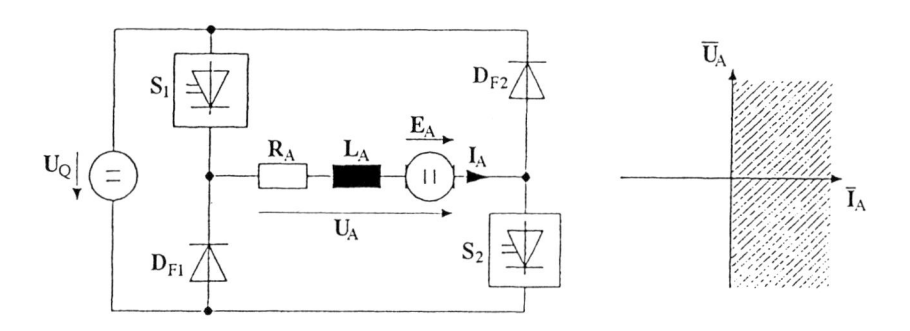

Fig. 4.13. Circuit for armature voltage reversal

Three different control methods will now be described:

1. Control method: if a $\overline{U}_A > 0$ is required, then S_2 is switched on continuously while S_1 operates in pulse width modulation. So one-quadrant motor operation is realized (section 4.1.1).

In order to obtain a current at $\overline{U}_A < 0$, S_1 must be switched off continuously and S_2 is operating in pulses width modulation. The DC-machine is in the fourth quadrant operation the DC-DC converter is a boost converter and reduces the armature current very quickly.

2. control method: simultaneous control. The two switches receive the same control pulses. A negative armature voltage is obtained $0 \leq a < 0.5$, and a positive for $0.5 < a \leq 1$ (continuous armature current assumed). Advantage: very easy control strategy. Disadvantage: no free wheeling operation is possible, i.e. constant succession between motor- and generator-operation.

3. Control method: this procedure is slightly more complicated and is therefore illustrated in a line diagram (Fig.4.14) (1 means: the valve conducts).

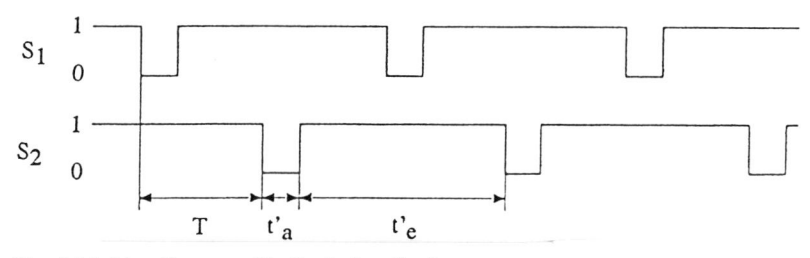

Fig. 4.14. Line diagram of 3. Control method

$$a' = \frac{t_e'}{T} - 1 \ .$$

The third method offers various advantages: the switches are loaded evenly and the switching frequency is lower compared to the other control strategies or at constant switching frequency the harmonic content is lower.

Example of application: lifting devices, winches.

Four-Quadrant Operation. By using four diodes and four controllable power semiconductors, two directions for $\overline{U}_A$ and $\overline{I}_A$ are possible (Figure 4.15).

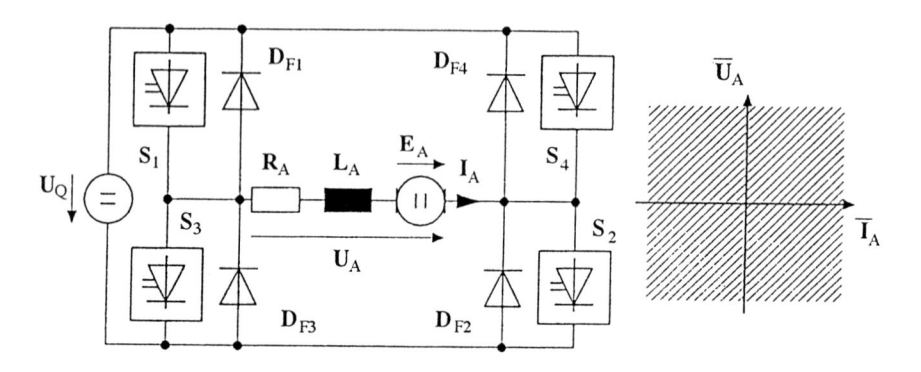

Fig. 4.15. Four-quadrant DC-DC-converter

There are two control strategies:

First control strategy: two switches close alternately: either S_1 and S_2 or S_3 and S_4. The pair of conducting switches must be switched off long enough (recovery time) before the other pair starts conducting to avoid short-circuiting the power supply.

If S_1, S_2 on-time is expressed by t_e and their off-time by t_a, with $T = t_e + t_a$ and $a = t_e/T$, then we obtain:

$$\overline{U}_A = a\, U_Q - (1-a)\, U_Q = (2 \cdot a - 1)\, U_Q$$

$$\overline{I}_A = \frac{1}{R_A} \left((2\, a - 1)\, U_Q - E_A \right).$$

Second control strategy: The diagonal pair of switches like S_1 and S_2 are controlled as in the case of two quadrant operation by armature voltage reversal. The switches in series S_3 or S_4 are switched in a push-pull operation. The equations for the armature voltage and armature current are:

$$\overline{U}_A = a' \ U_Q \qquad \text{with } a' \text{ from section 4.1.3}$$

$$\overline{I}_A = \frac{1}{R_A}\left(a' \ U_Q - E_A\right).$$

The armature voltage U_A is alternatively $+U_Q$ and zero for $\overline{U}_A > 0$ and U_Q and zero for $\overline{U}_A < 0$.

Comparison of the two control strategies: the first strategy is easier to implement and allows a measurement of the armature currents without potential separation: the armature current is always conducted on one of the lower branches. If these branches are connected to precision resistors to the negative terminals of the voltage source (which must be clamped on the zero potential of the control electronics), then the armature current can be measured by adding the voltage drops of the precision resistors.

The switching of U_A between positive and negative source voltage leads to greater ripple in the current and therefore to additional losses. Furthermore a higher switching frequency is required than in the second method, leading to increased switching losses.

4.2 Line Commutated Converters

In line commutated converters there is a conversion of power from an alternating or three-phase system to a direct voltage system. In this case the direct voltage follows the alternating voltage waveform or the AC-waveforms of the three-phase system. The shape of the current waveform is basically determined by the waveform of the AC-voltage, firing angle α and the load data. The definition *line commutated converters* or *converters with natural commutation* depends on the type of current commutation from one semiconductor to another. In this case the reactive power is supplied by the AC- or the three-phase AC- system in order to enable this commutation. This chapter, using the three-phase "Y" half wave circuit (chosen solely for explanatory purposes), illustrates the basic concepts of line commutated converters. In the sections 4.2.2 to 4.2.3 converters circuits and their data are shown, which are used generally.

4.2.1 Three Phase "Y" Half Wave Converter

Circuit: pulse number $p = 3$

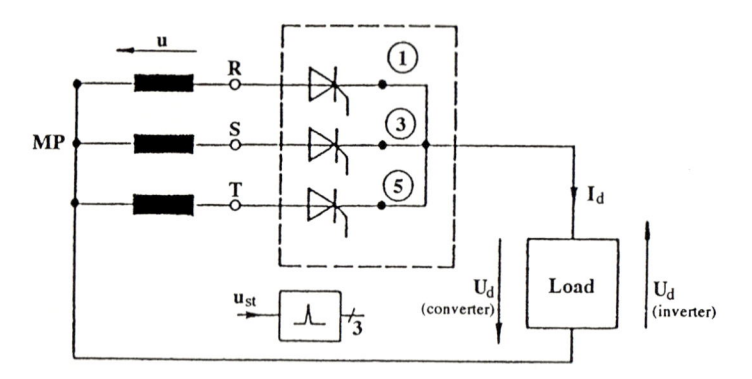

Fig. 4.16. Three phase ",Y" half wave converter (Principle)

Caution: The three-phase "Y" half wave circuit cannot be used as shown above, since there are DC-current components in the transformer windings which may saturate the iron core. The scheme shown in Fig.4.16 was chosen solely for explanatory purposes. In practice with a transformer and a three-phase ,"Y" half wave circuit, a transformer design must be used which avoids the DC-current components in the transformer windings (e.g. zig-zag windings).

Load R   Voltage waveforms, example ($\alpha = 60°$)   **Load R - L**

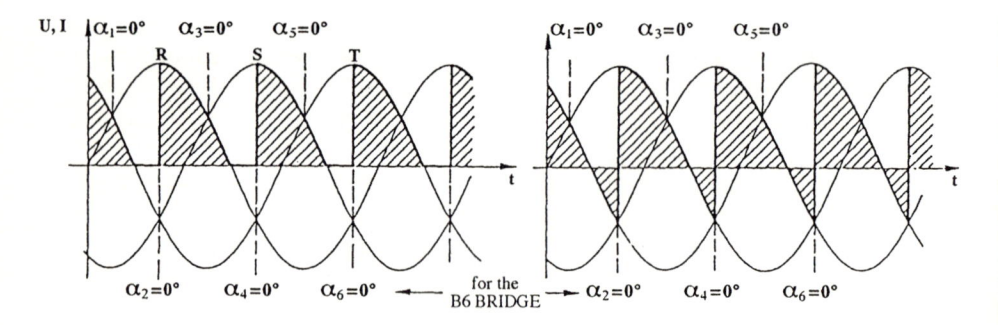

Fig. 4.17. Voltage waveforms at $\alpha = 60°$

The mean value of the direct voltage U_d is obtained from the shaded areas (voltage - radiant area) during the period $1/pF_N$. With $\alpha \leq 90$ the mean voltage is $U_d \geq 0$ and this type of operation is called the rectifier operation. The shapes of the waveforms clearly show that at the same control angle α that load has a considerable effect. In case (a) (load R), negative voltages cannot occur as a result of the valve. In case (b) (load R - L), owing to $U_L = L\, di/dt$, the inductor produces a voltage which maintains a positive current but a negative di/dt and consequently a positive voltage on the conducting semiconductor. So even negative voltage in the output voltage waveform can occur, as long as the flux is not zero or reversed in the inductor.

(a) ohmic load:

U_d, I_d : same waveform

discontinuous current

(b) Load R-L

with $T = \dfrac{L}{R} \gg \dfrac{1}{F_{mains}}$

→ continuous current

DC-current alternates from thyristor 1→3, 3 → 5, 5 → 1. This is called *Commutation*

The following applies: Load $R\text{-}L$; $\dfrac{L}{R} \gg \dfrac{1}{F_{mains}}$, ignoring commutation effects.

Ideal direct voltage: $\quad U_{di0} = 1.17\, U \qquad$ (for $\alpha = 0°$)

$$U_{di\alpha} = U_{di0}\, \cos(\alpha)$$

U: r.m.s. value of star voltage

Discontinuous - Continuous Operation. When the ohmic portion is dominant in the load, the so-called "discontinuous current" operation occurs. This means when the current becomes too low the thyristor blocks and the current is interrupted until the next thyristor is activated. Conversely, if the load circuit includes a large inductance (coil), then the current conduction of a thyristor normally ends when the next thyristor is fired. The load current thus remains roughly constant (due to the smoothing effect of reactor) and no longer reduces to zero.

In *discontinuous operation* with constant control angle α, the direct voltage $U_{di\alpha}$ is greater than in *continuous operation*. Moreover in the discontinuous range of operation the gain $\Delta I_d / \Delta U_{St}$ depends on the operating state and is generally much lower than in the continuous mode. In addition, in terms of control strategies, the time constant $T_A = L_A/R_A$ of the load is no longer effective. This creates considerable difficulties in terms of control, which can be avoided with special control concepts (adaptive current-control).

Commutation - Commutation Interval. There is another effect called commutation of the valves. The commutation is the transition of the current conduction from one thyristor to the next. Due to the limited di/dt of the current for the thyristors starting conduction, inductors must be inserted at the AC-side. These inductors prevent a sharp change of the current I_d from the conducting thyristor to the next on thyristor, for a brief interval during the transition (called commutation interval) during which two thyristors are conducting the current at the same time. Therefore during switching there is an overlap in the current conduction duration (commutation interval) of the thyristors involved in the commutation.

With the commutation interval $ü = f(I_d, \alpha)$, the output voltage U_d decreases as a function of the load current I_d, this can be explained by the following consideration. During commutation two thyristors conduct simultaneously, the two mains voltages are shorted with two inductors at the AC-side as loads. Since these are equal, the resulting direct voltage U_d, during the interval $ü$, is equal to half the star voltage.

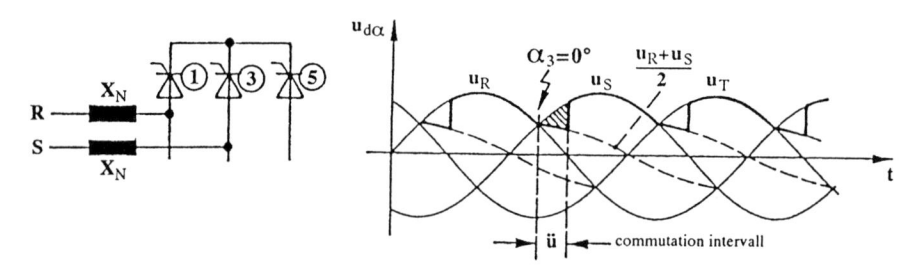

Fig. 4.18. Example: Commutation from thyristor 1 to thyristor 3

Compared with ideal switching without AC-side inductors, a voltage-radiant area is lacking (see hatching), the mean value of the direct voltage therefore is reduced and is lower compared to the no commutation interval. The duration of the commutation interval depends on the load current I_d, on the control angle α, on the rating of the AC-inductors and on the mains voltage. The AC- inductors are characterized by their relative short circuit voltage $u_{k\%}$:

$$u_{k\%} = X_N \frac{S_{Nmains}}{3\,U^2} 100\%; \quad S_{Nmains} = 3\,U I_{Nmains}$$

$$u_{k\%} = 5 \text{ to } 10\%$$

$$u_{k\%} = \frac{I_{Nmain} X_N}{U} 100\%$$

U: phase voltage

Due to the commutation interval $ü = f(I_d, \alpha, u_{k\%})$ we now have:

$$U_d = U_{di0}\cos\alpha - D_x = U_{di0}(\cos\alpha - d_x)$$

$$D_x = d_x U_{di0}; \quad d_x = \frac{D_x}{U_{di0}} = \frac{I_d}{2\sqrt{2} I_K}$$

D_x: inductive voltage drop of the direct voltage
d_x: relative voltage drop of the direct voltage

$$I_K : \frac{U_v}{2X_N}$$

With the mathematical equation between the rated current on the AC- side (I_{Nmains}) and on the direct current side (I_{dN}) of the three-phase "Y" half wave circuit, we can obtain d_x depending on I_d and I_{dN}:

$$I_{Nmains} = \frac{\sqrt{2}}{3} I_{dN} \quad \Rightarrow \quad d_x = \frac{\sqrt{3}}{2} u_{k\%} \frac{I_d}{I_{dN}}$$

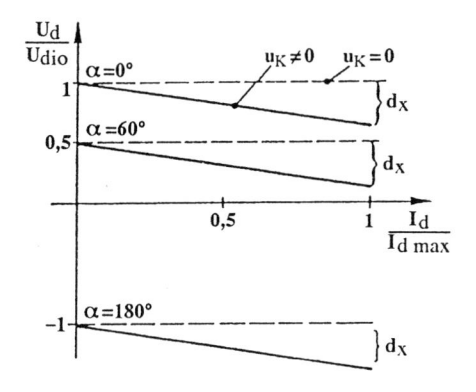

Fig. 4.19. Voltage drop caused by through commutation continuous current in the whole operating range

Inverter Operation-Commutation Failure. By controlling with $\alpha > 90°$, a negative mean value of the direct voltage is obtained from $U_{di\alpha} = U_{di0}\cos\alpha$:

$$U_{di\alpha} < 0: \text{ inverter operation}$$

with $90° < \alpha < 180°$ (150°)

Inverter operation is only possible with a voltage in the load circuit so that the voltage on the thyristors remains positive.

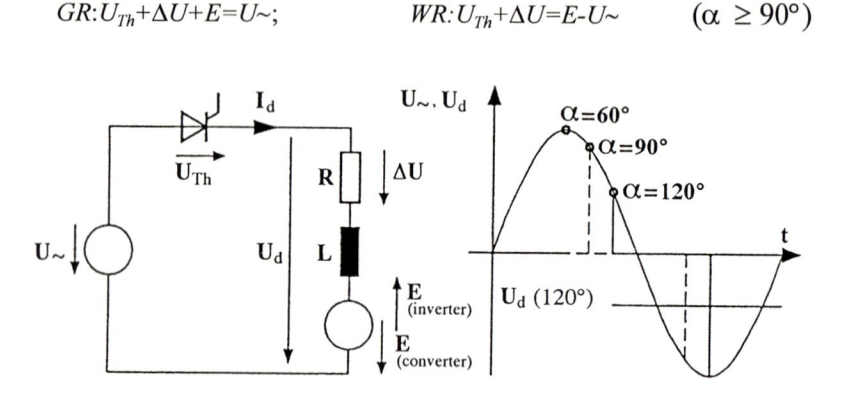

$$GR: U_{Th} + \Delta U + E = U_\sim; \qquad WR: U_{Th} + \Delta U = E - U_\sim \qquad (\alpha \geq 90°)$$

Fig. 4.20. Voltage polarity and operating behaviour in the rectifier and inverter operation

Control angle of $\alpha > 180°$ cannot be achieved since there is a negative voltage at the time of firing of the thyristor to be activated. As a result, commutation does not occur and the thyristor which was conducting current until that instant does not de-activate.

To prevent this effect, known as *Commutation failure*, a safety distance must also be maintained in relation to the limit angle $\alpha = 180°$. The reason for this is based on the fact that the commutation interval $ü$ and the protection time t_s of the thyristor to regain forward blocking capability must have elapsed before, starting with $\alpha = 180°$, the voltage of the conducting thyristor becomes positive again. If the commutation is not finished (commutation interval $ü$) before $\alpha = 180°$, the thyristor, conducting current up to then, will continue to conduct and the thyristor having been activated will go into the off-state again. In this operating state the counter-electromotoric force of a DC-machine for example and the output voltage of the converter will be added together starting from $\alpha = 180°$ and a very high load current will be the result, generally leading to damage in the converter and/or the load.

If the protection period t_s does not end before $\alpha = 180°$, the thyristor that was conducting current before can still not be able to block positive voltages, i.e. it can switch on again, without firing pulse but with positive voltage U_{AK}. The result is the same situation as described above. Therefore a safety distance must be maintained $\Delta\alpha = ü + \gamma$ in relation to $\alpha = 180°$.

$$\alpha_{max} = 180° - ü - \gamma \;.$$

In practice, a maximum control angle $\alpha_{max} = 150°$ is assumed to be safe.

4.2.2 Single-Phase Converters

(a) voltages and currents (continuous DC-current) single-phase, full-wave converter (common cathode) and centre tap transformer:

(2) ideal no load average voltage : $\quad U_{di0} = \dfrac{2\sqrt{2}}{\pi} U = 0.90\, U$

(maximum DC-voltage at $\alpha = 0°$)

(3) „inductive voltage drop" $\quad d_x = \dfrac{1}{\sqrt{2}}\, u_{k\%}\, \dfrac{I_d}{I_{dN}}$

(commutation)

(4) average voltage
without commutation: $\quad U_{di0} = U_{di0}\cos\alpha$
with commutation: $\quad U_d = U_{di0}(\cos\alpha - d_x)$

(5) peak reverse voltage: $\quad \hat{U}_T = 2\,\hat{U} = 3.14\, U_{di0}$

(6) r.m.s. current in the valves: $\quad I_T = \dfrac{1}{\sqrt{2}} I_d$

(7) average current in the valves: $\quad I_{TAV} = \dfrac{1}{2} I_d$

(8) harmonic ripple $(\alpha = 0°)\quad U_{ü} = 0.48\, U_{di0}$

(b) Voltages and current at single-phase bridge converter (continuous current)
(2) ideal no-load average voltage: $U_{di0} = U_{di0}$ as in 4.2.2a (max. DC voltage)

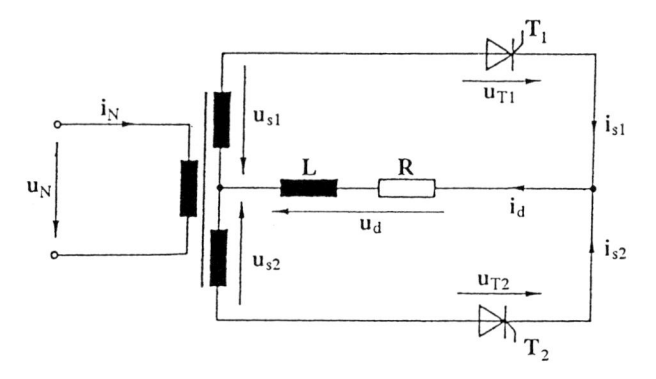

Fig. 4.21. single-phase full wave converter

(3) "inductive voltage drop": $d_x = d_x$ as in 4.2.2a
    (commutation)
(4) average voltage
    without commutation: $U_{di0} = U_{di0}$ as in 4.2.2a
    with commutation: $U_d = U_d$ as in 4.2.2.a
(5) peak reverse voltage: $\hat{U}_T = \hat{U} = 1.57\ U_{di0}$
(6) r.m.s. current in the valves: $I_T = I_T$ as in 4.2.2a
(7) average current in the valves: $I_{TAV} = I_{TAV}$ as in 4.2.2a
(8) harmonic ripple ($\alpha = 0$) $U_\ddot{u} = U_\ddot{u}$ as in 4.2.2a

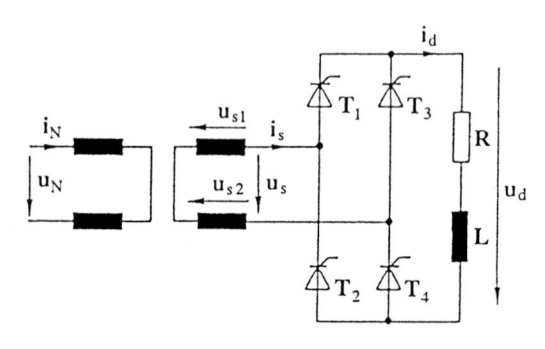

Fig. 4.22. Single-phase bridge converter

(c) Voltages and currents; single-phase bridge with common anodes diodes (continuous current)

(2) ideal no-load average voltage: $U_{di0} = U_{di0}$ as in 4.2.2a
    (maximum dc-voltage $\alpha = 0$)
(3) "inductive voltage drop"
    (commutation) $d_x = \dfrac{1}{4\sqrt{2}}\, u_{k\%}\, \dfrac{I_d}{I_{dN}}$
(4) average voltage
    without commutation: $U_{di0} = U_{di0}\, \dfrac{1+\cos\alpha}{2}$
    with commutation: $U_d = U_{di0}\left(\dfrac{1+\cos\alpha}{2} - d_x\right)$
(5) peak reverse voltage: $\hat{U}_T = \hat{U} = 1.57\ U_{di0}$

(6) r.m.s. current in the valves
(thyristors, diodes): $I_T = I_T$ as in 4.2.2a

(7) average current in the valves: $I_{TAV} = I_{TAV}$ as in 4.2.2a

(8) harmonic ripple ($\alpha = 0$) $U_{ü} = U_{ü}$ as in 4.2.2a

This solution has a low reactive power ($f(\alpha)$) at the line side

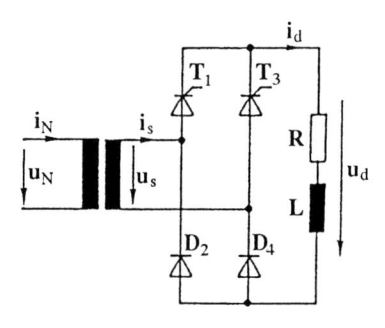

Fig. 4.23. Single-phase bridge converter (common anode diodes)

(d) Voltages and currents; single-phase free wheeling bridge converter (continous current)

(2) ideal no-load average voltage $U_{di0} = U_{di0}$ as in 4.2.2a
(maximum DC-voltage at $\alpha = 0°$)
(3) inductive voltage drop
(commutation)
durch commutation: $d_x = d_x$ as in 4.2.2 c
(4) average voltage
without commutation $U_{di0} = U_{di0}$ as in 4.2.2 c
with commutation: $U_d = U_d$ as in 4.2.2 c
(5) peak reverse voltage: $\hat{U}_T = \hat{U}$ as in 4.2.2 c

(6) r.m.s. current in the valves:

thyristors: $I_T = \sqrt{\dfrac{\pi - \alpha}{2\pi}} I_d$

diodes: $I_T = \sqrt{\dfrac{\pi + \alpha}{2\pi}} I_d$

(7) average current in the valves:

thyristors: $$I_{TAV} = \frac{\pi - \alpha}{2\pi} I_d$$

diodes: $$I_{TAV} = \frac{\pi + \alpha}{2\pi} I_d$$

(8) harmonic ripple ($\alpha = 0$) $U_ü = U_ü$ as in 4.2.2a

This solution has a low reactive power ($f(\alpha)$) at the line side and a free wheeling capability at the side.

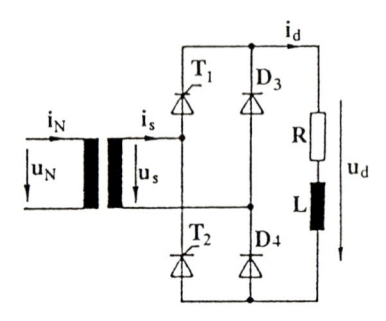

Fig. 4.24. Single phase freewheeling bridge converter

4.2.3 Three-Phase bridge-Converter

The three-phase bridge converter (a six pulse converter) can be considered as a series connection of the two three-phase "Y" half wave converters, the first with common cathodes and the second with common anodes.

Voltages and currents (continuous current) of a six pulse bridge converter

(1) line-to-line voltage: $U_v = \sqrt{3}\ U$ U: star-voltage
   (r.m.s. value)

(2) ideal no-load average voltage: $U_{di0} = \frac{3\sqrt{2}}{\pi} U_v = 1.35\ U_v$

   (maximum voltage at $\alpha = 0$)

$$= \frac{3\sqrt{6}}{\pi} U = 2\ 1.17\ U$$

   six pulse: connection of two converters (section 4.2.1)

(3) "inductive voltage drop"

   commutation: $d_x = \frac{1}{2} u_{k\%} \frac{I_d}{I_{dN}}$

(4) average voltage
without commutation: $U_{di0} = U_{di0} \cos\alpha$
with commutation: $U_d = U_{di0}(\cos\alpha - d_x)$

(5) peak reverse voltage: $\hat{U}_T = \hat{U} = 1.05\, U_{di0}$

(6) r.m.s. current in the valves: $I_T = \dfrac{1}{\sqrt{3}} I_d$

(7) average current in the valves: $I_{TAV} = \dfrac{1}{3} I_d$

(8) harmonic ripple ($\alpha = 0$) $U_ü = 0.042\, U_{di0}$
(low harmonic content)

three-phase common cathode circuit
Pulse number six

three-phase common anode circuit

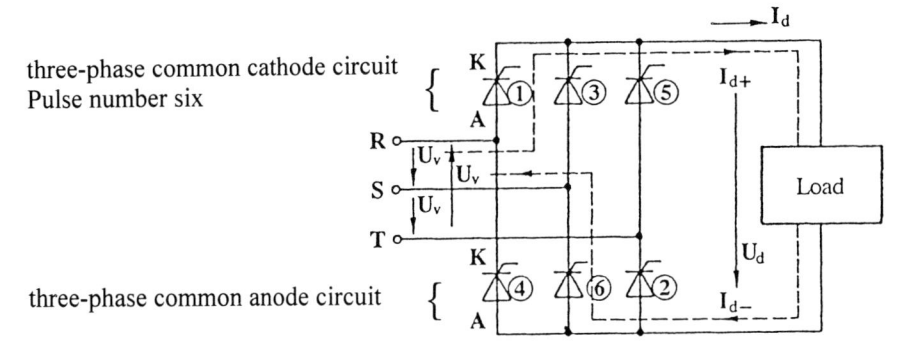

Fig. 4.25. Three-phase bridge converter

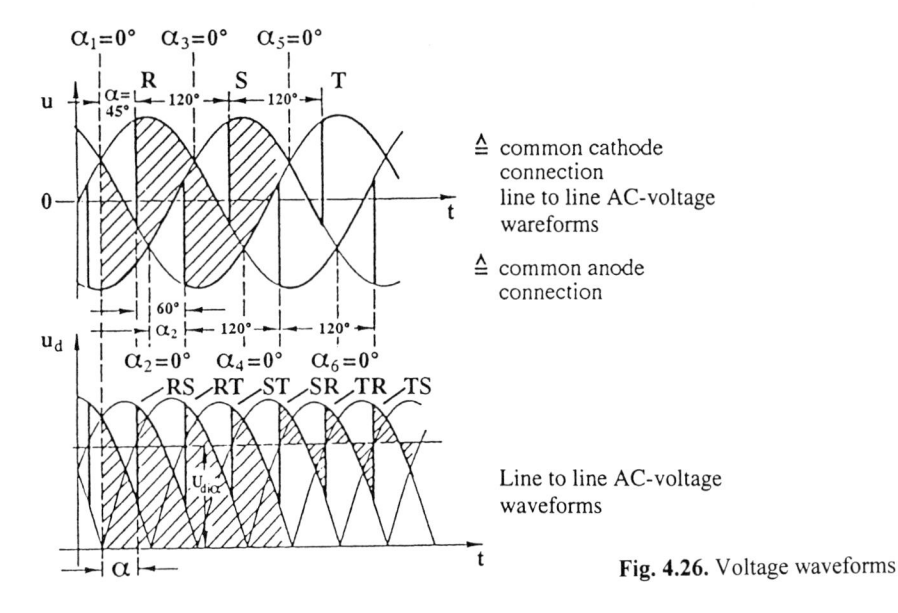

$\triangleq$ common cathode connection
line to line AC-voltage waveforms

$\triangleq$ common anode connection

Line to line AC-voltage waveforms

Fig. 4.26. Voltage waveforms

450   Dierk Schröder Part IV

4.2.4 Operating Limits - Converter and Machine

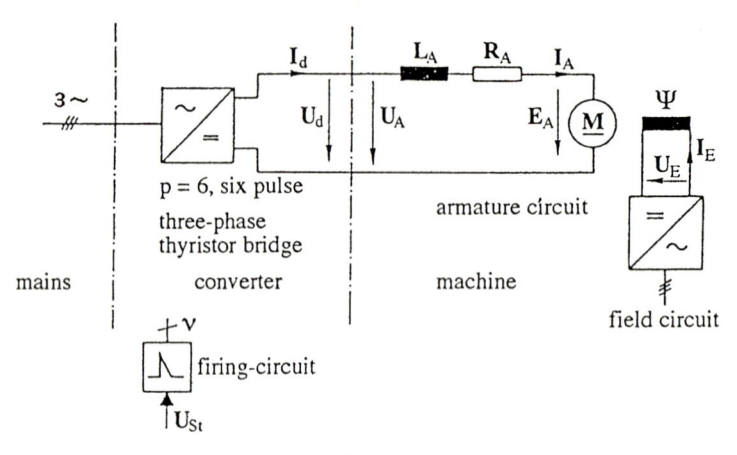

Fig. 4.27. System: converter- machine

(A) Armature control range $\Psi = \Psi_N$; $I_E = I_{EN}$.

(A1) Rectifier in first quadrant operation (rectifier operation, motor - right hand rotation)

Converter: $0° \simeq \alpha \simeq 90°$; $0 \simeq U_d \simeq U_{d\max\,(\alpha_{\min})}$

with :

$$U_d = U_{di0} \left(\cos\alpha - \underbrace{\frac{1}{2} u_{k\%} \frac{I_d}{I_{dN}}}_{d_x} \right) = U_{di\alpha} - D_x$$

Stationary equation of the machine characteristic:

$$N = \frac{U_A}{C_E \Psi} - T_{Mi} \frac{R_A}{C_E C_M \Psi^2}$$

with:

$$\Psi = \Psi_N, \quad T_{Mi} = C_M I_A \Psi_N .$$

Power supply by converter: $U_A = U_d$, $I_A = I_d$

$$N = \frac{U_d}{C_E \Psi_N} - I_d \frac{R_A}{C_E \Psi_N} = \frac{U_d}{C_E \Psi_N} - T_{Mi} \frac{R_A}{C_E C_M \Psi_N^2}$$

$$N = \frac{U_{di0} \cos\alpha}{C_E \Psi_N} - I_d \left(\frac{1}{2} u_{k\%} \frac{U_{di0}}{I_{dN}} + R_A \right) \frac{1}{C_E \Psi_N}.$$

The effects of the inductive voltage drop (converter) and the armature resistance (machine) voltage drop are added up.

Upper operating limit:

Rectifier with $\alpha = \alpha_{min}$ $U_d(\alpha_{min}) = U_{d\,max}(I_{A\,max})$

Cut off characteristic curve:

$$N = \frac{U_{d\,max}}{C_E \Psi_N} - I_d \frac{R_A}{C_E \Psi_N} = U_{di0} \frac{\cos\alpha - d_x}{C_E \Psi_N} - I_d \frac{R_A}{C_E \Psi_N}$$

$$N_N = \frac{U_{d\,max}}{C_E \Psi_N} - T_{MN} \frac{R_A}{C_E C_M \Psi_N^2}; \quad U_{d\,max} = f(I_d) = f(T_M)$$

Note: for good dynamic response provision must be made for a reserve capacity in the operating range of the converter. Therefore the converter is not used in stationary operations up to its maximum potential. Therefore in general: $\alpha_{min} > 0°$ and $U_{d\,max} < U_{di0}$. In the case of a four-quadrant converter with two three-phase antiparallel bridges, the minimum control angle α_{min} is limited to $\alpha_{min} \simeq 30°$ (symmetrically in relation to α_{max} for inverter operation, $\alpha_{max} = 150°$) (see section 4.2.1).

Converter
with inductive voltage drop

System
converter and DC-machine

$$U_{di\alpha} = U_{di0} \cos\alpha$$

$$U_d = U_{d\alpha} - D_x$$

Note: Two lines identified by —++— are parallel.

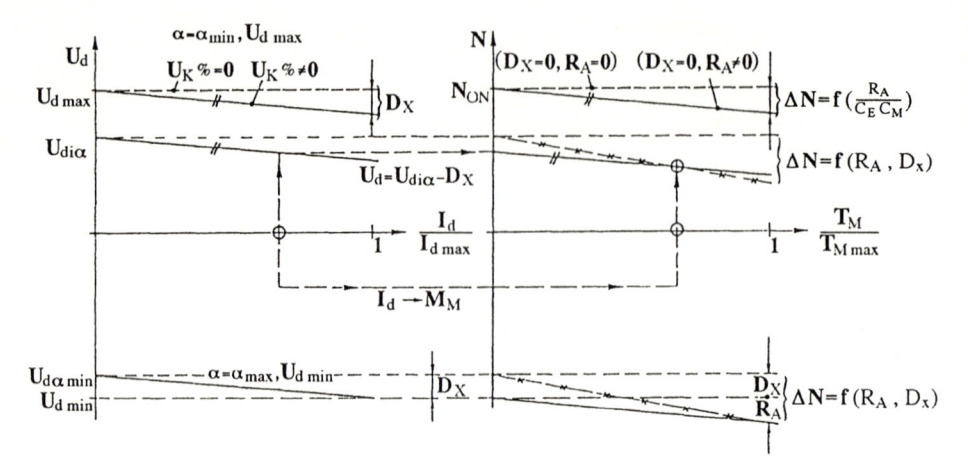

Fig. 4.28. Resulting effects of D_x and R_A on the speed

(A2) Converter-machine system in the fourth quadrant operation
(inverter operation, generator, left-hand speed).

$$\text{Converter: } 90° < \alpha \leq 150°; U_{d\min} \leq U_d < 0$$

$\alpha_{max} \approx 150°!$
the same characteristics as in A1 are obtained; in particular, the *lower limit* is given by:

$$\text{Converter with } \alpha = \alpha_{max} = 150°, U_{d(\alpha_{max})} = U_{d\min} < 0!$$

Limit characteristic:

$$N = \frac{U_{d\min}}{C_E \Psi_N} - I_d \frac{R_A}{C_E \Psi_N}$$

with $U_{d\min} = U_{di0} \cos\alpha_{max} - D_x = U_{d\alpha\min} - D_x$

F Field weakening

F1 Armature converter in first quadrant

Converter: $\alpha \simeq 30° = \text{const.}; U_d = U^*_{d\max}\big|_{I_d = I_A} = U_{di0} \cos\alpha - D_x$

$\alpha \neq 0$ reserve capability in order to conduct the current $I_{A\max}$!

$$N = \frac{U^*_{d\max}}{C_E \Psi} - I_d \frac{R_A}{C_E \Psi}; \quad T_M = C_M \Psi I_d$$

$$N = \frac{U^*_{d\max} - I_d R_A}{\dfrac{C_E}{C_M} \dfrac{T_M}{I_d}}; \quad \frac{C_E}{C_M} = 2\pi$$

$$N = \frac{\left(U^*_{d\max} - I_d R_A\right) I_d}{2\pi T_M}$$

for $I_A = I_{A\max} = I_{d\max}$ the fundamental relations in steady state are:

$$N \approx \frac{1}{\Psi}$$

$$T_M \approx \Psi$$

F2 Armature converter in fourth quadrant operation

Converter: $\alpha = 150°$; $U_d = U^*_{d\min} = U_{d\min}\big|_{I_d = I_A} = U_{di0} \cos\alpha - D_x$

$$N = \frac{U^*_{d\min}}{C_E \Psi} - I_d \frac{R_A}{C_E \Psi}$$

$$N = \frac{U^*_{d\min} - I_d R_A}{\dfrac{C_E}{C_M} \dfrac{T_M}{I_d}} = \frac{\left(U^*_{d\min} - I_d R_A\right) I_d}{2\pi T_M}$$

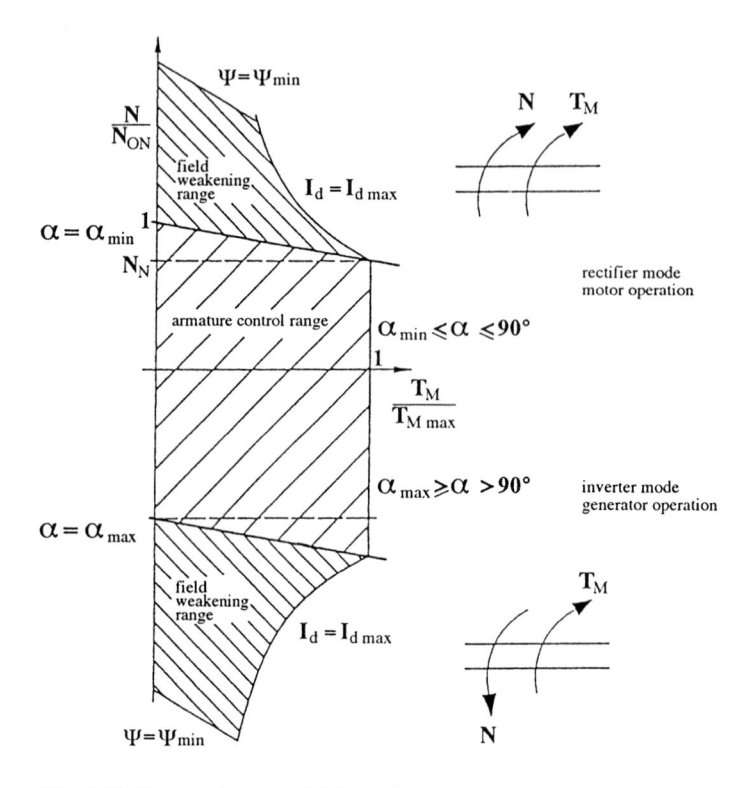

Fig. 4.29. Separately excited DC-machine operating ranges

4.2.5 Torque Reversal of DC-Machines Supplied by Converters

A multitude of DC-drives requires a reversal of torque either occasionally or under normal operating conditions. In contrast to DC-machines supplied by DC-generators, the direct current cannot change the polarity in a converter because of the valve effect of the semiconductors. Accordingly, a reversal of torque for these systems is only possible if particular steps are taken. The torque in a DC-motor is proportional to the product of armature current and flux. Since the flux is determined by the field current, the torque of a DC-motor may change direction either by *reversal of the armature current or by reversal of the field current* and thus of the flux.

$$+T = (+J_A)(+\psi)$$

$$-T = (+J_A)(-\psi) \quad \text{(field current reversal)}$$

$$-T = (-J_A)(+\psi) \quad \text{(armature current reversal)}$$

Torque Reversal by Reversing the Armature Current.
(a) Armature reversal with changeover switch:

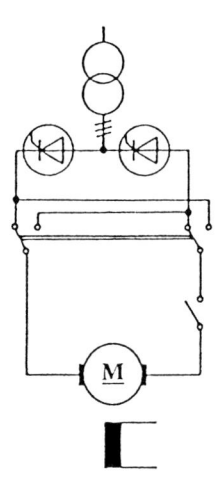

Fig. 4.30. Reversal of armature current with contactor

The simplest way of reversing the armature current in the machine is by using the armature changeover switch. This is a very inexpensive solution as only a two-quadrant converter is required. Since the mechanical switch is only activated in the absence of any current, pole reversing switches or air-break contactors can also be used for this purpose. However, mechanical switches are subject to more frequent failures due to the mechanically moving components and their wear. Therefore they are unsuitable for application with frequent torque reversal. Moreover, the *interval without armature current* derived from the switch-over *duration is from 100 to 500 msec.* Therefore this solution is only used at low power and only if the interval without current ("torque gap") can be accepted.

During switching of the contactor the control signal of the converter must be reversed simultaneously (e.g.: from rectifier- to inverter-mode) since otherwise the converter voltage U_d and the induced machine voltage E_A are in series and in the same polarity and produce a dangerous current.

(b) Four-quadrant armature converter: four-quadrant converters have valves for each direction of the current. A distinction is made between the cross connection and direct antiparallel converters. The cross connection is a circuit *with circulating current* and the second solution is *without circulating current.*

(c) Four quadrant converter without circulating current: the reversal of the armature current is performed electronically since a second antiparallel converter bridge accepts the negative load current. Of the two six pulse bridges in the example only one is always active, while the firing pulses of the other are blocked. The above mentioned "torque gap" during the reversal still exists because it is necessary to accept the recovery time of the valves to achieve the positive blocking capability. But the zero interval in the current reversal is only 1 ms to 6.6 ms.

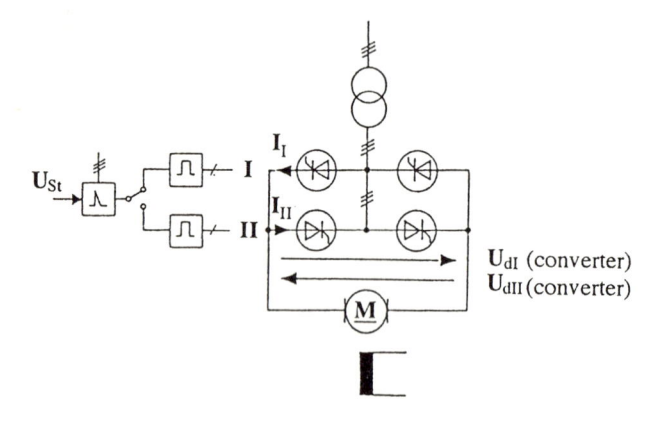

Fig. 4.31. Four quadrant converter without circulating current

Here too, in the bridge changeover, there must be a reversal from rectifier- to inverter-operation to avoid overcurrents. Since the current in the converters, in the event of low currents, operate in the discontinuous mode, the dynamic behaviour worsens if a non-adaptive current control is used so, at the time of current reversal, the reversal times are much longer and this in turn leads to much longer torque reversal times. Using an adaptive current controller, performance can be improved in the case of discontinuous armature current. In general, if the controller parameters are set exactly, current response times of less than 6 ms can be achieved for continuous armature current. If a good adaptive current controller is used, the current response time will increase to approximately 8 ms. If a fast zero current detection circuit is used, the zero current interval can be adjusted to less than 1.5 ms (depending on the recovery time t_q of the thyristors used). The zero current interval that actually occurs basically depends on the required current changeover instant and is therefore less than 3 ms. The difference in dynamics compared with the circulating current converters is so small that the latter is practically no longer used.

Due to its simple and inexpensive structure, the four quadrant converter without circulating current has become the standard solution in practical applications.

Torque Reversal by Reversing the Field Current. General Note: As a general rule the same circuits are used as for the armature circuit. However, since the excitation circuits have low power compared to the armature circuits, expenses are much lower. In the following examples it is assumed that the armature circuit is supplied by only one six pulse bridge (positive current, positive and negative voltage). All the methods based on field reversal are linked by the fact that during the reversal at $\Psi \approx 0$ a critical working condition arises in the machine unless counter-measures are taken. The explanation according to the stationary characteristic curve is:

$$n = \frac{u}{\psi} - t_M \frac{r_A}{\psi^2}.$$

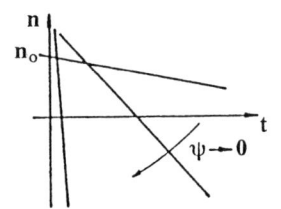

Fig. 4.32. DC-machine group characteristics with ψ as a parameter

In the case of a loadless machine ($t_w \to 0$) and constant armature voltage, the term $n_0 = \dfrac{u}{\psi}\bigg|_{\psi \to 0}$ tends toward infinity. Therefore the speed of the machine increases greatly and the machine is destroyed by the overspeed (centrifugal forces).

The counter-measure to prevent this effect entails zero-setting of the armature current by means of a corresponding control of the armature current and thus the motor's torque. This prevents an acceleration torque of the machine. This solution ensures stability but a torque gap also exists. The resulting delay time is in the 100 ms to 2s range, which cannot be considerably reduced due to the large inductance of the field winding which does not permit any rapid field dynamics, or only at the cost of an excessive voltage (dynamic overvoltage of the exciting voltage). Accordingly, in the case of field reversal, a good dynamic response cannot be achieved with this solution.

Moreover, it should be remembered that field reversal must always be related to a reversal of the control signal of the armature converter. In general, torque reversal by inverting the field is seldom used.

4.3 Control of DC-Drives (Current-Control, Speed Control)

Because of the poor dynamics response achieved by the flux control and the non-linear characteristic (hysteresis), the DC-machine is usually controlled by armature variables.

The most widely used method comprises:

Cascade control: inner control loop - current control;
                     superimposed control loop - speed control

The cascade control is frequently used for converter supplied drives, because:

(a) The DC-machine and the converter are sensitive to excessive armature currents I_A;
(b) the DC-machine is sensitive to excessive armature current variations di_A/dt;
(c) the DC-machine (and the converter) is sensitive to excessive armature voltages U_A;
(d) The armature current of the DC-machine represents an important state variable
$$\left(t_M = \psi\, i_A\right) ;$$
(e) the speed controller is superimposed onto the current controller and the control circuits can be activated step by step.

Conclusion: (speed controlled drives)

I: Limiting the armature current according to the amplitude and possibly the first derivation $\Rightarrow$ limiting and controlling the armature current
II: Limiting and regulating speed

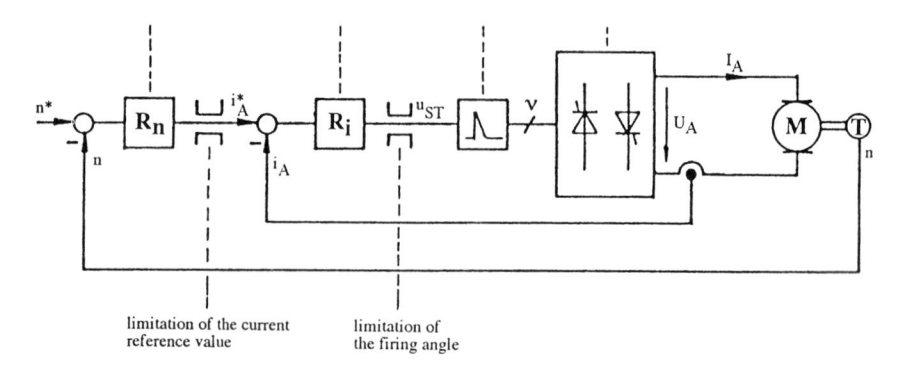

Fig. 4.33. Separately excited Dc-machine with cascade control

4 Converters and DC-Machine Control   459

Both the signal-flow graph in Chapter 3 for the DC-machine and the most important transfer functions in section 3.2 have been described. In this chapter we will now derive the basic principles for the control of the DC-machine. As explained, cascade control is frequently used (Fig.4.33). Fig. 4.33 shows that the speed control loop is superimposed onto the current control loop. In order to limit the current i_A in the DC-machine, the current reference value i_A^* is limited.

Advantages:

1. Reduction of the complexity of the transfer functions of the DC-machine→ simple controllers and optimization rules
2. Excellent disturbance rejection
3. The controlled signals can be limited and thus the real values
4. The effects of non-linear or discontinuous operating components in the control loop are reduced
5. Gradual putting into operation

4.3.1 Control of the Armature Current

Fig. 4.34 shows the current control loop. Without going into particular deductions for an approximation of the dynamic behaviour of the converter, and without inferring optimization rules, the essential points only are represented. The converter is regarded as a dead time with the gain v_{Str}. The dead time T_t depends on the converter used and is stated for example with $T_t = \dfrac{T_N}{2p}$ in the worst case scenario of line commutated converters.

According to Figure 4.34 the open current control loop comprises the transfer functions $G_{Str}(s)$ of the converter, the armature circuit with $G_{S1} = \dfrac{1}{r_A(1+sT_A)}$, the transfer function of the rotor $G_{S2} = \dfrac{1}{sT_{\Theta N}}$ and the feedback channel by the e.m.f.

Because of this feedback the following transfer function results:

$$G_S(s) = \frac{i_A(s)}{u_{St}(s)} = \frac{G_{Str}(s)\, G_{S1}(s)}{1 + G_{S1}(s)\, G_{S2}(s)}\bigg|_{\psi=1}$$

$$= v_{Str}\, e^{-sT_t}\, \frac{sT_{\Theta N}}{r_A(sT_{\Theta N} + s^2 T_{\Theta N} T_A) + 1}.$$

This transfer function is undesirable since it may contain a conjugate pair of poles. Therefore the design of the controller is difficult. Moreover, a controller with an integrating behaviour would take the order of the closed control loop to the third order and would have a negative effect on the dynamics of the closed control loop.

To avoid these disadvantages a disturbance feedforward of the e.m.f (dotted line) is recommended. Due to this the internal feedback branch is compensated with e_A. A transfer function $1/v_{Str}$ must be introduced additionally for this branch to ensure compensation of the e.m.f. and to prevent positive feed forward. If these conditions are observed the transfer function is reduced to (Figure 4.34):

$$G_S(s)\big|_{e.m.f.} = v_{Str}\, e^{-sTt}\, \frac{1}{r_A\left(1+sT_A\right)}$$

e.m.f. feedforward
(observer
to obtain e_A)

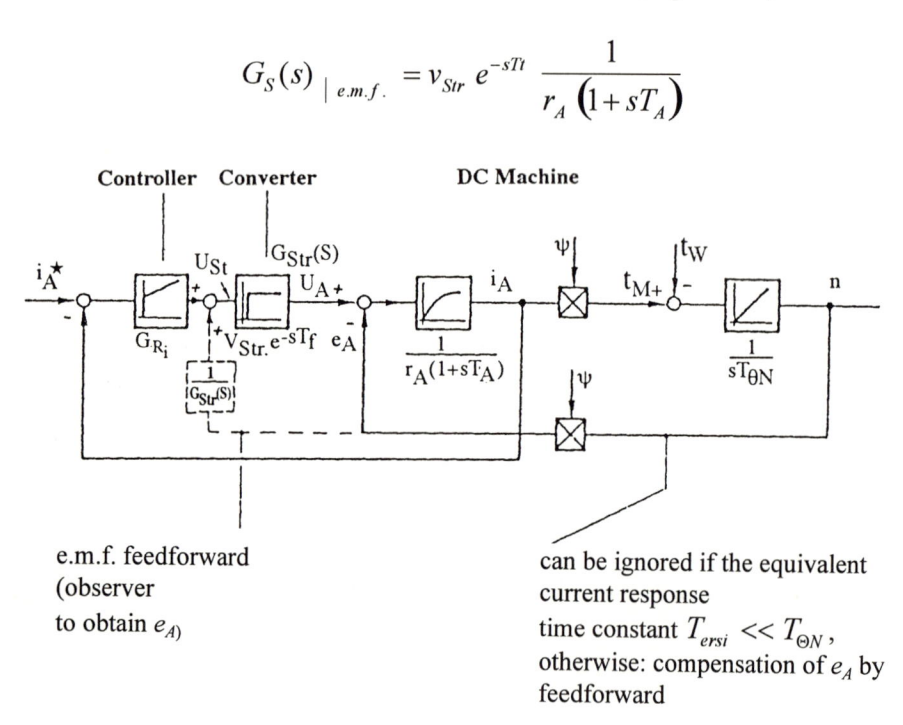

can be ignored if the equivalent current response time constant $T_{ersi} \ll T_{\Theta N}$, otherwise: compensation of e_A by feedforward

Fig. 4.34. Inner control loop - current control

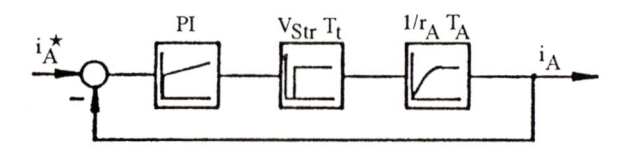

Fig. 4.35. Signal flow graph for the current control circuit with e.m.f. feed forward

In order to simplify the control loop considerably, the dead time is approximated with a first order delay (PT_1):

$$G_S(s)\Big|_{e.m.f.} = \frac{v_{Str}}{r_A} \frac{1}{(1+sT_t)(1+sT_A)} \qquad T_A > T_t .$$

We can now use the BO optimization criterion with these simplifications. In the case of a BO optimum, a PI-controller is used:

$$G_R(s) = K_R \frac{1+sT_R}{s} .$$

The optimization conditions are:

$$T_R = T_A$$

$$K_R = \frac{1}{2T_t} \frac{r_A}{v_{Str}} \quad ; \quad T_t = \frac{T_N}{2p} = \frac{1}{2pF_N} .$$

If the current control loop is optimized, and if no low-pass filter is introduced in the feedback channel to smooth the measured current, then the following variable command control transfer function is obtained:

$$G_{wi}(s) = \frac{1}{1+s\,2T_t + s^2\,2T_t^2} .$$

This transfer function has a complex conjugate pair of poles and damping $d = \frac{1}{\sqrt{2}}$. Figure 4.36 shows the transient response.

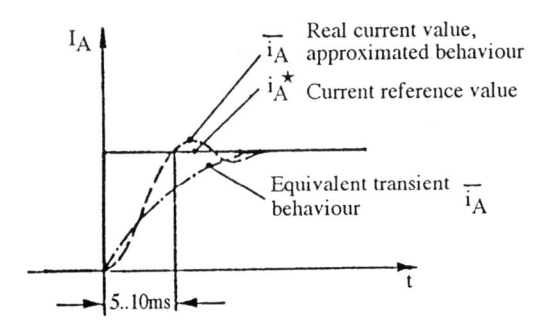

Fig. 4.36. Transient response of current control

To simplify the optimization of the superimposed control loop, generally an equivalent transfer function is chosen (Figure 4.3.5), which is a simplification of the real transfer function.

$$G_{wi}(s) = \frac{1}{1 + s\, 2T_t + s^2\, 2T_t^2} \quad \rightarrow \quad G_{wersi}(s) = \frac{1}{1 + sT_{ersi}}$$

$$\text{approximation} \qquad T_{ersi} = 2\, T_t$$

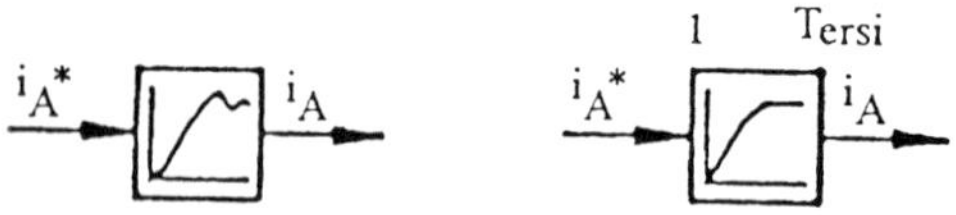

Fig. 4.37. Equivalent transfer function

In many current control loops with converters as the controlling element a first order delay (smoothing) is introduced in the feedback channel. This smoothing is regarded as a prudent measure, but if the sensor and shielding are constructed accurately it is not necessary as far as control is concerned; indeed, it may even be a disadvantage. Supposing a smoothing time constant T_G and $T_G \ll T_A$, BO-optimization would then be modified as follows:

$$T_R = T_A; \quad K_R = \frac{r_A}{v_{Str}} \frac{1}{2(T_t + T_G)} = \frac{r_A}{v_{Str}} \frac{1}{2T_t^*}$$

with $T_t^* = T_t + T_G$.

In this way T_{ersi} deteriorates to T_{ersi}^*. This greater time constant T_t^* must also be considered later in the optimization of the speed control loop, i.e. this dynamic behaviour is also reduced.

Briefly, to return to adaptive current control for discontinuous current. There is an endless number of possible configurations of the controllers and the identification circuits that cannot be discussed here. The basic purpose of adaptive current control is to achieve dynamics, comparable to the dynamics obtained for continuous current. Today this objective has almost been achieved as the detection of the discontinuous current range and the adaptation of the parameters in this range are acceptable from the point of control.

4.3.2 Speed Control

The current control loop (Figure 4.38) is considered solely as an equivalent function G_{wersi} when speed control is investigated. A PI-controller is used for the speed control to ensure stationary precision even with such disturbances as $t_w \neq 0$.

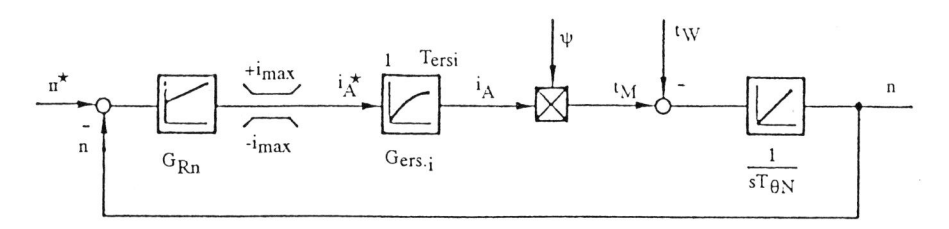

Fig. 4.38. Signal flow graph of the speed control loop

In the following considerations the rated flux ($\psi = 1$) will be assumed. Under these assumptions the following holds true:

$$G_{Sn}(s) = \frac{1}{1+T_{ersi}\ s}\ \frac{1}{sT_{\Theta N}}$$

$$G_{Rn}(s) = K_R\ \frac{1+sT_R}{s}$$

$$-G_0(s) = \frac{K_R}{s^2 T_{\Theta N}}\ \frac{1+sT_R}{1+s\ T_{ersi}}\ .$$

Due to the double integration in the open loop, the phase displacement angle ψ_0 of the open circuit begins with $\varphi_0 = -180°$. In order to stabilize the closed control loop $T_R > T_{ersi}$ must be chosen. The parameters of the controller must be selected according to the so-called symmetrical optimum (SO) rules:

SO - Optimization $\quad T_R = 4\ T_{ersi}$

$$K_R = \frac{T_{\Theta N}}{8\cdot T_{ersi}^2}$$

By choosing the control parameters according to SO, the speed variable command transfer function $G'_{wn}(s)$ is obtained:

$$G'_{wn}(s) = \frac{1+4sT_{ersi}}{1+4sT_{ersi}+8s^2 T_{ersi}^2 + 8s^3 T_{ersi}^3}\ .$$

If the reference signal is a step-function, the real value will respond with enormous overshoot. This overshoot is effected by the *4sT_{ersi}* in the numerator of $G'_{wn}(s)$. To ensure the desired transient response, the zero in the transfer function $G'_{wn}(s)$ must be compensated by a first order delay with a time constant $T_G = 4sT_{ersi}$ (Figure 4.39). The final result is as follows:

$$G_{wn}(s) = \frac{1}{1 + 4sT_{ersi} + 8s^2 T_{ersi}^2 + 8s^3 T_{ersi}^3}$$

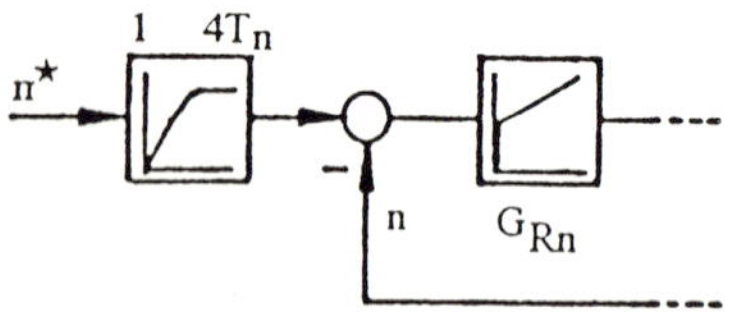

Fig. 4.39. Insertion of a first order delay into the reference channel

The rise time, including the effect of the first order delay in the reference channel is approximately $7\, T_{ersi}$.

$$G_{wn}(s) = \frac{1}{1 + s\, 4T_{ersi} + s^2\, 8T_{ersi}^2 + s^3\, 8T_{ersi}^3} \rightarrow G_{wersn}(s) = \frac{1}{1 + s\, T_{ersn}}$$

$$T_{ersn} = 4\, T_{ersi}.$$

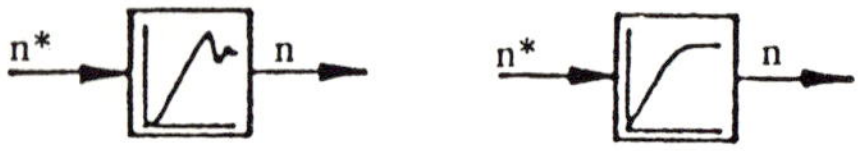

Fig. 4.40. Equivalent transfer function of the speed control loop

In Figure 4.38 the reference value of the current is limited. As we have already mentioned, this limitation is necessary both for the converter and for the machine (collector). However, it should be noted that due to the limitation the speed control loop is no longer closed. This leads to a different behaviour in the entire system compared to the closed system.

If the reference value reaches or exceeds the limiting value, only the limiting value will be effective as the current reference value. This means that with t_W constant a constant acceleration torque is effective during this interval. However, a constant acceleration torque leads to linear speed variation over time (Fig. 4.41).

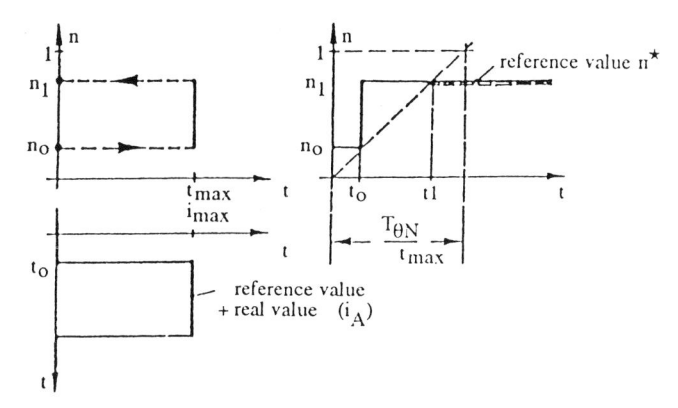

Fig. 4.41. Speed versus time in the case of armature current limitation

It should also be noted that - because of the open speed control loop during the interval of limitation - the transient response of the real speed is different compared to the behaviour based on a closed control loop.

The output signal of the speed controller will increase as a result of the integral action, even though the reference current value i_A^* remains constant on the limiting value. When the real speed has ultimately reached the reference value, the output signal of the controller will be much greater than in the case without limitation. The speed will therefore increase further until the proportional action of the speed controller exceeds the integral action and the acceleration torque changes sign.

The result is excessive overshoot of the real speed, while the control loop was open for the purpose of limitation. To prevent this, the integral action of the speed error in the speed controller must be suspended for the entire duration of the current reference value limitation.

Figure 4.41 also shows the effect of ψ variation. A flux of $\psi < 1$ will lead to a reduction in the torque in the field weakening range:

$$t_{Mi} = i_A \psi .$$

This torque reduction reduces the gain of the open speed control loop or acts comparable to it as an increase of the integral action factor $T_{\Theta N}$:

$$\frac{1}{sT'_{\Theta N}(\psi < 1)} = \frac{\psi}{sT_{\Theta N}(\psi = 1)} .$$

Therefore in the case of $\psi < 1$ the integral action time of the speed controller must be adapted to the modified gain. The simplest way to do this is a division by ψ in the speed controller which compensates the multiplication by ψ in the forward channel.

5 Three-Phase AC-Machines - Signal - Flow Graphs

Rotating field machines, i.e. the asynchronous and synchronous machines including permanent magnet excited synchronous machines (as well as the brushless DC-machines) are increasingly used as controllable drives which are adjustable in speed and torque. The first difficulty, before being able to understand AC-machines, is the three-phase system, i.e. there exist, for example, three voltages, currents and fluxes for the stator and rotor of the induction machine. To obtain a simpler description with just two quantities in a Cartesian system, the space vector theory must be used. But even using the space vector theory, the mathematical equations and related signal-flow graphs are complicated and non-linear. It is only possible with highly specific assumptions to simplify the system of equations and therefore the signal-flow graphs so as to help understanding and explaining the control concepts. A detailed account of the deductions of all of these considerations would go beyond the scope of this handbook. We will leave this task to special literature, such as the Springer series of *Electrical Drives I to IV*. Here we will simply illustrate the crucial steps in the deductions and show the results.

5.1 Space-Vector Theory - Phasors

In order to transform the three-phase system into a space-vector system, a crucial condition must be met, namely that only a symmetrical three-phase system may exist without neutral wire. In this case the assumption that the geometric sum of the voltages and currents is zero applies. This means that if two voltages or currents are known, then the third voltage or current can be calculated exactly. However, this also signifies that with just two known quantities, the third quantity involved in the three-phase system can be calculated each time. It is on this observation that the space-vector theory rests. To explain the basic method, the space-vector of the flux will be determined. In principle, it is known and assumed that the three symmetrical windings of the stator, for example, are supplied by the three voltages of the three-phase voltage system which lead to corresponding currents. In turn, the currents produce the magnetic field strength H, as well as the flux densities B and ultimately, by superposition also the total interlinked flux Ψ.

On the basis of the superposition of the three flux densities B, the definition of the space-vector will be illustrated. As we have said, for each flux density B a sinusoidal spatial distribution in the air gap is assumed. The basic definition of the space-vector theory is, that on the one hand in the time domain the geometric sum of the three flux densities is zero, this also holds for the currents and voltages, if the system is symmetrical and there is no neutral wire. On the other hand, in addition to considering the *phase* displacement of 120° electrically, account must also be taken of the *spatial* displacement of 120° electrically of the windings. This is shown in Fig. 5.1.

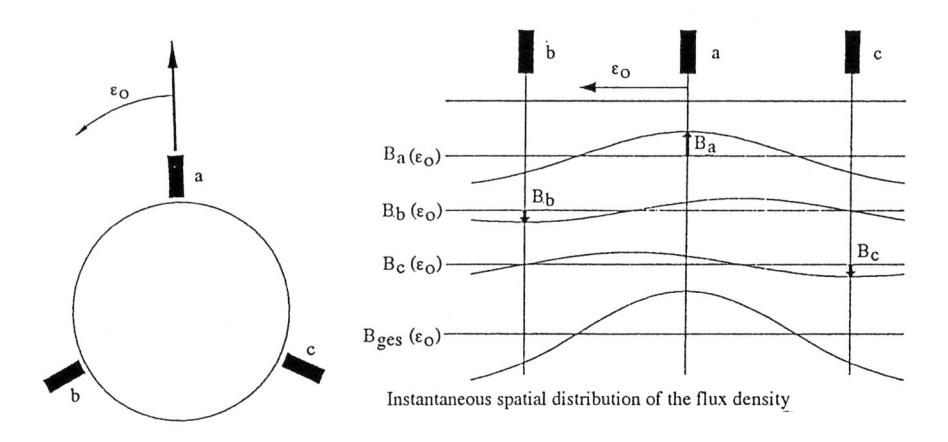

Fig. 5.1. Electrical and spatial displacement of field B

$$B_a(\varepsilon_0) = B_a \cos(\varepsilon_0) = \Re e\{B_a\, e^{j\varepsilon_0}\}$$

$$B_b(\varepsilon_0) = B_b \cos(\varepsilon_0 + 120°) = \Re e\{B_b\, e^{j\varepsilon_0}\, e^{j120°}\}$$

$$B_c(\varepsilon_0) = B_c \cos(\varepsilon_0 + 240°) = \Re e\{B_c\, e^{j\varepsilon_0}\, e^{j240°}\}$$

$$B_{tot}(\varepsilon_0) = B_a(\varepsilon_0) + B_b(\varepsilon_0) + B_c(\varepsilon_0) = \Re e\{(B_a + B_b\, e^{j120°} + B_c\, e^{j240°})\, e^{j\varepsilon_0}\}$$

With the operator $\underline{a} = e^{j120°}$ and $\underline{a}^2 = e^{j240°}$ the complex *space - vector or phasor* is defined:

$$\vec{B} = \frac{2}{3}\left(B_a + \underline{a}\,B_b + \underline{a}^2\,B_c\right).$$

From fig. 5.1 and the equations it is ascertained that owing to the phase and spatial displacement the resulting signal is no longer zero; however, instead we have a signal that rotates in the stationary state with constant amplitude and frequency.

$$B_{ges}(\varepsilon_0) = \Re\left\{\frac{3}{2}\,\vec{B}\,e^{j\varepsilon_0}\right\}.$$

The $\frac{2}{3}$ factor is comprehensible in the definition of the space-vector if a symmetrical three-phase voltage system is supplying a symmetrical load.

Thus we will have:

$$B_a = \hat{B}\,\cos(\Omega t)$$

$$B_b = \hat{B}\,\cos(\Omega t - 120°)$$

$$B_c = \hat{B}\,\cos(\Omega t - 240°)$$

$$\vec{B} = \frac{2}{3}\hat{B}\left(\cos(\Omega t) + \cos(\Omega t - 120°)\,e^{j120°} + \cos(\Omega t - 240°)\,e^{j240°}\right)$$

$$\vec{B} = \frac{2}{3}\left(\frac{3}{2}\hat{B}\,e^{j\Omega t}\right) = \hat{B}\,e^{j\Omega t}.$$

The space - vector can now be split into a real and an imaginary component, if we fix a cartesian coordinate system, for example, on the stator winding system; this coordinate system is said to be fixed to the stator - therefore it is spatially fixed - and has the components B_α and B_β:

$$\vec{B} = \hat{B}\,e^{j\Omega t} = B_\alpha + jB_\beta$$

$$B_\alpha = \hat{B}\cos\Omega t = \Re\{\vec{B}\} = B_\alpha$$

In the same way, B_b and B_c can be calculated. Instead of a spatially fixed coordinate system (fixed to the stator winding R for the real axis), rotating coordinate systems can also be used, for example, on the rotor windings or they can be defined optionally as required. Additionally space-vectors of different coordinate systems can be converted from one system to the other.

The space vector can also be differentiated. However, in this case care must be taken since, depending on the coordinate system chosen, the space-vector involved has an amplitude and possibly a rotating operator, so that differentiating must be made according to the amplitude and rotating operator (product rule).

As we have said, the details of the space-vector theory will not be described here in detail. The specialist literature should be consulted for further details.

5.2 General Rotating Field Machine

In the general rotating field machine one must start with the assumption that there is a three-phase winding system in the stator and a three-phase winding system in the rotor. Both winding systems are supplied by symmetrical and independent three-phase voltage systems. The following conditions apply to the general rotating field machine:

(a) saturation of the magnetic circuits will be neglected and the characteristic of magnetization will be linear;
(b) the spatial distributed windings will be replaced by concentrated windings;
(c) all the spatially distributed signals have a sinusoidal course in the air gap (for instance flux density) with the fundamental frequency;
(d) iron losses and the skin effect will be neglected, as will the friction and the torque of the ventilating system;
(e) the resistances and inductances will not depend on the temperature;
(f) the rotor parameters and signals will be expressed in terms of the stator.

The purpose of this chapter is to derive a general signal-flow graph of rotating field machines. This general signal-flow graph can then be used to deduce particular signal-flow graphs for synchronous and asynchronous machines.

The space-vector theory will be used to derive the signal-flow graph.

Fig. 5.2 shows the structure of the machine.

The following superscripts and subscripts are used:

superscripts:
S   phasor in a stator winding fixed coordinate system
L   phasor in a rotor winding fixed coordinate system
K   phasor fixed in an optional rotating coordinate system

subscripts:
1   Stator signal or parameter
2   Rotor signal or parameter

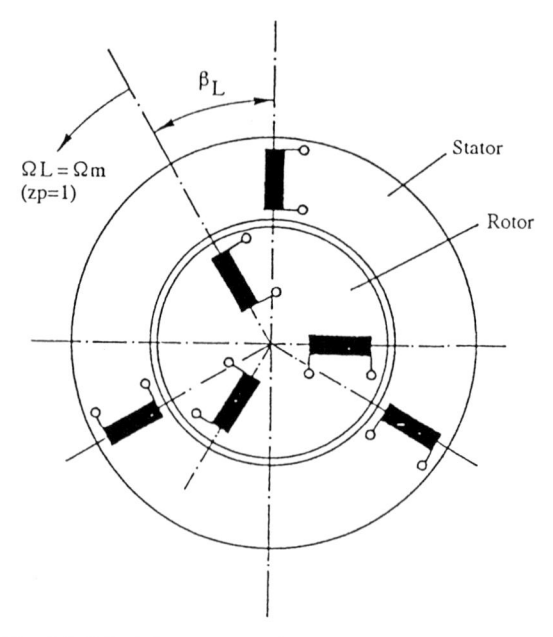

Fig. 5.2. Principle structure of a general rotating field machine

Components of the complex phasors:

α: real component
β: imaginary component
$\Big\}$ in stator winding fixed coordinate system

A: real component
B: imaginary component
$\Big\}$ in an optional rotating coordinate system

There are now four groups of equations:
Voltage equations:

$$\vec{U}_1^s = \vec{I}_1^s\, R_1 + \frac{d\vec{\Psi}_1^s}{dt}$$ voltage differential equations of the stator circuit

$$\vec{U}_2^L = \vec{I}_2^L\, R_2 + \frac{d\vec{\Psi}_2^L}{dt}$$ voltage differential equation of the rotor circuit.

In the general rotating field machine both of the winding systems will be supplied, i.e. the two supply voltage systems will produce currents in each winding system. These currents produce fluxes on the stator side and on the rotor side. In squirrel cage machines $\vec{U}_2^L = 0$, and in slip-ring-machines $\vec{U}_2^L$ or $\vec{I}_2^L$ may be applied.

In the flux interlinkage equations it must be remembered that, because of the slip frequency, the concentrated winding systems are in a time-variant angular position to each other. The time-variant angle between the angular position of the stator fixed phasor and the angular position of the rotor fixed phasor will be β_L.

This gives the *flux interlinkage equations* for:

$$\vec{\Psi}_1^S = L_1 \vec{I}_1^S + M e^{j\beta_L} \vec{I}_2^L$$

$$\vec{\Psi}_2^L = M e^{-j\beta_L} \vec{I}_1^S + L_2 \vec{I}_2^L$$

L_1: inherent inductance of the stator winding

L_2: inherent inductance of the rotor winding

M: maximum mutual inductance between stator and rotor

β_L: angle between stator fixed coordinate system and rotor fixed coordinate system

$M e^{j\beta_L}$: time-variant mutual inductance

The air gap torque is:

$$T_{Mi} = \frac{3}{2} Z_p \, \Im m \left[\vec{\Psi}_1^{*S} \vec{I}_1^S \right] = -\frac{3}{2} Z_p \, \Im m \left[\vec{\Psi}_2^{*L} \vec{I}_2^L \right]$$

Z_p: Pole pair number; $Z_p = \Omega_{el} / \Omega_m$

*: complex conjugate phasor

The last equation in the rotating field machine is the *mechanical equation*:

$$J \frac{d\Omega_m}{dt} = (T_{Mi} - T_W)$$

472   Dierk Schröder, Part IV

Ω_m: mechanical angular velocity of rotor
Ω_{el}: electrical angular velocity of rotor

$$\Omega_m = \frac{\Omega_{el}}{Z_p}; \quad \Omega_m = 2\pi N \left[\frac{1}{s}\right]$$

$$\Omega_{el} = \Omega_L = \frac{d\beta_L}{dt}.$$

Up until now all the equations have been represented in their own coordinate system. If we wish to develop a signal-flow graph for the rotating field machine, then the equations must be transformed into a common coordinate system. When we want a stator fixed coordinate system for the signal-flow graph, then the rotor variables must be converted to the stator system. Similarly, for a rotor fixed coordinate system the stator variables must be converted to the rotor system.

In the general rotating field machine it must be kept in mind that the angle β_L is time variant.

In the following deductions the stator fixed coordinate system will serve first as a reference basis. Without going into the detail of the mathematical deduction, the first voltage equation is:

$$\vec{U}_1^S = \vec{I}_1^S R_1 + L_1 \frac{d\vec{I}_1^S}{dt} + M \frac{d\vec{I}_2^S}{dt}$$

  ohmic   inductive   induced voltage
                     voltage drop   from the rotor current
                     stator side

If fluxes are used instead of currents, then the following holds:

$$\vec{U}_1^S = \frac{d\vec{\Psi}_1^S}{dt} + \vec{\Psi}_1^S \frac{R_1}{\sigma L_1} - \vec{\Psi}_2^S \frac{MR_1}{\sigma L_1 L_2}$$

$$\vec{U}_2^S = \frac{d\vec{\Psi}_2^S}{dt} - j\Omega_L \vec{\Psi}_2^S + \vec{\Psi}_2^S \frac{R_2}{\sigma L_2} - \vec{\Psi}_1^S \frac{MR_2}{\sigma L_1 L_2}.$$

As has been said, the stator fixed coordinate system is aligned on the winding axis R of the stator, which is spatially fixed. For this reason all the variables

$\vec{U}^S, \vec{I}^S, \vec{\Psi}^S$ are sinusoidal in stationary state in the time domain, and therefore, for example:

$$\vec{U} = |U| \sin\Omega_1 t \stackrel{\wedge}{=} |U| e^{j\Omega_1 t}.$$

If we now assume a rotating coordinate system that rotates with the angular velocity Ω_1 and coincides with the R-winding axis at time $t = 0$, then only the amplitude of the voltage phasor is seen. Therefore, if all the stator fixed equations are transformed into a optional coordinate system rotating with Ω_1, then in the stationary state all the machine signals in this coordinate system appear as vectors fixed in phase displacement to each other. Specifically, the voltage vector is located on the real axis of this optional coordinate system.

This method is convenient in terms of control, since all the signals are constant and are DC-signals in the stationary state. Moreover, by considering the equations of the machine in a coordinate system rotating with Ω_1 and assuming the stationary state, all the derivations can be set $\dfrac{d}{dt} = 0$, so that algebraic equations will remain in place of differential equations.

The first voltage equation that was transformed to the K-coordinate system reads as follows:

$$\vec{U}_1^K = \frac{R_1}{\sigma L_1} \vec{\Psi}_1^K - \frac{MR_1}{\sigma L_1 L_2} \vec{\Psi}_2^K + \frac{d}{dt}\left(\vec{\Psi}_1^K\right) + j\Omega_K \vec{\Psi}_1^K \ .$$

All the other equations are subject to the same transformation; the torque equation and the mechanical equation are independent from the coordinate system. If, as a last purely mathematical step, all the equations are manipulated into the state space form as far as possible, and if they are decomposed to the real and imaginary components, then the signal-flow graph can be drawn.

Therefore, the following applies by means of an example:

$$\frac{d\vec{\Psi}_1^K}{dt} = -\frac{R_1}{\sigma L_1}\left(\vec{\Psi}_1^K - \frac{M}{L_2}\vec{\Psi}_2^K\right) - j\Omega_K \vec{\Psi}_1^K + \vec{U}_1^K$$

thus:
Real signal component:

$$\frac{d\Psi_{1A}}{dt} = -\frac{R_1}{\sigma L_1}\left(\Psi_{1A} - \frac{M}{L_2}\Psi_{2A}\right) + \Omega_K \Psi_{1B} + U_{1A}$$

and imaginary signal component:

$$\frac{d\Psi_{1B}}{dt} = -\frac{R_1}{\sigma L_1}\left(\Psi_{1B} - \frac{M}{L_2}\Psi_{2B}\right) - \Omega_K \Psi_{1A} + U_{1B} .$$

In this way the signal-flow graph of the general rotating field machine can be drawn for impressed voltages (Fig. 5.3). If the machine has a squirrel cage rotor then $U_{2A} = U_{2B} = 0$. If the machine is a slip-ring machine, then U_{2A} and U_{2B} can also be impressed.

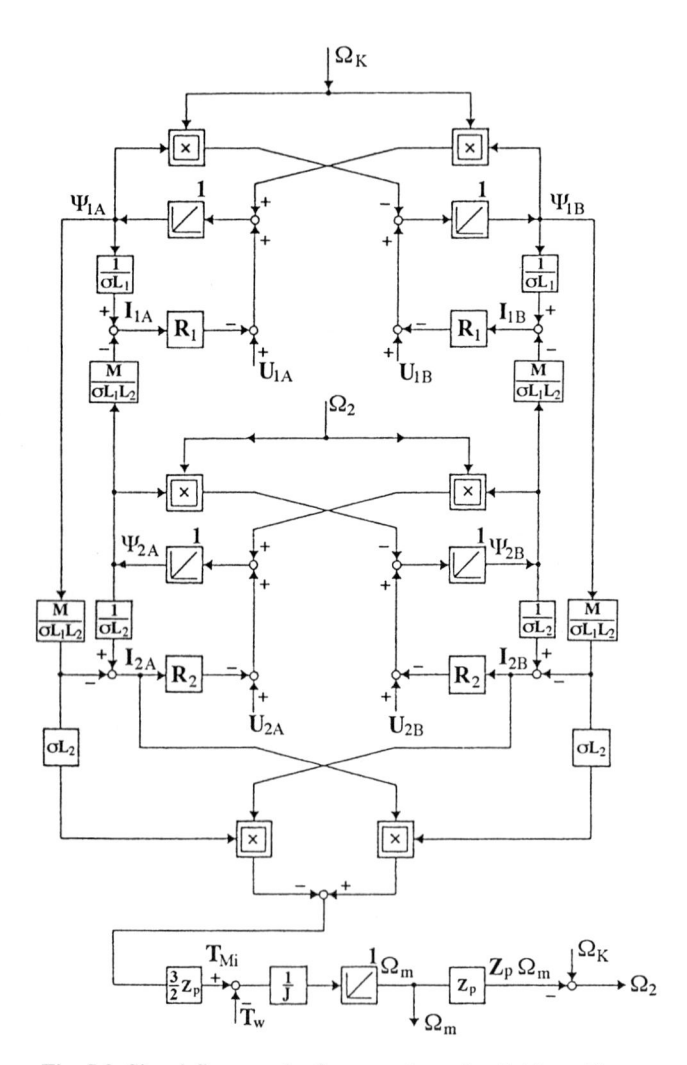

Fig. 5.3. Signal-flow graph of a general rotating field machine

Fig. 5.3 shows the signal-flow graph of the general rotating field machine in the complex A - B system.

However, the rotating field machine is usually connected to a three-phase voltage system. Fig. 5.4 shows a block diagram of this machine connected to a three-phase voltage system.

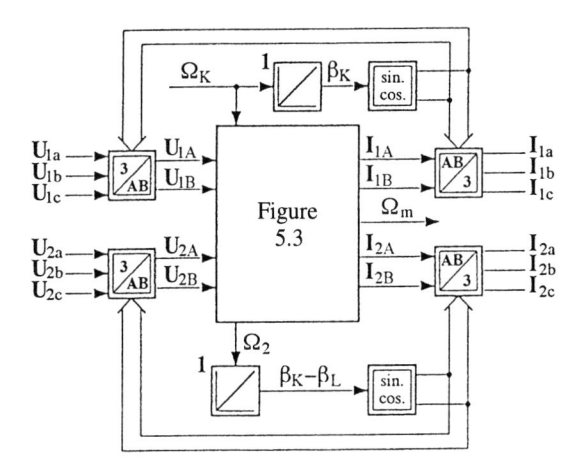

Fig.5.4. Block diagram of an induction machine in the three-phase voltage system

Fig.s 5.5 and 5.6 show the transformation of the three-phase system to the A- B system and vice versa.

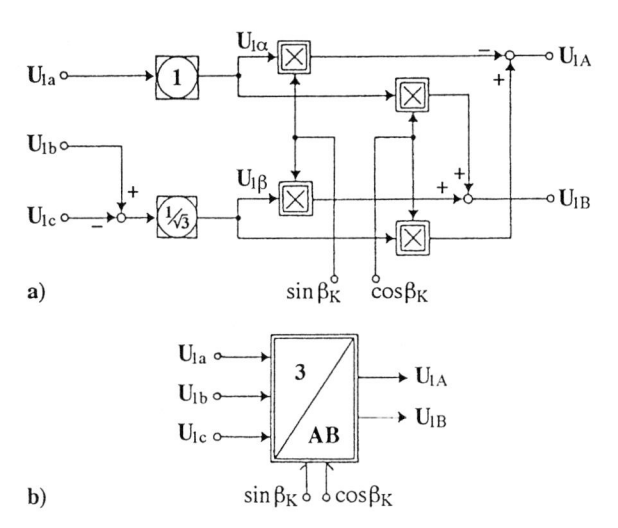

Fig. 5.5. Transformation of a three-phase voltage system to the A - B System
(a) structural diagram; (b) block diagram

a)

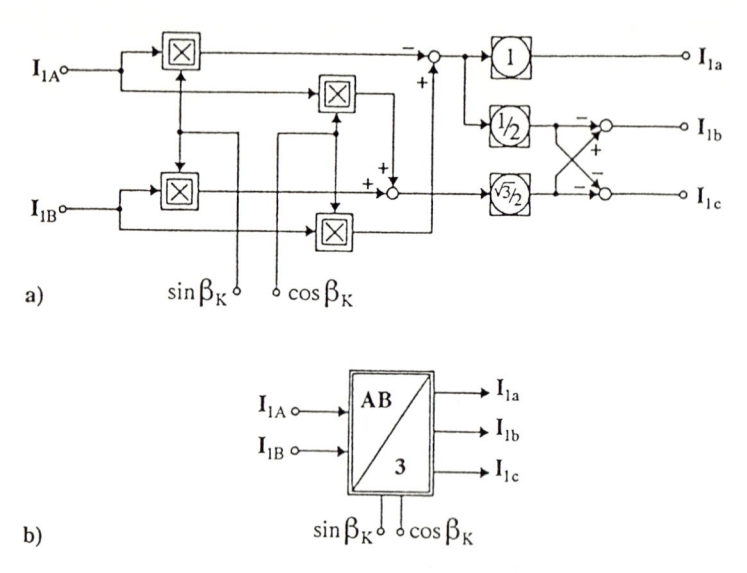

b)

Fig. 5.6. Transformation of currents of the A - B System in three-phase currents
(a) structural diagram; (b) block diagram

In this way the signal-flow graph of the general rotating field machine is known. From the signal-flow graph or state equations it will be seen that the rotating field machine is a non-linear system of the fifth order. The state space variables $\vec{\Psi}_1^K$ and $\vec{\Psi}_2^K$ and Ω_m using complex notation, or Ψ_{1A}, Ψ_{1B}, Ψ_{2A}, Ψ_{2B} and Ω_m using a real and imaginary notation.

The correcting variables are $\vec{U}_1^K$, $\vec{U}_2^K$ and $\Omega_2 = \Omega_K - Z_p \Omega_m$ with Ω_k for the dynamic angular frequency of the stator circuit (Ω_k to Ω_1 in steady state).

Additionally, it should be noted that multiplications are constantly present between state space variables and correcting variables. The mathematical treatment of such a system is complicated. Linearization at the working point is not applicable since the machine is to be used for the entire operating range. Accordingly, a control method has to be found to control the complex system and permit a type of control similar to that of the separately excited DC-machine.

Similarly, we can obtain a signal-flow graph for the machine with impressed currents. However, this will be done later since first the various control strategies must be deduced.

5.3 Induction Machine in the Steady State Supplied from the Mains

5.3.1 Mains and Converter Power Supply

This chapter examines the torque-characteristics T_{Mi} of the induction machine when the machine is running in steady state, thus connected to a mains having a constant voltage and frequency but a variable load. Many methods for open-and closed-loop control of the induction machines supplied by inverters also start from the basic premise of a quasi-stationary operation.

Our examination will be concerned with a machine with a squirrel cage rotor, thus $U_{2A} = U_{2B} = 0$.

If we assume stationary operation $\frac{d.}{dt}$ can be set $\frac{d.}{dt} = 0$ in the differential equations in the coordinate system K and $\Omega_k = \Omega_1$. In this case, the phasor $\vec{U}_1^K$ is time invariant and is therefore constant in amplitude and frequency.

The flux and torque equations are algebraic and remain unchanged even in steady state. This clearly simplifies the equations. Now follows a second step, truly essential for the simplification that will be increasingly used in the following chapters.

For instance, let us suppose that R_1 is $R_1 = 0$, this is true for machines having medium to high power ratings. Fig. 5.3 shows that the integrators have no feed-back channels in this case.

The third essential premise is that the three-phase power supply system is connected to the machine terminals, so that the voltage phasor $\vec{U}_1^K$ is identical to U_{1B}. This means that $U_{1A} = 0$ and $U_{1B} = U_1 = $ constant

Therefore, if $U_{1A} = 0$ and $U_{1B} = U_1$ and $d/dt = 0$ and $R_1 = 0$, Fig. 5.3 or the stator voltage equations indicate that:

$$\Psi_{1A} = \frac{U_1}{\Omega_1}$$

$$\Psi_{1B} = 0 \ .$$

With $R_1 = 0$, the stator flux is therefore constant and independent of the load. The block diagram of the induction machine in fig. 5.7 may now be considerably simplified, because one can describe all transfer functions through their steady state behaviour alone.

Fig. 5.7 shows the rotor signal-flow diagram.

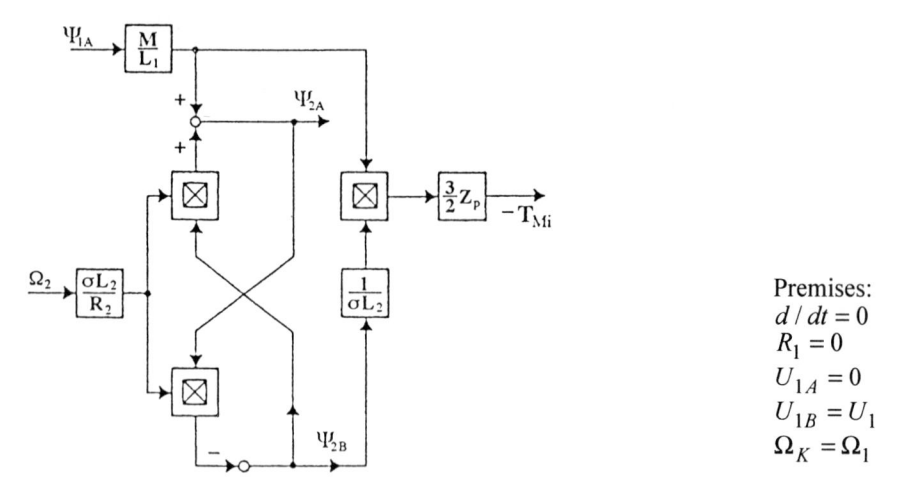

Fig. 5.7. Block diagram of the IM with $\Psi_{1B} = 0$ and steady state operation

Premises:
$d/dt = 0$
$R_1 = 0$
$U_{1A} = 0$
$U_{1B} = U_1$
$\Omega_K = \Omega_1$

The interdependencies shown in fig. 5.7 allow us to calculate the steady state torque of the motor $T_{Mi} = f(\Psi_{1A}, \Omega_1)$:

$$T_{Mi} = -\frac{3}{2} Z_p \frac{M}{\sigma L_1 L_2} \Psi_{1A} \Psi_{1B}$$

or

$$T_{Mi} = +\frac{3}{2} Z_p \frac{M^2}{\sigma L_1^2 L_2} \Psi_{1A}^2 \frac{1}{\frac{\Omega_2}{\Omega_{2K}} + \frac{\Omega_{2K}}{\Omega_2}}$$

with

$$\Omega_{2K} = \frac{\Omega_2}{\sigma L_2}.$$

In this equation we must now insert the ratio between Ω_2 and the angular frequency of the machine Ω_m or $\Omega_L = Z_p \Omega_m$. To do so, we insert the machine slip as an auxiliary variable.

Slip and pull-out slip. The slip gives the relative difference in the angular speed Ω_m of the IM and the synchronous angular speed (ideal angular speed during no-load operation) Ω_0- also called Ω_s.

$$S = \frac{\Omega_0 - \Omega_m}{\Omega_0} = 1 - \frac{\Omega_m}{\Omega_0}$$

with

$$\Omega_0 = \frac{\Omega_1}{Z_p} \text{ and } \Omega_m = \frac{\Omega_L}{Z_p} :$$

$$S = \frac{\Omega_1 - \Omega_L}{\Omega_1} = \frac{\Omega_2}{\Omega_1} .$$

Two types of operation are especially significant:
No-load operation: $\quad s = 0$
Zero speed: $\quad s = 1$.

The name "synchronous speed" means that the IM-rotor rotates synchronously with the stator fixed phasors of $\vec{U}_1, \vec{\psi}_1$. In an IM, this is only possible at ideal no-load operation, because at $\Omega_2 = 0$ no torque can be produced.

The torque-speed characteristic of the IM shows, as indicated below, another important point - the *pull-out torque* T_K. The corresponding slip is called *pull-out slip* s_K:

$$s_K = \frac{\Omega_{2K}}{\Omega_1} = \frac{R_2}{\Omega_1 \sigma L_2} .$$

Kloss Equation and the Pull-Out Torque. Using this definition, we may indicate torque as:

$$T_{Mi} = \frac{3}{4} Z_p \frac{M^2}{\sigma L_1^2 L_2} \left(\frac{U_1}{\Omega_1}\right)^2 \frac{2}{\frac{s}{s_K} + \frac{s_K}{s}}$$

or

$$T_{Mi} = T_K \frac{2}{\frac{s}{s_K} + \frac{s_K}{s}} = T_K \frac{2 s s_K}{s^2 + s_K^2} .$$

This is the *Kloss equation*.

$$T_K = \frac{3}{4} Z_p \frac{M^2}{\sigma L_1^2 L_2} \left(\frac{U_1}{\Omega_1}\right)^2$$

is a constant value known as the *pull-out torque* T_K.

Discussion of the Torque Speed Characteristic. Fig. 5.8 shows the non-linear torque-speed characteristic behaviour for the IM according to the Kloss equation.

The characteristic is symmetrical to the synchronous point of operation. Near this operating point the IM shows a shunt characteristic like that of the separately excited dc-machine. This is the usual operating range.

The maximum torque that the machine can produce is the pull-out torque T_K at the pull-out slip s_K. A load torque greater than T_K produces an overload and thus an instable operating condition. This must be avoided at all cost.

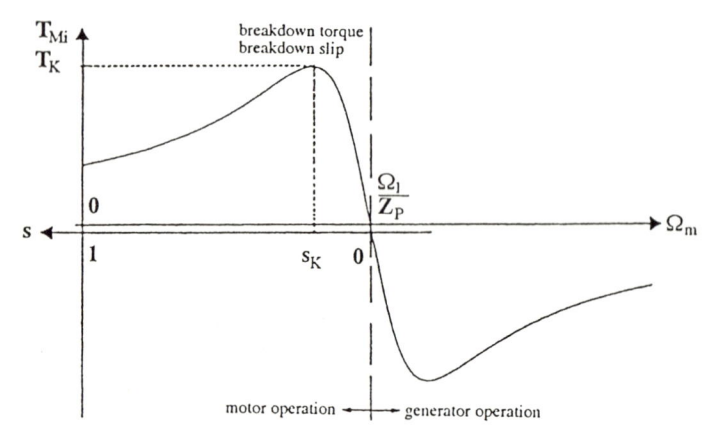

Fig. 5.8. Torque speed characteristic for the IM with $R_1 = 0$

Linearized Characteristic. Within the usual operating range near to the synchronous speed, it is generally sufficient to consider the linearized characteristic. For $|s| \ll s_K$ one obtains the approximation:

$$T_{Mi} \simeq 2T_K \frac{s}{s_K} = 2T_K \frac{\Omega_2}{\Omega_{2K}}$$

and after an additional manipulation the angular speed:

$$\Omega_m = \frac{1}{Z_p} \left(\Omega_1 - T_{Mi} \frac{\Omega_{2K}}{2T_K} \right).$$

The analogy with respect to the equation of the speed-torque characteristic in the separately excited dc-machine can be easily recognized here.

Modifying the Torque-Speed Characteristic. Without an inverter for power supply, one can modify the characteristic externally only by means of Z_p (pole pairs) or R_{2v} (external rotor resistors), since $U_1 = U_{mains}$ and $\Omega_1 = \Omega_{mains}$ are fixed premises.

External rotor resistor R_{2v}. Due to external rotor resistors (assumption: wound rotor slip-ring induction machine), one obtains an effect similar to that of the DC-machine:

$$R_2 = R_{20} + R_{2v} \quad ; \quad s_K = \frac{R_2}{\Omega_1 \sigma L_2}.$$

The characteristic flattens, but the synchronous speed remains unchanged. The pull-out torque also remains constant, while the pull-out slip increases (Fig. 5.9) (flattening effect). This method reduces the efficiency and has been largely replaced by inverter power supply.

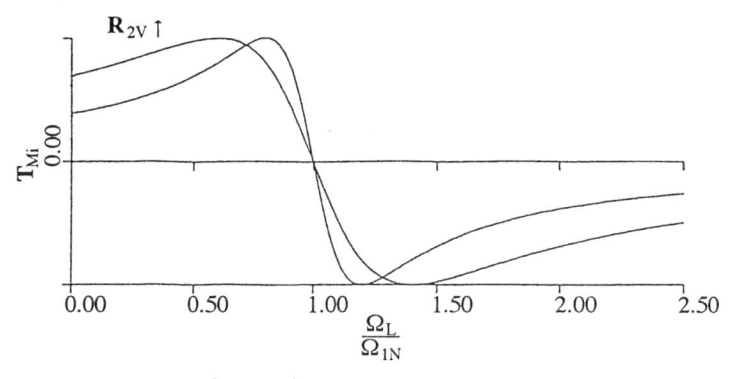

Fig. 5.9. $T_{Mi} = f(\Omega_L, R_2)$ with $\Omega_L = Z_p \Omega_m$

Supply of the AC-machine by inverters. With the power supply of an inverter the stator voltage U_1 and the stator frequency Ω_1 are adjustable. With Ω_1 in particular, one can adjust the synchronous speed of no-load operation. Similar to the DC-machine, one must attempt to keep the interlinked flux constant within the armature control range.

$$\vec{\Psi}_1 = \Psi_{1N} = \frac{U_1}{\Omega_1} = const.$$

There should be no attempt to have an interlinked flux much greater than nominal flux with U_{1N} and Ω_{1N}, since otherwise the inductances would become saturated. Therefore there are two operating ranges, similar to the DC-machine: the armature control range and the field weakening range.

Armature control range. The armature control range has the stator frequencies:

$$0 < \Omega_1 \leq \Omega_{1N}$$

(Ω_{1N}: rated stator frequency of the machine) and therefore the speed range:

$$0 \leq \Omega_m \leq \frac{\Omega_{1N}}{Z_p}.$$

The following hold for the constant stator flux:

$$\frac{U_1}{\Omega_1} = const. = \frac{U_{1N}}{\Omega_{1N}} = \Psi_{1N}$$

$$U_1 = U_{1N} \frac{\Omega_1}{\Omega_{1N}}.$$

In this way the entire characteristic will be shifted along the frequency axis (speed axis) (Fig. 5.10). This corresponds to the shifting in the armature control range of the DC-machine.

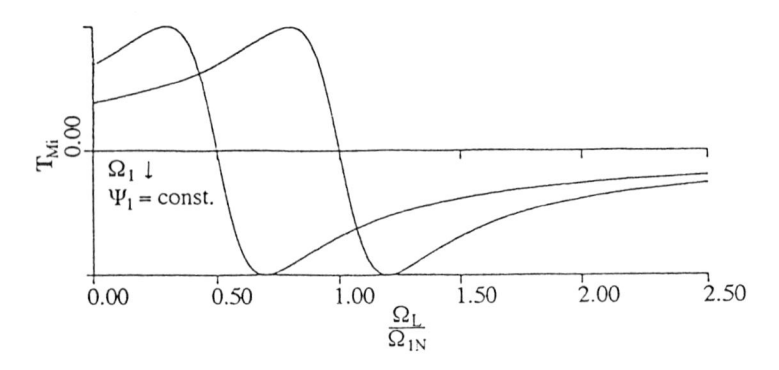

Fig. 5.10. Family of characteristics of the IM in the armature control range

One can also achieve the same effect by applying an adjustable counter-electromotoric force to the rotor terminals of a wound rotor machine. This is used in subsynchronous cascade drives (wound rotor induction machine with slip recovery) (Chapter 7.2).

The limit of the armature control range is reached with $\Omega_1 = \Omega_{1N}$ since the stator voltage U_1 should not increase beyond its rated value U_{1N}.

Field weakening range. If a higher speed than in the armature control range is required, only Ω_1 may be increased, while $U_1 = U_{1N}$ is kept constant. The stator flux therefore will decrease. As Ω_1 rises, the synchronous speed also increases and the speed-torque characteristic becomes increasingly flat. Simultaneously, due to the quadratic relation of the torque-voltage ratio, the pull-out torque is considerably reduced (Fig. 5.11).

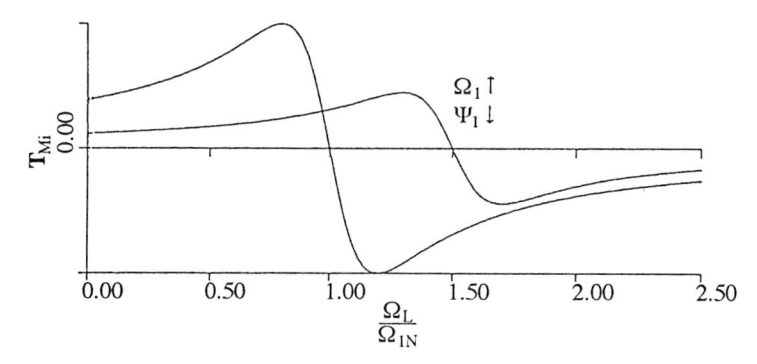

Fig. 5.11. Group of characteristics of the IM in the field weakening range

5.4 Single-Phase Equivalent Circuit in Steady-State Operation

5.4.1 Equivalent Electrical Circuit for the IM

During steady-state operation, the phasors in the optional coordinate system K are constant in amplitude and phase (frequency). With this premise, they can be considered as complex vectors in the time domain.

It is possible to develop an equivalent electrical circuit for these complex vectors that show the electrical behaviour of the IM in the steady-state.

The stator circuit and rotor circuit are coupled inductively. That is why the electrical behaviour of the IM in the stationary state is related to that of a three-phase transformer, and one can develop a similar equivalent electrical T circuit.

(a) The stator and rotor inherent inductances may be further split into mutual inductances and leakage inductances:

$$L_1 = L_{1H} + L_{\sigma 1} = L_{1H}\left(1+\sigma_1\right)$$

$$L_2 = L_{2H} + L_{\sigma 2} = L_{2H}\left(1+\sigma_2\right)$$

(b) Then with the transformation ratio $ü$, the rotor side parameters are transformed to stator side parameters (values '):

$$\ddot{u} = \frac{L_{1H}}{M} = \frac{M}{L_{2H}}$$

$$\underline{U}'_2 = \underline{U}_2 \ddot{u} \qquad \underline{I}'_2 = \frac{\underline{I}_2}{\ddot{u}}$$

$$R'_2 = R_2 \ddot{u}^2 \qquad L'_2 = L_2 \ddot{u}^2$$

$$M' = M\ddot{u}^2 = L_{1H} = L'_{2H} \qquad \text{(equal interlinked flux)}$$

For $\underline{U}'_2 = 0$ (squirrel cage machine) this is equal to:

$$\underline{U}'_2 = R'_2 \underline{I}'_2 + j\Omega_2 L'_{\sigma 2} \underline{I}'_2 + j\Omega_2 L'_{2H}(\underline{I}_1 + \underline{I}'_2) = 0.$$

(c) The inductive voltage drops related to Ω_2 in the rotor circuit are transformed to Ω_1, and are so adapted to the stator frequency:

(d) with $s = \dfrac{\Omega_2}{\Omega_1}$ and $\underline{I}_\mu = \underline{I}_1 + \underline{I}'_2$ (magnetising current) we obtain the machine equation for the equivalent electrical single-phase circuit of the IM:

$$\underline{U}_1 = R_1 \underline{I}_1 + j\Omega_1 L_{\sigma 1} \underline{I}_1 + j\Omega_1 L_{1H} \underline{I}_\mu$$

$$\frac{\underline{U}'_2}{s} = \frac{R'_2}{s} \underline{I}'_2 + j\Omega_1 L'_{\sigma 2} \underline{I}'_2 + j\Omega_1 L'_{2H} \underline{I}_\mu = 0$$

from which we obtain the equivalent electrical single-phase circuit shown in Figure 5.12.

The equivalent electrical circuit thus obtained is based on the physical variables and parameters of the machine. From the standpoint of calculation, Blondel's leakage coefficient σa is sufficient in order to consider the leakage coefficients σ₁ and σ₂.

$$\sigma = 1 - \frac{1}{(1+\sigma_1)(1+\sigma_2)}.$$

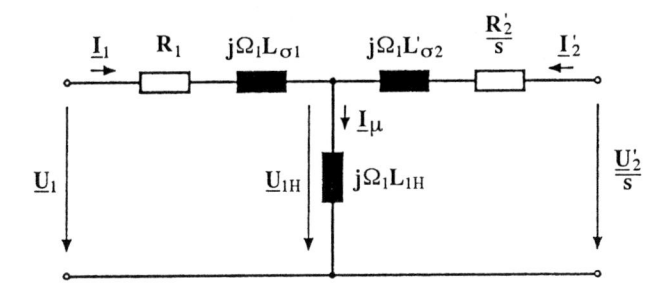

Fig. 5.12. Equivalent electrical single-phase circuit of the IM with stator and rotor-leakage inductances

The electrical diagram shown in Fig. 5.12 is used, for example, in steady-state operations in order to calculate the stator voltage U_1 as a function of the speed, and torque in controllable drives.

It is also possible to manipulate the equivalent electrical circuit shown above so that the leakage inductances are concentrated only on the stator side or only on the rotor side. Another use is in calculating the stator current locus as a function of voltage, slip and torque of the IM and its parameters. This provides the well known Heyland- or Osanne-locus. The steady-state behaviour of the AC-machine may now be easily described in order to gain a basic understanding of this rotating field machine.

5.5 Induction Machine Supplied by Inverter

If the induction machine is supplied by an inverter, one may then freely choose the supply voltage of the IM in the converter range. This chapter will help us determine the control conditions and the signal-flow graphs resulting from the specified control conditions. Through these control conditions, we can determine the behaviour of the induction machine and load of the inverter during a static and dynamic operating situation.

Generally speaking, in the IM we must distinguish between three control strategies: the control based on constant stator flux, the control based on constant rotor flux, and the control based on constant air gap flux. Generally, the flux amplitudes remain constant at their rated value in the armature control range. In the field weakening range, the amplitudes reduce correspondingly to the increasing speed.

We must also distinguish between voltage source inverters, in which the output voltages are impressed in terms of amplitude and frequency, and current source inverters, in which output currents are impressed in terms of amplitude and frequency.

Current source inverters may be real source inverters or voltage source inverters which additionally have a current control.

5.5.1 Control Strategies Based on Constant Stator Flux

In the previous chapter, induction machine in steady-state operation supplied from the mains, we assumed that $R_1 = 0$, the stator flux Ψ_{1A} must be
$\Psi_{1A} = \Psi_{1N}$, $\Psi_{1B} = 0$ and thus $U_{1A} = 0$, $U_{1B} = U_1$. These assumptions were made in order to show the characteristic properties of the IM in steady-state operation as easily as possible.

If the IM is now supplied by an inverter, we must ensure that Ψ_{1A} is $\Psi_{1A} = \Psi_{1N}$ and that $\Psi_{1B} = 0$ with a suitable open or closed control. The essential principle in the open or closed control is that the real axis of the optional coordinate system K, in which the machine equations had been deduced, is oriented on the stator flux.

IM with $R_1 = 0$. The orientation of the flux Ψ_{1A} and the voltage U_{1B} are illustrated in Fig. 5.13.

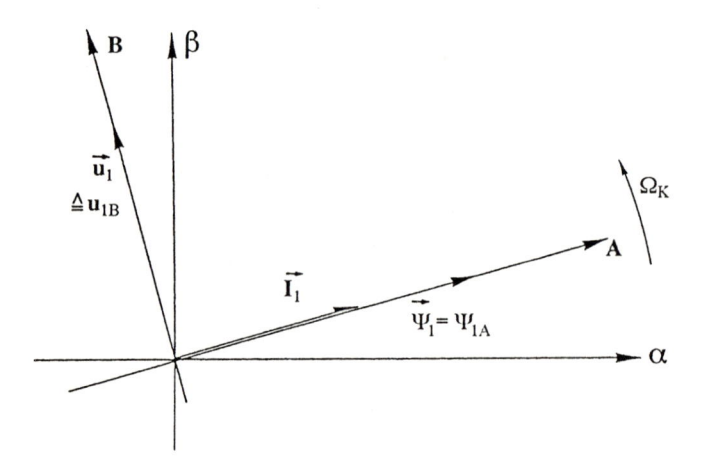

Fig. 5.13. Orientation of the stator flux ($R_1 = 0$)

If - as shown in Figure 5.13 - $R_1 = 0$, then the stator currents I_{1A} and I_{1B} do not produce any ohmic voltage drop and thus, as shown in Figure 5.3, the torque T_{Mi} also may not have any feedback effect on the stator voltages.

5 Three-Phase AC-Machine Signal-Flow Graphs

Thus the statements developed in the previous chapter, with $R_1 = 0$, they also hold for stator voltages and fluxes with inverter supply:

with	is	or
Ψ_{1A}	$U_{1A} = 0$	$\dfrac{d\Psi_{1A}}{dt} = U_{1A}$
$\Psi_{1B} = 0$	$U_{1B} = \Omega_K \Psi_{1A}$	$\dfrac{d\Psi_{1B}}{dt} = 0.$

If we observe these control conditions for the stator side, only the signal-flow graph of the rotor remains (see Figure 5. 3).

It is essential that with $\Psi_{1B} = \dfrac{d\Psi_{1B}}{dt} = 0$ only Ψ_{1A} is coupled to the rotor side from the stator side, and that based on the premise $R_1 = 0$ the rotor fluxes no longer have a feedback channel to the stator fluxes and the stator voltages. The torque T_{Mi} may therefore be controlled through Ψ_{1A} and Ω_2. If - as assumed in section 3.3 - Ψ_{1A} is constant, then the torque may be controlled only through Ω_2.

In steady-state operation, the equations given in section 5.3 for the torque TMi hold. The same considerations remain valid.

Since, as mentioned earlier in section 5.3, the inverter may control the stator voltage and stator angular frequency Ω_1 freely, any speed and torque may be set in the operating range of the inverter and IM, through setting U_1, Ω_1 and Ω_2. The following holds: $\Omega_2 \ll \Omega_{2K}$:

$$\Omega_K = Z_p \Omega_m + \Omega_2$$

$$\Omega_2 \approx \frac{2}{3} \frac{1}{Z_p} \frac{R_2 L_1^2}{M^2} \frac{1}{\Psi_{1A}^2} T_{Mi} \ .$$

Meaning that the speed torque characteristic in Figure 5.12 may be shifted to any desired operating point by means of $\Omega_0 = \Omega_K / Z_p$

This also determines the basic behaviour of the IM supplied by inverters in steady-state operation, with $\Psi_{1B} = \dfrac{d\Psi_{1B}}{dt} = 0$ and $R_1 = 0$.

Induction machine with $R_1 \neq 0$. In fact is $R_1 \neq 0$. Therefore the rotor fluxes produce voltage drops at the stator side and influence the stator fluxes.

Corresponding to Figure 5.3, with $\Psi_{1B} = \dfrac{d\Psi_{1B}}{dt} = 0$ and $\Omega_2 \ll \Omega_{2K}$, the following holds for steady-state operation:

$$U_{1A} = I_{1A} R_1 \approx \Psi_{1A} R_1 \left[\frac{1}{\sigma L_1} - \frac{M^2}{L_1^2} \frac{1}{\sigma L_2} \right] = \Psi_{1A} \frac{R_1}{L_1}$$

$$U_{1B} \approx \Psi_{1A} \Omega_K + \frac{M^2}{L_1^2} \frac{R_1}{R_2} \Psi_{1A} \Omega_2 .$$

Therefore the stator voltage U_{1A} must be kept constant in steady-state operation which is determined by Ψ_{1A} and by the machine parameters. The voltage U_{1B} is determined, on the one hand, by Ψ_{1A} and Ω_K and on the other hand by Ψ_{1A} and Ω_2 (influence of the required torque).

The voltage equations may also be given according to the torque and speed. Ω_K and Ω_2 may be substituted with the known equations:

$$\Omega_K = Z_p \Omega_m + \Omega_2$$

$$\Omega_2 \approx \frac{2}{3} \frac{1}{Z_p} \frac{R_2 L_1^2}{M^2} \frac{1}{\Psi_{1A}^2} T_{Mi} .$$

The speed is directly correlated to Ω_m:

$$\Omega_m = 2\pi N \left[\frac{1}{s} \right] .$$

The voltage equations may be written as follows:

$$U_{1A} \approx \frac{R_1}{L_1} \Psi_{1A}$$

$$U_{1B} \approx \frac{2}{3} \left(R_1 + R_2 \frac{L_1^2}{M^2} \right) \frac{T_{Mi}}{Z_p \Psi_{1A}} + Z_p \Psi_{1A} \Omega_m$$

$$\left| \vec{U} \right| = U_1 = \sqrt{U_{1A}^2 + U_{1B}^2}$$

The groups of characteristics in Figure 5.14 have been calculated with the parameters of a real IM. Three groups of characteristics have been drawn for different torques.

5 Three-Phase AC-Machines – Signal-Flow Graphs

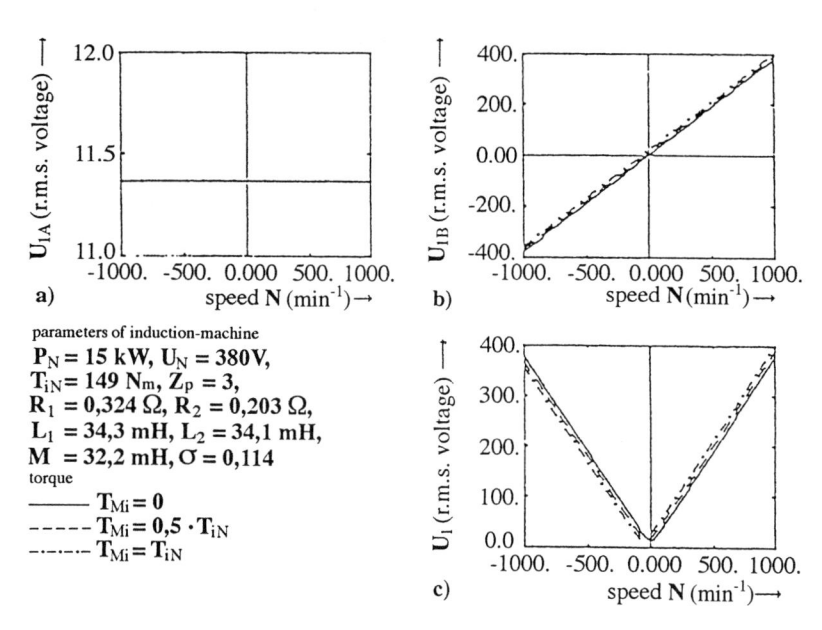

a) U_{1A} (r.m.s. voltage) vs. speed N (min⁻¹)

b) U_{1B} (r.m.s. voltage) vs. speed N (min⁻¹)

c) U_1 (r.m.s. voltage) vs. speed N (min⁻¹)

parameters of induction-machine
$P_N = 15$ kW, $U_N = 380$V,
$T_{iN} = 149$ Nm, $Z_p = 3$,
$R_1 = 0{,}324\ \Omega$, $R_2 = 0{,}203\ \Omega$,
$L_1 = 34{,}3$ mH, $L_2 = 34{,}1$ mH,
$M = 32{,}2$ mH, $\sigma = 0{,}114$
torque
——— $T_{Mi} = 0$
- - - - - $T_{Mi} = 0{,}5 \cdot T_{iN}$
—·—·— $T_{Mi} = T_{iN}$

Fig. 5.14.a) b) c) Group of steady-state characteristics of a real induction machine with constant stator flux

The results in Fig. 5.14 demonstrate that U_{1A} is approximately constant, regardless of the torque T_{Mi} and relatively small in the stationary state compared to U_{1B} (a). On the contrary, U_{1B} (b) shows a clear linear dependency from $Z_p\Omega_m$, and there is a clear influence from the torque T_{Mi}. Furthermore the ohmic resistances, especially R_1, are important at low speeds. As a result we get $|U_1|$, shown in Figure 5.14 (c).

Since $|U_{1A}| \ll |U_{1B}|$, $\vec{U}_1$ shall remain approximately in the direction of the B-axis of the optional coordinate system K.

After deducing the steady-state operation behaviour with the assumptions

$$\Psi_{1A} = \Psi_N, \quad \Psi_{1B} = \frac{d\Psi_{1B}}{dt} = 0,$$

and inverter supply, one can then immediately determine the signal-flow graph of the induction machine from Figure 5.8 (Fig. 5.15 for a squirrel cage machine). This signal-flow graph also describes the dynamic behaviour.

We must remember that, when assuming a constant stator flux $\psi_{1A} = \psi_N$ and a power supply from an inverter, the signal-flow graph is not as obvious as in the separately excited DC-machine, either in the case of impressed voltage, or with of impressed current.

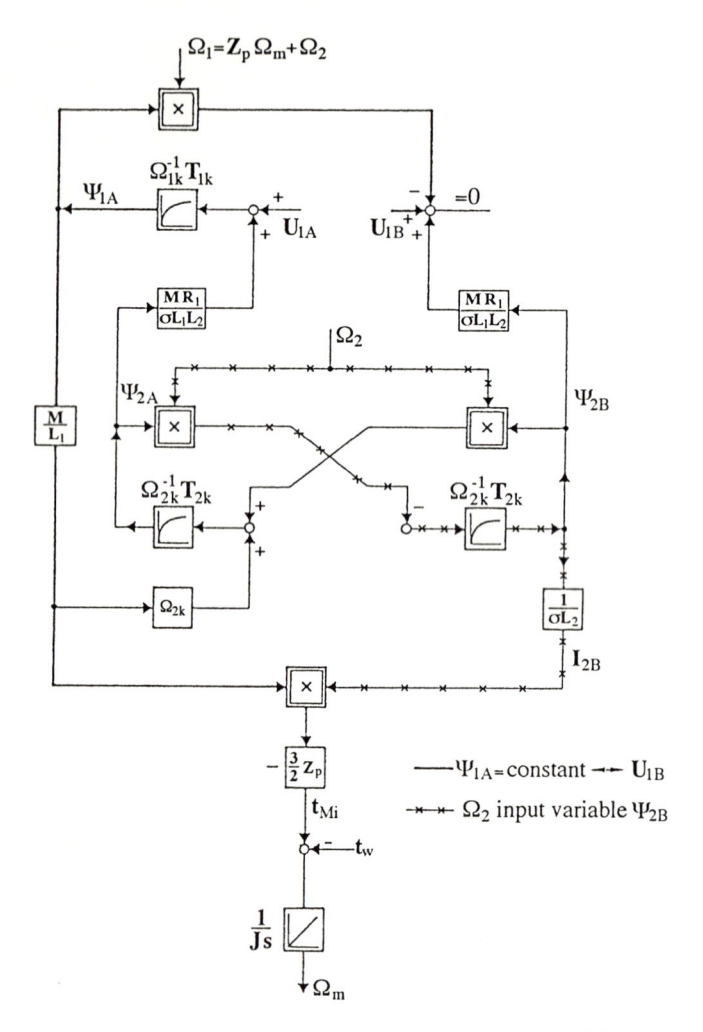

Fig. 5.15. Signal-flow graph for the induction machine with constant stator flux and impressed voltages

5.5.2 Control Strategies Based on Constant Rotor Flux

Instead of an orientation of the optional coordinate system K on the stator flux, we may also fix it to the rotor flux. In this case the real axis of the system K is fixed to the direction of $\Psi_{2A} = \Psi_{2N}$ and thus $\Psi_{2B} = 0$. In the following considerations, we assume a squirrel cage machine $(U_{2A} = U_{2B} = 0)$. From the signal-flow graph in Figure 5.3, with $\Psi_{2A} = \Psi_{2N}$, $\dfrac{d\Psi_{2B}}{dt} = 0$ and $\dfrac{d\Psi_{2A}}{dt} = 0$

thus in steady-state operation for the flux Ψ_{2A} (armature control range), we can derive:

$$\Psi_{2A} = \Psi_{1A} \frac{M}{L_1}$$

or

$$\Psi_{1A} = \Psi_{2A} \frac{L_1}{M}$$

and therefore $I_{2A} = 0$.

In this case, the signal-flow graph (armature control range) may be further simplified as shown in Figure 5.16.

The graph immediately shows the differences with respect to the example based on constant stator flux. The torque T_{Mi}, in the armature control range with $\Psi_{2A} = $ constant and rotor flux orientation $\Psi_{2B} = \frac{d\Psi_{2B}}{dt} = 0$ may be controlled directly through Ψ_{1B}.

However, in order to ensure $\Psi_{2B} = \frac{d\Psi_{2B}}{dt} = 0$, we must maintain

$$\Omega_2 \Psi_{2A} = -R_2 I_{2B} = \frac{R_2 M}{\sigma L_1 L_2} \Psi_{1B} = \frac{M}{L_1} \Omega_{2K} \Psi_{1B}$$

and therefore

$$\Omega_2 = \frac{M}{L_1} \frac{\Omega_{2K}}{\Psi_{2A}} \Psi_{1B} \quad .$$

The resulting torque is:

$$T_{Mi} = +\frac{3}{2} Z_p \frac{M^2}{L_1^2} \frac{\Psi_{1A}}{\sigma L_2} \Psi_{1B} \quad .$$

The stator voltage equations in steady-state operation result as functions of Ω_m and of the torque T_{mi}:

$$U_{1A} = \frac{R_1}{\sigma L_1} \left(\frac{L_1}{M} - \frac{M}{L_2} \right) \Psi_{2A}$$

$$- \frac{4}{9} \frac{R_2 \sigma L_1 L_2}{Z_p^2 M} \frac{T_{Mi}^2}{\Psi_{2A}^3} - \frac{2}{3} \frac{\sigma L_1 L_2}{M} \frac{T_{Mi}}{\Psi_{2A}} \Omega_m$$

$$U_{1B} = \frac{2}{3} \frac{R_2 L_1 + R_1 L_2}{Z_p M} \frac{T_{Mi}}{\Psi_{2A}} + Z_p \frac{L_1}{M} \Psi_{2A} \Omega_m$$

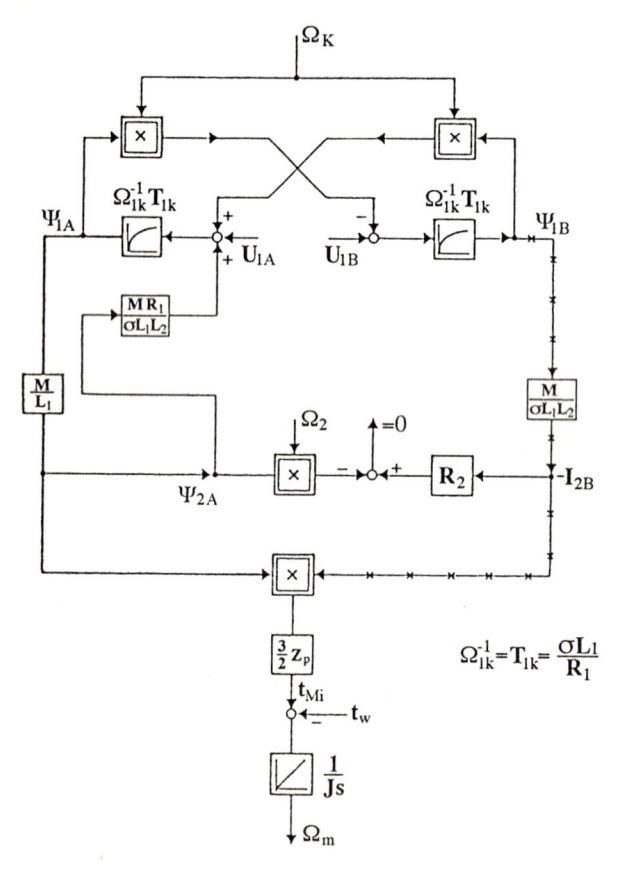

Fig. 5.16. Signal-flow graph of the induction machine, rotor flux orientation $\left(\Psi_{2B} = \dfrac{d\Psi_{2B}}{dt} = 0\right)$ and armature control range (Ψ_{2A} = constant.)

The groups of characteristics in Figure 5.17 are obtained by using the parameters from a real induction machine:

Figure 5.17 a) clearly shows that, with constant rotor flux Ψ_{2A}, the effects on U_{1A} are considerably greater than at constant stator flux, and depend on the torque. The dependency of U_{1B} (Fig. 5.17 b) corresponds to that of the constant stator flux, and depends essentially on the speed.

5 Three-Phase AC-Machines – Signal-Flow Graphs

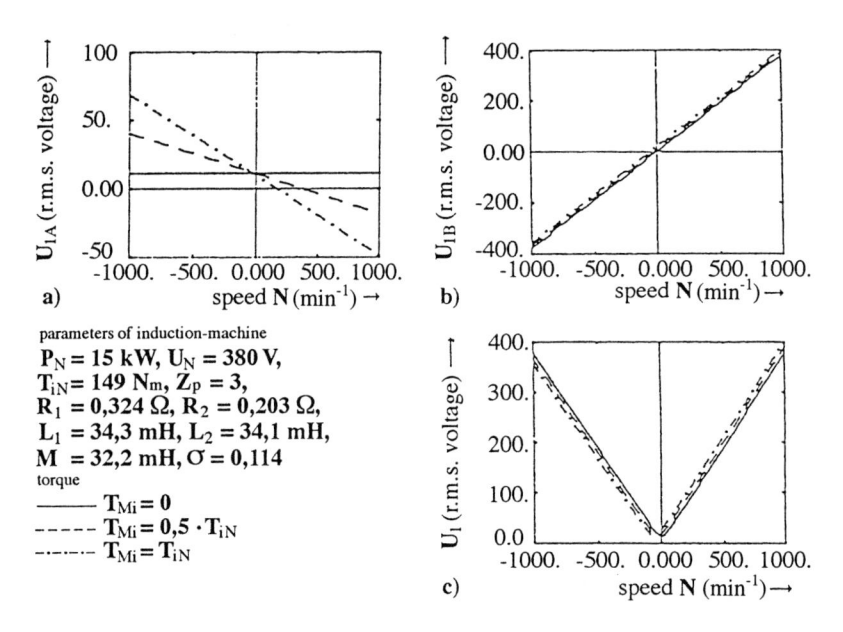

a) U_{1A} (r.m.s. voltage) vs. speed N (min⁻¹)

b) U_{1B} (r.m.s. voltage) vs. speed N (min⁻¹)

c) U_1 (r.m.s. voltage) vs. speed N (min⁻¹)

parameters of induction-machine
P_N = 15 kW, U_N = 380 V,
T_{iN} = 149 Nm, Z_p = 3,
R_1 = 0,324 Ω, R_2 = 0,203 Ω,
L_1 = 34,3 mH, L_2 = 34,1 mH,
M = 32,2 mH, σ = 0,114

torque
——— $T_{Mi} = 0$
– – – $T_{Mi} = 0{,}5 \cdot T_{iN}$
–·–·– $T_{Mi} = T_{iN}$

Fig. 5.17. a) b) c) Groups of characteristics in the steady-state operation for a real induction machine with constant rotor flux

Power Supply from an Inverter with Impressed Current. Instead of a power supply with impressed voltages, we may also have a power supply by an inverter with impressed currents.

The signal-flow graph in Figure 5.3 reveals the following relations for $\Psi_{2B} = 0$:

$$\Psi_{1B} = \sigma L_1 I_{1B}$$

$$I_{2A} = \frac{1}{\sigma L_2}\Psi_{2A} - \frac{M}{\sigma L_1 L_2}\Psi_{1A}$$

$$I_{2B} = -\frac{M}{\sigma L_1 L_2}\Psi_{1B} = -\frac{M}{L_2}I_{1B} \quad.$$

If we insert this relation in the torque equation, we obtain:

$$T_{Mi} = \frac{3}{2} Z_p \frac{M}{L_2} I_{1B} \Psi_{2A} \quad.$$

It is apparent that, for an orientation on the rotor flux, T_{Mi} is controlled without delay by means of the current component I_{1B} if Ψ_{2A} remains constant. The following is obtained for the control of Ψ_{2A}, from the basic equations (with $\Psi_{2B} = 0$):

$$\frac{d\Psi_{2A}}{dt} + \frac{R_2}{L_2}\Psi_{2A} = M\frac{R_2}{L_2}I_{1A} .$$

Ψ_{2A} may therefore be controlled with a first order delay and the time constant $T_2 = \dfrac{L_2}{R_2}$ by the current component I_{1A}:

$$\Psi_{2A}(s) = M I_{1A}(s) \frac{1}{1+sT_2} \quad \text{with } T_2 = \frac{L_2}{R_2} .$$

Figure 5.18 shows the signal-flow graph.

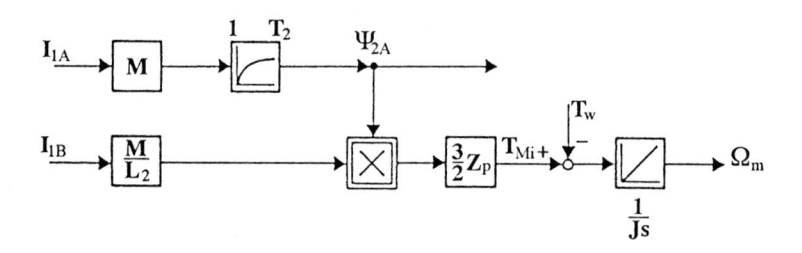

Fig. 5.18. Signal-flow graph of the induction machine with rotor flux orientation and impressed stator currents

Figure 5.18 shows that with impressed stator currents and $\Psi_{2B} = 0$ we get a signal-flow graph similar to that of a separately excited DC-machine. The stator current I_{1A} generates the flux, while the stator current I_{1B}, without delay, produces the torque. The advantage of these assumptions is the simplicity of the signal-flow graph on the one hand but this simplicity however results in a greater sensitivity to variations of the parameters during open- and closed-loop control on the other hand.

6 Synchronous Machine

In the following chapters signal-flow graphs of synchronous machines will be deduced and, from these, also the control strategies. To avoid going into too many details for the signal-flow graphs, the following simplifying assumptions apply:
- The state of saturation of the machine will be considered constant, but may vary in a longitudinal and transverse direction.
- The influence of skin effects on the conductors will not be considered.
- Losses in the iron will be neglected.
- Only the reciprocal damping of the basic magnetic fields (simple number of pole pairs) in the magnetic gap will be considered.
- The asymmetries of an irregular or incomplete damping cage may be considered in the form of asymmetrical resistances and inductances in the twin-coil compensating winding of the damper.
- The excitation axis must be aligned with the centre of one mesh of the damper or with the centre of one damper bar.
- The machine's main and mutual inductances may vary in the longitudinal and transverse directions.
- The stator has a three-coil symmetrical winding that can be converted to an equivalent two-coil winding, rotating with the rotor.
- A magnetic coupling between the branch winding and damper cage through the transverse inductance (if the two windings are arranged in common slots) may if necessary be considered by means of high mutual inductance M_{ED}.
- The three-phase power supply system is fixed and does not contain any neutral component.
- The parameters on the rotor side are converted to the stator side.

6.1 Synchronous Machine with Salient Pole Rotor

6.1.1. Machine with Salient Pole Rotor with Impressed Stator Voltage Signal-Flow Graph

A machine with a salient pole rotor will be supposed for the deduction of the signal-flow graph. In this case the rotor is comprised of salient poles and is therefore mechanically asymmetrical. Only the excitation winding of the synchronous machine is placed there (Fig. 6.1).

The presence of a symmetrical three-phase stator winding system is crucial for deriving the equations for a signal-flow graph. This three-phase winding system can be described, as in the chapters for the induction machine, by phasor equations with a coordinate system oriented to the stator windings.

As in the case of general rotating field machines, the following phasor voltage equation also holds true for the stator winding system:

$$\vec{U}_1^S = R_1 \vec{I}_1^S + \frac{d\vec{\Psi}_1^S}{dt} \qquad S: \text{stator windings fixed coordinate system}$$

As illustrated in Fig. 6.1, there is an angular difference ϑ between the stator coordinate axis α and the d coordinate axis fixed to the salient pole rotor.

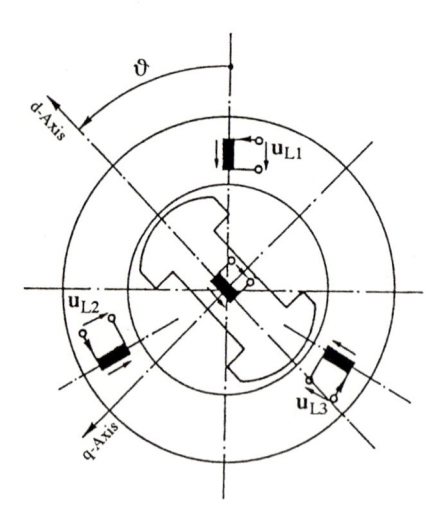

Fig. 6.1. Synchronous machine (machine with salient pole rotor) without damper winding (lumped winding systems)

Therefore:

$$\vartheta = \vartheta_0 + \int_0^\tau \Omega_L dt$$

where ϑ_0 is the initial value of the angle and Ω_L is the electrical angular velocity of the salient pole rotor seen from the stator fixed coordinate system.

As in section 5.2, the second step is choosing a common coordinate system K for the stator- and the excitation-winding systems. For this machine with a salient pole rotor it is reasonable to direct the optional oriented coordinate system K, as shown in Fig.6.1, in the excitation direction of the salient pole rotor. The real axis is called the d axis, and the imaginary axis the q axis.

When transforming the stator voltage equation, it should be taken into account that both the amplitude of the flux Ψ_1 and the position in relation to the coordinate system K are time-variant. Accordingly, in differentiating the flux the product rule must be applied, given that the differentiation must be carried out on both the time variant amplitude and the position. Therefore we get:

$$\vec{U}_1^K = R_1 \vec{I}_1^K + \frac{d\vec{\Psi}_1^K}{dt} + j\Omega_L \vec{\Psi}_1^K \qquad \text{where } \frac{d\vartheta}{dt} = \Omega_L .$$

The second term in the preceding equation describes the induced voltage as a result of variation in amplitude, while the third term describes the induced voltage due to variations in position.

The equation can be broken down into its d and q components:

$$U_d = R_1 I_d + \frac{d\Psi_d}{dt} - \Omega_L \Psi_q$$

$$U_q = R_1 I_q + \frac{d\Psi_q}{dt} + \Omega_L \Psi_d$$

where $\Omega_L = Z_p \Omega_m$ Ω_m: mechanical angular velocity of the rotor.

A comparable system of equations was also obtained for the stator system of the rotating field machine, since U_d is generated by the rotational induced voltage $\Omega_L \Psi_q$ and similarly U_q is generated by the rotational induced voltage $\Omega_L \Psi_d$.

Only the representation of the excitation circuit is now required in order to obtain the signal-flow graph. Similarly, the following applies:

$$\vec{U}_E^L = R_E \vec{I}_E^L + \frac{d\vec{\Psi}_E^L}{dt} .$$

The rotor fixed coordinate system L corresponds to the flux oriented coordinate system K, because of the orientation of the excitation winding. The superscript can be ignored since at this stage all the equations are present in the same system of coordinates.

$$U_E = R_E I_E + \frac{d\Psi_E}{dt} .$$

As in the case of general rotating field machines, the flux interlinkage between stator and rotor must be described.

In the salient pole machine the inductances of the axes d and q are different. The inductances of the stator windings are L_d and L_q, the inductances of the excitation windings is L_E, and the mutual inductances between stator and rotor are M_{dE} and M_{qE}.

It can be derived from what has been discussed so far that in the case of the salient pole rotor machine without damper winding, only a flux interlinkage on the d axis through M_{dE} is possible. Therefore:

$$\Psi_d = L_d I_d + M_{dE} I_E$$

$$\Psi_q = L_q I_q$$

$$\Psi_E = L_E I_E + M_{dE} I_d .$$

The inductances of the d and q axes can be separated into leakage and main inductances. On the d axis the main inductance corresponds to the mutual inductance.

$$L_d = L_{\sigma d} + L_{hd} = L_{\sigma d} + M_{dE} \qquad L_q = L_{\sigma q} + L_{hq} .$$

Since, as in the case of the general rotating field machine, the torque T_{Mi} and the mechanical motion equation are independent from the coordinate system used, the following equations can be taken from section 5.1:

$$T_{Mi} = \frac{3}{2} Z_p \left(\Psi_d I_q - \Psi_q I_d \right) .$$

The equation for the torque should again be interpreted for the machine with a salient pole rotor. Introducing Ψ_d and Ψ_q in the equation we obtain:

where $$\Psi_d = L_d I_d + M_{dE} I_E$$
$$\Psi_q = L_q I_q$$

$$T_{Mi} = \frac{3}{2} Z_p \left(M_{dE} I_E I_q + \left(L_d - L_q \right) I_d I_q \right) .$$

6 Synchronous Machine

It follows from the equation of the torque that the first term is the result of the coupling of the excitation flux with the impressed stator current I_q. The second term describes a torque independent of the excitation flux. If for instance we set $I_E = 0$ and we have a machine with salient poles, then it is possible to generate a torque due to the unsymmetry $L_d \neq L_q$, this is the reluctance torque (second term). In the case of a non-salient pole rotor there is $L_d = L_q$ and the second term no longer appears. For a machine with a salient pole rotor and without damper winding, we can transform:

$$T_{Mi} = \frac{3}{2} Z_p \left(\left(M_{dE} I_{\mu d} + L_{\sigma d} I_d \right) I_q - L_q I_d I_q \right)$$

where $I_{\mu d} = I_d + I_E$.

At this point the equations must be normalized. The reference values for the stator correspond to the rated machine variables:

$$U_{Norm} = \sqrt{2}\, U_{effN}; \quad I_{Norm} = \sqrt{2}\, I_{effN}; \quad T_N = \frac{1}{2\pi f_N}.$$

The following normalized values are obtained:

$$\Psi_{Norm} = T_N\, U_{Norm}; \quad Z_{Norm} = R_{Norm} = \frac{U_{Norm}}{I_{Norm}};$$

$$L_{Norm} = \frac{\Psi_{Norm}}{I_{Norm}} = T_N \frac{U_{Norm}}{I_{Norm}}$$

$$\Omega_{Norm} = \frac{1}{T_N}: \text{ electrical}; \qquad \Omega_{0N} = \frac{1}{T_N\, Z_p}: \text{ mechanical};$$

$$N_{0N} = 2\pi \Omega_{0N} \qquad T_{Norm} = \frac{3}{2} \frac{U_{Norm} I_{Norm}}{\Omega_{0N}}.$$

Inductances and reactance at rated frequency are equal in the normalized case. Example:

$$l_d = \frac{L_d}{L_{Norm}} = \frac{2\pi f_N L_d}{Z_{Norm}} = x_d .$$

The normalized mechanical and electrical angular velocities and the rotor speed are also equal:

$$n = \frac{N}{N_{0N}} = \omega_m = \frac{\Omega_m}{\Omega_{0N}} = \omega_L = \frac{\Omega_L}{\Omega_{Norm}}.$$

With these reference values and the preceding equations we obtain the normalized equations and from these the signal-flow graph shown in Fig. 6.2.

$$T_N \frac{d\psi_d}{dt} = n\, \psi_q + u_d - r_1\, i_d$$

$$T_N \frac{d\psi_q}{dt} = -n\, \psi_d + u_q - r_1\, i_q$$

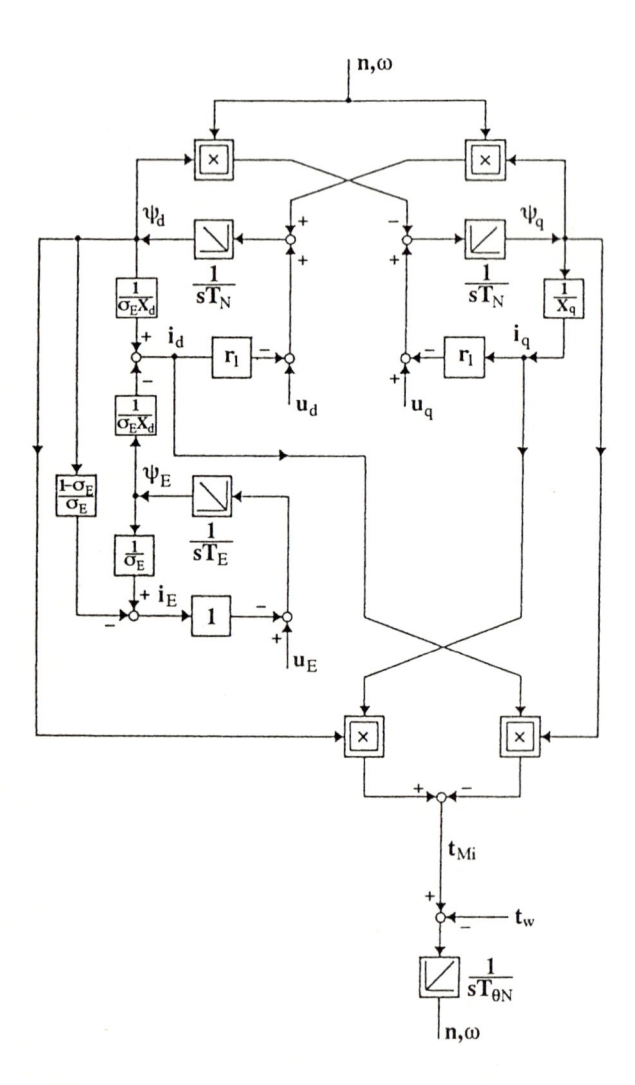

Fig. 6.2. Signal-flow graph of a salient pole rotor synchronous machine

$$T_E \frac{d\psi_E}{dt} = u_E - i_E$$
$$t_{Mi} = \psi_d i_q - \psi_q i_d$$
$$T_{\Theta N} \frac{dn}{dt} = t_{Mi} - t_W .$$

The transformation of the coordinate systems for example from the three-phase system to the d - q system is depicted in Figure 6.4, or from the speed n to $\sin \vartheta$ and $\cos \vartheta$ functions is shown in Figure 6.5.

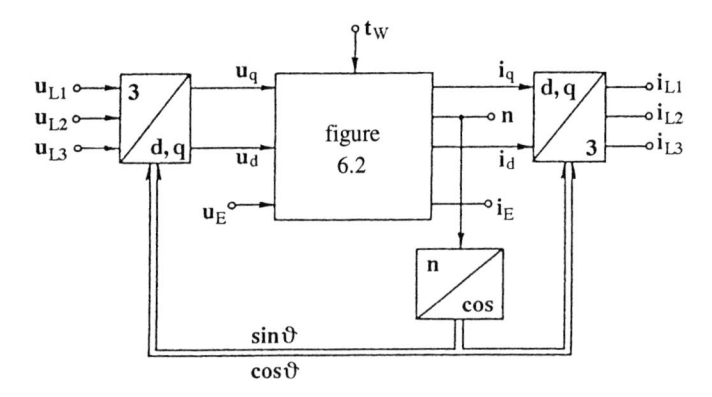

Fig. 6.3. Lock diagram of synchronous machine with impressed stator voltages

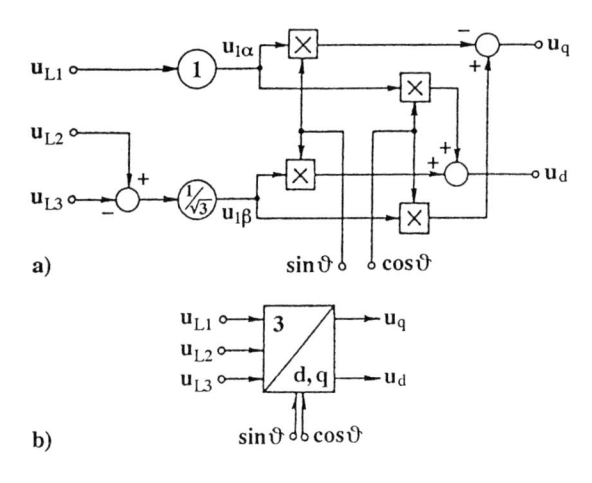

Fig. 6.4. Transformation of the three-phase voltages u_{L1}, u_{L2} and u_{L3} to the voltages u_d and u_q of the d and q axes of the synchronous machine.(a) structural diagram; (b) block diagram

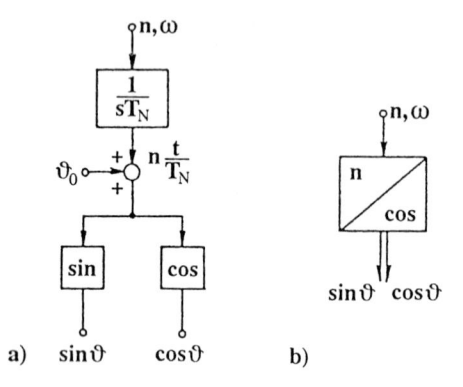

Fig. 6.5. Transformation of n speed to $\cos\vartheta$ and $\sin\vartheta$. (a) structural diagram; (b) block diagram

6.2 Synchronous Machine with Non-Salient Pole Rotor

Unlike the salient pole rotor machine, the non-salient pole rotor machine has a rotor which is cylindrical and therefore has a symmetrical structure in the d and q axes and this avoids the differences between the inductances:

$$L_1 = L_d = L_q; \quad L_h = L_{hd} = L_{hq} \quad .$$

Consequently, the different time constants due to the d and q axis are also equal. For reasons of symmetry the normalised flux equations are simplified in the system oriented on the d-q coordinates as follows:

$$\Psi_d = L_1 I_d + M_{dE} I_E$$

$$\Psi_q = L_1 I_q$$

$$\Psi_E = L_E I_E + M_{dE} I_d \quad .$$

Because of the rotational symmetry no reluctance torque will be generated and the torque equation is simplified to:

$$T_{Mi} = \frac{3}{2} Z_p \left(M_{dE} I_E I_q \right) .$$

The mechanical equation remains:

$$T_{\Theta N} \frac{dn}{dt} = t_{Mi} - t_W \quad .$$

Therefore the signal-flow graph for the non-salient pole rotor synchronous machine can be transferred from that of the synchronous machine with salient pole rotor. Compared with the normalised equations of the salient pole rotor synchronous machine, the following can be written:

$$x_1 = x_d = x_q .$$

To obtain an equivalent electric circuit we can, as in the case of the induction machine, interpret the phasors in the steady-state operation as complex vectors:

$$\underline{U}_1 = U_d + jU_q; \quad \underline{I}_1 = I_d + jI_q; \quad \underline{I}_2 = I_E .$$

The following complex equation is obtained:

$$\underline{U}_1 = R_1 \underline{I}_1 + j\Omega_L \underline{I}_1 + j\Omega_L M_{dE} I_E$$

with $\underline{U}_p$, the synchronous electromotoric force $\underline{U}_p = j\Omega_L M_{dE} I_E = jX_h I_E$ and $\underline{U}_h$, the transient internal voltage $\underline{U}_h = \underline{U}_p + j\Omega_L L_h \underline{I}_1 = \underline{U}_p + jX_h \underline{I}_1$:

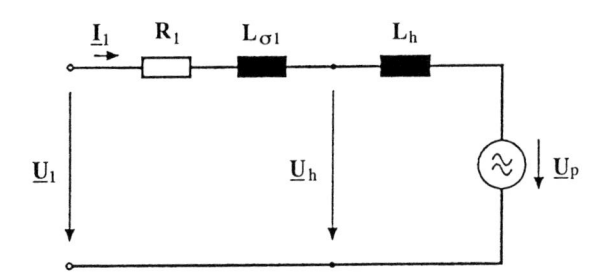

Fig. 6.6. Equivalent electric circuit of a non-salient pole rotor machine in steady state operation

The equivalent electric circuit is shown in Fig.6.6.

The voltage of the synchronous electromotoric force is a function of Ω_L and I_E and can be measured at the stator terminals when no stator voltage is applied.

Conversely, if the excitation winding is not excited $(I_E = 0)$, the vectors of $\underline{U}_1$ and $\underline{I}_1$ and therefore the parameters R_1 and X_1 can be determined.

If $I_E \neq 0$ and $\underline{U}_1$ is connected to the stator terminals, then

$$\underline{U}_1 = R_1 \underline{I}_1 + j\Omega_L L_{\sigma1} \underline{I}_1 + jX_h \underline{I}_1 + jX_h I_E .$$

The interlinked flux previously obtained from the signal-flow graph is also obtained in the equivalent electric circuit by:

$$\underline{U}_h = jX_h\left(\underline{I}_1 + \underline{I}_E\right).$$

6.3 Non-Salient Pole Rotor Synchronous Machine - Control Conditions

The synchronous machine now has to be controlled to obtain a behaviour similar to the separately excited DC-machine.

The control conditions will be derived for steady-state operation, therefore $d(.)/dt = 0$ are set for the derivations. Now the results in section 6.2, which led to the signal-flow graph in Fig. 6.6, can be used again.

These considerations have shown that with $I_E = 0$ then also $\vec{U}_p = 0$ and that for the stator voltage $\vec{U}_1$ an ohmic-inductive load circuit remains. Conversely, if $\vec{U}_1 = 0$, then the electromotoric force $\vec{U}_p$ can be measured at the terminals.

The flux interlinkage previously derived from the signal-flow graph again results in the transient internal voltage $\vec{U}_h$. With $I_q = 0$ it follows:

$$\underline{\vec{U}}_1 = R_1 I_d + j\Omega_L L_{\sigma d} I_d + jX_h I_{\mu d}$$

with $I_{\mu d} = I_d + I_E$ and $\left|\vec{U}_h\right| = X_h I_{\mu d}$.

If we wish to keep $\left|\vec{U}_h\right|$ constant, $I_{\mu d}$ must be kept constant also. This can be achieved therefore by means of I_d and I_E.

At constant speed and with $I_q = 0$, the current displacement of the current I_d can be controlled between capacitive behaviour (overexcited synchronous machine) and inductive behaviour (underexcited synchronous machine). The control of this behaviour at the terminals is achieved by the control of the exciting current I_E. The synchronous machine can therefore be used to compensate reactive power (Fig.6.7).

The relation already found for the rotating field machines is also valid, i.e. the linear increase of the terminal voltage versus the increase in speed N with constant $I_{\mu d}$ or constant flux.

After discussing the idling operation, the next step is to study control of the torque T_{Mi}. The following equations apply for the torque T_{Mi} and the stator current $\vec{I}_1$

$$T_{Mi} = \frac{3}{2} Z_p M_{dE} I_E I_q$$

$$\vec{I}_1 = I_d + j I_q$$

capacitative behaviour,
overexcited synchronous machine

$$\left(\vec{I}_1 = I_d \right)$$

inductive behaviour
underexcited synchronous machine

$$\left(\vec{I}_1 = I_d \right)$$

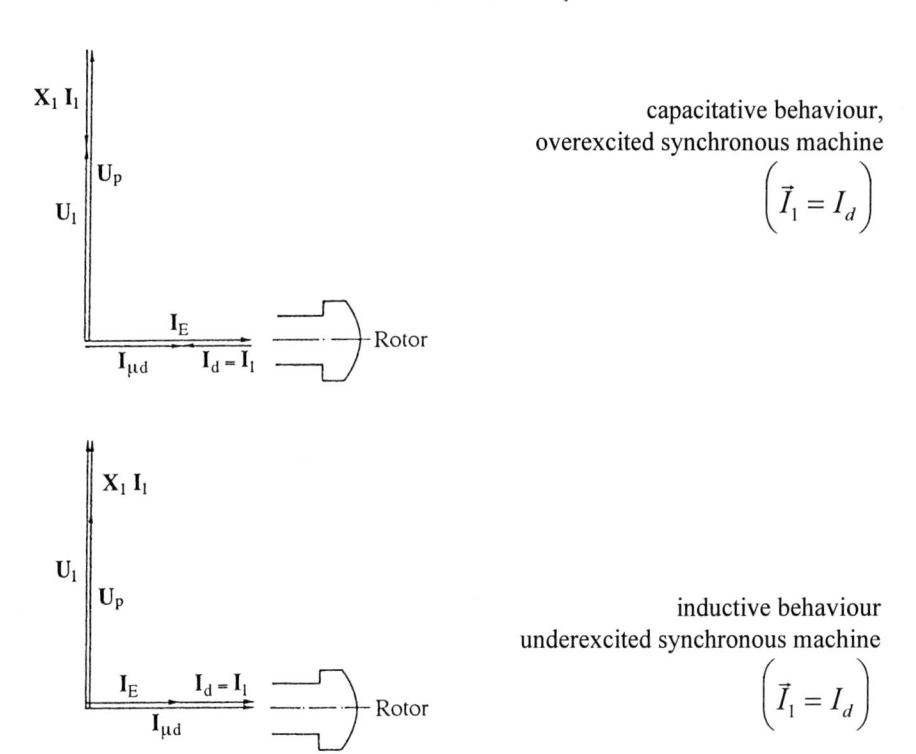

Fig. 6.7. Vector diagrams for $R_1 = 0$ (ideal idling)

This means that at $T_{Mi} \neq 0$, the current I_q is necessary in the stator current $\vec{I}_1$. Consequently a voltage drop $jX_h I_q$ is produced and thus the voltages $\vec{U}_p$ and $\vec{U}_h$ are no longer in phase. The following angles are obtained from Fig. 6.8.

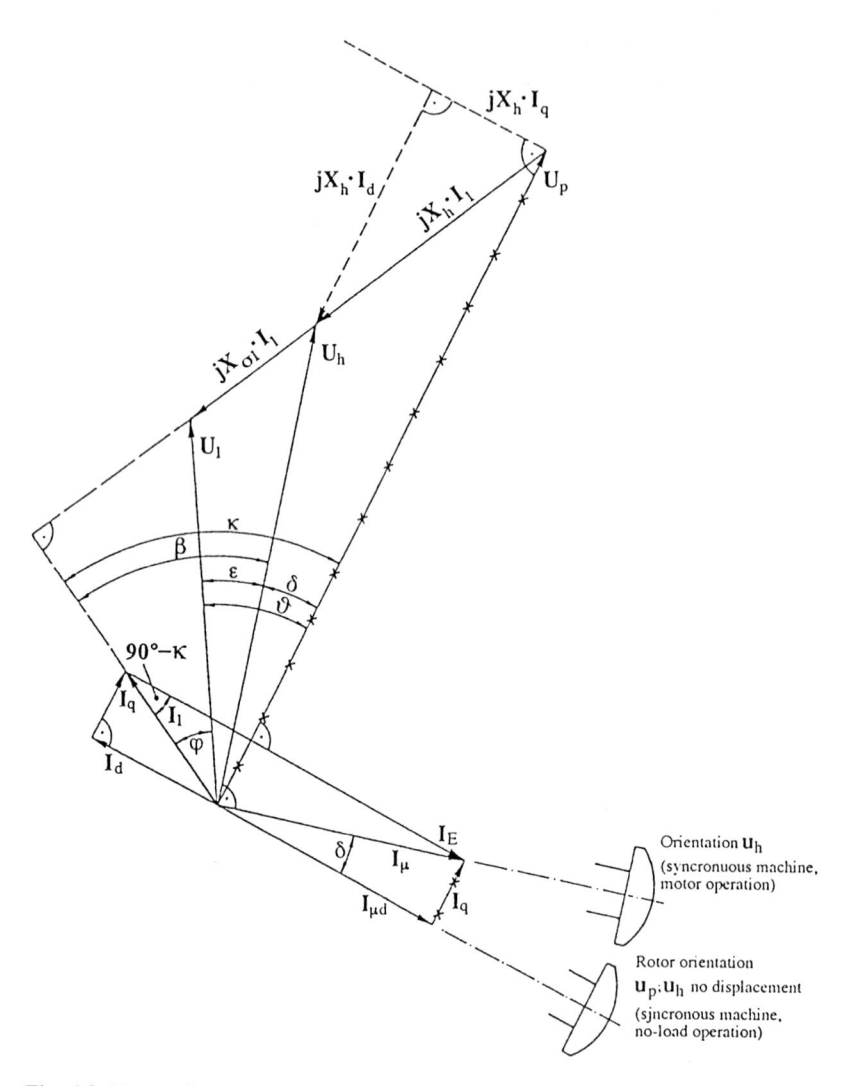

Fig. 6.8. Vector diagram of overexcited synchronous machine ($R_1 = 0$)

$\vartheta \angle (\vec{U}_1, \vec{U}_p)$,   rotor displacement angle, load angle

$k: \angle (\vec{I}_1, \vec{U}_p)$,   shaft control angle

$\varphi: \angle (\vec{U}_1, \vec{I}_1)$,   phase angle

6 Synchronous Machine

When the machine operates under load, the rotor will change position by the rotor displacement angle ϑ in relation to its no load position (motor operation, leading $\vartheta < 0$; generator operation, leading $\vartheta > 0$).

With $\vec{I}_\mu = I_d + I_E + jI_q$ the following equations are obtained from the vector diagram:

$$\left|\vec{I}_\mu\right|^2 = I_E^2 + \left|\vec{I}_1\right|^2 - 2I_E \left|\vec{I}_1\right| \sin \kappa$$

$$\delta = \arcsin\left(\frac{\left|\vec{I}_1\right|}{\left|\vec{I}_\mu\right|} \cos \kappa \right)$$

$$\vartheta = \delta + \varepsilon$$

$$P = 3\left|\vec{U}_1\right|\left|\vec{I}_1\right| \cos \varphi \qquad \text{active power}$$

$$\vec{U}_1 = jX_1\vec{I}_1 + \left|\vec{U}_p\right| e^{-j\vartheta}$$

$$\vec{I}_1 = -j\frac{U_1}{X_1} + j\frac{\left|\vec{U}_p\right|}{X_1} e^{-j\vartheta} = \frac{\left|\vec{U}_p\right|}{X_1} \sin \vartheta + j \frac{\left|\vec{U}_p\right|\cos\vartheta - U_1}{X_1}$$

and therefore: $\vec{I}_1 = f(\vartheta)$.

Circle with centre $-j\dfrac{U_1}{X_1}$ and radius $\dfrac{\left|\vec{U}_p\right|}{X_1}$, angle of rotation ϑ (Fig. 6.9).

The real part of the stator current $\vec{I}_1$ is:

$$\mathfrak{Re}(\vec{I}_1) = \frac{\left|\vec{U}_p\right|}{X_1} \sin \vartheta .$$

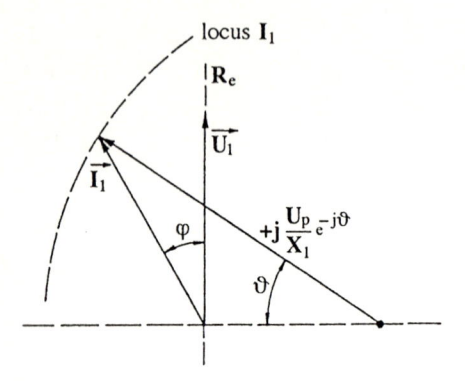

Fig. 6.9. Locus of $\vec{I}_1$, $R_1 = 0$

Therefore the active power is:

$$P = 3\, U_1\, \frac{|\vec{U}_p|}{X_1}\, \sin\vartheta$$

and the torque:

$$T_{Mi} = \frac{P}{\Omega_L} = \frac{3\, U_1\, |\vec{U}_p|}{X_1\, \Omega_L}\, \sin\vartheta\ .$$

Consequently the torque is a function of ϑ and reaches the maximum value with $\vartheta = \pm\pi/2$.

7 Inverter Drives

As described in Chapter 5 (Induction machine) and Chapter 6 (Synchronous machine) the speed and torque are controllable in the rotating field machines. To achieve this, a power electronic actuator (inverter) must convert the three-phase voltage system of the mains with fixed frequency and voltage either to a three-phase voltage system (inverter with impressed voltage) with variable amplitude and frequency of the voltages, or to a three-phase current system (inverter with impressed current) with variable three-phase amplitude and frequency of the currents. As mentioned above, an inverter with impressed voltage can be transformed to an inverter with impressed current by a current control loop. The inverters have a much more complex circuitry than those of line commutated inverters used for DC-machines. For this reason a wide range of solutions was used in the past.

7.1 Cycloconverter

A cycloconverter, for instance, generally uses a line commutated converter for each phase. The advantage of this solution is a well known technique. However, the frequency range of this inverter is close to zero and thus in the low frequency range $F_1 \leq 0.5$ (0.3) F_N. Consequently only relatively low speeds are achievable with this solution.

The drive supplied by a cycloconverter is able to supply high torque and power at moderate speeds around zero (four-quadrant operation).

Fig. 7.1. shows the circuit diagram of the control circuit which is used to impress the stator currents to the rotating field machine. As described in chapter 4.3 "D.C.-machine control", the electromotoric voltages U_{hi} of the rotating field machine are used to eliminate the disturbing variables in the current control loops.

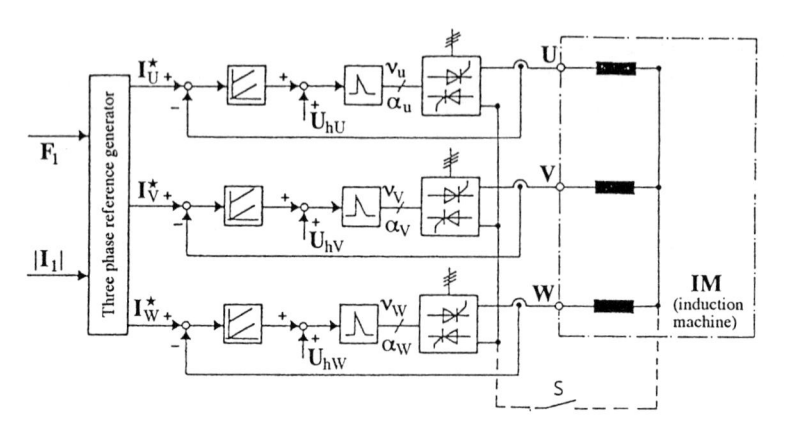

Fig. 7.1. Control circuit to produce a three-phase current system (DSQ: source for three-phase current reference variables)

The current controllers are adaptive to ensure the same dynamic responses both for discontinuous current and continuous current. The DSQ block represents the source of the required three-phase reference values of the current, the DSQ input variables are the stator frequency F_1 and the amplitude of the stator currents $|I_1|$. The DSQ output values are the three reference values for the currents.

Fig. 7.2. illustrates a quasi-stationary control method (control of the slip frequency). The slip frequency control loop is the inner cascade loop of the speed control loop. The reference values of the stator currents are controlled according to the slip frequency F_2. Starting with the rotor equation of the squirrel cage machine, we get:

$$\vec{U}_2^K = R_2 \vec{I}_2^K + j\Omega_2 L_{\sigma 2} \vec{I}_2^K + j\Omega_2 L_{2H}\left(\vec{I}_1^K + \vec{I}_2^K\right) = 0$$

with:

$$\vec{I}_1^K + \vec{I}_2^K = \vec{I}_\mu^K$$

we obtain:

$$\vec{I}_\mu^K\left[1 + j\Omega_2 \frac{L_2}{R_2}\right] = \vec{I}_1^K\left[1 + j\Omega_2 \frac{L_{\sigma 2}}{R_2}\right]$$

and therefore, after normalization to $I_{\mu N}$

$$\left|\frac{I_1}{I_{\mu N}}\right| = \left|\frac{I_\mu}{I_{\mu N}}\right| \sqrt{\frac{1 + \Omega_2^2\left(L_2 / R_2\right)^2}{1 + \Omega_2^2\left(L_{\sigma 2} / R_2\right)^2}}.$$

This equation is implemented in the non-linear block (Fig. 7.2.) and therefore the required reference value of the stator current $|I_1|$ is obtained from Ω_2. It should however be remembered that we started only with the quasi-stationary equations of the induction machine. Moreover, account must be taken of the fact that R_2 may vary considerably as a result of temperature, and L_2 because of field weakening. This will lead, in the case of deviations from the rated values, to substantial deviations in the effective values and poor dynamic performance.

Moreover it is vitally important that a digital frequency measurement is used instead of an analogue measurement of the speed.

Since the cycloconverter is mainly used for high power drives and in a low speed range (e.g. tubular mills), other solutions will not be discussed here. Another type of circuitry with cycloconverters is the *wound rotor induction machine* with slip-power recovery (Scherbius drive).

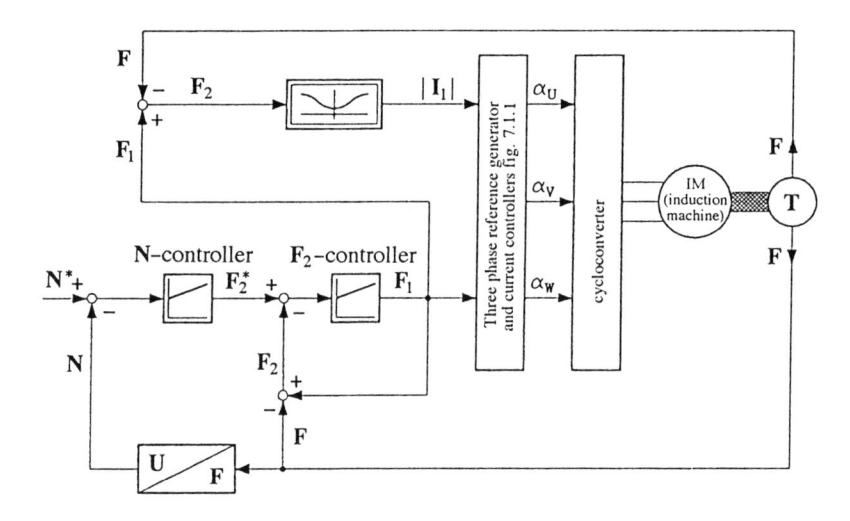

Fig. 7.2. Quasi-stationary control of slip frequency for the asynchronous machine

7.2 Wound Rotor Induction Machine With Slip-Power Recovery (Subsynchronous Cascade; Scherbius Drive)

The Scherbius drive uses an induction machine with three-phase windings in the stator and three-phase windings in the rotor. The windings in the rotor are connected to slip-rings. The structure of the drive-system is shown in Fig. 7.3. A non-controlled converter (diode bridge) is connected to the slip-rings of the wound rotor induction machine. This three-phase diode bridge will rectify the rotor voltages with frequency and amplitude proportional to the slip. The reactor with the inductance L_z absorbs the voltage difference due to the different instantaneous values of the voltages U_{ZG} and U_{zW}, thus smoothing the direct current I_z.

The line commutated converter, which usually is a three-phase thyristor bridge, supplies the effective power transmitted by the DC-intermediate circuit back to the line (energy recovery).

Fig. 7.4. shows the effects of a current-dependent converter voltage and a current-independent counter voltage (Fig. 7.5.) in the rotor circuit of an induction machine on the torque - speed characteristic. As can be seen, in the Scherbius drive the original behaviour of the machine is maintained throughout the speed range.

P_1: : stator input power

P_2: : rotor input power

P_{mech} : mechanical output power

I_z: : intermediate direct current

L_z:: reactor inductances

U_{lz}: : DC-voltage drop of the reactor $\overline{U}_{Lz} \approx 0$

U_{zG}: : diode rectifier, DC-voltage

U_{zW}: : thyristor inverter, DC-voltage

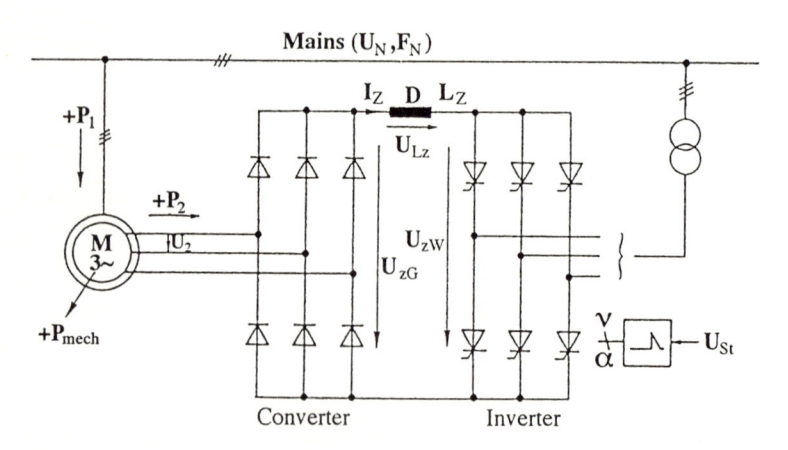

Fig. 7.3. Schematic circuit diagram of the Scherbius drive

7 Inverter Drives

Since the rating of the diode bridge is designed only for the maximum slip-voltage that may occur, the Scherbius drive differs from all the other types of inverter drives by the fact that it is sufficient to install a diode converter and a thyristor converter (in inverter operation) only with the maximum-slip-power corresponding to the speed control range actually required starting from the synchronous speed.

If this speed control range is less than 100 %, starting from stand still is obtained by means of the starting rheostat. The use of a transformer may be necessary for the thyristor converter at the mains side.

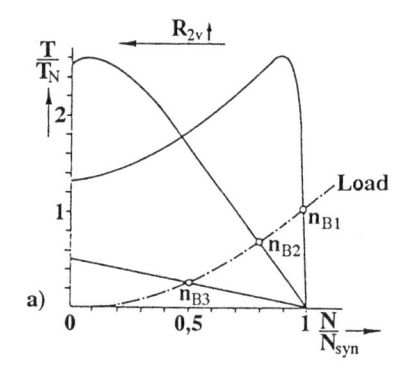

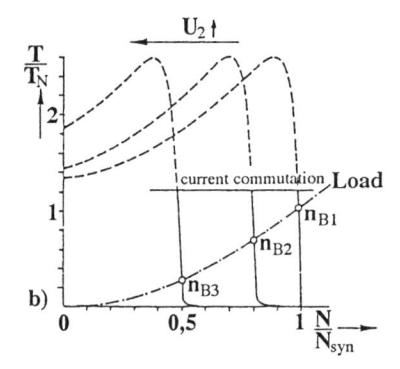

Fig. 7.4. Family of characteristics for different rotor resistances R_2, corresponding to a current-dependent converter voltage

Fig. 7.5. Family of characteristics for Scherbius drives with different converter voltages U_{zG}

Electrical relations. The electrical relations can be derived from the following considerations.

The induction machine consumes from the mains the active power P_1, which is transmitted to the stator windings. After deducting the stator losses (losses in copper and iron), the air gap power P_δ remains, which is transmitted to the rotor. After deducting the rotor losses (losses in copper and iron, through friction and ventilation) from this power P_δ, the following mechanical power is available on the shaft:

$$P_{mech} = (1-s) P_\delta \ .$$

The remaining rotor power P_2 is:

$$P_2 = s P_\delta \ .$$

The power P_2 is therefore a function of the slip and will be zero when $s = 0$ (synchronous speed). As illustrated in Chapter 5.4, the characteristic of the torque can be modified by varying the rotor resistance R_2. In the case of the Scherbius drive, the power P_2 is not converted to heat in the resistors, but about $U_z I_z$ is recovered from the mains. The following holds true

$$P_2 \simeq U_z I_z$$

$$s \Omega_{syn} T_M \simeq s P_{\delta z} = P_2$$

$$U_{zG} \simeq \frac{3}{\pi}\sqrt{2}\, s\, U_{20} \quad U_{20}: \text{idle rotor voltage}\;.$$

and therefore:

$$I_z \simeq \frac{\Omega_{syn} T_M \pi}{3\sqrt{2}\, U_{20}}$$

$$I_{2\sim} = \sqrt{\frac{2}{3}}\, I_z \simeq \frac{\Omega_{syn} T_M \pi}{3\sqrt{3}\, U_{20}}$$

and:

$$U_{zW} = U_{diW} \cos\alpha_{WR} \simeq \frac{3}{\pi}\sqrt{2}\, s\, U_{20} = U_{zG}$$

$$s \simeq \cos\alpha_{WR}\;.$$

The voltage U_{zG} is a function of the slip and, moreover, can be varied only by modifications of the design of the induction machine (U_{20}). The current I_z is controlled by $U_{zW} = U_{zG} - U_{LZ}$. The rotor current $I_{2\sim}$, and consequently the power P_2, can therefore be controlled through I_z. Thanks to this virtually any P_2/P_δ ratio can be set in the operating range and thus the mechanical power

$$P_{mech} = \left(1 - \frac{P_2}{P_\delta}\right) P_\delta = P_\delta - P_2$$

by control of the line commutated converter in inverter operation.

Limitations exist due to the fact that with $s = 0$ the voltage becomes $U_2 = 0 = U_{zG}$. At the same time, in respect of the rotor resistance R_2, there is a power loss in the rotor. In addition, the losses in the converters must be covered. It can be derived from both the boundary conditions that the synchronous speed cannot be included in the operating range and no braking operation is possible by control of the firing angle of the thyristor converter.

Control aspects. Control of the Scherbius drive is very simple as long as only steady state and quasi-stationary transients are considered (Fig. 7.6.). The firing angles of the line commutated converter, which are always in the inverter mode, produce a controllable counter voltage in the rotor, independent of the intermediate current and proportional to the slip. Therefore the rotor currents are also controlled. The torque developed by the drive is directly proportional to this current and thus to the intermediate current I_z.

The current I_z is controlled by the current control loop similar as in the DC-machine and is limited to the maximum value. The superimposed speed control loop presets, according to the load conditions, the required current reference value I_z^* and therefore P_2.

When $N < N_{syn}$ the induction machine can operate only as a motor. Therefore the Scherbius drive is always a one-quadrant system. For this reason only positive torque for accelerations and thus positive speed variations can be controlled. In the case of reduction in the required speed value, this must be accomplished by the load. It should however be noted that, during idling or running under minimum load, the speed reduction will necessarily occur under certain conditions since the parasitic currents induced in the rotor even with open slip-rings are sufficient to form a large enough torque to cover friction in the bearings, air ventilation, etc.

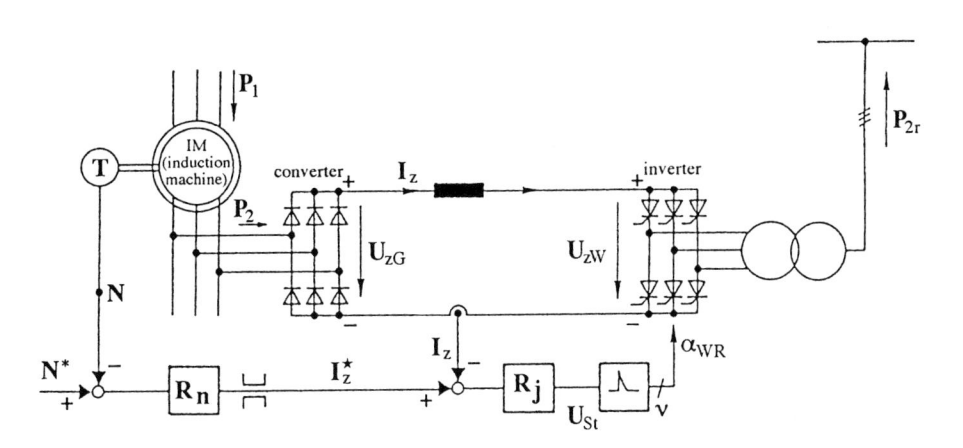

Fig. 7.6. Circuit diagram for the control of a Scherbius drive

In these cases the application of a further load torque or the disconnection of the stator from the mains is necessary. If we also wish to brake, this can generally be achieved by changing the rotation of the phase sequence at the stator. In this case the converters must however be rated for the double standstill rotor voltage, regardless of the speed control range required. This operation can be performed economically only for low power drives.

Operating range. The previous considerations show, the Scherbius drive offers advantages in all applications that require a speed control range close to the synchronous speed. In this range only the slip-power $P_2 = s \cdot P_\delta$ must be recovered by the converters. Consequently, the converter rating diminishes as the required slip range and therefore the speed control range reduces. Outside this range resistors must be used to operate the drive. This is particularly true during the start-up phase. This type of drive has, for instance, been used successfully in the operation of pumps and fans. The equations may be transferred symmetrically to the synchronous speed. At speed within $N < N_{syn}$ the induction machine can only operate as a motor. If $N > N_{syn}$, the induction machine operates as a generator. This is for example given in combustion engine test-beds. Because of the slip-rings and the rheostat start-up costs, the acceptance of the Scherbius drive has gradually diminished.

7.3 Double Fed Induction Machine

The Scherbius drive is a simple example of a double feed induction machine. The advantage of this solution with a limited speed control range is the simplicity of the converter circuitry and control and the efficiency of the system. To avoid limitation of the speed control range, an inverter can supply by the slip-rings the rotor windings. This inverter may, for instance, be an inverter with impressed voltages or impressed currents.

If, in the case of all impressed voltage inverter, the inverter is for example a cycloconverter - providing a voltage system for low output frequencies, the synchronous speed can be included in the operating range.

7.4 The Load-Commutated Inverter Drive

The load-commutated inverter drive is another type of circuit consisting of a mains-side converter STR I, as a power supply system, an intermediate coil D, a load-commutated inverter STR II, and a synchronous machine (Fig. 7.7.).

The power supply STR I is a line commutated converter supplied by the mains with a fixed voltage U_N and frequency F_N. By changing the firing angle α_I, we can obtain either a positive DC-voltage U_{zI} (GR), thus $30° < \alpha_I \leq 90°$ or a negative DC-voltage U_{zI} (WR), thus $90° < \alpha_I \leq 150°$. The *maximum firing* angle is limited

7 Inverter Drives 517

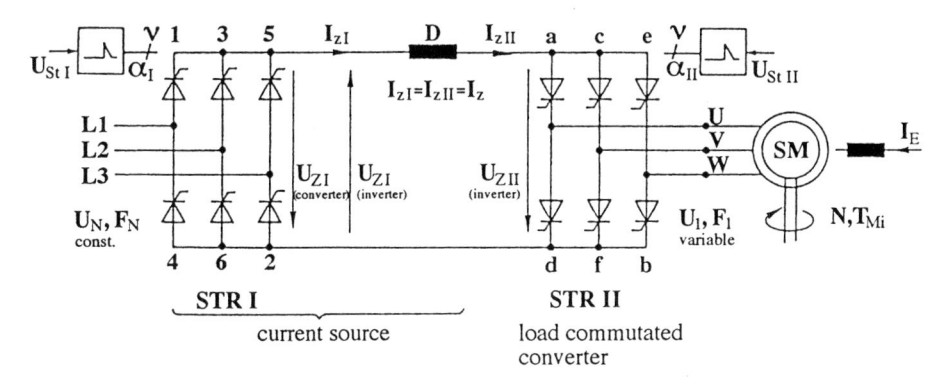

Fig. 7.7. Circuit diagram of a load commutated inverter drive

to 150° in inverter operation due to the recovery time of the thyristors. Due to the valve effect of the thyristors, however, only one current direction I_{zI} is possible.

In the intermediate circuit there is a reactor D between the converters STR I and STR II. Through the intermediate reactor D, the converters STR I and STR II are decoupled in instantaneous voltages. The average voltages, however, remain equal neglecting the ohmic voltage. The line commutated converter STR I and the reactor D, build a controllable current source with variable intermediate DC-current I_z.

This controllable intermediate DC-current I_z is now switched by the load commutated converter STR II (through two thyristors) to two coils of the synchronous machine, and a spatial current block is thus produced. By cyclical switching of the thyristors of the load commutated inverter, we can therefore produce a discontinuously rotating current phasor, which has six fixed spatial positions.

Since the rotor winding is excited by a DC-current, a torque is produced. The rotor thus follows the six spatial phasors. Therefore depending on the firing frequency of the load commutated converter STR II, the rotor will follow with the given speed of the firing frequency the six spatial current phasors and thus the speed is controllable. This means that the speed n of the rotor can be controlled through the firing frequency.

Now, it is not yet clear how the current can be switched from one thyristor to another. Assuming that the rotor winding is excited with the current I_E and that it has the speed n, a three-phase voltage system will be generated at the terminals of the synchronous machine proportional to the speed n. Commutation is possible thanks to this voltage system. Since the commutation voltage is provided by the load (synchronous machine), this type of operation is called "load commutation".

Control aspects. We must now deduce the control conditions for the converters STR I and STR II in relation to the actual power flow.

Fig. 7.1 shows that during steady-state operation, $U_{zI} = U_{zII} = U_z$, if R_D of the reactor is set to zero.

If U_{zI} is therefore in the rectifier mode, U_{zII} must be in the inverter mode and this will result in a power flow from the mains to the synchronous machine; the synchronous machine is in motor operation.

Contrary, if U_{zI} is in the inverter mode, U_{zII} must be in the rectifier mode, this will result in an active power flow from the synchronous machine to the mains, the synchronous machine is in the generator mode and the energy is recovered by the mains.

Thus only by controlling the firing angle α_I and α_{II} it is possible to achieve both motor operation and braking (operation of the synchronous machine).

Controlling the load commutated inverter, one must take into account - in order to avoid commutation failure in the inverter mode - that the control angle is chosen below $\alpha_{II} < 180°$. One will generally use $\alpha_{II\,max} = 150°$ and, for reasons of symmetry, $\alpha_{II\,min} = 30°$ as for DC-drives and four-quadrant operation. To reduce the reactive power requirements of STR II and the synchronous machine to a minimum, fixed control angle for α_{II} are used; $\alpha_{II} = 150°$ (motor operation) or $\alpha_{II} = 30°$ (generator operation), depending on the operating mode.

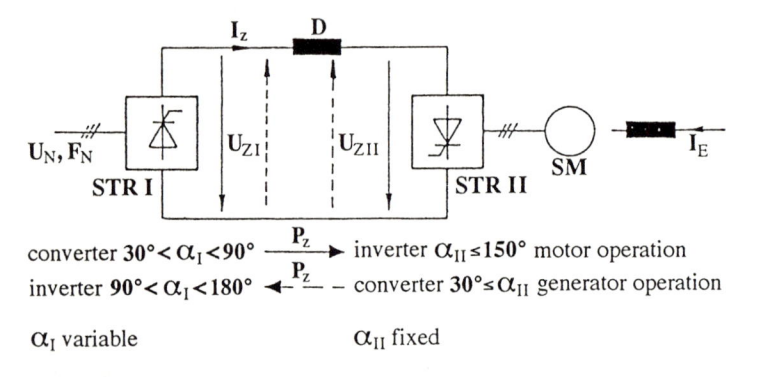

Fig. 7.8. Operating modes of the load commutated inverter drive

Due to the fixed control angle α_{II} there will be, at increasing speed n of the synchronous machine, an increasing intermediate voltage U_{zII}. Since $U_{zI} = U_{zII}$, we must therefore adjust the firing angle α_I in correspondence to the speed of the synchronous machine. For standstill $\alpha_I = 90°$, for the maximum speed in motor operation $\alpha_I = 30°$ and for maximum speed in generator operation $\alpha_I = 150°$, thus α_I uses the full range.

$$\left|U_{zI}\right| = \left|U_{zII}\right|$$

$$U_{diI}\cos\alpha_I - D_{xI}I_z = -U_{diII}\cos\alpha_{II} + D_{xII}I_z$$

$$U_{diII} = f(n, I_E)$$

$$\alpha_{II} = 150° \mid 30°$$

$$I_z = f(T_{Mi}) .$$

Speed reversal. By changing the sequence of the firing pulses for STR II, one my also change the direction of the rotation. A reversal of the direction of the rotation is therefore possible without additional components in the power circuitry - a valuable advantage.

Starting. The above mentioned conditions lead us to the conclusion that at the speed $n = 0$ of the synchronous machine, the voltage are also $U_1 = U_h = 0$. In the event of a low speed ($|n| = 0 \ldots 0,1$), the synchronous machine cannot provide the reactive power for the load commutated inverter STR II. Thus the load commutated inverter drive does not induce the speed range of $|n| \approx 0$.

In order to enable starting without too many additional components of the power circuitry, a solution of the intermediate timing control may be used. This means that STR I will be in the rectifier mode for a short period of time and two thyristors of STR II will also be fired. In the next period of time for STR I, the adjacent thyristors of STR II will be fired, producing discontinuously rotating spatial phasors with a low frequency.

Torque harmonics. At the beginning of this chapter we mentioned that the spatial fixed current phasors in the synchronous machine rotates only discontinuously, due to STR II. From the six-pulse inverter STR II, each winding is supplied by a current block of electrically 120° positive and a current block of electrically 120° negative. These current blocks produce the spatially fixed stator current phasors, which are switched every 60° electrically. During the period of current conduction of two thyristors (no load commutation of STR II), the current phasor is fixed spatially, but the rotor moves with a constant speed N. This results in a time-variant displacement between the stator current phasor and the centre position of the continuously rotating flux, produced by I_E.

In the vector diagram this results in an effect that the armature current vector covers an angle of 60° electrically and returns to its starting position during commutation.

The following equations describes the power consumption:

$$P_1 = 3 U_1 I_1 \cos\psi$$

and thus:

$$\overline{T}_{Mi} = \frac{3 U_1 I_1}{\Omega_L} \cos\psi .$$

Since the phase angle between U_1 and I_1 changes by ±30° electrically during the on-time of the two thyristors, the air gap torque T_{Mi} will have the following equations.

$$\Delta T_{Mi} = \frac{3\,U_1\,I_1}{\Omega_L} \cos(\varphi + 30° - \Omega t) \qquad 0° \leq \Omega t \leq 60°; \ p = 6.$$

As has already been mentioned, phase angle φ will be for example approximately 30° (this corresponds to α_{II}) at $\ddot{u} = 0$. Thus the torque pattern is also fixed (Fig. 7.9.).

These torque harmonics in the air gap torque are harmonics of the order $6k$, $k = 1, 2, 3, \ldots$. If the mechanical system, coupled to the drive, has mechanical resonance frequencies which coincide with torque harmonic frequencies, considerable harmonic excess torque may occur in the mechanical system. One possible remedy at low frequencies is to change the shape of the current waveshape at the load side to reduce the torque harmonics.

Another remedy to reduce torque harmonics, which is often used, is to increase the pulse code p of the inverter and therefore the windings groups of the machine.

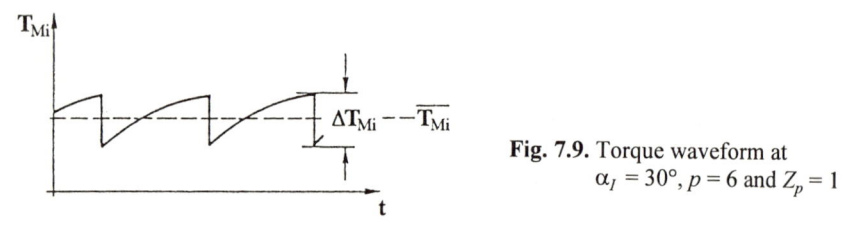

Fig. 7.9. Torque waveform at $\alpha_I = 30°$, $p = 6$ and $Z_p = 1$

For example, to increase the pulse code p to 12, the synchronous machine must have two three-phase winding systems spatially shifted by 360°/12 = 30° with respect to one another.

In addition, we must include two half-power supply systems at the mains. Thanks to the provision $p = 12$, the torque harmonics having an odd order k of the two spatially shifted systems will compensate each other; but the torque harmonics having an even order remain.

Control aspects. As already described, the converter I shall be a current source for the load commutated converter and for the synchronous machine. The current I_z will be controlled by means of a current control of the converter I. In the inverter II, the stator frequency for the synchronous machine will be controlled by adjusting the frequency of the firing pulses of the thyristors in the inverter II. In order to achieve a behaviour by the synchronous machine equal to that of the separately excited DC-machine, i.e. in order to avoid torque instability of the synchronous machine, according to the equation

$$T_{Mi} = \frac{3\,U_1\,U_p}{X_1\,\Omega_1} \sin \vartheta$$

the rotor displacement angle ϑ must always remain below 90° electrically. Similarly the angle between U_1 and U_p must also be below 90°. Since the torque $T_{Mi} = f(\vartheta)$ - in non-salient pole rotor synchronous machines without a damper coil is controlled by I_q, we can ensure that this rotor displacement control condition is observed by limiting the stator current I_1 electrically.

For inverter the load commutation must be assured. Generally, this is obtained, for instance, with $\alpha_{II} = 150°$ (30°). This information may be obtained from the zero transitions of the phase voltages of the synchronous machine, and is called "machine self control".

From Fig.s 6.4.2 and 6.4.4 in section 6.3 we can conclude that in the event of torque variations, the current I_μ, must remain constant. The following was valid:

$$I_\mu^2 = I_E^2 + I_1^2 - 2 I_E I_1 \cos(90°-k)$$

where I_1 is direct proportional to I_z. Thus, in the event of a variation in I_1, with $I_\mu = constant$, we must adapt the excitation current I_E.

Thus a control circuit may be depicted for quasi-steady-state operation of the synchronous machine (Fig. 7.10.).

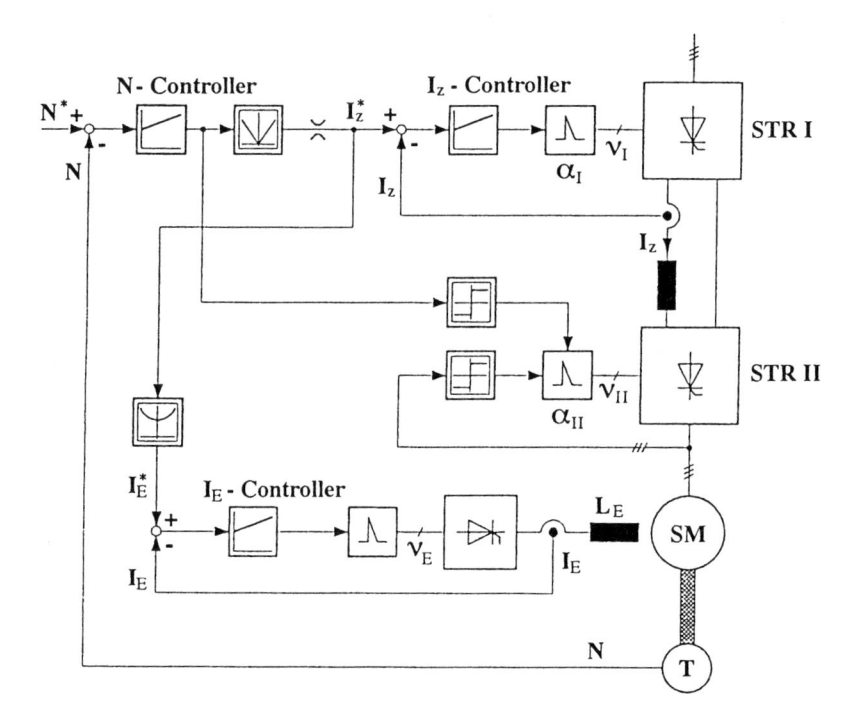

Fig. 7.10. Principle control diagram for a quasi-steady-state control of the load commutated inverter drive

Fields of application. As we have already explained the load commutated inverter drive is a four-quadrant drive for high power applications with high speeds and good static and dynamic characteristics. There are limitations: zero and low speeds. Thanks to these properties, this drive may be used for pumps, fans or rolling mills.

7.5 Forced Commutation Current Source Inverter with Thyristors

Basic system behaviour. Fig. 7.11. shows the principle circuit diagram of this drive system.

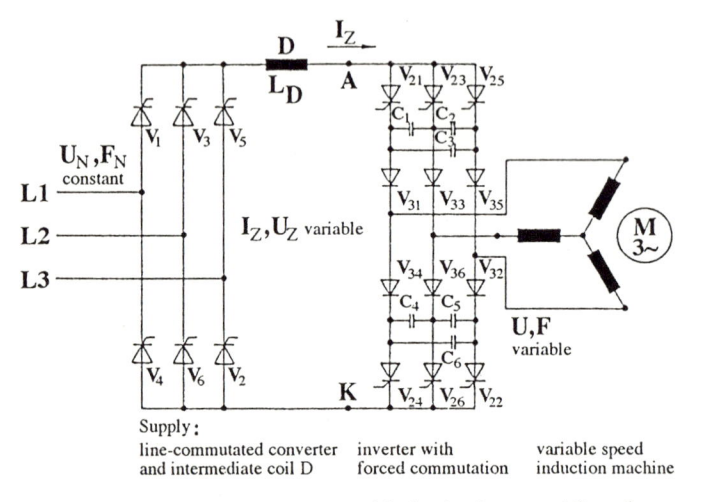

Fig. 7.11. Current source inverters with the thyristors and forced commutation

The circuit diagram looks very similar to the load commutated inverter drive. The power sources are the same in both circuits. These are current sources with controllable current I_z. The intermediate voltage U_z can be controlled according to the load conditions on the machine side. While in the load commutated inverter drive the machine-side inverter is commutated by the load, thus the reactive power is supplied from the synchronous machine, the asynchronous machine needs circuits for forced commutation to switch off the thyristors in the inverter. These commutation circuits are the commutation condensers from $C1$ to $C6$ and the commutation diodes from V_{31} to V_{36}.

The basic operating behaviour of this drive corresponds in principle to the operating behaviour of the load commutated inverter drive, thus:

- spatially fixed current pulses are applied to the induction machine windings, which are switched to different windings according to the switching status of the inverter;

7 Inverter Drives

- a discontinuously rotating spatial current phasor is produced in the stator;
- the stator frequency for the induction machine is set through the firing frequency of the inverter;
- the induction motor drive is in motor operation when the current source converter is in rectifier mode;
- braking can be achieved by a reversal of the power flow, thus the current source is in the inverter mode;
- by reversing the firing pulse sequence for the inverter, the direction of the rotation of the current phasor of the stator is reversed;
- as in the load commutated inverter drive, torque harmonics are produced in the air gap torque.

We have thus described the major features of this system. The differences are mainly affected by the commutation, which here is of a forced commutation type, and we shall refrain from going into further details. One essential aspect of this type of forced commutation is the fact that the commutation condensers are always charged in such a sequence that it is possible to switch the thyristors in either a positive or negative rotational direction. Thus the operating range for low speeds is included, and in the event of very low stator frequencies some undesirable torque harmonics may be eliminated in the air gap torque, using the pulse width modulation method for the current pulses, if a correct position and duration of the current pulse sequences are chosen.

Control aspects. The drive system has two control inputs, as in the load commutated inverter drive. The first control input is given by the current source. This can be used to adjust the intermediate current I_z and thus the machine current. The second control input is the firing frequency for the thyristors in the forced commutated inverter.

Fig. 7.12. shows a schematic diagram of the most simple control for this type of drive.

For the constant stator flux, the following idealized equations had been valid in no-load operation:

$$\Psi_{1A} = \Psi_1 = constant \qquad \Psi_{1B} = 0$$
$$U_{1A} = 0 \qquad U_{1B} = \Omega_1 \Psi_1$$

and therefore

$$U_1 = \sqrt{U_{1A}^2 + U_{1B}^2} \approx F_1.$$

In the control diagram shown in Fig. 7.12. this proportional relationship between stator frequency and stator voltage is used.

Thus the reference value of the speed directly determines the stator frequency. Simultaneously, the reference value of the speed is used as the idealized reference value for the stator voltage, and with this information the stator voltage can be controlled. The stator voltage control loop is superimposed to a current control loop for the mains-side converter (current source).

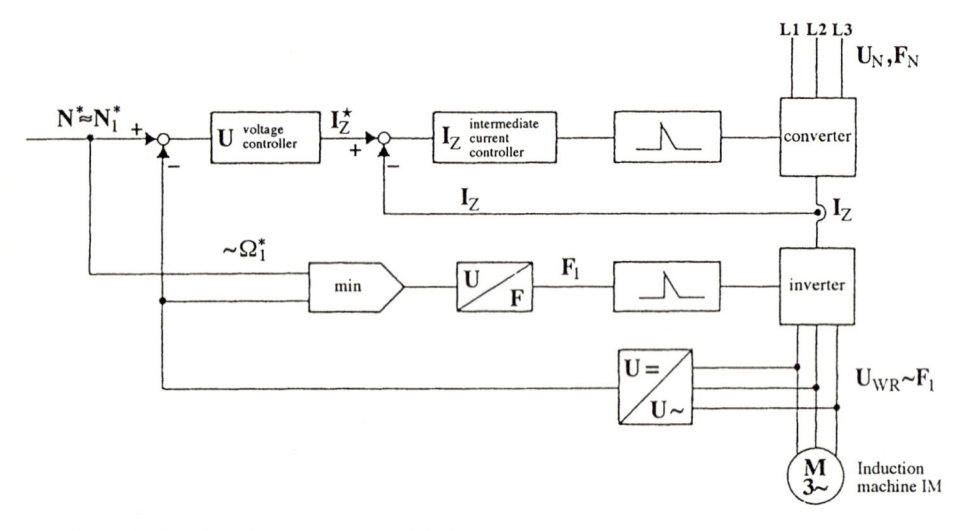

Fig. 7.12. Quasi-stationary control of the induction machine

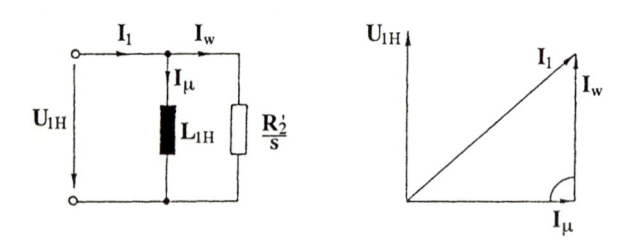

Fig. 7.13. Equivalent electrical diagram and vector diagram of the induction machine

Due to the current source the stator currents are impressed and since the leakage reactance of the rotor is neglected, we can simplify the equivalent electrical diagram of the induction machine.
Considering:

$$I_\mu X_{1H} \simeq \frac{I_W R_2'}{s}$$

with :

$$s = \frac{F_2}{F_1}$$

or:
$$F_1 s = F_2 = \frac{I_W R_2'}{2\pi L_{1H} I_\mu} = K_2 I_W$$

$$F_2 \approx I_W$$

and
$$F_1 \simeq N_L + F_2 .$$

Because there is no speed sensor, we must accept that the real speed of the induction machine is different to the reference value N^* and varies according to the load and the slip. In case of overload, the stator frequency will be reduced by minimal control. This type of stator frequency- and stator current-control is often used in drives without high dynamic requirements, such as pumps and fans. This is especially true if the induction machine is built for severe environmental conditions - such as, for instance, danger of explosion - and thus additional sensors, such as an additional speed sensor, are not allowed.

Fig. 7.13 shows a variation of the control strategy with a speed sensor. This solution may also not be used for drives with high dynamic requirements. But since the real speed is sensed, we can use a speed controller. The output signal of the speed controller is a signal corresponding to the required torque component. Since only quasi-stationary control procedures have been considered here, and we use a current impression, we can neglect the voltage drop on the stator side. This will also hold approximately for the voltage drop at the leakage reactance of the rotor side, since F_2 is small with respect to F_1 in the normal operating range. In this case, the stator current I_1 is the magnetizing current I_μ, and the current that produces the torque I_w:

$$I_1 = \sqrt{I_\mu^2 + I_W^2} .$$

Generally speaking, the same considerations are valid here as when explaining Fig. 7.2 or 7.13. This equation is used as a characteristic curve in the control path between the output of the speed controller and the current reference value. A current controller for the mains-side converter (current source) is the last detail.

Further developments for current source inverters. The introduction of power semiconductors which can be switched on and off, has led to the circuit shown in Fig. 7.14.

In this circuit the supply circuit at the mains-side is again a current source. The load side (induction machine side) inverter uses power semiconductors, which can be switched on and off. It should be indicated that these power semiconductors at the load-side need positive and negative blocking capability.

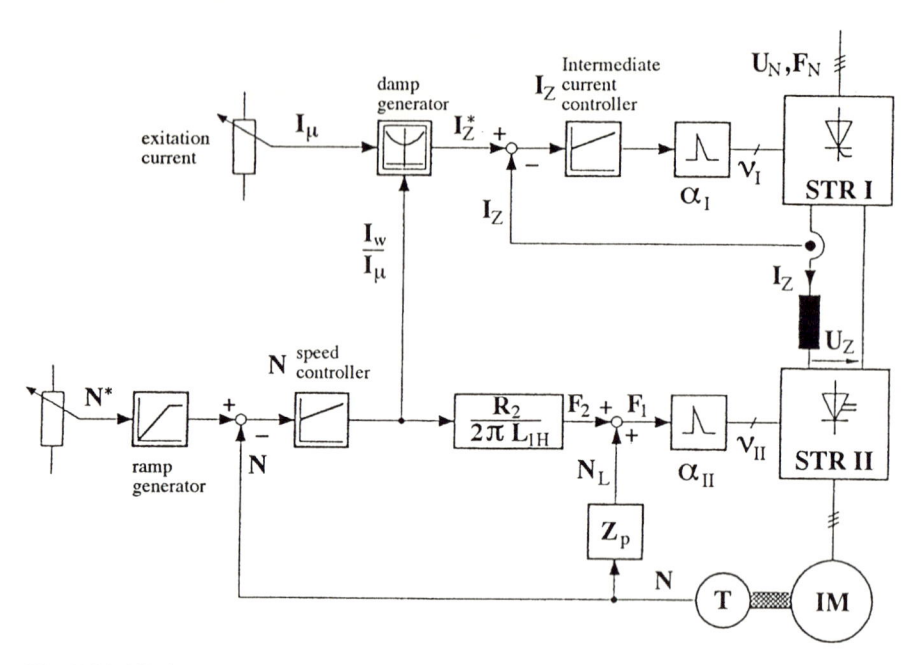

Fig. 7.14. Slip/current control of the induction machine

But switching these switches off without a decoupling circuit - the capacitor-bank - between inverter and electrical machine is not possible due to the inductances of the machine (switching off a current in an induction machine produces high voltages). By this decoupling capacitor bank the power-semiconductors can be switched on and on very frequently compared with the normal conduction period of 120° electrically in the load commutated inverter drive or the forced commutated current source inverter with thyristors. Therefore pulse width modulation can now be used. As a result of this pulse width-modulation, the capacitor bank is charged and discharged by current pulses, producing a symmetrical three-phase voltage system with a low harmonic content. The induction machine has ohmic-inductive behaviour and thus the stator currents have a lower harmonic content as the voltages. There are control strategies available, which control the voltage phasor and the current phasor at the load side. Therefore this type of inverter drive is excellent.

Fields of application. Both current source inverters make full four-quadrant drives possible, which therefore also include zero speed operation. The static and dynamic characteristics are essentially determined by the type of control, thus the user can select the most suitable type of control according to his requirements.

In the first solution according to Fig. 7.11, the fact that inexpensive power semiconductors may be used is an advantage. However, in this solution, a great

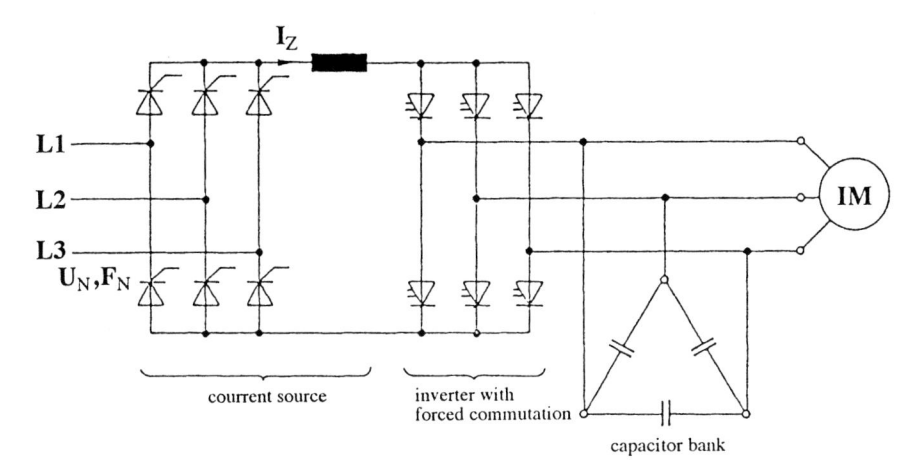

Fig. 7.15. Current source inverter with sinusoidal voltages and currents

field weakening range may be achieved only if the thyristors, diodes and commutation condensers are rated for this application. These limitation do not exist in the solution shown in Fig. 7.15.

In both solutions, the protection is easy to achieve, because a short circuit of a current source is a normal operation in pulse width modulation of current source inverter. This is a firm advantage compared to voltage source inverters.

Another important advantage is the design of the induction machines. Since the current is impressed to the induction machine, it is not necessary, as for inverters with impressed voltage, to limit the peak current, and thus the harmonic content in the current, in order to ensure current switch-off capability of the valves. The peak current must be limited in converters with impressed voltage by increasing $L_{\sigma l}$ or inserting a reactance between inverter and machine. In both cases however, with $T_K \approx M^2/\sigma L_1$, there is a reduction of the maximum breakdown torque and an increase of the breakdown slip s_K; thus the machine may no longer be as overloaded, as in the installation of converters with impressed current. Therefore drives with impressed current are used in many general applications.

7.6 Voltage Source Inverter with Intermediate Direct Voltage (VSI)

The voltage source inverter with intermediate direct voltage circuit and forced commutation has achieved an extraordinary expansion in use thanks to the availability of power semiconductors, which can be switched on and off. Furthermore these power semiconductors frequently only have the blocking capability of positive anode-cathode voltages. However, since in these inverter there is always a diode anti-parallel to the power semiconductor, this avoids any limitation.

Generally speaking, one must distinguish between VSI with variable or constant intermediate direct voltage. The difference are illustrated in the following two sub-chapters.

7.6.1 VSI with Variable DC-Voltage

Fig. 7.16 Depicts the principle electrical circuit diagram of the VSI with variable intermediate direct voltage.

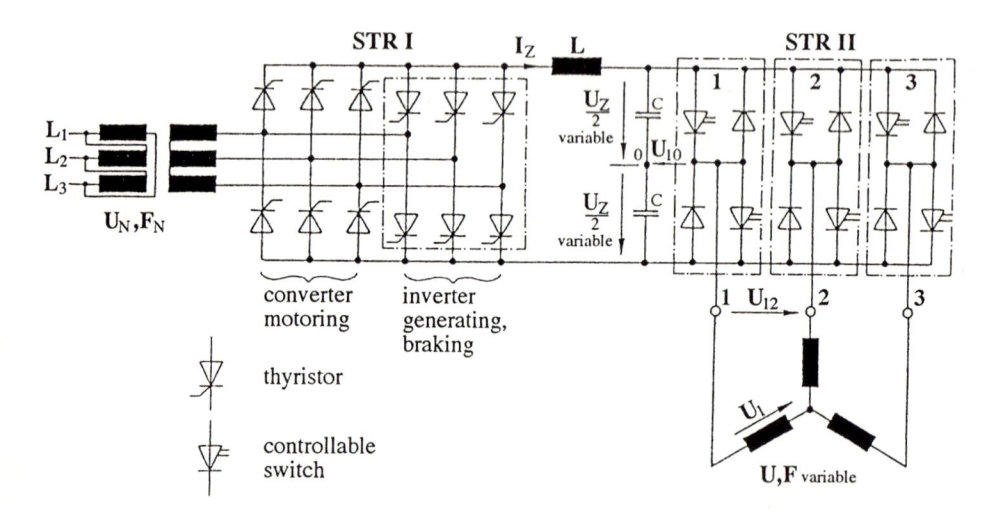

Fig. 7.16. Principle electrical circuit diagram of a VSI with variable DC-voltage

The controllable converter STR I converts the mains voltage system with a fixed voltage U_N and frequency F_N into a controllable intermediate direct voltage U_z. The voltage U_z may be regulated, as in direct current operation, through the control angle α_I.

$$U_z = U_{di} \cos \alpha_I - D_x .$$

The direct voltage U_z, which is constant in a specific point of operation but variable for the whole operation range, is converted by the inverter STR II into a three-phase voltage system having a variable amplitude and a variable frequency of the voltage for the rotating field machine.

This function may be seen in fig. 7.17..

7 Inverter Drives

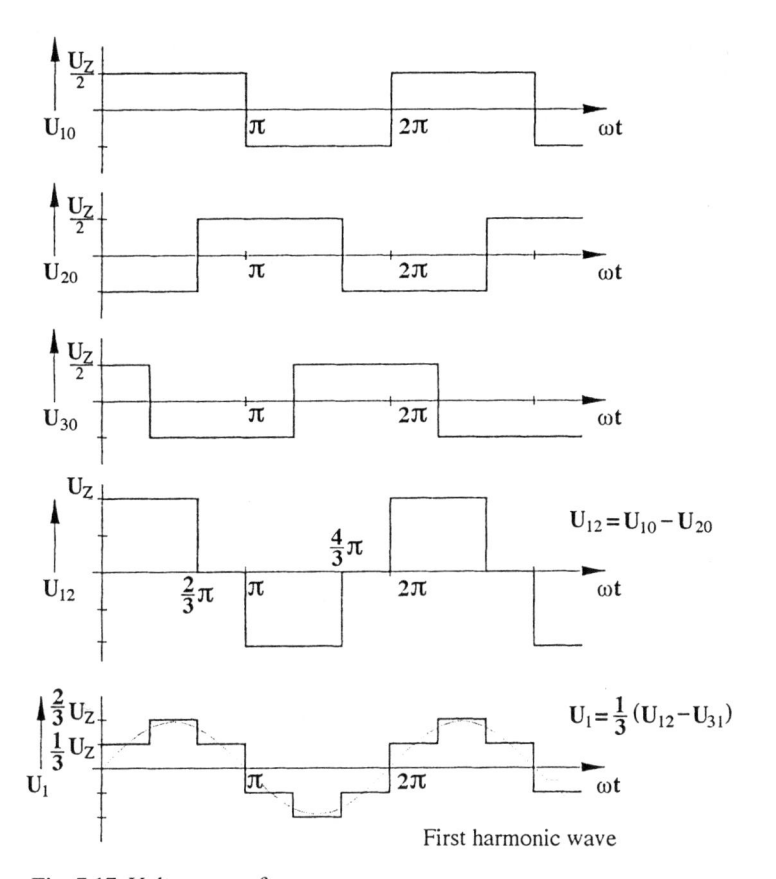

Fig. 7.17. Voltage waveforms

If the systems 1 - 3 of the inverter are controlled with the desired output frequency (fundamental frequency) and with the desired phase sequence, one gets the three voltages U_{10} to U_{30}.

By superposition the interlinked voltages (voltages between the terminals of the load) are achieved:

$$\text{e.g.:} \quad U_{12} = U_{10} - U_{20}$$

and the phase voltages

$$\text{e.g.:} \quad U_1 = \frac{1}{3}(U_{12} - U_{31}).$$

In this solution the switching sequence with only the fundamental frequency is advantageous: thus each inverter valve is activated only once per cycle. This leads to small switching losses. However, the dynamics of the control at the mains-side converter - controlling the intermediate direct voltage relatively slowly - is a disadvantage.

The VSI with intermediate variable direct voltage U_z is therefore suitable to supply rotating field machines with variable voltages and frequency. The voltage may be adjusted with the controllable thyristor bridge STR I, and the frequency is adjusted through the forced commutated inverter STR II.

In the event of a torque reversal of the induction machine, the intermediate direct current I_z must reverse its direction. In order to ensure the braking function as well, an anti-parallel converter must also be included (dotted outline).

Another version is shown in Fig. 7.18.

In Fig. 7.20, through the diode bridge, the three-phase mains voltages are converted into a constant direct voltage U_{zl}. This constant direct voltage U_{zl} is then converted into a variable DC-voltage $U_{z2} < U_{zl}$ by means of the buck converter. Finally, the variable direct voltage U_{z2} is converted into a three-phase voltage system with variable voltage U_1 and variable frequency F_1 (inverter with fundamental frequency control).

In this case, no energy recovery is possible due to the diode bridge at the mains and the buck converter. When the induction machine is in the generating mode (braking), the voltage U_{z2} will therefore increase and should consequently be limited. This voltage limitation generally takes place through a pulsed braking rheostat.

7.6.2 VSI with Constant Intermediate Direct Voltage

The elementary electrical circuit diagram of a VSI with a constant intermediate direct voltage is depicted in Fig. 7.19. The power supply from the mains-side is a diode bridge, thus the intermediate direct voltage is constant. During braking of the induction machine, a controllable thyristor bridge must be placed anti-parallel to the diode bridge. Since the maximum control angle in the inverter mode of the thyristor bridge is approximately $\alpha = 150°$, the maximum direct voltages are different between the diode bridge and the thyristor bridge.

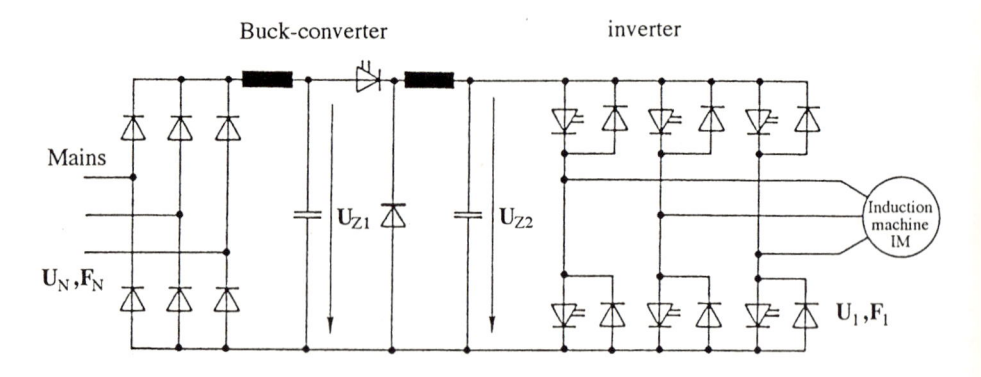

Fig. 7.18. VSI with constant intermediate direct voltage and buck converter

7 Inverter Drives   531

The thyristor bridge must therefore be supplied by a different voltage from the one used for the diode bridge (transformer). The circuitry of the forced commutated or self commutated inverter is equal to the circuitry of the inverter in the previous chapter.

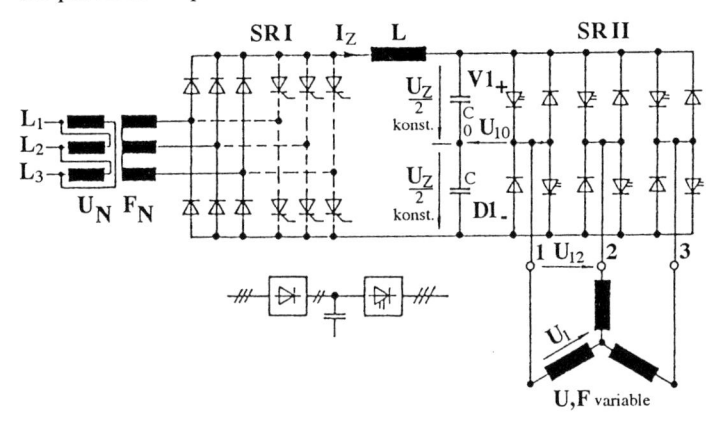

Fig. 7.19. Elementary electrical circuit diagram of the VSI with a constant intermediate direct voltage

The output voltage of the inverter may be regulated in various ways.

Modulation strategies. On-off controller with hysteresis. An easy method for controlling the VSI is on-off control. Fig. 7.20 shows the elementary control diagram and the waveforms of the output signals for an on-off current control with an ohmic-inductive load. Depending on the difference between the current reference value and the actual value measured, the direct voltage of the supply will be switched on or off, so that the current remains within a tolerance band given by the hysteresis band of the on-off controller. The switching frequency and thus the on- and off-times of the power switch may responds freely depending on the operating point and the load parameters.

The on-off control with a hysteresis band is extremely simple and has excellent dynamic behaviour. At the same time however this behaviour has a number of disadvantages that may seriously limit its application. The switching frequency does not have any distinct harmonics but generally a continuous harmonic spectrum. This disadvantage is particularly important when the switching frequency is not much higher than the desired fundamental frequency of the control.

In particularly interesting three-phase circuits with counter-electromotoric forces, the three on-off current controller - with a free star connection of the load - are no longer independent of each other. Here the sum of the instantaneous currents must always be zero. This leads to an extension of the tolerance bands and to an increase of the mean switching frequencies, but the good dynamic behaviour is maintained.

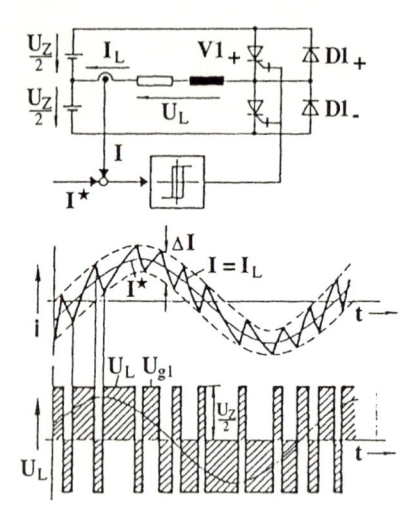

Fig. 7.20. On-off current control of one phase of a VSI

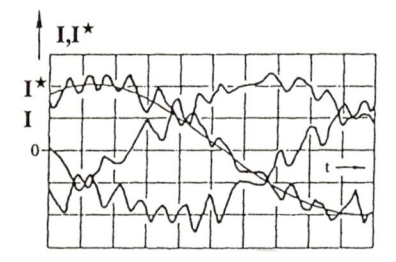

Fig. 7.21. Three-phase VSI with on-off current control

Fig. 7.21 shows the motor currents in an experimental set-up with three independent on-off controllers which show the discussed behaviour. More favourable is the direct phasor modulation strategy, which avoids this effect.

Pulse width modulation. The objective of the pulse pattern generated by the pulse width modulation process is to produce a desired output voltage at a presumed constant intermediate direct voltage. The fundamental component of the output voltage should be controllable as widely as possible and the harmonic content should be as small as possible. There are different pulse width modulation strategies, one is described in fig. 7.22.

Basically, a high frequency triangular sampling signal (e.g. triple modulation frequency) is compared with a rectangular fundamental frequency signal. The fundamental frequency signal can have a duration of 120° or 180° electrically for each half cycle. A sinusoidal fundamental frequency signal can also be used instead of the rectangular fundamental frequency signal. The difference is that a greater amplitude of the fundamental component is generated at the output with a rectangular modulation signal, while the sinusoidal modulation signals produce a smaller harmonic content. (By adding synchronised harmonics with the order $3 \cdot n$, $n = 1, 2, 3$, a transition between the two strategies can be obtained.)

The triangular signal and the fundamental frequency signal can be synchronised or unsynchronised. In the latter case, in addition to the desired fundamental output signal, harmonics with lower frequencies than the desired fundamental frequency are generated; accordingly, this type of strategy can only be applied if the control can suppress these undesired effects of low frequency components. The amplitude variation of the fundamental frequency signal is accomplished by a variation in the amplitude of the fundamental frequency signal. Different frequencies of the triangular modulation signals are necessary, when the maximum switching frequency of the power semiconductors is not high compared to the desired output frequency.

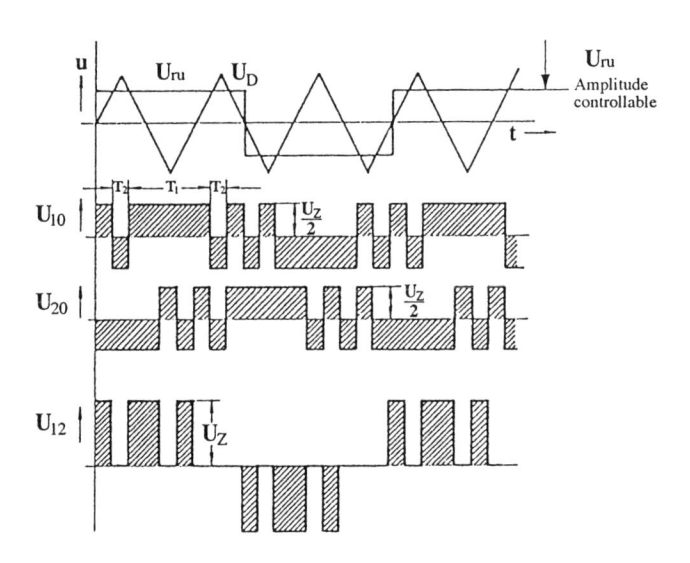

Fig. 7.22. Pulse width modulated VSI, generation of the output voltage. Sampling of a rectangular reference voltage (with variable amplitude) with a triangular signal (triple modulation frequency).

It should be noted that due to a switching of the modulation signal in frequency the output voltage of the inverter before and after the switching is not the same in amplitude and phase, which results in undesired transients.

Modern modulation strategies are optimized in real time (on-line optimized).

8 Basic Considerations for the Control of Rotating Field Machines

In the chapters featuring the signal-flow graphs of rotating field machines and control, the basic control assumption was a supply with voltage- or current-phasors. Moreover, it was shown that dynamic behaviour may be readily understood only, if an optional coordinate system K, for example, is oriented on the flux (stator or rotor). It was also pointed out that in the actual situation a rotating field machine generally has symmetrical three-phase stator windings and that this system of windings is supplied by an inverter, which provides a three-phase, symmetrical current or voltage system, variable in amplitude and frequency.

This combination of facts produces the following problem that on the one hand the signal-flow graphs were based on the K-coordinate system while on the other hand the stator windings, and with them the inverter must be considered in the S-coordinate system supplying the stator windings. This means that all control signals in the K-coordinate system in steady state are DC-signals, while in the S-coordinate system they are sinusoidal signals with stator frequency Ω_K. Consequently it is necessary to have a coordinate transformation for the control signals oriented on the flux, to signals oriented on the stator windings. This transformation was achieved in the synchronous machine by means of the rotor orientation to obtain the angle of rotation between the K- and S-coordinate system. In the asynchronous machine the determination of the angle is much more complex since the position, for example, of the flux is not directly available.

In order to use the deducted signal-flow graphs, a distinction must be made between two different control strategies. With the first strategy the frequency Ω_2 and indirectly the torque is controlled, while the flux has only an open loop control. This method is known as "decoupling". In the second strategy both the torque and the flux are controlled in a closed loop. This method is known as "field orientation". Both of these methods will be explained in principle.

8.1 Decoupling

As has been discussed, in the case of decoupling the frequency Ω_2 is controlled in a closed control loop, due to this, indirectly, the torque is also indirectly controlled, while the flux is only controlled in an open control loop. The phrase,"indirect control of the torque" means that in the case of incorrect control of the flux, the torque is also affected. Figure 8.1 shows the elementary signal flow graph of the decoupling control.

It must be noted that reference values for $\psi_{1(2)}^*$ and Ω_2 are provided as input signals for the decoupling circuit. In the decoupling circuit an inverse model of the rotating field machine is realized which calculates the stator voltages or -currents in a Cartesian coordinate system K and the stator frequency as reference values for the inverter. The dashes indicate estimated signals as a result of parameter uncertainties. It is important in this solution that, because of the open loop flux control, on the one hand the real value of the flux (amplitude and position) is not required, but that on the other hand the static and dynamic couplings of the signals in the rotating field machine are considered.

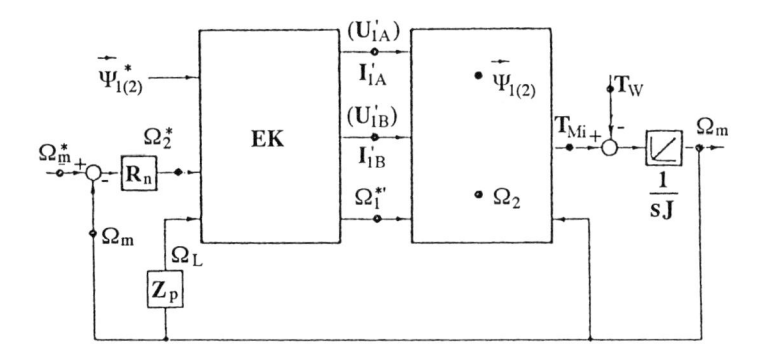

Fig. 8.1. Elementary signal flow graph of the decoupling control

If it is supposed that the flux has been set exactly by the open loop control, then the considerations known for control of the torque T_{Mi} can be used to understand the following figures $\Psi_{2A} = const.$ and $\Psi_{2B} = 0$:

$$T_{Mi} = \frac{3}{2} Z_p \frac{M}{\sigma L_1 L_2} \Psi_{2A} \Psi_{1B}$$

$$\Omega_2 = \frac{M}{L_1} \frac{\Omega_{2k}}{\Psi_{2A}} \Psi_{1B}$$

and therefore

$$T_{Mi} = \frac{3}{2} Z_p \frac{1}{\sigma L_2 \Omega_{2k}} \Psi_{2A}^2 \Omega_2$$

i.e. the torque is - in the case of Ψ_{2A} - controllable through Ω_2. Ω_2 is represented as output signal of the speed controller then- on the assumption that $\Psi_{2A} = const.$ - the torque T_{Mi} is controlled in a closed loop control. It should be noted that the flux is only open loop controlled. This means that in the presence of different parameters in the rotating field machine on the one hand and in the decoupling circuit on the other hand, or in the case of amplitude errors or delays in the inverter, clear differences will arise between the real flux and the reference value: to draw attention to this fact the output signals of the decoupling circuits EK are identified by the superscript dash (estimated quantity - may be incorrect).

The remaining problem is how, starting from Fig. 8.1, the inverter is actually controlled.

Figures 8.2 to 8.4 show possible solutions.

The stator voltage phasor in the coordinate system K has been calculated in the section 5.2:

$$\vec{U}_1^{*K'} = \vec{\Psi}_1^{*K}\left[j\Omega_K' + s\right] + \frac{R_1'}{\sigma' L_1'}\left[\vec{\Psi}_1^{*K} - \frac{M'}{L_2'}\vec{\Psi}_2''^{K}\right].$$

This stator voltage phasor must be impressed to the stator windings of the induction machine in the desired operating point. The essential consideration for decoupling is: to design a decoupling circuit in such a way, that in the desired operating point - parameters $\Omega_K = Z_p\Omega_m + \Omega_2$ and $\vec{\psi}_1^K$ for example the stator voltage phasor $\vec{U}_1^K$ is the output signal. This consideration results in the very important equation:

$$\vec{U}_1^K = \vec{\Psi}_1^K\left[j\Omega_K + s\right] + \frac{R_1}{\sigma L_1}\left[\vec{\Psi}_1^K - \frac{M}{L_2}\vec{\Psi}_2''^{K}\right].$$

The result of this consideration is $\vec{U}_1^{*K} = \vec{U}_1^K$. The signals calculated by the decoupling circuit EK are marked by a dash, and the reference values by an asterisk. If all the parameters of the induction machine and the model are exactly equal and also the initial conditions, e.g. zero, then $\vec{\Psi}_1^{*K} = \vec{\Psi}_1^K$ and $\vec{\Psi}_2''^{K} = \vec{\Psi}_2^K$ must be equal in static and dynamic state. Consequently the reference value of the stator flux $\vec{\Psi}_1^{*K}$ will directly control $\vec{U}_1^K$ and thus $\vec{\Psi}_1^K$

However, it should be noted that the inverter between the decoupling circuit and the stator windings of the induction machine, which is not shown in Figure 8.1,

8 Basic Considerations for the Control of Rotating Field Machines 537

must generate a signal corresponding to $\vec{U}_1^{*K'}$ perfectly in order to guarantee no errors between $\vec{\Psi}_1^{*K}$ and $\vec{\Psi}_1^{K}$ (in amplitude and phase). In this way the basic method of decoupling is described.

The implementation of this method is shown in Fig. 8.2.

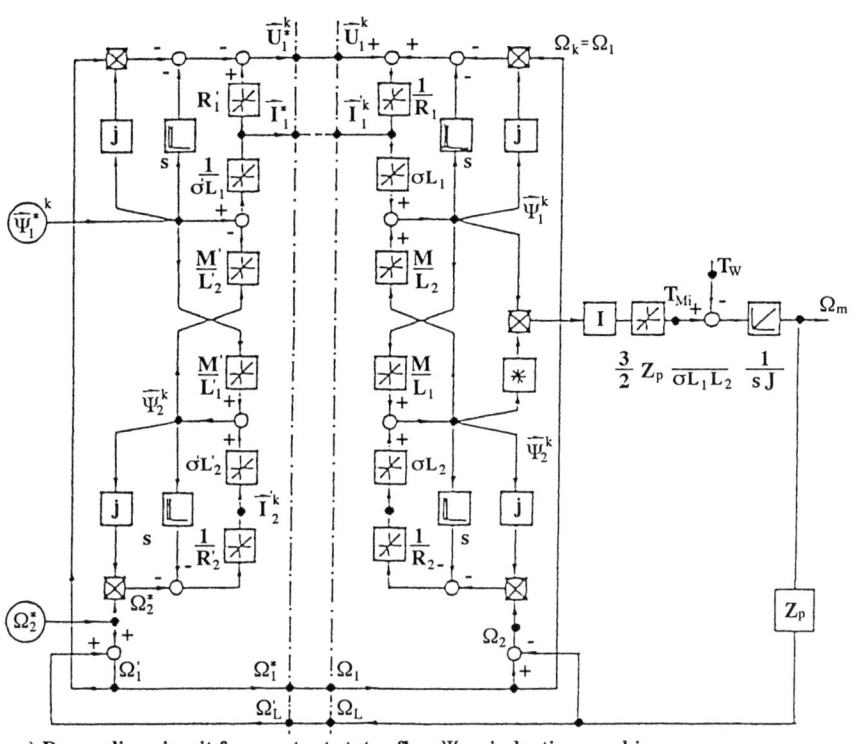

a) Decoupling circuit for constant stator flux Ψ_1 induction machine

Fig. 8.2. Induction machine with decoupling control for constant stator flux Ψ_1 and impressed stator voltages (- - - impressed stator currents)

On the left-hand side of this figure is depicted the signal-flow graph of the decoupling circuit (last but one equation) taken from Figure 8.1, and on the right-hand side the induction machine in a complex representation (last equation) with the assumption of a constant stator flux $\left|\Psi_1^K\right|_0$. In practical realizations of the decoupling circuit EK, a separation must be made, as in section 5.2 et seq., between the real and the imaginary part of the signals:

$$U_{1A}^{*'} = \frac{R_1'}{\sigma'L_1'}\left[\Psi_{1A}^* - \frac{M'}{L_2'}\Psi_{2A}''\right] + \frac{d\Psi_{1A}}{dt} - \Omega_K'\Psi_{1B}^*$$

$$U_{1B}^{*'} = \frac{R_1'}{\sigma'L_1'}\left[\Psi_{1B}^* - \frac{M'}{L_2'}\Psi_{2B}''\right] + \frac{d\Psi_{2B}}{dt} - \Omega_K'\Psi_{1A}^* \ .$$

The value $U_1^{*'} = \sqrt{U_{1A}^2 + U_{1B}^2}$ is the amplitude of the stator voltage phasor and $\Omega_1^{*'} = \Omega_K'$ is the angular frequency.

A general application is shown in Fig. 8.3. If, as has generally been supposed thus far, it is desired that $\Psi_{1A}^* = \Psi_N$ and $\Psi_{1B}^* = 0$, then the brokenline path of the signal no longer applies in the signal-flow graph with Ψ_{1B}^*. If only the armature control range is assumed, i.e. if it is established that $\Psi_{1A}^* = \Psi_N$ = constant, then the differentiating transfer-function with the Laplace parameter s can be omitted too.

The result is an extremely simple signal-flow graph for torque- and flux control. In the same way the known equations in Chapter 5.2 et seq., can be used to design decoupling circuits for the rotor flux as a reference value in the open loop control or when the stator currents are impressed.

Figure 8.4 shows the same structure as in Fig. 8.1. However, the Cartesian signals of the stator voltage phasor or the stator current phasor are transformed by a "Cartesian/polar" coordinate conversion:

	cartesian	polar
	U'_{1A}, U'_{1B}	$\left|U_1^{*'}\right|, \gamma_u^{*'}$
or		
	I'_{1A}, I'_{1B}	$\left|I_1^{*'}\right|, \gamma_i^{*'}$.

The conversion of the coordinates means that both the complex voltage or the complex current signals are transformed to the amplitude and the phase.

To obtain the real control signals for the inverter in amplitude and frequency, the value of the amplitude signal $\left|U_1^{*'}\right|$ is used directly. The frequency signal $\Omega_1^{*'}$ is obtained by an additioned dynamic signal with $\dfrac{d\gamma_u^{*'}}{dt}$ with $\gamma_u^{*'} = \arctan \dfrac{U_{1B}^{*'}}{U_{1A}^{*'}}$ and Ω_1. This gives the resulting reference value of the stator frequency.

With regard to the solution with the decoupling circuit, a behaviour similar to that in a separately excited DC-machine may be achieved, if the excitation current is open loop controlled and the armature current controlled in a closed loop.

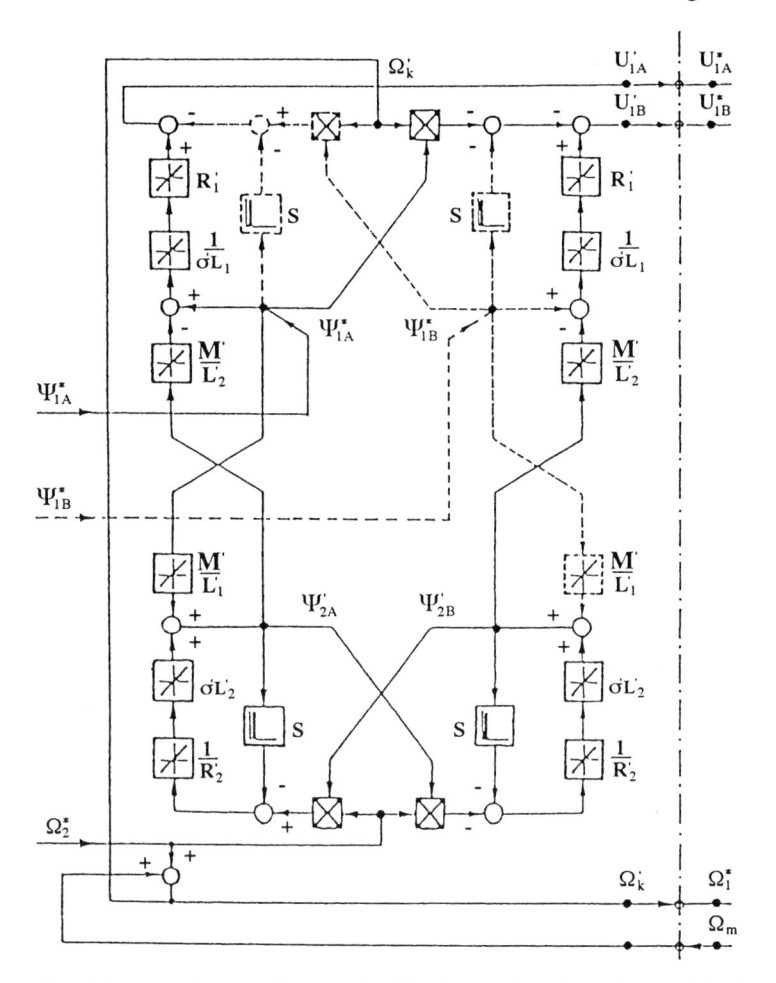

Fig. 8.3. General signal-flow graph of the decoupling circuit for control of stator flux Ψ_{1A}

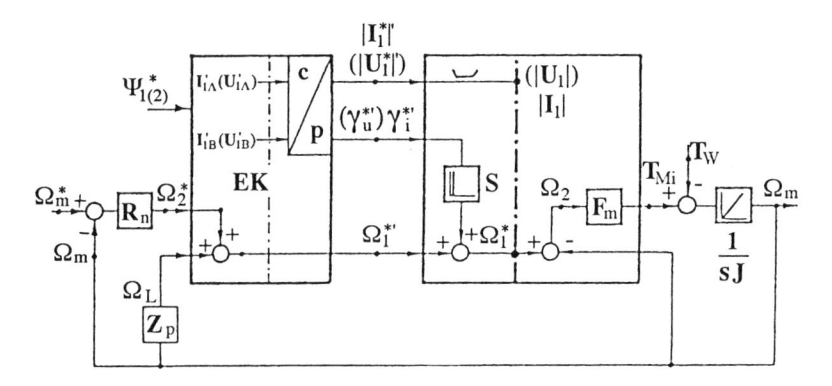

Fig. 8.4. Basic structure of variable speed drive including inverter control

8.2 Field Orientation

The basic assumption for the decoupling strategy was to ignore the orientation of the flux. When the flux orientation is chosen as the starting point for control of the rotating field machine, the phasor of the flux Ψ_1 or Ψ_2 must be known. Two possibilities arise (Fig. 85):
1. The flux space vector is measured - this is the direct field orientation method
2. The flux space vector is estimated - this is the indirect field orientation method

Figure 8.5 shows a general scheme of both possibilities for rotor flux orientation.

It can be seen from Fig. 8.5, there are two closed loop control circuits, if the field orientation is used: The first control circuit includes the speed controller and the control circuit for I_{1B}. The second control circuit contains the flux controller with the real value of the flux (measurement or model), and the incorporated control circuit for I_{1A}. In both control circuits a vector rotation circuit VD+ is

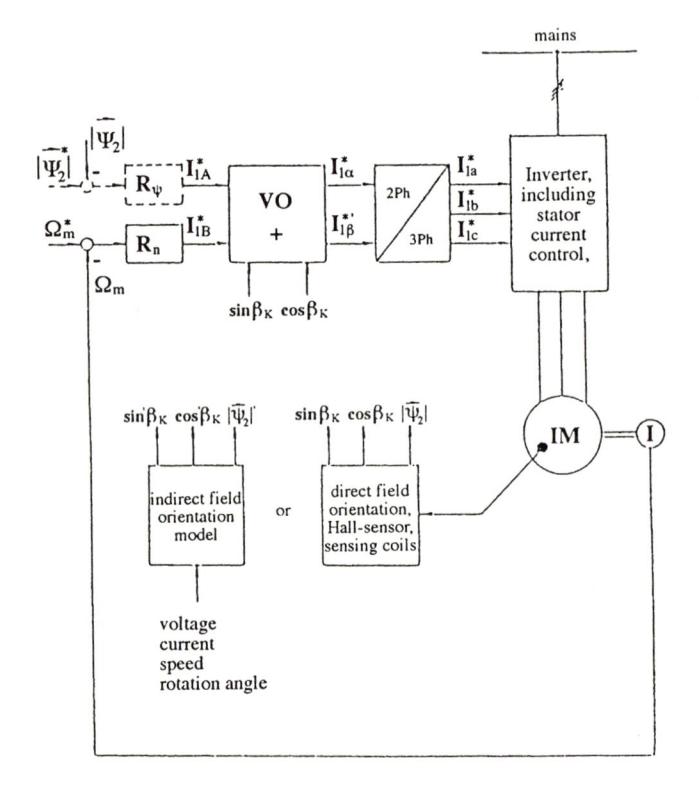

Fig. 8.5. Simplified structure of field-oriented speed control, control of the stator current in the stator fixed coordinate system

8 Basic Considerations for the Control of Rotating Field Machines 541

necessary, which is required for the transformation of the signals from the K coordinate system of the control loop to the S coordinate system of the inverter and machine. This transformation of the K coordinate system to the S coordinate system is necessary, as explained above, in order both to guarantee control of the currents I_{1A} and I_{1B} in the K coordinate system and also to provide the converter and therefore the stator windings of the rotating field machine with the stator voltages and currents in the stator winding fixed coordinate system.

First of all, it must be supposed that the flux orientation and with it the angle β_K is exactly known. Thanks to this assumption the upper part of Fig. 8.3 can be explained as example the equations of the torque T_{Mi}.

$$T_{Mi} = \frac{3}{2} Z_p \frac{M}{L_2} I_{1B} \Psi_{2A}$$

and of the rotor flux

$$\Psi_{2A} = M I_{1A}$$

(boundary conditions Ψ_{2A} = const., Ψ_{2B} =0) are repeated, then - thanks to the assumption of the control of the fluxes Ψ_{2A} and of the current I_{1B} - the torque T_{Mi} is controlled exactly too.

The basic difficulty in achieving the field orientation control is the determination of the flux phasor according to amplitude (absolute value) and phase. As stated above, the determination can be made by measurement or by a model. Since however in the case of measurement (direct method) a sensor must be incorporated in the rotating field machine, and since it is sensitive to disturbances, the direct method is generally not used.

Using the indirect method, the flux phasor must be estimated in a model with the available signals of the rotating field machine, like the stator currents and voltages and the speed. A more detailed representation of the indirect field orientation strategy is shown in Fig. 8.6.

The upper part of Fig. 8.6 corresponds to the upper part of Fig. 8.5. The bottom part of Fig. 8.6 shows, with slightly greater precision, one of the many variations of the indirect field operation. It is of fundamental importance that the stator currents I_{1a} - I_{1c} and the machine's speed, for example, may be used as available signals.

The three-phase stator currents represented in a stator fixed coordinate system are transformed in a first transformation stage into the currents $I_{1\alpha}$ and $I_{1\beta}$ (stator fixed coordinate system). And now the real problem of field orientation begins. With the currents $I_{1\alpha}$ and $I_{1\beta}$ the estimated currents I'_{1A} and I'_{1B} in the K coordinate system are calculated by means of the vector rotation VD- circuit. The problem lies in the fact that to calculate the angle β'_K, , which is necessary as signal in both the vector rotation circuits VD- and VD+, the estimated currents I'_{1A} and I'_{1B} are again required, as well as a model, the speed and an integration.

This means that the determination of the amplitude and orientation of the flux occurs in a very complex closed loop non-linear control circuit. It also means that any errors in determining the speed, as well as in the model, the integration and the trigonometric functional diagram $\sin\beta'_K$ and $\cos\beta'_K$, lead to the errors in determining the orientation and amplitude $\left|\Psi''_{2A}\right|$ of the flux.

However, ally errors in the estimation of the flux orientation immediately leads to an incorrect separation of the components of the stator currents which generate the flux and torque and may lead to undesirable machine operating conditions.

Due to this extremely high precision standards must be demanded, especially in the estimation of the orientation of flux. Moreover, parameter identifications are necessary to ensure that the time variant parameters of the rotating field machine are not a further source of errors.

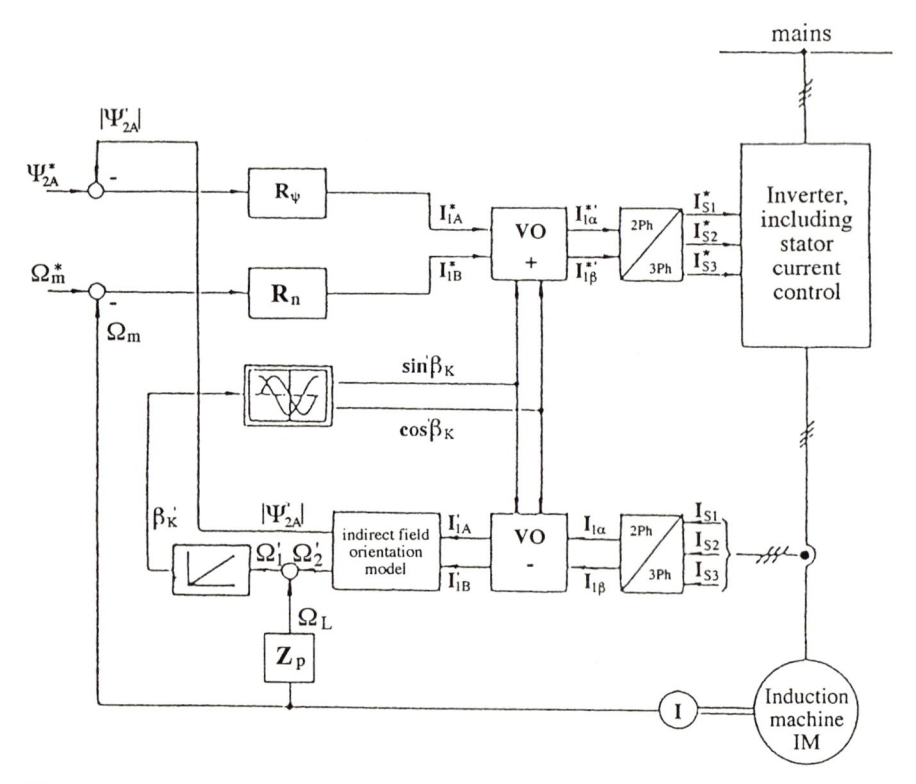

Fig. 8.6. General representation of the indirect field-oriented control using the current model

At this stage there is no chance to go into further detail or discuss the decoupling strategies and field orientation. This is a matter of control technique and is dealt with in detail in the specialist literature, as for example in *Electrical drives II, control of drives* in the Springer-Verlag.

9 Permanent Magnet Machines

9.1 Permanent Magnet Synchronous Machine

In permanent magnet excited rotating field machines the structure is broadly the same as the synchronous machine, i.e. there is a stator with the symmetrical three phase winding system and a rotor, which in this case has a permanent magnet instead of the excitation winding.

A basic distinction must be made between two types of construction: on the one hand the permanent magnet synchronous machine and on the other hand the brushless DC-machine. In the permanent magnet synchronous machine the stator is supplied with a sinusoidal current or voltage, while in the brushless DC-machine rectangular voltages or currents pulses are used. Basically, the mathematical treatment of the first type is simpler, since in deducing the signal-flow graph one can use the system of equations as for the synchronous machine and the equation for the excitation winding does not apply. Given that this machine is generally constructed symmetrically in the rotor, the signal-flow graph is simplified in a manner similar to the non-salient pole rotor machine (e.g. no reluctance torque) (Fig. 9.1).

Of importance of this machine is the fact that ψ_{PM} is fixed through the permanent magnet. Field weakening operation is in principle possible with $I_d < 0$, but the result is a considerable increase of the stator current and therefore of the inverter current too. It is therefore useful in the armature control range to establish $I_d = 0$, in order to use the inverter and the stator windings only for the current I_q that generates the torque.

As in the synchronous machine the d-q axes are fixed to the rotor, i.e. the d axis is oriented on Ψ_{PM} and with it I_d is oriented on Ψ_{PM} too. The q axis is electrically leading by 90°. This rotor fixed coordinate system rotates with Ω_L and therefore an angle β_L can be defined between the d axis and the stator-fixed coordinate system too.

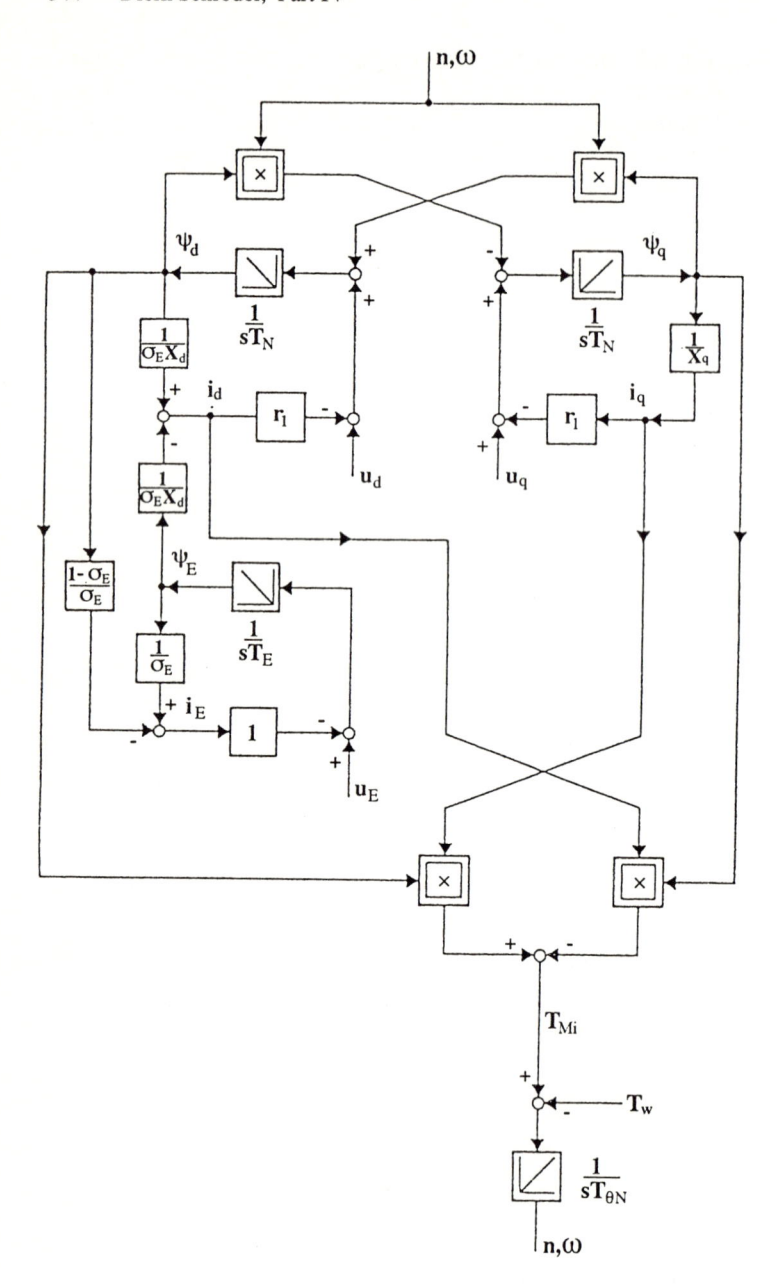

Fig. 9.1. Signal-flow graph of the permanent magnet synchronous machine

As discussed for the induction machine, a simplified and linear control strategy is achieved by an orientation of an optional coordinate system K to the flux. In the case of a PM synchronous machine the stator current I_d may be fixed to $I_d = 0$. With this the following equation for T_{Mi}, which has no reluctance effects (in $I_d = 0$ and $\Psi_d = \Psi_{PM}$), holds true:

$$T_{Mi} = \frac{3}{2} Z_p \Psi_{PM} I_q$$

i.e. by impression of I_q the torque can be controlled directly. If the signal-flow graph shown in fig. 9.1 is used with the additional condition $I_d = 0$ as a prior assumption for the design of the control system, then it will be seen that for the PM machine the control conditions for the voltages are broadly similar to those of the synchronous machine. From the signal-flow graph (Fig. 9.1) it can be seen that for $I_d = 0$ we obtain:

$$\Psi_d = \Psi_{PM}$$

$$\Psi_q = L_1 I_q$$

$$\frac{d\Psi_d}{dt} = 0 = U_d + \Omega_L L_1 I_q$$

$$\Rightarrow U_d = -\Omega_L L_1 I_q .$$

With this control condition $I_d = 0$ maintained.
Moreover it follows:

$$\frac{d\Psi_q}{dt} = U_q - R_1 \frac{\Psi_q}{L_1} - \Omega_L \Psi_{PM}$$

and then

$$T_1 \frac{d\Psi_q}{dt} + \Psi_q = T_1 (U_q - E)$$

with

$$T_1 = \frac{L_1}{R_1}$$

With these control conditions the signal-flow graph in Fig. 9.1 is simplified and Figure 9.2 ensues (including the reluctance effects $L_d \neq L_q$ but $I_d = 0$).

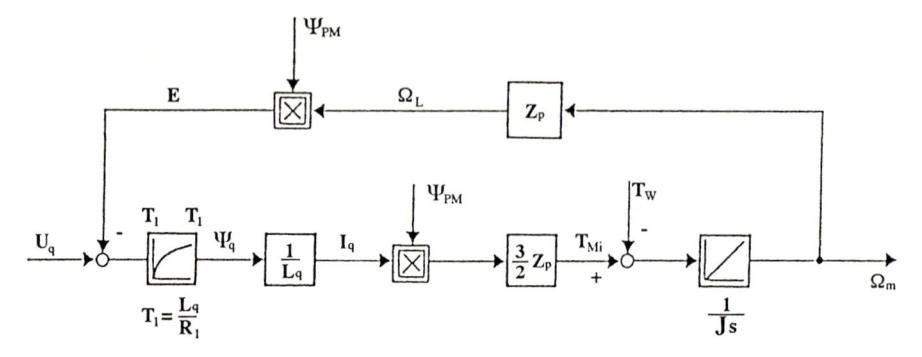

Fig. 9.2. Simplified signal flow graph of PM machine with $I_d = 0$

With this assumption the PM machine has for $I_d = 0$ a signal-flow graph roughly equal to the DC-machine. Therefore the same control strategies can be applied as for the separately excited DC-machine.

The flux model and coordinate transformations from a flux-fixed coordinate system to a stator-fixed coordinate system (or vice versa), which were necessary in the induction machine (in the presence of incorrect models), can now be replaced by the clear transformations of coordinates. The rotor position is sensed (sensor is necessary) and results in the blocks $e^{+j\beta_L}$ and $e^{-j\beta_L}$, which is an enormous advantage !

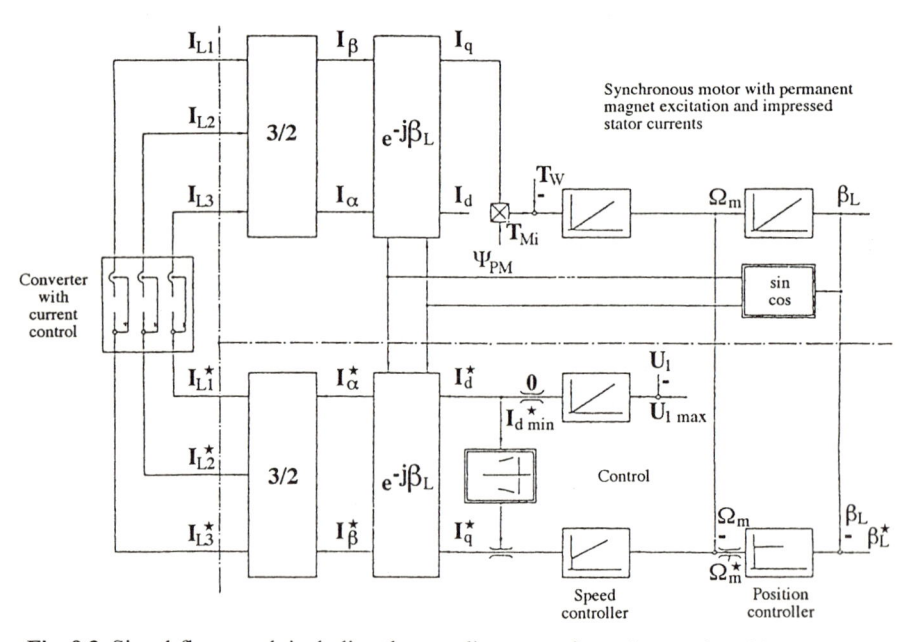

Fig. 9.3. Signal-flow graph including the coordinate transformations and position control ($Z_p = 1$)

Fig. 9.3 shows the signal-flow graph extended with the additional signal processing components. With regard to the PM-machine, it is an advantage that the angle β_L can be measured easily, since the spatial position of the flux Ψ_{PM} is fixed by the permanent magnets and with it to the rotor. The following questions remain to be answered with regard to the application of the current. Owing to the structure of the current control circuit, the same difficulties emerge in the current controller as in the separately excited DC-machine. Specifically, in the case of small currents the undesirable discontinuous current may arise, which has markedly negative effects, especially in dynamic behaviour. This deterioration of the dynamic behaviour results in all amplitude reduction of the current phasor and with it to a shift in orientation, leading overall to a substantial reduction in the available torque. Accordingly, measures to prevent discontinuous currents vital. Moreover, the electromotoric force E in an extremely dynamic signal as in the separately excited DC-machine, a disturbance variable compensation should be used.

9.2 Brushless DC-Machine

As was stated in the previous chapter, a rectangular current impression may be considered instead of the sinusoidal stator current and this results in the brushless DC-machine. In this solution it is important that, in contrast to the application hitherto assumed of three simultaneous stator currents, only two phases are supplied, each with current blocks of 120° electrically. The analysis of this machine is not so simple since account must be taken of both the fundamental and the harmonics. Because of this difficulty the space vector method is not applicable as easy as might be supposed.

These machines are analysed favourably in a time domain using the state space method; in this way the inductance variations that vary with time and depend on position are also taken into account. Because of these difficulties, reference should be made to the specialist literature rather than going into particular detail here. However, for the purpose of separation, several typical results should be mentioned: Abbreviations: re = rectangular; sin = sinusoidal.

In the case of rectangular supply the torque harmonics must be observed as these torque harmonics may disturb the mechanical system.

$$I_{re} = 1.22 I_{sin}$$

α: half of the electrically elongation of the magnet at the rotor surface

$$M_{re} = 0.86 M_{sin} \quad \text{with} \quad \alpha = \pi/2$$

$$M_{re} = 1.11 M_{sin} \quad \text{with} \quad \alpha = \pi/3$$

Part V

Total Quality Management and Continuous Improvement

Prof. Haijme Yamashina
Doctor of Engineering in Mechanical Engineering
Master of Science in Industrial Engineering
n and Operations Research

1 Outline of Total Quality Management

1.1 Definition of Quality

Quality has been defined in a variety of ways by many experts in quality who have received much recognition. They may be categorized in the following four groups:

(1) From the viewpoint of the user:

   (a) Quality is a predictable degree of uniformity and dependability, at low cost and suited to the market - Deming
   (b) Quality if fitness for use - Juran
   (c) Quality is the totality of features and characteristics of a product or service that bear on its ability to satisfy stated or implied needs - (ISO 9000), etc.

(2) From the viewpoint of the maker

   (a) Quality is conformance to requirements - Crosby etc.

(3) From the viewpoint of both of the user and the maker

   (a) Quality is in its essence a way of managing the organization - Feigenbanm
   (b) Quality is a systematic approach to the search for excellence. (Synonyms : productivity, cost reduction, schedule performance, sales, customer satisfaction, team work, the bottom line).
   The American Society for Quality Control Poster, etc.

(4) From the viewpoint of the society

  (a) Quality is the loss imparted by the product to society from the
      time the product is shipped. - (Taguchi), etc.

Obviously, there are as many definitions of quality as there are experts, protagonists, authors of quality books and quality organizations. Here it is sufficient to take the first viewpoint and define quality in the following way:

Quality is a characteristic property to be the object of valuation for determining whether or not a product or service is satisfying its purpose - (JIS: Japanese Industrial Standards).

Quality control (QC) can be defined in the following way :

QC is a system of means to economically produce products or services which satisfy customers' requirements.

Quality has various characteristics. The quality of gear boxes for general power transmission and changer in speed, for example, includes such quality characteristics that cover the maximal power transmission, speed changing ratios, shift movement smoothness, noise level, power loss, weight, shape, length, dimensions of key components, starting and initial characteristics, maintenance factor, life, appearance, etc. Such requirements are often incorporated in technical specifications. But technical specifications may not necessarily mean that the intended result of a product or service will be consistently met if there happen to be any deficiencies in the specifications or in the organizational system to plan, design and produce the product or service. Therefore, there are two aspects in dealing with quality, i.e. management and technical. To explain the first, it is worth having a comprehensive view of total quality management in Japan which is dealt with in Paragraph 1.2. Paragraph 1.3 addresses to technical aspects.

1.2 Brief History of Total Quality Management in Japan

1.2.1 Introduction of SQC (Statistical Quality Control)

Recently, the need for quality has become a matter of supreme importance. This first originated and, ironically, ignored for more than two decades by American and European industry, but taken up very seriously by Japanese industry. Consequently, the main facts in the emergence of ideas pertaining to quality have flourished in Japan and the West still lags behind.

After World War II, the Japanese launched a new nationalistic drive for growth, pursuing economic rather than military goals. But, at the beginning, the Japanese faced a serious problem of overcoming the reputation of Japanese goods having poor quality.

1 Outline of Total Quality Management

In the occupation period, one of the troubles the American army faced was many breakdowns in Japanese telecommunication systems. Also, they often did not function properly. There were large differences of quality among Japanese telecommunication devices. Therefore, the American occupation forces advised the Japanese electrocommunication industry to adopt the newly developed concepts of statistical quality control (SQC); this was the first adoption of SQC in Japan.

In 1950, the Japanese Industrial Standards (JIS) system was established. For items specified by the government, this system enabled Japanese manufacturing companies to mark products with the JIS label when the government acknowledged the factory producing the products to have carried out SQC properly. This helped SQC to grow among Japanese manufacturing companies eventually.

In the same year, the U.S. quality expert Dr. W. Edwards Deming was invited to speak by the Japanese Union of Scientists and Engineers (JUSE), which was founded in 1946, and he held an eight-day seminar on SQC for QC engineers and a one-day seminar for top management. What he taught was the importance of taking four steps: Plan-Do-Check-Act in problem solving, which is often referred to as the Deming Cycle, the viewpoint of statistical variability and the concept of process control with the usage of control charts, etc. His visit was a revelation for the Japanese and many factories began to use SQC tools such as control charts. But it was not an easy task for them to introduce SQC and many problems relating to personell emerged :

(1) Skilled workers who were used to relying on their own experience and intuition showed mental objections towards statistical methods.
(2) Technical, operational and inspection standards necessary to manage a factory properly were either rarely available or not complete. When manufacturing staff tried to establish such standards, many people objected to them, saying, "There are too many factors to be identified in such standards. " Many skilled workers insisted that they could do their jobs without such standards."
(3) Proper data, needed to carry out SQC, in many cases were not available.
(4) Even when some data were available, the methods used to collect data were incorrect, therefore the available data were not usable.
(5) When measuring equipment to collect data was installed, there were workers who thought it was being used to assess them, they sabotaged it.

These problems were gradually overcome through continuous efforts by the Japanese management.

1.2.2 Necessity of Quality Assurance

In the history of Japanese quality assurance, quality was, at the beginning, assured based on inspection. This quality assurance was called *inspection oriented quality assurance*. This meant that inspection had to be carried out strictly in order to assure quality. This is referred to as the first stage of quality assurance which focuses mainly on inspection of final products, involves only the inspection and quality departments. Their main functions are to prevent defective products from being released from the company. In other words : "Quality assured by sorting out good and bad products." But this method of quality assurance had many problems :

(1) Inspectors actually do not build in quality.
(2) The responsibility of quality assurance does not belong to the inspection division, but to the design and production divisions. Thus, unless design and production divisions work at solving quality problems, there will not be any real improvement of quality.
(3) It is liable to take too much time to feedback information from the inspection division to the production division.
(4) As the production speed is increased, it becomes impossible to make inspection by operators properly.
(5) There are often many items which cannot be guaranteed by inspection.

If defectives are actually produced one after another during processing, it is no use carrying out strict inspection because that is not solving the actual quality problem unless inspection is made to prevent any recurrence. When sufficient effort is made not to produce defectives by monitoring and controlling the quality factors of the processes, unnecessary expenses for inspection can be saved. Thus, the Japanese abandoned this quality assurance system after they started to build in quality based on process control. This quality assurance was called *process oriented quality assurance*. This is referred to as the second stage of quality assurance which focuses on process control, involves the shop floor, the subcontractors, the purchasing department, the production engineering department and even the business department. Hundred per cent good quality is pursued by investigating process capability carefully and by controlling the production process properly. In other words : "Build-in quality at the process". Some Japanese manufacturing companies stated that they were not interested in buying products from the company which sold good products, but were interested in buying products from the company which had good production processes." This concept is still important at present, but as the customers' requirements have become more stringent, even this quality assurance system has proved insufficient. For instance, problems concerning reliability, safety, bad design, bad selection of material, etc., cannot be solved solely by the manufacturing division.

1.2.3 Toward Total Quality Management

In 1954, another U.S. quality expert Dr J. M. Juran was invited by JUSE to give lectures on quality management. With this opportunity, Japan became more aware of the need to bring quality to management and Japanese quality control began to be transferred from technically oriented QC on the shop floor into quality assurance which focuses on all levels of management.

Quality assurance carried out only at the shop level has obvious limitations: the best time to implement quality assurance effectively is at the early design stage of developing a new product as shown in Figure 1.1.

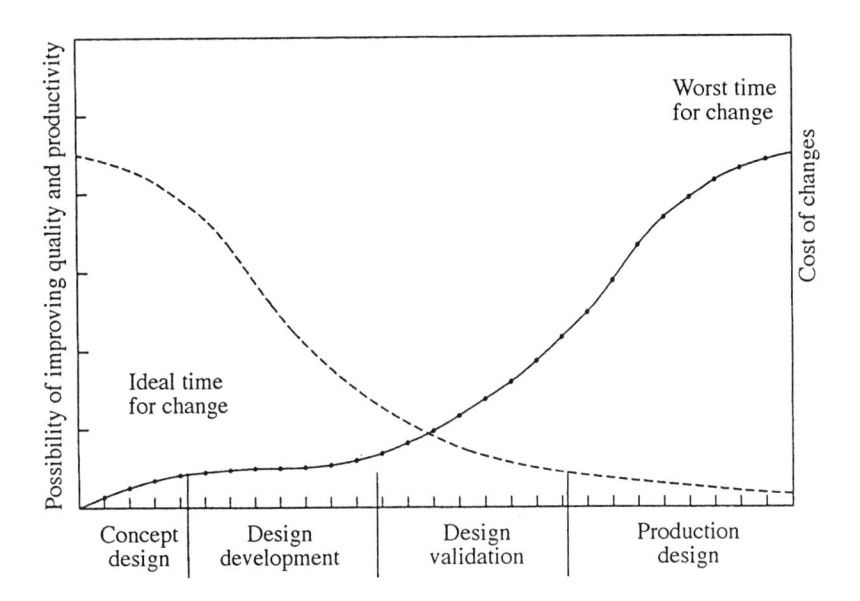

Fig.1.1. The Best Time to Implement Quality Assurance Effectively

It illustrates how much greater and cheaper the possibilities of improving product quality and reducing product cost are than during production. This method of quality assurance was called *quality assurance in new product development*. This is referred to as the third stage of quality assurance, which necessitates the participation and cooperation of everyone in the company involving top management, managers, supervisors and workers in all areas of corporate activity such as market research, research and development, product planning, design, preparations for production, purchasing, vendor management, manufacturing, inspection, sales and sales after services as well as financial control, personnel administration and training and education.

In fact, failure at the introductory stage of new products results in total failure of the production process. The visit of Dr Juran to Japan was an epoch-making event for the Japanese in developing quality control into Total Quality Management which gradually grew among Japanese manufacturing companies in the late 1950s (Fig. 1.2).

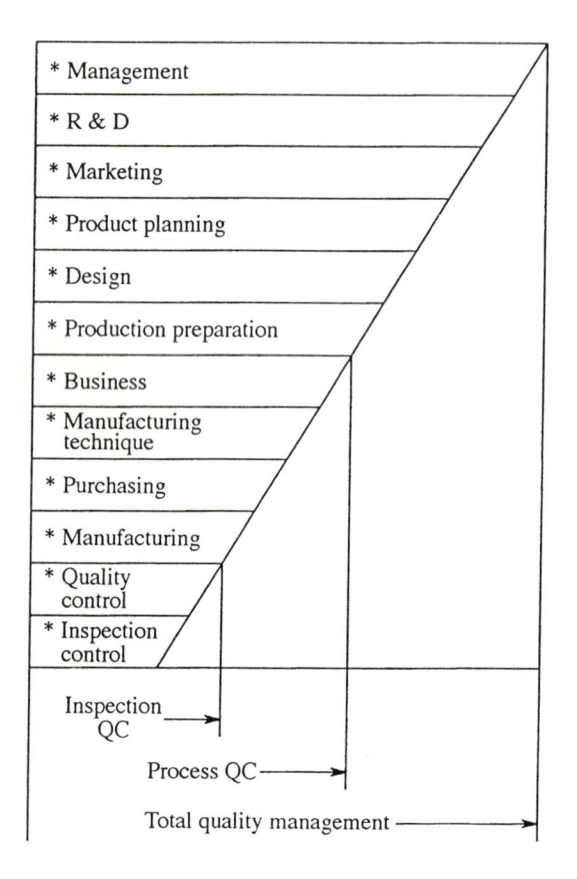

Fig. 1.2. Development of the Japanese Quality Assurance

1.3 Importance of Production Techniques to Build-in Quality in new Innovative Product Development

1.3.1 Product and Process Technology

During the 1970 and 1980, Western companies sent endless study groups to assess Japanese business success. One way of looking at the continuing evolution of the Japanese approach is to consider the techniques that have been adopted in the West, such as SQC, Total Quality Management, Just in Time, Total Productive

1   Outline of Total Quality Management   557

Maintenance, etc. They are certainly important, and Japanese firms are continuing to refine and perfect them. But confining the study to the mechanics of production risks discounting the importance of more subtle, long-term factors of organization and human resources. Fashionable production techniques are not a short cut; the necessary skills based on good education and training and organizational ability based on total quality management can only be built into the company over time. Without exception the techniques evolved used by Japanese manufacturers are founded on the increasingly sophisticated organization of production techniques. Figure 1.3 shows the growing range of production techniques and disciplines that have been integrated into manufacturing. They are:

MD	Manufacturing Department
PTS	Production Techniques Section to improve quality, value added per employee and to shorten lead times
PTD	Production Techniques Department
PTC	Production Techniques Centre
PTH	Production Techniques Headquarters
R&D of PT	R&D of Production Techniques

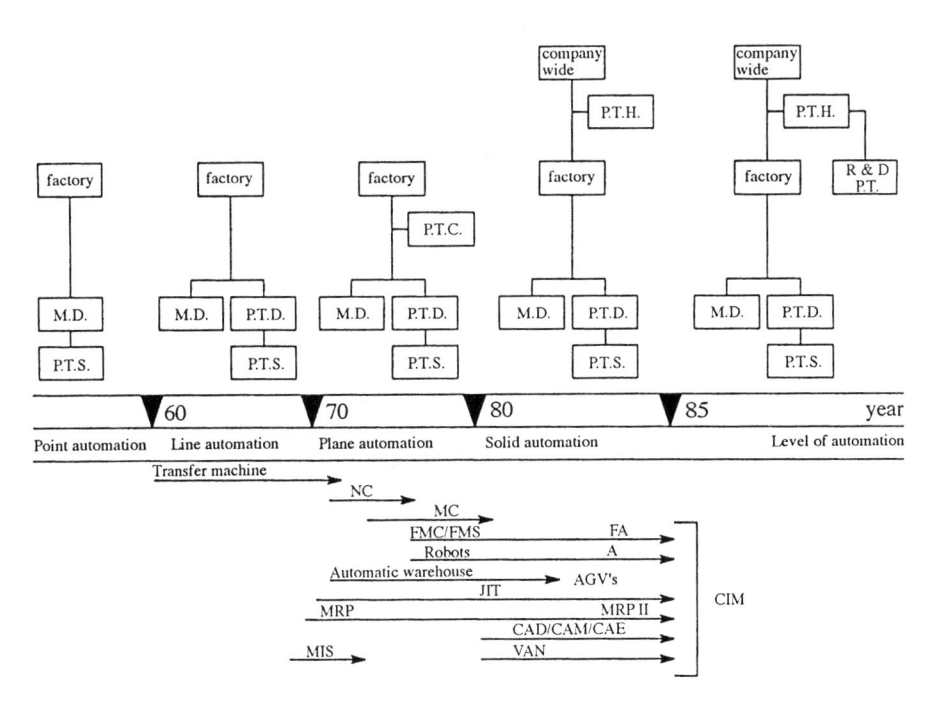

General trend of organizational change

Fig. 1.3. Increasingly Sophisticated Organization of Production Techniques

To understand what this means, it is necessary to step back and look at the wider manufacturing picture. Unlike firms in the West, the Japanese have always believed that products and the processes by which they are made are two sides of the same coin. Process technology involving good quality control is as important as product technology. The two go together. In the same way, they also believe that creative development goes together with, and is as important as, creativity in invention. Table 1.1 shows five examples of important consumer products which were developed by Japanese firms after having been originally created in the West. Far from mere "copycatting", as some Western businessmen like to think, Japanese companies consider these as textbook examples of commercial creativity.

Table 1.1 Invention and Development

Item	Originator	Developer
Transistor radio	Regency	Sony
VTR	Ampex	Sony, Victor
TV	RCA	Matsushita
Rotary engine	Vanchel	Mazda
CD	Philips	Sony

It is true that since the Second World War Japan has been able to license or buy patents for attractive products from the West. This is especially the case for consumer goods. For that reason, in the pursuit of competitive manufacturing, its companies have been able to focus more attention and channel more resources into production techniques than their Western counterparts.

1.3.2. A Closer Look at the Production Process

To be able to manufacture attractive products at an attractive price with excellent quality, companies need not just plant and equipment but different production staff. Japanese firms identify the key categories as :

(a) basic research scientists and engineers;
(b) applied research engineers;
(c) product development and design engineers;
(d) process engineers;
(e) process improvement engineers;
(f) operators(workers).

Japanese companies generally concede that in basic research they are inferior to the West, but applied research and product development, they are even. In process

1 Outline of Total Quality Management 559

and process improvement, however i.e. on the factory floor the advantage shifts decisively to Japan as is shown in Figure 1.4. In particular, Japanese engineering skills have been applied more consistently and intensively than in the West to three crucial areas: quality control, value added per employee, and the shortening of lead times.

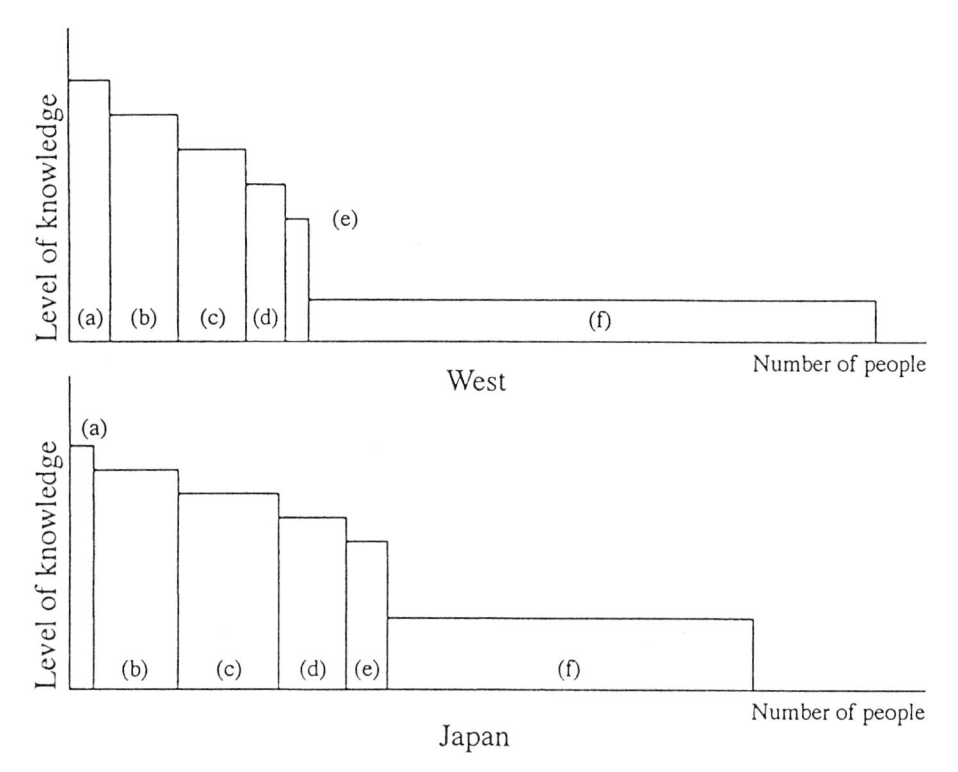

Fig. 1.4. Different types of production staff

These policies are reflected in company organization. Comparing companies of similar size and business area in Japan and the West, the Japanese firm uses more engineers in applied research, product development and design, process development and process improvement and far fewer operators for daily work. It is not just that there are more engineers in the applied areas: in each category the level of knowledge required is higher than in the West. Think of the strength of all of those production people pursuing excellent quality and competitive manufacturing based on Total Quality Management.

1.3.3 Matching Manufacturing to Market Changes

The effects of any disparity in human resource organization become more, not less, significant as the demands on manufacturing increase to match changes in the markets. Take the example of consumer durable air-conditioners. Up to, say, 1974, the time of the first oil crisis, one air-conditioning unit per house was the Japanese norm. The market was demand-led, and manufacturers needed only to mass-produce standardized designs to meet it. All markets eventually become saturated, and evidence of stalling demand for standard air-conditioners began to appear in the mid-1970. To maintain and increase sales, companies had to create new markets. In time, they succeeded in persuading increasingly affluent consumers that they needed an air conditioner for each room. This meant that manufacturers had to develop a variety of products to meet a diversity of customer needs different shapes and sizes of room, wall, floor or ceiling mounting or risk losing market share to broader-range suppliers. The production-led years gave way to a marketing-led period. The combination of the marketing-led trend led not only to an air-conditioner for every room, but an air-conditioner to match each individual's personality, lifestyle and taste; in short, a separate product for each customer. Avant-guarde companies now see themselves in the middle of this "customer-oriented" period.

1.3.4 Costs of Diversification

While diversification undeniably increases sales, it also carries costs. It requires more work in design. In addition, if production cannot cope with changing market requirements quickly enough, it can lead to waste as a result of mismatches between production and sales overproduction and inventory overstocking of products in low demand, together with inability to supply products in great demand. In short, it is vital for the manufacturing company to provide customers with the right products with excellent quality at the right time at the right price in other words, to implement just-in-time manufacturing, which requires essential support of good quality assurance. For conventional manufacturing systems and quality control methods, diversified demand, rapid changes in demand and shorter product life-cycles are the most difficult of all conditions with which to deal. All these conditions intensified in the 1980s. In the second half of the 1980s, a variety of manufacturers responded to growing competition by launching a barrage of new products with ever shorter life cycles. All the signs are that this trend of multiple product variations with shorter and shorter lives will increase in the 1990s. So manufacturing effort will become even more complex.

Table 1.2 shows the complexities and problems raised by these trends. The implications are that unless product variety and quality can be built in at the design stage, it is difficult to manage competitively. Excellent production techniques are not an optional extra. Think of production as an iceberg. The

1 Outline of Total Quality Management

Table 1.2 Manufacturing Headaches: Diversity, Changing Demand, Model Changes

Unfavorable conditions	Problems	Major issues to be considered
Changes in demand	* Decrease in competitiveness of personnel and facilities	* Flexibility * Pursue of less people
Diversified demand	* Lower quality * More space * More complex control	* Built-in quality at the process * High investment effectiveness
Frequent model	* Decrease in investment effectiveness	* High space utilization * Rapid capability at the start of production
New products	* Increase in re-building costs of old facilities, etc.	* Mix production * Less maintenance cost

innovative, diversified products which customers see are the visible tip as shown in Figure 1.5. Below the waterline is the necessary infrastructure of, in descending order, applied research, production techniques, and right at the

New innovative products

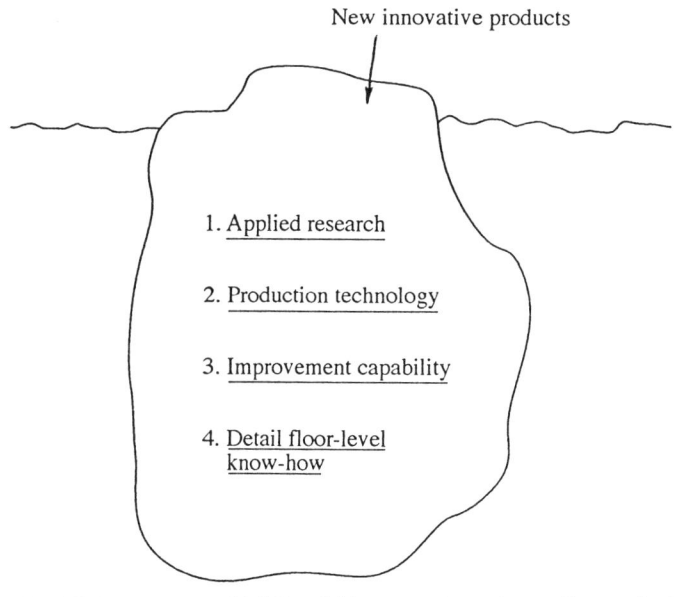

1. Applied research

2. Production technology

3. Improvement capability

4. Detail floor-level know-how

Fig. 1.5. Importance of 1,2,3 and 4 to support new innovative product development

bottom, the base on which everything else floats, detailed shop floor production knowhow, all of which require good quality control activities and must be integrated from the viewpoint of Total Quality Management as illustrated in Figure 1.2.

1.3.5. Mobile Competitive Edge

As markets and manufacturing have evolved, the focus of the competitive battleground has also shifted. As Figure 1.6 shows, when customer needs were largely standardized and only changing slowly, manufacturers could compete on quality and price. As variety and the pace of change increased, the winning weapon was high-variety, small-lot production with excellent quality. In the 1990s, speed and variety have converged again in time-based competition plus customer-oriented production. The important point is that the production techniques to which Japan has attached great importance play a key role in both customer-oriented production and time-based competition.

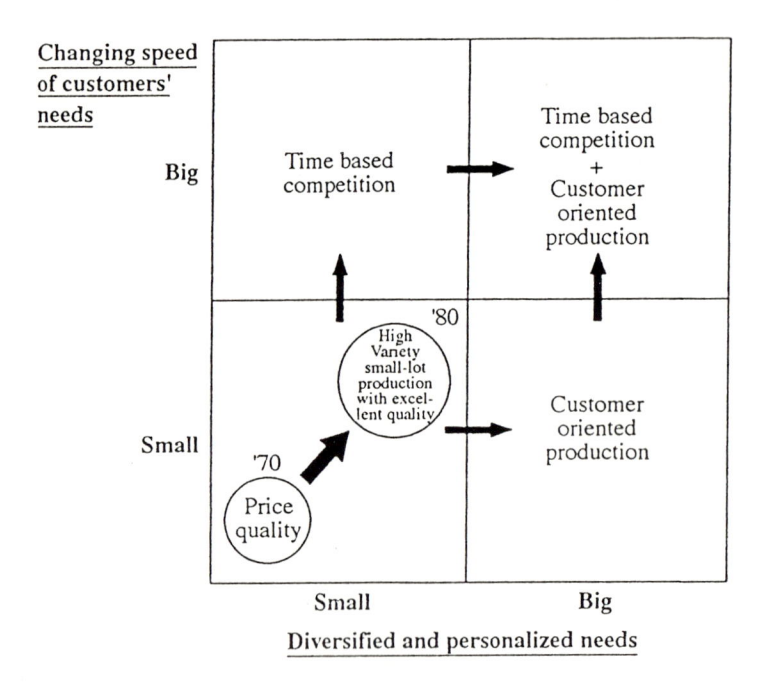

Fig. 1.6. The shifting competitive edge (Boston Consulting Group 1990)

Figure 1.7 demonstrates in general how companies have been changing their organization and subjects of interest based on the requirements of the market. The following abbreviations are used in Figure 1.7.

1 Outline of Total Quality Management   563

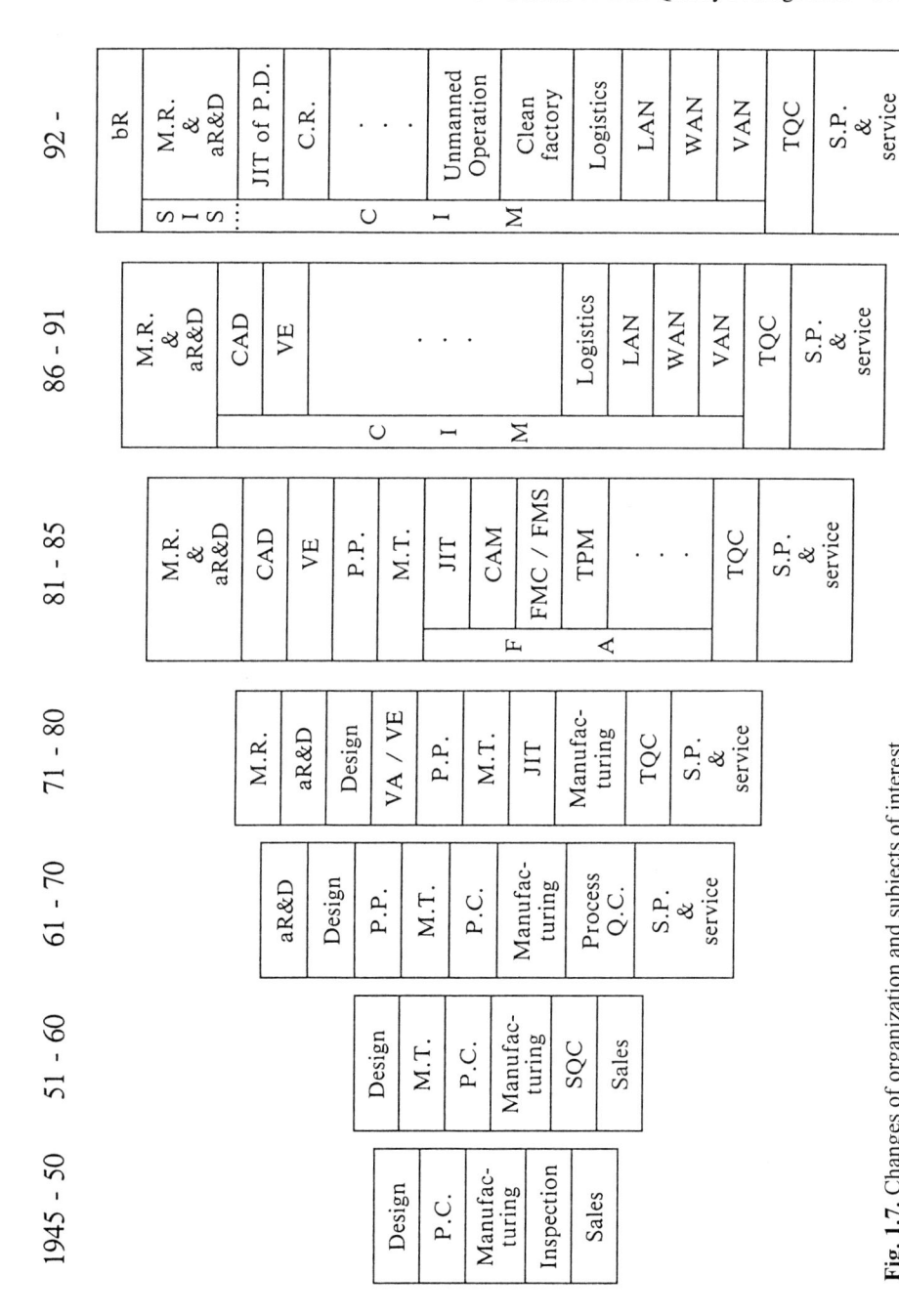

1945 - 50	51 - 60	61 - 70	71 - 80	81 - 85	86 - 91	92 -
Design	Design	aR&D	M.R.	M.R. & aR&D	M.R. & aR&D	bR
P.C.	M.T.	Design	aR&D	CAD	CAD	M.R. & aR&D
Manufacturing	P.C.	P.P.	Design	VE	VE	JIT of P.D.
Inspection	Manufacturing	M.T.	VA / VE	P.P.	...	C.R.
Sales	SQC	P.C.	P.P.	M.T.		...
	Sales	Manufacturing	M.T.	JIT		Unmanned Operation
		Process Q.C.	JIT	CAM / FMC / FMS	Logistics	Clean factory
		S.P. & service	Manufacturing	TPM	LAN	Logistics
			TQC	...	WAN	LAN
			S.P. & service	TQC	VAN	WAN
				S.P. & service	TQC	VAN
					S.P. & service	TQC
						S.P. & service

Fig. 1.7. Changes of organization and subjects of interest

PC	Production control.
PT	Production techniques.
SQC	Statistical quality control.
R&D	Research and Development.
PP	Preparation for production.
Process QA	Process oriented quality assurance.
SP	Sales promotion.
MR	Market research.
VA/VE	Value analysis / value engineering.
JIT	Just in time.
TQM	Total quality management.
FMC/FMS	Flexible manufacturing cells/flexible. Manufacturing systems.
TPM	Total productive maintenance.
FA	Factory automation.
CIM	Computer integrated manufacturing.
LAN	Local area network.
VAN	Value added network.
JIT of PD	Just in time of product development.
CE	Concurrent engineering.

Looked at in this light, it is not too much to say that Japan's economic success stems in large measure not from culture, consensus or copying, as the West would beleive, but from its organizational flexibility. The lessons which Western companies need to learn are at this level, too: not so much the techniques themselves, as the organization of human resources to integrate them based on the total quality management.

2 The QC Viewpoint

2.1 Customer Orientation

QC should be oriented toward customers' satisfaction and be carried out in order to produce the products with the quality that can meet customers' requirements. Therefore, they must be investigated thoroughly, and based on investigation, a proper product should be designed, produced and sold. Thus, the basic concept of quality management is to control quality in every sense.

If quality is interpreted in just a narrow sense, it is interpreted as the quality of the product. But in a wider sense, it means the quality of work, that of service, information, processes, departments, personell - workers, technicians, managers, or administrators, systems, the company, policies, etc. In other words, quality management is continuous improvement of all processes through quality to satisfy customers.

It should be noted that however good the quality may be, if the price is too high customers will not be satisfied. That is, without considering the price, quality cannot be defined. This is important in planning or designing quality. In other words, QC is not possible without considering the price, profit and cost management. From the QC viewpoint, products with the right level of quality should be supplied in a proper quantity at a reasonable price. Therefore, based on QC, a firm should run not only on cost, price and profit but also on quantity such as manufacturing quantity, selling quantity, inventory, etc. and delivery time. Thus, in total quality management in which all members of the firm including the managers, supervisors, workers as well as the executives participate, quality management should be carried out along with cost management and quantity control.

2.2 Customers' Quality Requirements

2.2.1 Quality Requirements and Substitutional Characteristics

In carrying out quality management, first of all it is necessary to grasp what the quality required is in the real sense, to select the measuring method for it, and then to determine the quality level. After this, its substitutional characteristics which are supposed to influence the quality should be carefully selected. Then a question arises : how far is it necessary to analyze substitutional characteristics in order to satisfy the quality characteristic in the real sense? In other words, the relationship between the quality characteristic in the real sense and its substitutional characteristics must be analyzed statistically. Figure 2.1 shows the relationship between the quality characteristic in the real sense and its substitutional characteristics.

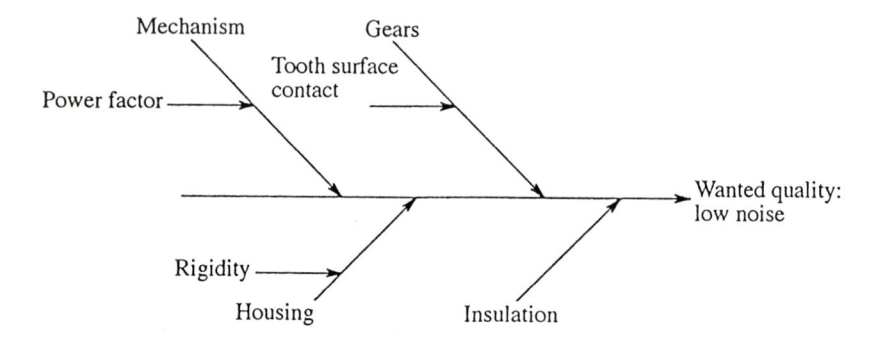

Fig. 2.1. Wanted quality and its substitutional characteristics

The following three steps are important in identifying customers' requirements. This process of analysis is called quality analysis or quality function deployment.

(1) Determine the quality characteristic in the real sense.
(2) Determine the measuring method or testing method for it.
(3) Find out the substitutional characteristics and understand the relationship between the quality characteristic in the real sense and its substitutional characteristics.

2.2.2 How to Express Quality

After a quality characteristic to satisfy a customer's requirement is determined, how can it be expressed properly? Customers' requirements are not always easily expressed in concrete physical terms. Depending on the analysis, producers can interpret and handle them in various different ways.

The following are the important points in expressing quality.

(1) Determine the unit for guaranteeing quality. This becomes an issue in guaranteeing quality for articles not encounted.
(2) Determine the measuring method. Without a precise measuring method, the clear definition of quality is impossible.
(3) Determine the degree of importance for each quality characteristic.

For many quality characteristics, it is important to find the precedence of their importance. Generally, quality problems can be classified into the following three categories :

(1) Fatal quality problems: quality characteristics concerning human life or safety, for example a gear of the gear box becomes loosened.
(2) Severe quality problems: quality characteristics which influence the performance of the product a great deal. for example the gear box does not function properly.
(3) Minor quality problems: quality characteristics which do not influence the performance of the product but are not accepted favourably by customers, for example the noise of the gear box is too high.

This priority analysis is extremely important in carrying out QC. These quality characteristics, such as defectives or bad quality, are called "backward quality", while the characteristics that can be selling points such as compactness or low noise level are called "forward quality." The latter should be studied in the precedence of importance.

In a similar way various quality characteristics can be divided into essential quality and attractive quality in the following way :

(a) Essential quality: customers find the product normal if this quality is satisfied but find it unpleasant if this quality is not met.
(b) Attractive quality: customers find the product attractive if this quality is satisfied but accept it even if this quality is not met.

In order to make the products the customers really want, such nature of quality must also be carefully investigated as in the case of forward quality.

(4) Unify the interpretation of bad quality. This is because between producers and customers or even within the same company, people tend to interpret defectives or bad quality in different ways.
(5) Exposure of latent defects. The apparent defectives that are expressed in terms of numbers in a plant or a company are only a small part of total

defectives or waste. In other words, waste in a broader sense must amount to ten to hundred times the latent waste. To reveal these latent waste is the basic concept of quality management. Quite often only these defective products that cannot be used and will be thrown away are defined as defectives, but these products must be carefully checked in a much stricter sense. For example, the products which did not satisfy quality standards and have been reworked or the products which needed adjustment should also be defined as defectives.

Incidentally, the percentage of perfectly good products without any adjustment or rework through all the processes including the final assembly, against the total products, is called "straight going rate." The products which need adjustment or rework will be liable to create more breakdowns in the market eventually. Therefore, design and manufacturing processes should be controlled so as to achieve a direct going rate between 95% and 100%.

Thus, strictly speaking, any company has latent waste and latently defective processes. In carrying out quality management, the definition of waste should be clearly defined and waste or latently defective processes should be investigated and proper countermeasures must be taken.

(6) Consider quality from the statistical point of view. Customers only accept the products of stable and uniform quality. Therefore, the quality of a group should be studied from the statistical point of view as the investigation of its statistical dispersion is important for controlling the manufacturing processes.

(7) Quality of design and quality of conformance. The difference between the two, quality of design and quality of conformance, means the defects or the necessity of rework. If quality of conformance is improved, cost should be reduced. Those who are not familiar with QC tend to think QC only require more money and less productivity. If they take QC as being inspection, repeated inspection obviously increases cost. If the quality of design is to be improved, it also requires more cost. But with better quality of conformance, defects, rework and adjustment will decrease. As a result, lower cost and higher productivity can be gained. Furthermore, if the quality of design matches customers' requirements, the amount of sales will increase, which results in more effective mass production and rationalization, with the consequence that the cost can be further reduced. The reason Japanese goods are competitive in the international market is because of this interrelated effect of these two qualities, quality of design and quality of conformance.

In order to win international competition, Japanese companies have improved the quality of their design. This has required further cost, but by improving quality of design manufacture and by improving processes thereafter such as the quality of conformance and the reduction elimination of defects overall cost can be reduced.

2.3 Management through Quality to Satisfy Customers

2.3.1 Problems in Conventional Management

Management means all the activities necessary for performing a job efficiently, achieving an objective and meeting a set target. This encourages workers to perform jobs by working hard is known as "mental management." With this kind of management, it is impossible to manage the whole company effectively. Defectives or failures that should have been the responsibility of the workers on the shop floor are only one fourth or one fifth of the whole. Most of them are attributed to management or staff. With this mental management, all the responsibilities of producing defectives and causing failures are liable to be attributed to workers on the shop floor.

It is very common in conventional management to find various problems such as:

(1) Many of the management theories used are too abstract and not practical.
(2) Management is based on experience and intuition and not based on facts and data.
(3) Management policies tend to be capricious based on mere casual ideas.
(4) There are no scientific or logical methodologies used in management.
(5) There are no analytical methods or management methods based on statistics.
(6) The president, executives, managers, supervisors and workers are not educated sufficiently enough on quality and quality control.
(7) Although there are specialists in each discipline, they do not cooperate to solve company-wide problems.
(8) All members of the company do not participate in finding the means to achieve company targets.
(9) Strong sectionalism normally exists and departments tend to fight each other for authority.

2.3.2 Management by rotating the Plan-Do-Check-Act cycle

When managing anything, it is important to follow the four steps of Plan-Do-Check-Act cycle, the so called Deming Cycle as shown in Figure 2.2 in the following way:

Step 1 : Prepare a plan (Plan).
Step 2 : Implement the plan (Do).
Step 3 : Check the results (Check).
Step 4 : Take actions based on the findings of Step 3 (Act).
The following explains each step in more detail:

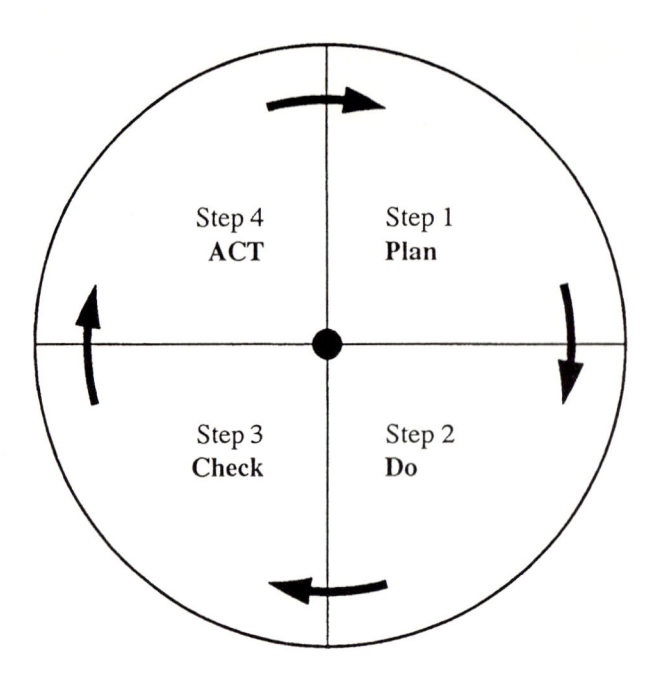

Fig. 2.2. Plan-Do-Check-Act cycle

Step 1 : Prepare a plan

(1) Determine objectives and targets. This can be done based on the company policy. In order to decide on the policy, its ground work and the data on which it is based must also be clear. In other words, management must be run based on facts and data. Without the company's policy, targets cannot be set. This policy must be determined by the top management, but the analysis of data and collect them are the responsibility of managers and staff.

The policy must be decided based on a comprehensive viewpoint. If there are defectives of between 40% and 50%, it is understandable to decide on a policy "to reduce defectives." But when there is a policy to "assure the production quantity" at the same time, it will cause confusion on the shop floor. Therefore, policies or targets must be based on priority with a maximum of ten to twenty items.

With policies clarified, targets can be determined easily. They must be clarified by concrete numbers in terms of, say, manpower reduction, quality improvement, cost reduction, profit increase, quantity, delivery lead times, etc. Without concrete policies, good and effective control cannot be made.

(2) Decide on the methods to achieve targets; i.e. Formulate, observe and utilize standards. If only objectives and targets are ascertained and the methods to attain them are not determined, management with this level of planning will result in mere mental management. If the management only decides on the target to reduce the defective rate to 1% and only pushes production personell to reach that figure, very little can be achieved.

2 The QC Viewpoint 571

Scientific and logical methods must be decided and once the methods have been determined, they must be standardized so that all people concerned can follow them strictly.

The following are most likely to happen in standardization.

(a) Product and process engineers tend to set too detailed technical standards or regulations without considering the opinions of the personell on the shop floor who have to follow them in practice. There are engineers who tend to make their own standards by desk work and let the people on the shop floor suffer from many inconveniences due to the lack of practicality.

(b) There are people who interpret "control" as setting many regulations to restrict others.

Standardization without the right objectives results in irregular operation, less efficiency and less respect for humanity.

Figure 2.3 shows the cause and effect diagram which highlights the process factors that contribute towards a particular result. The quality characteristic shown on the right is the result or objective required. Causes are shown at the end of the branches in the diagram. The grouping of these causes is called a process. A process represents not only the manufacturing process but design, purchasing, sales, personnel, accounting work, etc.

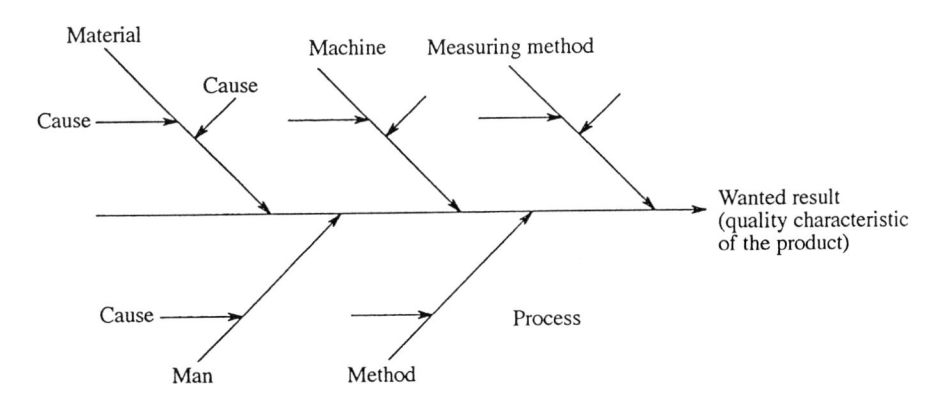

Fig. 2.3. Cause and effect diagram

Quality control can be effectively conducted by identifying and controlling causes for an effect. This is called "process control" as mentioned in the previous chapter. Quality control produces good quality products by controlling the processes is called "control by processes" or "prior control". Quality control for taking measures after checking the results is called "control by results" or "posterior control".

There are many causes. An effect can easily have ten or twenty causes as can any work or process. It is practically impossible to control all the causes. However, really important causes or those that influence a result a great deal are few. According to the Parete principle, standardization of five to ten important causes can bring about a great effect. This explains the importance of being aware of priorities; finding major causes; i.e. attack priority problems and address them properly.

The method applicable for searching possible causes is to first speak with knowledge of the facts try to collect frank opinions of all the people involved by brain storming. Second. analyse the opinions by statistical methods and prove them by the data scientifically and logically. This is called process analysis. The result drawn by this method should be comprehensible enough for everybody to understand. At this point, it is strongly advisable to draw a sketch to illustrate the relationship between a cause and the effect to identify a real cause without spending much time in brain storming.

It is also important to clarify what to do in case of abnormality. It must be made clear (a) who should at what extent at the time and how far (concerning authority) and (b) who should give orders and take responsibility and institute radical countermeasures to ensure that the same abnormality is not repeated.

Step 2 : Implement the plan

(1) Educate and train the people concerned. Even if good operation and technical standards are set, it is not enough just to have them distributed among the people concerned. They will not read them or, even if they are read, they cannot really understand the basic concept behind the standards or the way to use them. Therefore, it is important to educate the people who are to use them.

By education, it does not only mean group education by lectures which should in fact be only one third or one fourth of total education. Superiors or leaders must train their personell on the job. This is called "on-the-job training."

(2) Implement the plan. If the aforementioned steps are taken, management should be properly.

Step 3 : Check the results

The most important issue in management is the rule of exception (abnormality). It is all right to leave everything as it is when operations are following standards, moving smoothly and accomplishing targets step by step. But when an exception takes place, it is necessary to detect it and take countermeasures against it. Therefore, to identify and check exceptions, the basic policies, objectives, targets and standards must be clearly known. Some managements try to check very strictly without any policy, objective or target, but they are never able to convince their personell with this kind of attitude.

The detection of abnormality can be made in the following way :

(a) Check the causes. The first step is to see whether the causes are under good control. In other words, it must be checked whether all the causes are controlled according to the standards in design, purchasing or manufacturing processes, etc.

For this purpose, it is necessary to oversee the workshop carefully. It is not enough just to walk into the workshop but to watch and check everything with regard to its standards or regulations. As there are frequently many causes, it is impossible for one person to check everything. A check list is useful for checking priority causes efficiently and to prevent previous and possibly future problems. Checking of causes is the responsibility of lower management.

(b) Check process and management by results. Another method of checking may be achieved by the results of operations. By checking such things as number of suggestions, quality, quantity, delivery lead time or cost, it is possible to check the present situation of the process, operations and management. Checking by results is the responsibility of higher management.

A bad result indicates that there are some abnormal problems somewhere in the processes. Therefore, by finding the real cause for the bad result and putting it under control, the process can be controlled properly.

The check items of the process and management are called control items. Foremen should have five to twenty control items while personell from sectional managers up to the president of the company should have twenty to fifty items.

It must be understood that to check by results and to check results are two different things. As for quality, the process or management can be checked through quality control. To check quality there must be an inspection and check the process and management through quality and to check actual quality are completely different. By focusing on quality, the process can be controlled properly and good products can be produced.

As there is an infinite number of causes, effects such as quality, production quantity, or cost, always fluctuate. That is, by analyzing the distribution of results statistically, exceptions or abnormalities can be discovered. In other words, by paying attention to dispersion and identifying its causes, abnormalities can be discovered.

Step 4 : Take actions based on the findings of Step 3

An important point in taking necessary countermeasures is the prevention of any recurrence. It is not that abnormality should be eliminated by adjusting other causes but that the very cause which has created the abnormality should be controlled or eliminated. Adjustment and prevention of recurrence are two different concepts with different countermeasures.

The following points should be taken into consideration :

(a) Failure due to workers' responsibility comprises one fourth or one fifth of the total number of failures. So the management should never get angry about these failures. Otherwise, the real facts may become hidden behind false data and false reports. Comprehensive company atmosphere in which workers can report failures to their superiors or colleagues with ease and all the members can cooperate in prevention of recurrence should be developed.

(b)If there are many unknown causes, these indicate that the concept of process control is not thoroughly understood or put into practice properly.

(c)The result of a countermeasure must be checked and by referring to the cause, it must also be checked as to whether the preventive measure is working properly against further recurrence. Short-term and long-term checking are needed.

(d) Control does not mean maintaining present conditions. Even if recurrence of the same problem has been prevented absolutely, further continuous improvement should be carried out.

The QC viewpoint is that it enables personell to make continuous improvement in order to reduce or eliminate all the waste in all processes and all areas of operation through quality to satisfy customers continuously.

3 Quality Assurance

3.1 Basic Quality Assurance Policy

To assure quality in order to satisfy and attract customers, there must be a basic quality assurance policy as discussed in the previous chapter. Each company should have its own quality assurance policy established properly for its organization and its products. The following gives sample examples of a quality assurance policy of a Japanese manufacturing company :

(1) Quality must be foremost in every aspect of the company policy.
(2) Customers' satisfaction can only be met by supplying the products with world class quality.
(3) Quality must be achieved through participation and cooperation by all personell in the company.
(4) The quality assurance system must preclude all potential problems.

3.2 Quality Assurance Organization

Many manufacturing companies are comprised of corporate departments and product divisions, each division being independent of the others. In such cases, the company head office normally has a quality assurance department, which administrates the overall quality assurance activities throughout the company, often serves as the total quality management promotion centre and performs company-wide quality management activities.

The manager of the quality assurance department is responsible for overall control of the quality assurance system of the company and often for establishing the basic policy and objectives to be followed in the operation of the system. If necessary, he also revises quality assurance procedures.

The main activities of the quality assurance department includes the following :

(1) Establishment of company-wide quality targets and follow-up of their results.
(2) Checking and approval of the quality evaluation results made by a product division for its new product from the customers' viewpoint.
(3) Taking necessary actions for serious quality difficulties and to prevent these problems from recurring.
(4) Development of advanced techniques for evaluation of reliability and safety.
(5) Company-wide training of QC methods
(6) Promotion of QC circle activities, etc.

Alongside the production forces in each product division or in its plant, a quality assurance section is provided as shown in Figure 3.1. These sections obviously should form a closely interwoven network operating under the quality assurance department in the corporate office.

The general manager of the product division is responsible for the quality of products manufactured in his division.

The main activities of the quality assurance section includes the following :

(a) promotion of total quality management activities throughout the division;
(b) establishing product quality targets and following up their results;
(c) summation of quality assurance activities at every phase from design to production in the product division;
(d) test and evaluation of new products from the customer's viewpoint;
(e) control of inspection activities;
(f) control of the activities related to disposition of market claims;
(g) collection and summation of product quality information, etc.

Figure 3.2 illustrates the roles of each department in a company for quality assurance.

Normally quality assurance functions from the sales department or from the service department, if there is only one as in the case of Figure 3.2, the following must be covered:

(1) Control of activities for disposition of quality claims.
(2) Summation of quality claims.
(3) Collection of field quality information, etc.

The purchasing department is responsible for assuring the quality of purchased materials and parts (cf. section 4.2.5).

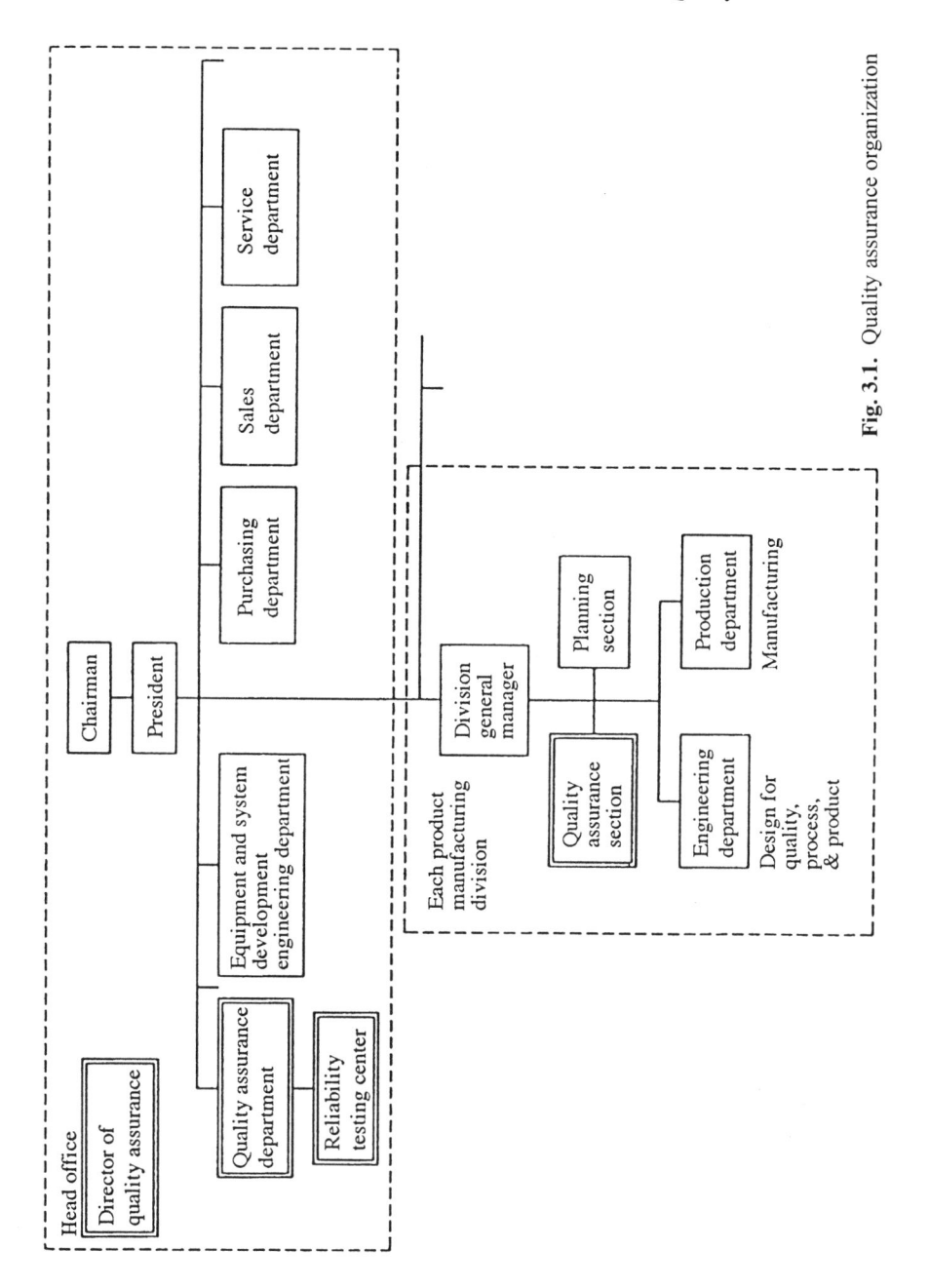

Fig. 3.1. Quality assurance organization

	Head office		Each product manufacturing division
QA department	Establishment of company-wide quality targets and follow-up of their result	**Planning section**	Product planning
	QC training, etc.	**Engineering department**	Development and designing of the new product
			••• responsible for the quality of design
Purchasing department	Promotion of quality assurance of suppliers	**Manufacturing department**	Preparation for mass-production
	Purchasin of parts and materials		Manufacturing
	••• responsible for the quality of the purchased goods		••• responsible for the quality of conformance
Service department	Collection of product quality information on the field	**QA section**	Promotion of quality assurance
			• quality evaluation
			• product inspection

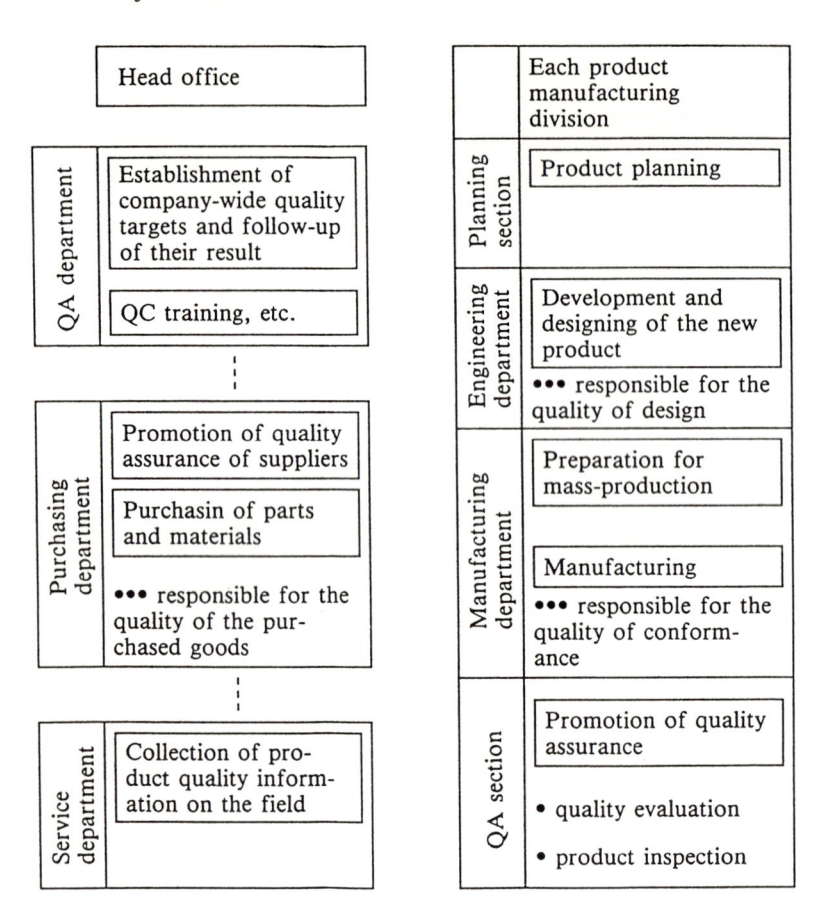

Fig. 3.2. Roles of each department for quality assurance

3.3 Quality Assurance Process

The quality assurance of a new product comprises of various steps.

Figure 3.3 shows an example of the quality assurance process. In this case, a quality assurance meeting is held at the end of each phase to determine whether or not the current new product development can move to the next phase. The important point in this quality assurance meeting is the presence of the senior board member in charge of the product division and he must take responsibility for the quality of the product. When dealing with an important new product, the executive vice president should participate and take responsibility.

```
Product planning
    │
    ├── 1st QA meeting      ★ Planning the product that
    ▼                          will satisfy customers.
Product design
    │
    ├── 2nd QA meeting      ★ Realization of quality targets
    ▼                       ★ Important quality specifications
Prototype design/
manufacture/ test/
evalutation
    │
    ├── 3rd QA meeting      ★ Confirmation of design quality
    ▼                       ★ Evaluation of degree to which
Production                     quality targets have been
preparation                    achieved.
    │                       ★ Stipulation of quality control
    │                          criteria
    ▼
Trial production
test/ evaluation
    │
    ├── 4th QA meeting      ★ Securing machine and process
    ▼                          capacity
                            ★ Confirmation of inspection pro-
                               cess and methods
Mass production             ★ Confirmation of manufacturing
(early-stage)                  quality and process capability
    │
    ├── 5th QA meeting      ★ Striving for quality in manufac-
    ▼                          ture
Mass production             ★ Ensuring the quality of incoming
                               materials and parts
                            ★ Confirmation of the quality of
                               the new product
```

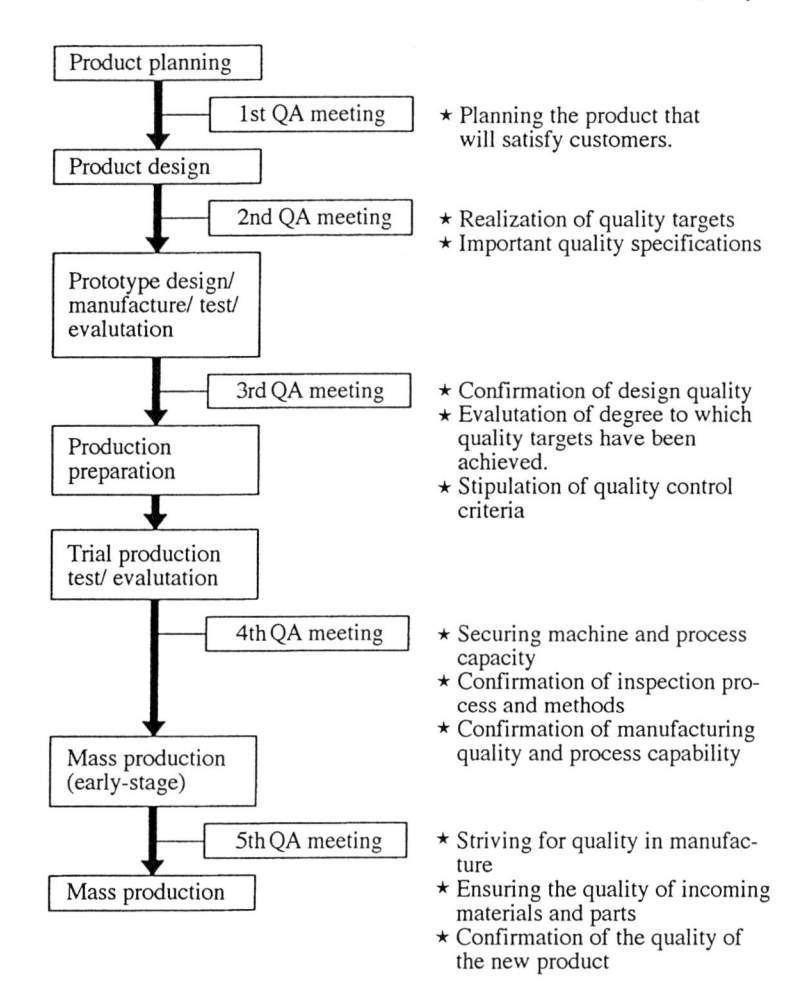

Fig. 3.3. Example of quality assurance process

3.4 Quality Assurance System Diagram

Figure 3.4 illustrates how the quality of a new product must be assured right from the stage of collecting customers' information until the start of production. This diagram shows just the main framework of a quality assurance system diagram and many world class manufacturers have made continuous improvement to refine and perfect it to a more detailed diagram.

In the product planning phase, various pieces of information such as customers' information, competitors' information, quality problems of previous models, new research findings, etc. must be collected to obtain information in

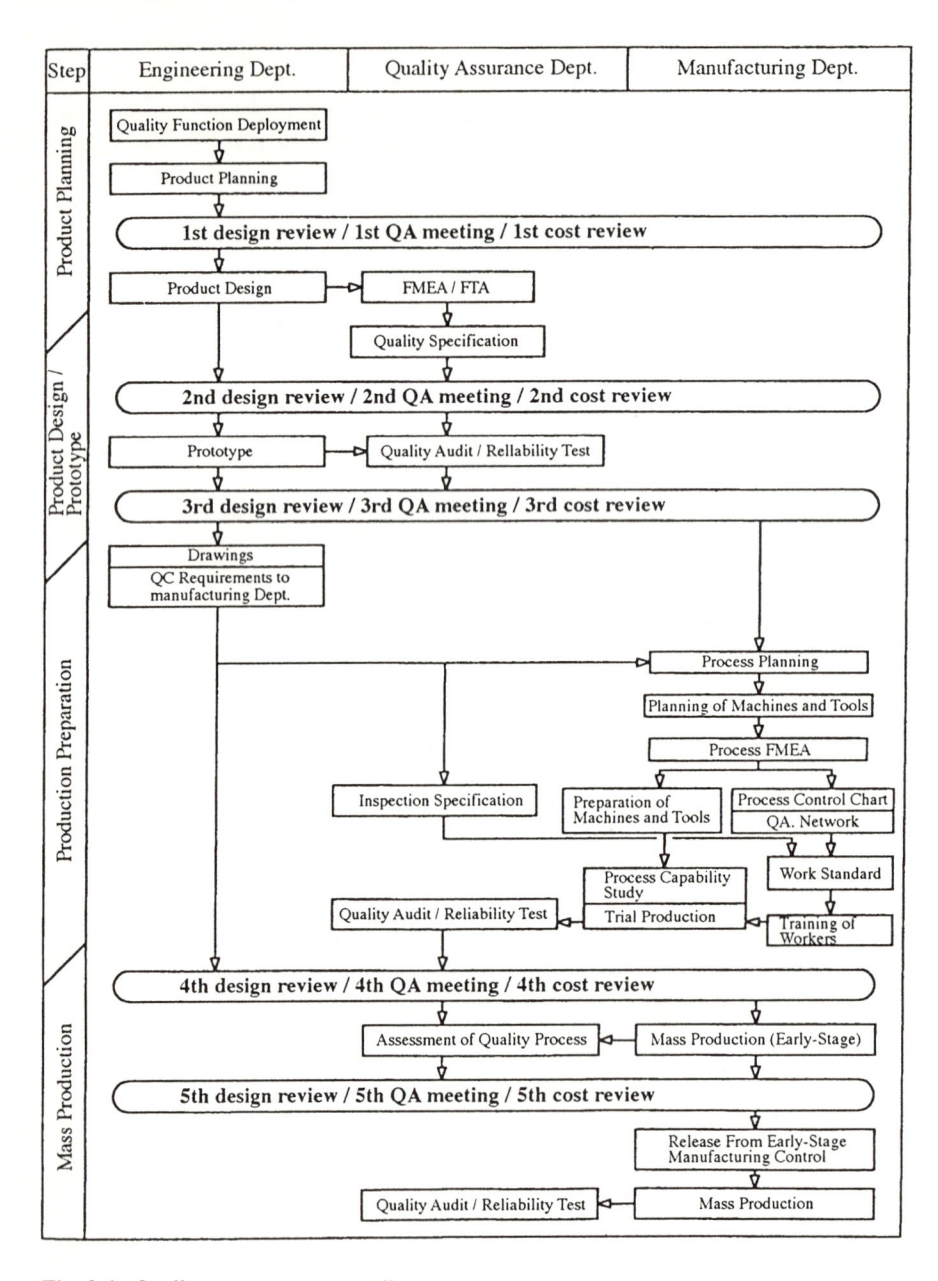

Fig. 3.4. Quality assurance system diagram

order to determine the customers' requirements as shown in Figure 3.5. Based on this investigation, quality function deployment (QFD), should be utilized to ensure the product meets the customers' requirements. Quality deployment converts customers' requirements into substitute characteristics and establishes the design quality of a final product. There are several sequential phases to QFD.

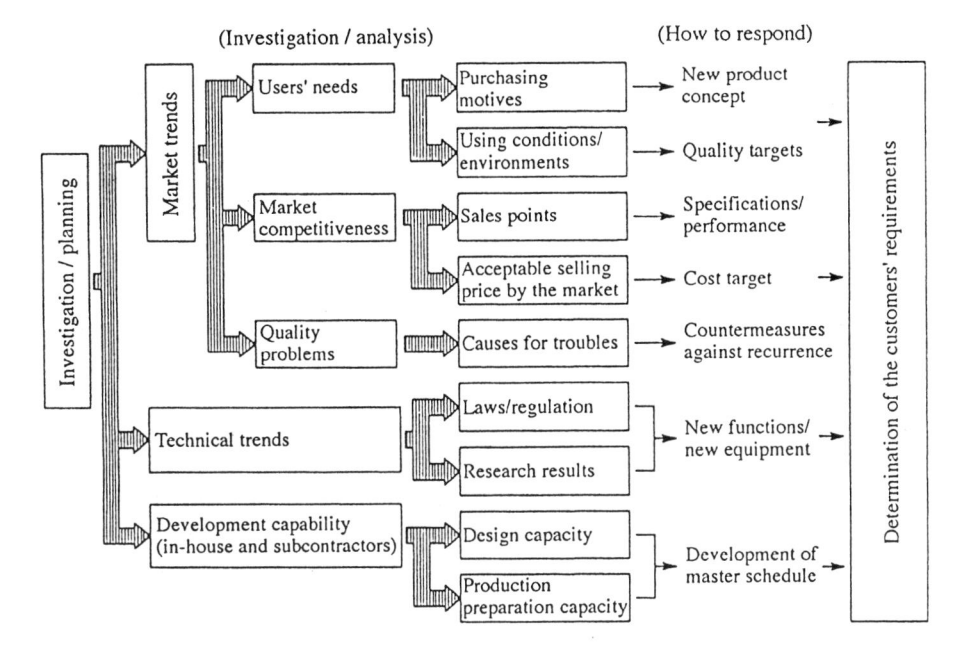

Fig. 3.5. Determination of the customers' requirements

The first phase is the conversion of customers' requirements from the customers' demands into quality characteristics. The second phase uses a quality planning matrix to define the characteristics of critical components. The third phase plans the process and identifies the critical processes to achieve the planned quality characteristics. The last phase plans production to ensure that the customers' requirements are met. For details refer to Akao [1].

The presence of the quality assurance department in this figure means that they need to check the quality of design and conformance from their own viewpoint and to be assured that there will not be any problems at later stages. In other words, the quality assurance department is responsible for detecting as many quality problems as possible at an earlier stage in order that they are not passes on to the succeeding stages.

In the design phase in order to assure the quality of design, various tools such as design reviews, quality assurance meetings, failure mode and effect analysis (FMEA), fault tree analysis (FTA), quality audit and reliability testing should be properly utilized as shown in the figure.

Quality audit is normally used to evaluate whether the quality assurance system has been properly operated and each phase of the new product development has been completed as planned. If any problem has been found through quality audit, proper countermeasures are to be taken.

Upon the completion of the prototypes, reliability testing is conducted to see whether they will meet the specified customers' requirements.

It is very important that at the time of preparations for production the product engineering department specifies quality requirements clearly and conveys important engineering information to the manufacturing department.

In the phases of preparations for production and manufacturing, design reviews, quality assurance meetings, process FMEA, QA network, process capability assessment, quality audit and reliability testing must be undertaken in order that the quality of conformance may be assured.

It is very important that in quality audit the staff concerned of the product engineering department participate and evaluate production process capability together with other staff.

The process control chart specifying control items, measuring methods, acceptance criteria, inspection frequency, etc. must be prepared by the manufacturing staff so that quality during the process is assured. Then the original drawing of the part takes place to manufacture and process control chart, the responsible foreman can prepare the operation standard for each process. It should be noted that for continuous improvement, the process control chart as well as the operation standard must be refined continuously.

4 Main Quality Control Activities

4.1 Quality Control at the Product Design Stage

4.1.1 Quality Engineering

Product design has the greatest impact on product quality, and there are three factors which influence the quality of products :

(1) External noise factors due to operating environment variables such as temperature, humidity, vibration, contamination, etc.
(2) Internal noise factors due to deterioration.
(3) Variation among units due to differences in parts and/or materials.

The purpose of quality engineering, which has been developed originally by Taguchi[2], is to design a product that is robust with respect to these three factors. Robustness means that the product's functional characteristics are not sensitive to variation caused by these factors. In order to achieve robustness, quality control must be made from the product design phase through the preparation phase for production to the manufacturing phase. In order to improve the quality of products so that robustness is achieved, the following three kinds of design must be used:

(1) System design to develop a prototype design and to determine materials, parts and the assembly method.
(2) Parameter design to select the levels of design parameters of each element so that the functional deviations of the product are minimized.
(3) Tolerance design to specify narrow tolerances for the deviations of design parameters in relation to the levels determined by the parameter design if the reduction in variation of the functional characteristic achieved by parameter design is insufficient.

Table 4.1 Reliability Grade Classification

Reliability grade \ Parts	Unused	NEW PARTS Different conditions	NEW PARTS Similar conditions	NEW PARTS Same conditions	SIMILAR PARTS Different conditions	SIMILAR PARTS Similar conditions	SIMILAR PARTS Same conditions	SAME PARTS Different conditions	SAME PARTS Similar conditions	SAME PARTS Same conditions
AAA	○									
AA		○								
A			○		○					
B				○		○		○		
C							○		○	
D										○

4.1.2 FMEA and FTA

In order to design a reliable product, the product must be assessed from the various engineering viewpoints. FMEA and FTA are powerful tools for this purpose, and enable designers to detect potential quality problems in the early design stage in spite of the lack of reliability data.

FMEA lists every part of the product and predicts the effect, the frequency and severity of its potential problems which may cause a failure of the product and highlights priority parts. The causes of failures of such priority parts must be carefully studied and necessary corrective countermeasures should be taken care of before the drawings of these parts are released.

Table 4.1 gives a reliability grade classification for various parts. FMEA should be made for those grades from A to at least B level. The result of FMEA will give guidance in establishing control limits on the control chart used during manufacturing. If the quality distribution is approximately a normal distribution with a target value at the centre, control limits can be set in the following way :

(1) Defectives that have to be avoided at any cost:
    Control limits = target value $\pm 5\sigma$
    where, σ is the standard deviation of the quality distribution
(2) Defectives that bring out substantial problems:
    Control limits = target value $\pm 4\sigma$
(3) The remainder:
    Control limits = target value $\pm 3\sigma$

For a possibly serious failure of the product, FTA should be carried out. FTA shows up on a critical failure of the product and analyzes its causes down to quality problems of constituent parts.

4.2 Quality Control at the Preparation Stage for Production

4.2.1 Quality Engineering

There are also three design steps to be followed at the preparation stage for production as in product design. These three steps deal with variation among units:

(1) System design to determine the manufacturing processes that can produce the product within the specified limits and tolerances at minimal cost.
(2) Parameter design to determine the operating levels of the manufacturing processes so that variability in product parameters is minimized.
(3) Tolerance design to determine the ranges allowed for changes in operating conditions and other variables.

4.2.2 Process FMEA

In order to make reliable processes to produce the products, the processes must be assessed from various production engineering viewpoints. Process FMEA serves this purpose. Process FMEA lists every process and its parts of the process and predicts the effect, frequency and severity of its potential problem which may cause a quality problem to the product or a failure of the process, and highlights priority processes or parts of the processes.The causes of product quality problems or the failures of the processes must be carefully investigated and the necessary corrective countermeasures against them should be taken care of before equipment is set up.

4.2.3 Process Capability Assessment

In order to conduct quality control during the process effectively, it is very important and effective to survey and assess process capability and utilize the results to monitor and control the process. Process capability is a measure of the inherent variability of the process and can be determined indirectly by measuring the product uniformity.

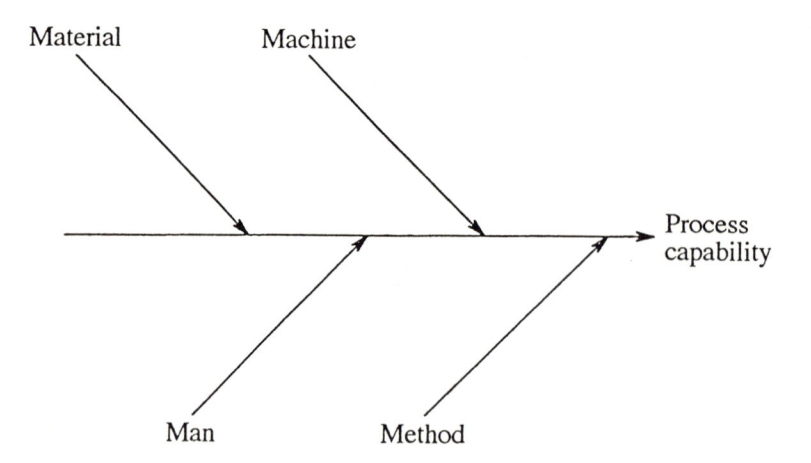

Figure 4.1 shows four major items which influence process capability. They are material, machine, man and method, often abbreviated to 4M.

Fig. 4.1. Four major items which influence process capability

If the quality characteristic is assumed to follow a normal distribution, where $\pm 3\sigma$ includes 99.73% of the population, process capability is defined in the following way :

Process capability = $\pm 3\sigma$ or 6σ

4 Main Quality Control Activities

while process capability index can be calculated as follows:

Case (a) : Two-sided, upper and lower limits of standard
$$C_p = (u_t - l_t) / 6\sigma$$
Case (b) : One-sided, upper limit of standard
$$Cp = (u_t - x) / 3\sigma$$
Case (c) : One-sided, lower limit of standard
$$Cp = (x - l_t) / 3\sigma$$
where u_t is the upper tolerance limit
l_t is the lower tolerance limit
x is the mean value of the distribution

In case where x deviates from the target value, the following C_{pk} should be used:
$$C_{pk} = (1-k)(u_t - l_t) / 6\sigma$$

Table 4.2 Process based on the value of Cp

Cp	Distribution	Judgement	Action
$Cp \geq 1.67$	l_t σ u_t / $\bar{x}$	Process capability is more than enough.	Simplification of process control and cost reduction can be considered in certain cases.
$1.67 > Cp \geq 1.33$	l_t σ u_t / $\bar{x}$	Process capability is sufficiently high.	Ideal condition. Maintain it.
$1.33 > Cp \geq 1.00$	l_t σu_t / $\bar{x}$	Process capability is not sufficiently high, but is adequate.	Control process properly and maintain it in a control state. Defects may results if Cp approaches 1. Take action if needed.
$1.00 > Cp \geq 0.67$	l_t σu_t / $\bar{x}$	Process capability is not sufficient.	Defects have been generated. Screening inspection and process control and Kaizen will be required.
$0.67 > Cp$	l_t σu_t / $\bar{x}$	Process capability is very low.	Cannot satisfy quality. Quality must be improved, cause must be pursued and emergency actions must be taken. Reexamine standards.

where k is the degree of deviation

$$k = \frac{(u_t + l_t)/2 - x}{(u_t - l_t)/2}$$

If $k > 1$, let $C_{pk} = 0$

Table 4.2 demonstrates the process.

During the phase of preparations for production, $Cp \geq 1.33$ must be achieved.

Figure 4.2 shows the important fact that product and process development must combine together during process capability assessment. In other words, the quality information acquired through various stages in the quality assurance system should be fed back to the product engineering department to modify the product. It must also be fed forward to improve the production process and the quality control system.

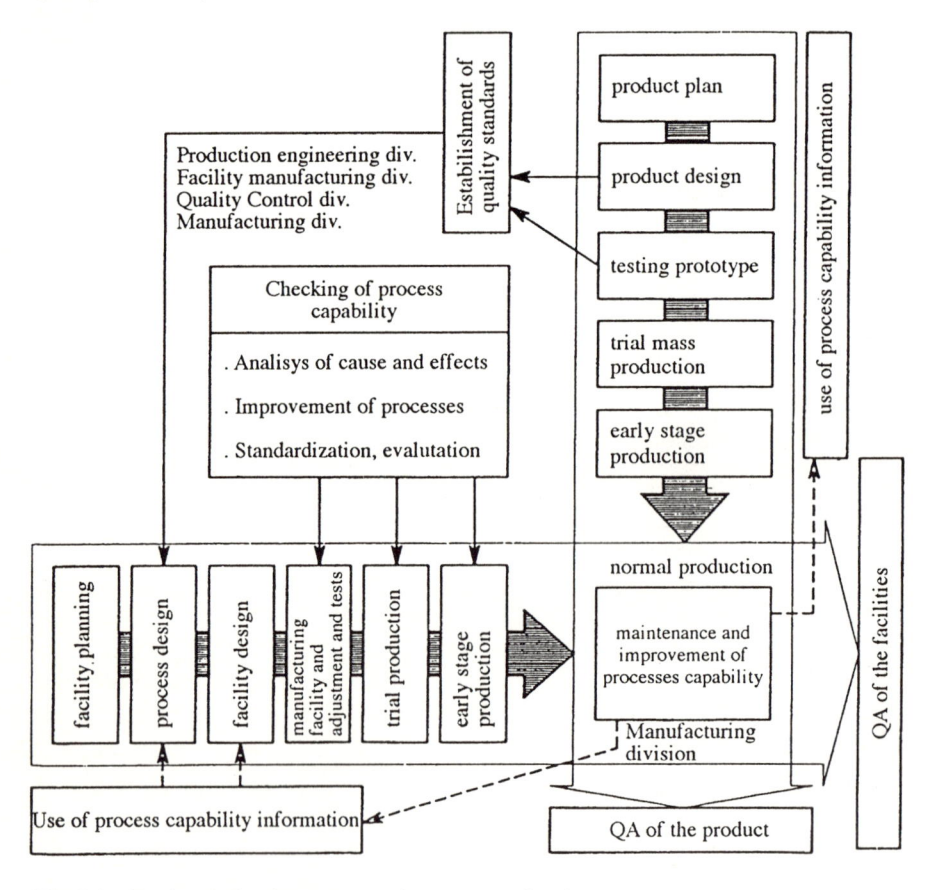

Fig. 4.2. Product development and process development must go together through process capability assessment

4.2.4 Fool-Proof System

Human beings make mistakes however hard they are trained not to do so. The following are typical mistakes caused by operators:

(1) Operators make mistakes in processing workpieces: non-processed workpieces, reversely processed workpieces, or over-processed workpieces.
(2) Absent-mindedness by operators, gives rise to fluctuation in part sizes or mis-positioning of parts.
(3) Due to missing or incorrectly fitted parts at the time of fixing parts to jigs and/or machines, operators produce scratches, non-acceptable appearance, non-processed, or parts reversely attached.

Causes of mistakes

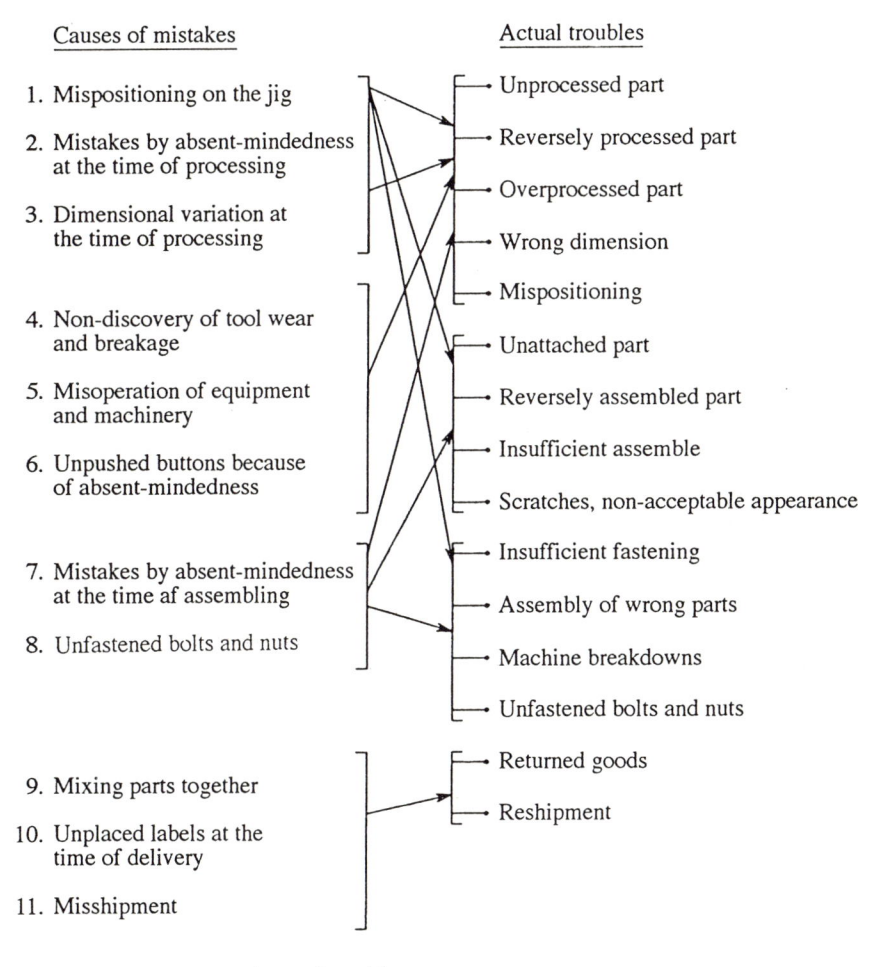

1. Mispositioning on the jig
2. Mistakes by absent-mindedness at the time of processing
3. Dimensional variation at the time of processing
4. Non-discovery of tool wear and breakage
5. Misoperation of equipment and machinery
6. Unpushed buttons because of absent-mindedness
7. Mistakes by absent-mindedness at the time af assembling
8. Unfastened bolts and nuts
9. Mixing parts together
10. Unplaced labels at the time of delivery
11. Misshipment

Actual troubles

- Unprocessed part
- Reversely processed part
- Overprocessed part
- Wrong dimension
- Mispositioning
- Unattached part
- Reversely assembled part
- Insufficient assemble
- Scratches, non-acceptable appearance
- Insufficient fastening
- Assembly of wrong parts
- Machine breakdowns
- Unfastened bolts and nuts
- Returned goods
- Reshipment

Fig. 4.3. Causes of mistakes and troubles

(4) Operators forget to attach some parts, assemble mischosen parts, assemble workpieces reversely, or assemble workpieces incompletely at the time of assembly.
(5) Operators forget to fix or fasten small parts, such as bolts and nuts.
(6) Due to the mixture of different kinds of parts, operators assemble them wrongly, thus causing breakdowns or breakage of machines. This most frequently happens with small parts, such as screws.
(7) At the time of delivery, operators forget to attach labels, name tags, etc. or put wrong ones on.

Figure 4.3 shows the various causes of mistakes and the resulting troubles.

Instead of setting an acceptable level of mistakes and sticking to it, speciality items, i.e fool-proof devices that can alert an operator to easily corrected but potentially troublesome defects without unnecessarily shutting down the process, must be developed and installed to the processes where operators are liable to make mistakes.

Even if a certain mistake rarely takes place, if it is caused by carelessness, it must be eliminated by using fool-proof devices.

4.2.5 Quality Control of Purchased Goods

Figure 4.4 shows how the purchasing department can be involved at the stage of preparations for production to assure the quality of purchased goods.

As shown in the figure, engineers of the purchasing department review the submitted data from a supplier, visit and investigate the process capability of its production process, evaluate the goods with various measurements and tests and also check its maintenance system.

If everything is satisfactory, inspection of incoming goods from the supplier can be omitted during the following production stage.

4.3 Quality Control at the Daily Production Stage

4.3.1 Statistical Process Control (SPC) by Using the Control Chart

The variations of quality characteristics in production should be maintained for the given tolerances because they directly indicate any deviation from the target value of the quality of design.

The important concept of quality control at this stage is:"Never send defectives or mistakes on to the next process."

Fig. 4.4. The involvement of the purchasing department at the preparation stage for ▶ production

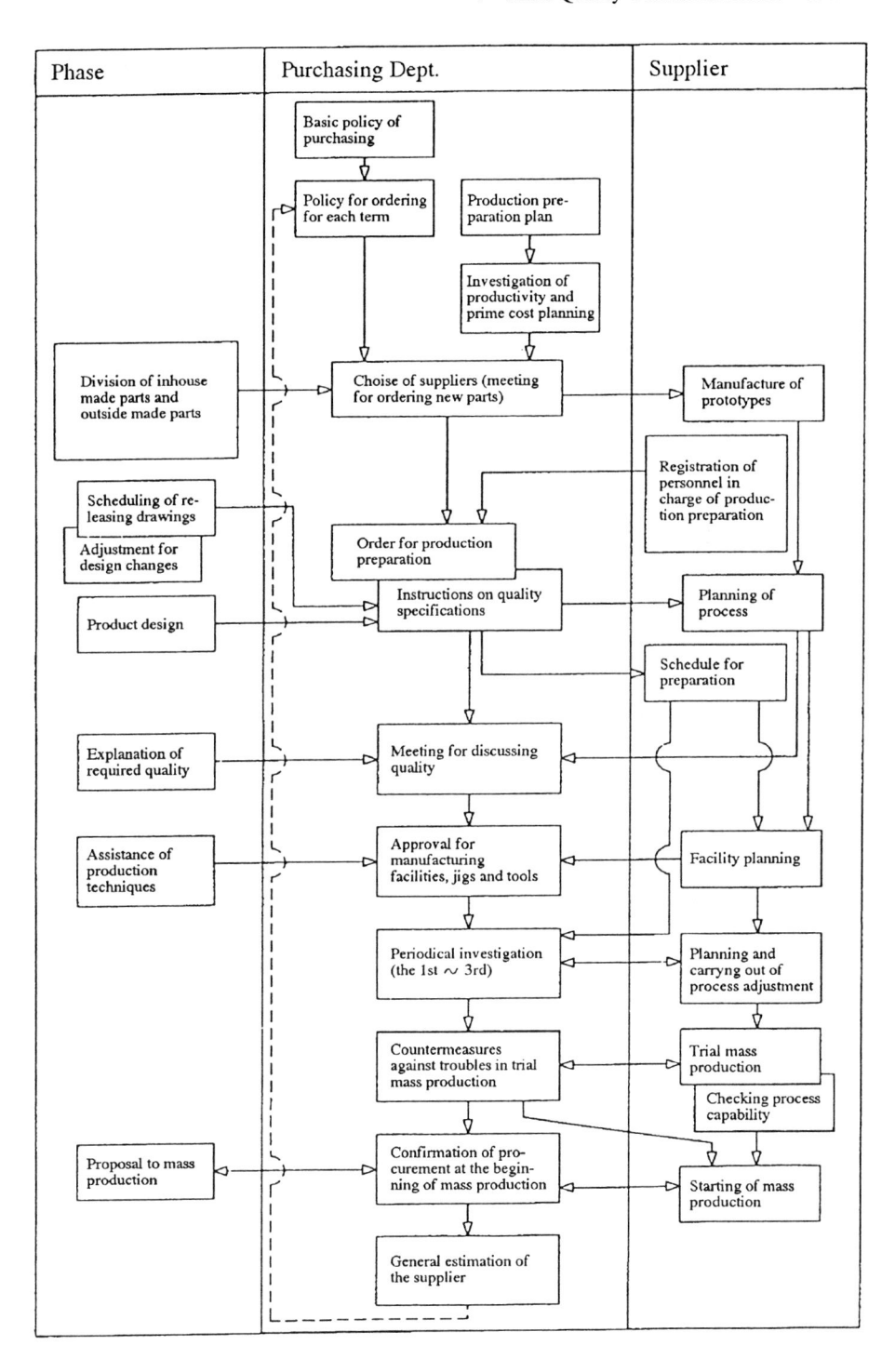

4 Main Quality Control Activities   591

Phase	Purchasing Dept.	Supplier
	Basic policy of purchasing	
	Policy for ordering for each term — Production preparation plan	
	Investigation of productivity and prime cost planning	
Division of inhouse made parts and outside made parts	Choise of suppliers (meeting for ordering new parts)	Manufacture of prototypes
		Registration of personnel in charge of production preparation
Scheduling of releasing drawings	Order for production preparation	
Adjustment for design changes	Instructions on quality specifications	Planning of process
Product design		Schedule for preparation
Explanation of required quality	Meeting for discussing quality	
Assistance of production techniques	Approval for manufacturing facilities, jigs and tools	Facility planning
	Periodical investigation (the 1st ~ 3rd)	Planning and carryng out of process adjustment
	Countermeasures against troubles in trial mass production	Trial mass production
		Checking process capability
Proposal to mass production	Confirmation of procurement at the beginning of mass production	Starting of mass production
	General estimation of the supplier	

This is sometimes paraphrased in the following:"The next process is your customer."

To achieve this, the process must be under control and every aspect of inspection must be carried out in one way or another.

Thus, in daily production, the process should be carefully monitored to see whether the process is kept under control by

SQC: through the X-R control chart and histograms, the following three must be confirmed :

(1) The central value of the histogram is the target value of the quality characteristics.
(2) The variations of the quality characteristics are in the given tolerances.
(3) The process capability index C_p is kept.

4.3.2 Every Aspect of Inspection

In order not to produce defectives, inspection must be made to prevent any recurrence of mistakes. As shown in Figure 4.5, inspection can be classified into two types; the first being the one used. In other words, the aim of inspection is not to select defectives from good products but eliminate defectives.

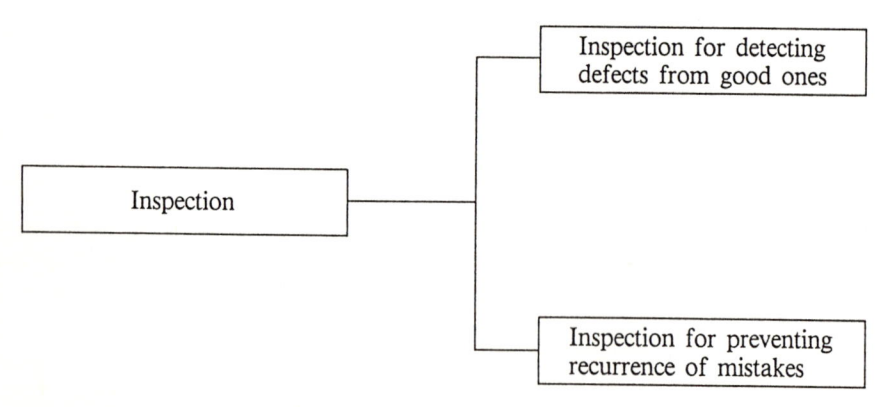

```
                          ┌─────────────────────────────┐
                          │ Inspection for detecting    │
                          │ defects from good ones      │
                          └─────────────────────────────┘
┌──────────────┐         
│  Inspection  │─────────┤
└──────────────┘         
                          ┌─────────────────────────────┐
                          │ Inspection for preventing   │
                          │ recurrence of mistakes      │
                          └─────────────────────────────┘
```

Fig. 4.5. Two types of inspection

Defectives must be discovered, in principle, immediately after they have taken place. Thus every aspect of inspection must be theoretically carried out in the production process. The problem is how to do it economically.

The essentials of preventing the recurrence of mistakes are:

(1) Workers must be instilled with the notion that each one of them is a quality-control inspector.

(2) If they spot a faulty item in the production process, they should be encouraged to shut down the whole process to fix it.
(3) Workers should tally the defect, analyze it, trace it to the source, make corrections, and then keep a record of what happens afterwards.

For item (3), theso called method of *5 why's,* should be utilized. The cause of the defect must be pursued by asking at least five times *WHY* did it take place, and the countermeasures must be taken.

Inspection can be classified into two types, according to the place where it is carried out as shown in Figure 4.6.

```
                            Before the process

                        ┌─────────────────────┐
                        │  source inspection  │
                        └─────────────────────┘

                        --- defect preventive


  ┌────────────┐
  │ inspection │
  └────────────┘


                        At and after the process

                        ┌─────────────────────┐
                        │  result inspection  │
                        └─────────────────────┘

                        --- defect finding

                            ┌─ at the process
                            ├─ at the next process
                            │  at the inspection of the production
                            ├  system
                            └─ at the final inspection process
```

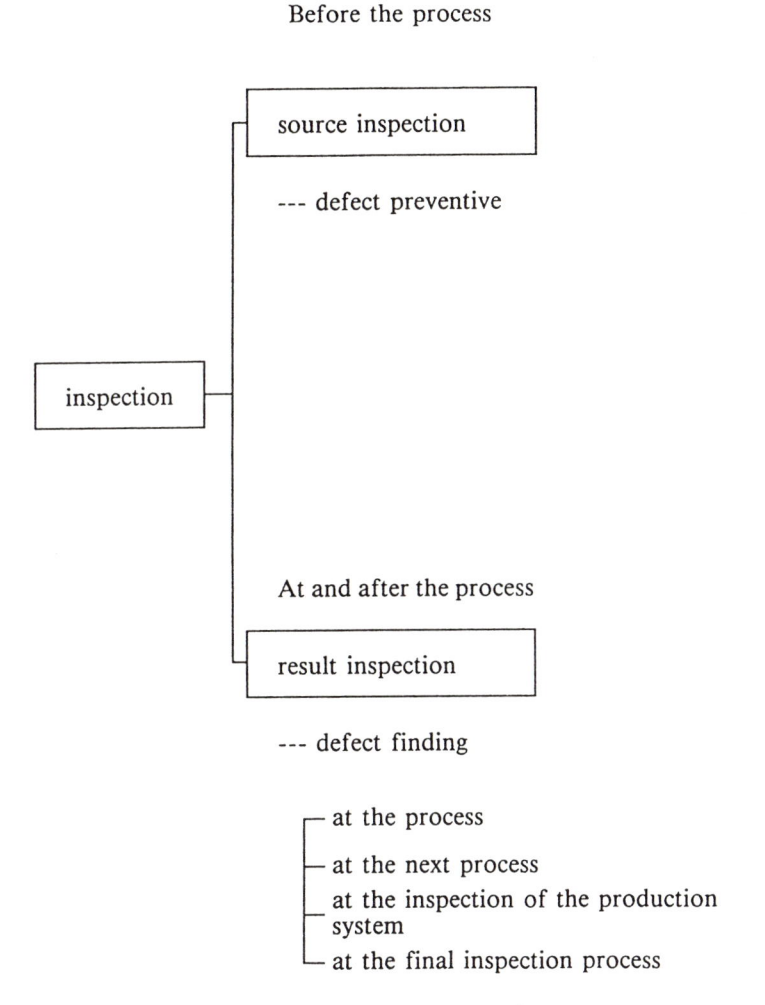

Fig. 4.6. Two types of inspection according to the place

Source inspection means detecting the cause of a mistake in advance before it takes place and preventing it from happening. For instance, to keep the production system going, all the parts fed to the system must be the correct ones. That means they must be checked totally before they are put into the system in order to prevent system stop page. This inspection is part of source inspection.

But in some cases, it is difficult to carry out source inspection, in which case result inspection must be carried out. This means that inspection is done after production; this is known as a defect finding inspection.

Result inspection can be classified into four types of inspection, as shown in the figure above. The earlier a defect is found, the better it is because proper countermeasures can be taken earlier. Thus it is very important to carry out inspection within the process.

To inspect all the items just at final inspection must be avoided. Inspectors cannot build-in quality. The important principle is that workers must build in quality during their work.

In machine processing of drills, reamers, or taps, due to the inverse proportion to the increased number of processed parts, the hole diameter becomes smaller. So considering that the quality of parts may change as time passes sampling inspection of the parts during processing can fulfil the same function for every inspection piece. That is, if the 50th product turns out to be a good one by inspection, then the 1st to the 49th products are assumed to all be good and if the 100th product is inspected as good, then the 51st to the 99th products are taken to be as good. Even when the 100th product is found out to be defective, by checking from the 51st product up to the 100th product, the defectives can be discovered. In the usual sampling inspection method, products are picked at random from a lot for sampling inspection and based on the statistics of the result, quality check is carried out.

The number of workpieces between one inspection and the next is called the quality inspection interval or quality inspection number. The quality inspection interval must be determined very carefully by tool life.

Quality control also has a great deal to do with traceability. That is, if the production process is badly organized, the manufactured goods do not come out according to the sequence of their operations but become mixed. Then even if a defective is found, it is difficult to see at what process the defective has been produced. In this sense, in single processes as well as parallel processes, parts must be inspected with the sequence of their processing.

Figure 4.7 shows one example of improving quality control. The production system has two machines (No. 1 to No. 2) in the firts process, and all workpieces processed by each machine become mixed and are transferred to the second process. If a defective is discovered on the conveyor to the second process, all the workpieces already mixed on the conveyor must be sorted out into good ones and defectives. Also all two machines must be investigated in order to find the one which has produced the defective. Thus it will be a difficult process; it is better to feed only good workpieces to the second process after going through every piece inspection.

No. 1 automatic machine

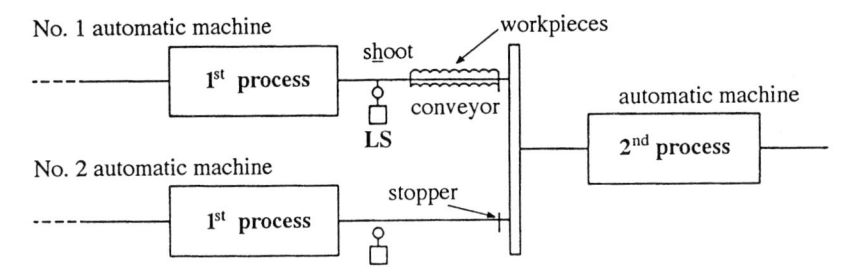

```
----[ 1st process ]
```
shoot, workpieces, conveyor, LS

```
[ 2nd process ]  automatic machine
```

No. 2 automatic machine

```
----[ 1st process ]
```
stopper

Fig. 4.7. Example of improving quality control

4.3.3 QA Network

In actual situations quality checking or inspection of a quality item at the process concerned is not sufficient for prevention of releasing defectives from the manufacturing system. Therefore, double checking or in certain cases triple checking in various ways within the system for important quality items is needed.

Figure 4.8 shows a matrix, called a QA network, to consider such a way of checking. This method is useful for avoiding missing or incorrectly fitted parts, assembly of mischosen parts, etc., and guaranteeing quality.

PROCESS / Important quality items	1	2	3	10
problem A	○	□		X
problem B		○	△	X
problem C		□		X
problem D			X	□
problem E	▽			

(Notes)

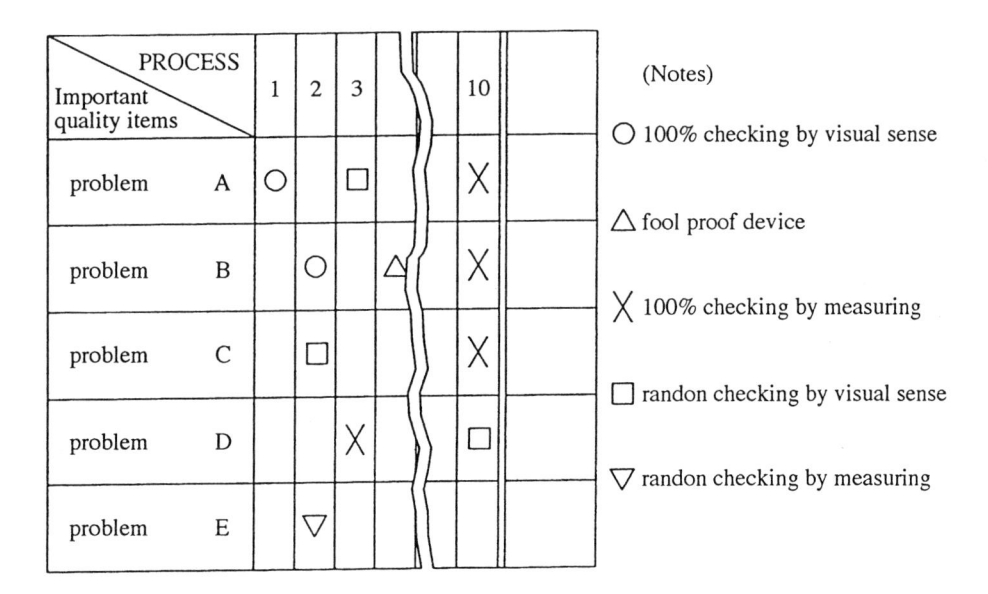

○ 100% checking by visual sense

△ fool proof device

X 100% checking by measuring

□ randon checking by visual sense

▽ randon checking by measuring

Fig. 4.8. QA matrix

4.3.4 Establishment of the Standard Operation

Standard operation is the base for a worker to follow when he or she operates machines. There are three main factors for standard operation, i.e. cycle time, sequence of operations, standard number of in-process inventory. Cycle time means standard time spent to produce one part (or product) by the manufacturing system. The sequence of operations is the one in which the worker does operations such as unloading a part from one machine, loading a new part to the machine and transferring the unloaded part to the next machine, etc. The standard number of in-process inventory is the minimal work in process necessary for carrying out the operations. This includes the workpiece in process on the machine.

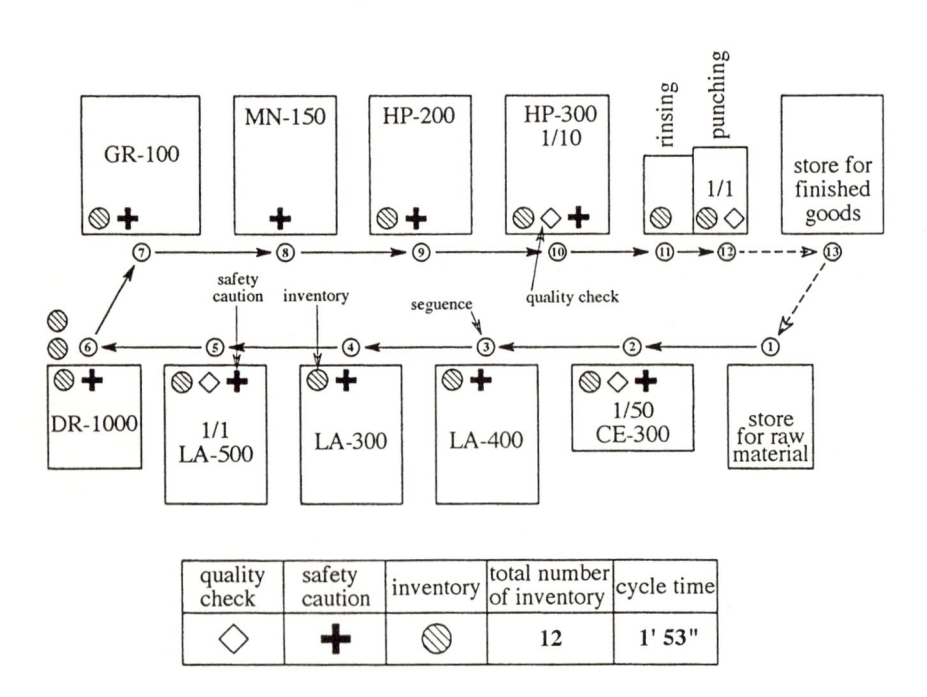

Figure 4.9 shows one example of standard operation. The worker starts operation (1), where he collects necessary material, and goes to machines from (2) to (12) and leaves the completed unit to the pallet marked by (13). At machines (2), (3), (4), (5), (6), (7), (8), (9) and (10), he needs to pay attention to safety. At machines (2), (5) (10) and (12), he has to inspect quality. The numbers 1/50 at machine (2) means that he needs to check the workpiece every 50 times. The system needs two more stocks between machines (6) and (7) besides the ones on machines (6) and (7).

In this way everything can be made clear.

Fig. 4.9. Example of standard operation

If the standard operation has been established properly, it must assure zero defect. If it has been strictly followed by the worker, planned results should be achieved and no human accidents or material damages should take place. Thus, if defectives have been found out, it means either

(1) The worker has not followed the standard operation, or:
(2) Materials or parts fed to the system have not been the correct ones, or:
(3) Machines, equipment, moulds, or jigs must have broken down.

This makes it easy to trace possible causes for defectives.

4.3.5 Role of the Foreman

The role of the foreman is to meet the production volume requirement, assure quality and improve the manufacturing system and the area for which he is responsible. For this purpose, the first thing he should do is to arrange the system and the area in such a way that abnormality can immediately be seen visually. The following methods are often used to achieve this.

1. Five S's   (*Seiri*, organization; *Seiton*, tidiness, *Seiso*, purity, *Seiketsu*, cleanliness and *Shitsuke*, discipline)
2. Straight and wide gangway
3. Limiting machine height for visuality
4. Determination of the standard numbers of parts (or products) and work in process to keep.
5. Standardized stores of materials and parts
6. Use of FIFO (First In First Out) principle
7. Paging mechanism
8. Conveyor stop button for each operator in assembly
9. Display panel
10. Dedicated area for storing the defectives

To build-in quality during this process, it is important that workers are motivated to make up good quality products. For this reason, the foreman must be in close contact with his workers in such a pleasant way that they will work willingly and if a query regarding quality takes place, he must be informed of it and should take the necessary countermeasures against it.

4.3.6 Quality Maintenance

In most factories today, the center of production has changed from labor to equipment and quality has much to do with equipment. There are two aspects for ensuring quality of the equipment. They are technical and management as follows:

1. Technical aspects : Establishment of the conditions for zero defect:
   (1) Are the conditions clear?
   (2) Are they easily set?

2. Management aspects : Controlling to maintain the conditions for zero defect:
   (3) Are they easily maintained as time progresses?
   (4) When they change, can the changes be recognized easily?
   (5) When they change, can they be easily restored?

The equipment must be maintained to ensure zero defect by providing good answers to the above five questions. PM analysis is a useful tool for establishing the conditions for zero defect. For details refer to Shirose[3].

This kind of maintenance is called quality maintenance and is a very important aspect to ensure for zero defect.

4.3.7 QC Circle Activities

To manufacture products that satisfy the customers, it is important to make them up on the shop floor when they are manufactured. For this purpose, quality consciousness and motivation of workers are vital. In other words, in this highly competitive market, poor performance by operators can no longer be afforded. The QC circle means a small group to perform quality control activities voluntarily within the workshop to which they belong.

The basic aims of QC circle activities can be summarized as follows

1. To involve workers and make them quality conscious
2. To provide workers with the opportunities for self development of both the knowledge and skills required to carry out their jobs satisfactorily.
3. To display these human capabilities fully and eventually to create infinite possibilities.
4. To create a rewarding work environment
5. To contribute to the improvement and development of the company.

4.3.8 Quality Control of Purchased Goods

Table 4.3 shows quality control situations between the customer and the supplier.

There are eight levels :

Level 1.
Since no inspection is made either in the supplier nor in the inspection department of the customer, the manufacturing department must inspect every piece in order not to send defectives to the production system.

Level 2.
Since it is normally too costly to carry out inspection of every piece in the manufacturing department, the inspection department takes care of this work. The drawback of this level of control is that the supplier does not try to improve his quality.

Table 4.3 Quality control situations between the customer and the supplier

Level	Supplier Manufacturing department	Supplier Inspection department	Customer Inspection department	Customer Manufacturing department
1	–	–	–	100% inspection
2	–	–	100% inspection	
3	–	100% inspection	100% inspection	
4	–	100% inspection	sampling inspection or check inspection	
5	100% inspection	sampling inspection	sampling inspection or check inspection	
6	process control	sampling inspection	check inspection or no inspection	
7	process control	check inspection	check inspection or non inspection	
8	process control	no inspection	no inspection	

Level 3.
In principle, the supplier must inspect every piece for the customer. Even if the supplier does do this, if the inspection method is not proper, or if the customer cannot trust the supplier's inspection, then the customer still has to inspect every piece.

Level 4.
When the customer reaches the level where the customer can trust the supplier, then the customer carries out sampling inspection in the inspection department.

Level 5.
The responsibility of quality assurance obviously belongs to the producer. So if this principle is applied to the supplier, the manufacturing department itself must inspect every item. Then the inspection departments of the supplier and the customer only need sampling inspection.

Level 6.
If the manufacturing department of the supplier continues only to inspect every item, then it does not lead to the reduction of defects. Obviously the manufacturing department must carry out process control properly in order to eliminate defects. But if process capability is not sufficient and defects still kept continue to be produced, inspection is then necessary for every item.

When the manufacturing department can carry out process control properly and, if necessary, inspect every piece, then the inspection department of the supplier only needs sampling inspection. The inspection department of the customer may need no inspection or check inspection. The latter is the small sampling inspection.

Level 7.
When the process control and inspection of every item in the manufacturing department of the supplier reaches the level where they can be fully trusted, then the inspection department of the customer only needs check inspection.

Level 8.
Ideal state. At this level no inspection is required in the inspection department of the supplier because the quality of the products produced by the manufacturing department is completely trust-worthy.

It goes without saying that the supplier should endeavor to reach level 8 to satisfy the customer.

4.4 Necessity of Education and Training on the QC Tools

Solid education and training for all the personnel is one essential part of developing capable human resources in total quality management. Education and training are investments in people that yield multiple returns if they are given to the personnel at the right moment. The content and organization of introductory materials on quality may not vary much from company to company. Table 4.4

Table 4.4 A list of QC Tools

Personnel		Subject	Contents
ENGINEER	Basic	Introductory course.	• Concepts of QC • Sampling techniques • Sampling inspection
	Intermediate	QC basic course	• Estimation and tests of hypotheses • Design of experiments (analysis of variance, orthogonal arrays, etc.) • Correlation analysis (simple correlation analysis, multiple correlation analysis) • Regression analysis (simple regression analysis, multiple regression analysis) • Orthogonal polynomials • Binomial probability paper • Simple analytic methods • Multivariate analysis techniques (Principal-component analysis, factor analysis, clustering and discrimination, quantification Types I - IV, etc.) • Optimization methods (simplex method, Box-Wilson method, EVOP, etc.)
		QC intermediate course	• The seven new QC tools (relation diagrams, systematic diagrams, matrix diagrams, affinity diagrams, arrow diagrams, process decision program charts, matrix data analysis)
		QC related methods course	• IE techniques • VE techniques • OR techniques • Idea-generating strategies
		Reliability engineering course	• Reliability data analysis • Concepts of reliability design • FTA, FMEA, Weibull probability paper, cumulative hazard paper, etc. • Failure analysis
	Advanced 1 for assist. manager	QC seminar for assist. manager	• TAGUCHI method • S/N ratio (parameter design, tolerance design)
	Advanced 2 for manager	Reliability engineering course	• Reliability control for manager
WORKER	Basic	Introductory course	• Quality consciousness 5S's • PDCA cycle • QC circle activity
	Intermediate	Intermediate QC Education and training course	The seven QC tools (cause-and-effect diagrams, pareto diagrams, graphs, check sheets, histograms, scatter diagrams, control charts)
	Advanced 1 for foreman	QC education and training for foreman	• Instruction of 7 tools for QC • Process capability study method • Problem solving procedures
	Advanced 2 for assist. manager	QC circle facilitator course	• Method to facilitate QC circle activities • Group discussion

summarizes a list of QC tools with which personnel at different levels must become intimately acquainted.

References

1. *Quality Function Deployment*, Yoji Akao (Ed), Productivity Press, Cambridge, Mass., 369pp.
2. *Introduction to Quality Engineering*, G. Taguchi, Asian Productivity Organization
3. *TPM for Workshop Leaders*, K. Shirose, Productivity Press, Cambridge, Mass.

Author and Subject Index

Dudley, Darle W. 1
Schröder, D. 19, 59
Sprengers, Jacques 367
Yamashina, Haijme 549

Abrasive wear 168
Acceleration 22, 23, 31
Addendum 80
Adhesion 40
Alternate stress 64
Amplitude 306
Angular speed 27
Annealing 70
Aperiodic behaviour 416
Application factor 173
Approach path 89
Armature control range 450, 482, 543
Armature converter 452, 453
Armature current 406, 425, 459
Armature current reversal 436, 455
Asynchronous behaviour 388
Austempering 72
Axial force 131
Axial module 137
Axial pitch 136

Backlash 164
Ball bearing 254, 255
Base circle 75, 84
Base tangent length 147
Basic rack 80, 92, 93
Belt 38, 256
Belt conveyor 348
Bending 50, 207
Bending factor 51
Bending moment 50
Bending stress 50
Bevel gear 124, 338, 343
Bevel gear factor (contact) 196
Bevel gear factor (root) 198
Blank 161

Bolt 288-290
Boost-converter 428, 435
Braking operation 435
Braking torque 47
Breakdown slip 402
Breakdown torque 402
Breaking stress 63, 290
Bridge converter 445, 447, 448
Bridge crane 346
Bronze 73, 201, 204
Brushless DC-mashine 547
Buck converter 426-428
Buckling 54

Carburizing 72
Case hardened steel 173, 180, 183, 189
Cast iron 70
Center distance 86
Chain conveyor 349
Chordal thickness 148
Clutch couplings 251
Coefficient of elasticity (contact) 48
Coefficient of elasticity (transverse) 49
Coefficient of friction 34
Coefficient of torsion 53
Commutation 442
Commutator 381
Composite radial deviation 159
Composite tangential deviation 158
Compression 48
Contact pattern 167
Contact ratio 88
Contact ratio factor (contact) 182
Contact stress 169, 180, 195
Continuous operation 376, 441
Continuous running duty 372
Control 426, 458, 486, 490, 534
Converter 426
Conveyor 348
Convolution 408
Cooling methods 370
Coordinate system 469, 470, 496
Coupling 247

Author and Subject Index

Crane 347
Cumulated pitch deviation 154
Current control 458
Current source converter 523
Current source inverter 485, 522
Curvature radius 76, 90
Curving length factor 193
Cyclo converter 509
Cylindrical roller bearing 260
Cylindrical wheel 96

Decoupling 535
Dedendum 80, 93, 136
Denominator polynominal 392
Deviation 147, 151, 154, 157, 158
Dip lubrication 281
Discontinuous operation 441
Disk 41
Double row self-aligning ball bearing 259, 318
Driving machine 386, 393
Driving power 386
Dynamic factor 174, 192, 270
Dynamic viscosity 275

E.M.F. feed forward 460
Eccentricity 161
Efficiency 124, 138, 322
Elastic coupling 250
Elasticity module, coefficient of elasticity (transverse) 48, 49, 169
End relief 123
Endurance 215, 216
Endurance factor 215
Endurance limit 63, 216
Energy 23
Equivalent electrical circuit 485, 503
Equivalent power 299, 300

Face load factor (contact) 175
Face load factor (root) 175
Face width 94
Failure Method Effect Analysis (FMEA) 581, 585
Failure Tree Analysis (FTA) 581, 585
Farm machines 353
Field circuit 413
Field current 456
Field orientation 534, 540
Field weakening 418, 422, 452, 483
Finite elements calculation 285
First-in First-out (FIFO) 597
First-quadrant 452
First-quadrant operation 450

Flexion vibration 304
Flux interlinkage equations 471
Forced vibration 303
Form factor 224
Four-quadrant-operation 438
Fourier transformation 307
Frequency 306
Fretting corrosion 243
Friction angle 35
Friction gaskets 272

Galton's curve 65
Gear motor 301
Gear units 288, 289, 295
Gearbox 286, 287, 289, 295, 298, 326
Generator operation 401, 480, 518
Gravity center 21
Grease 273, 274
Grinding 340, 344
Grinding wheel 341
Groove 222

Half wave converter 440
Helical gear 100
Helical wheel 100
Helix angle factor (contact) 182
Helix angle factor (root) 189
Helix form deviation 158
Helix slope deviation 158
Helix total deviation 158
Hob 336, 337
Honing 340, 343
Housing 284
Hub 239-241

Imbalance 30
Individual pitch deviation 151, 154
Individual pitch measurement 151
Induction machine 397, 477, 485
Inductive voltage drop 443
Inertia 28
Interference 84
Intermittent periodic duty 373, 375, 376
Inverter 509, 516, 522, 527
Inverter operation 443
Involute 75, 91

Just-in-Time 556, 564

Key 238
Keyway 221
Kinematic viscosity 275, 276
Kloss' equation 479

Laplace domain 392, 407
Life factor (contact) 183, 197
Life factor (tooth root) 189, 199
Line of action 79
Line-commutated converter 439, 512
Line-to-line voltage 448
Losses 314, 316, 317
Lower deviation 235

Magnetising current 484
Main stress 56
Martempering 72
Materials 69
Materials wear factor (worm) 203
Mechanical energy 23
Meshing frequency 308
Meshing interference 107
Middle zone factor 197
Miner's rule 68, 200
Mineral lubricant 273
Modified center distance 86
Modulation strategies 531
Module 79-81
Moment of inertia 25, 31
Multiple quadrant operation 434

Needle roller bearings 254
Nitriding 73
Noise 302, 309, 310
Non-reversing 139
Non-salient pole rotor 399, 502
Normal chordal 148
Normal contact ratio 106
Normal pressure 119
Normalization 409
Notch factor 219
Notch gradient 219
Notch sensibility factor 219
Nuts 288, 290

Open-loop speed control 420
Operating limits 450, 451
Operating point 390
Operating range 516
Overhanging load 211
Overlap ratio 106

Periodical stress 63
Permissible contact stress 182, 196
PI-controller 461, 463
Pinion cutter-tool 335
Pitch circle 87
Pitch cylinder 116
Pitch diameter 87

Pitting 168
Plan-Do-Check-Act 570
Planet 142
Planetary 142
Point of contact 115
Potential energy 23
Power 27
Precision 161, 164
Pressure angle 78
Profile measurements 154
Pull-out slip 478
Pull-out torque 479
Pulley 38
Pulsation 306
Pulse width control 430
Pulse width modulation 532
Pulse-frequency control 431

Quality 551
Quality Assurance (QA) 554, 564, 575
Quality Control (QC) 555, 565
Quality Function Deployment (QFD)
   581
Quenching 71
Quotient 136

Rack shift factor 84, 119
Rack tooth profile 80
Radial deep groove ball bearings 258
Radial force 117
Ratio 117
Recess path 113
Rectifier operation 450
Reference center distance 87
Reference circle 81, 87
Reference cylinder 94, 116
Reference diameter 81, 87
Reference list 94
Reference thickness 95
Reliability factor 216
Reluctance torque 499
Rigid coupling 248
Roller 40, 43, 44
Roller bearing 254, 255
Roller conveyor 350
Root circle 79
Root cylinder 93
Root diameter 81
Rotating field machine 397, 469
Rotor flux orientation 492
Roughness 237
Roughness factor 216, 315
Running in 177
Runout 161

Author and Subject Index

S-N curve 66, 67, 217
Safety factor 181, 225
Salient-pole machine 498
Salient-pole rotor 398, 496
Scherbius drive 511
Scoring 282
Screw 290
Screw conveyor 351
Seals 270
Sensitivity factor 190
Series resistor 423
Shaft 207
Shaft distorsion 231
Shaving 342
Shearing 49
Shearing load 209
Short-time duty 373
Shortening 110
Single contact 90
Single pair tooth contact factor 181, 182
Size factor 181, 199
Skiving 340, 343
Sliding 110
Sliding speed 110
Slip 400, 478
Slip-ring machine 474
Space-vector theory-phasors 466
Speed control 420, 458, 463
Spline 245
Spray lubrication 281
Squirrel-cage rotor 398, 400, 474
Stability 390, 391
Standards 369, 371
Starting torque 47
Static capacity 257
Static permissible stress 62, 214
Statistical Quality Control 552, 564
Stator flux orientation 486
Steel 70
Stress concentration factor 218, 224
Stress correction factor 186, 188
Stress reduced to the reference stress 225
Subsynchronous cascade 511
Surface factor 190
Synchronous behaviour 389
Synchronous machine 397, 403, 495
Synthetic lubricant 274

Taper roller bearing 263

Tempering 73
Thermal power 314
Thermic Treatment 70
Three-phase bridge converter 448
Time response 425
Tip circle 79
Tip cylinder 92
Tip diameter 81
Tip thickness 83
Tooth backlash 164
Tooth bending strength 185, 198
Tooth form factor 188
Tooth root stress 198
Torque reversal 454, 455
Torque speed characteristic 402, 480
Torsion 53, 212
Torsion vibration 305
Total Quality Management (TQM) 552, 555, 564
Traction 48
Transient response 414
Transverse contact ratio 106
Transverse load distribution factor 190, 193
Transverse module 100
Two-quadrant operation 435

Undercut 85, 94, 269
Upper deviation 235

Variators 359, 360
Vibration 302
Virtual gear 195
Viscosity 275-278

Voltage equation 470, 496
Voltage source converter 426
Voltage source inverter 527

Wear 168
Wear prevention 203
Work hardening factor 185, 197
Worm 135, 338
Worm gear 135
Worm profile 139
Worm wheel 135, 137, 338

Zone factor 181

Springer-Verlag and the Environment

We at Springer-Verlag firmly believe that an international science publisher has a special obligation to the environment, and our corporate policies consistently reflect this conviction.

We also expect our business partners – paper mills, printers, packaging manufacturers, etc. – to commit themselves to using environmentally friendly materials and production processes.

The paper in this book is made from low- or no-chlorine pulp and is acid free, in conformance with international standards for paper permanency.

Printing: Saladruck, Berlin
Binding: Buchbinderei Lüderitz & Bauer, Berlin